Studies in Computational Intelligence

Volume 490

Series Editor

J. Kacprzyk, Warsaw, Poland

For further volumes:
http://www.springer.com/series/7092

Shengxiang Yang · Xin Yao
Editors

Evolutionary Computation for Dynamic Optimization Problems

 Springer

Editors

Shengxiang Yang
School of Computer Science and Informatics
De Montfort University
The Gateway
Leicester LE1 9BH,
United Kingdom

Xin Yao
School of Computer Science
University of Birmingham
Edgbaston
Birmingham B15 2TT,
United Kingdom

ISSN 1860-949X   ISSN 1860-9503   (electronic)
ISBN 978-3-642-44843-0   ISBN 978-3-642-38416-5 (eBook)
DOI 10.1007/978-3-642-38416-5
Springer Heidelberg New York Dordrecht London

Printed on acid-free paper

Springer is part of Springer Science+Business Media (www.springer.com)

To our families

Preface

Evolutionary computation (EC) represents a class of optimization methodologies inspired by natural evolution. During the past several decades, evolutionary algorithms (EAs) have been extensively studied by the computer science and artificial intelligence communities. As a class of stochastic optimization techniques, EAs can often outperform classical optimization techniques for difficult real-world problems.

Due to the properties of ease-to-use and robustness, EAs have been applied to a wide variety of optimization problems. Most of these optimization problems tackled are stationary and deterministic. However, many real-world optimization problems are subjected to dynamic environments that are often impossible to avoid in practice. For example, the objective function, the constraints, and/or environmental conditions may change over time due to many reasons. For these dynamic optimization problems (DOPs), the objective of an EA is no longer to simply locate the global optimal solution, but to continuously track the optimum in dynamic environments. This poses serious challenges to classical optimization techniques as well as conventional EAs. However, conventional EAs with proper enhancements are still good choices for DOPs. This is because EAs are inspired by principles of natural evolution, which takes place in the ever-changing dynamic environment in nature.

Addressing DOPs has been a topic since the early days of EC and has only received increasing research interests over the last two decades due to its challenge and its importance in practice. A number of events, e.g., edited books, journal special issues, symposia, workshops and conference special sessions, have taken place, which are relevant to the field of EC for DOPs. A variety of EC methods for DOPs have been reported across a range of application backgrounds in recent years. This motivated the edition of this book. This book aims to timely reflect the most recent advances, including benchmark test problems, methodologies, theoretical analysis, and relevant real-world applications, and explore future research directions in the field.

We have a total of 17 chapters in this book, which cover a broad range of topics relevant to EC in dynamic environments. The chapters in this book are organized into the following four categories:

- Part I: Fundamentals
- Part II: Algorithm Design
- Part III: Theoretical Analysis
- Part IV: Applications

Part I: Fundamentals

During the last two decades, researchers from the EC community have developed a variety of EC approaches to address DOPs and evaluated them on many benchmark and real-world DOPs under different performance measures. Part I of the book consists of four chapters, which review the developments in terms of test and evaluation environments, methodologies, and challenges, and lay the foundations for the research field of EC for DOPs.

Chapter 1, contributed by Yang *et al.*, first introduces the concept of DOPs and reviews existing dynamic test problems (including both benchmark and real-world DOPs) that are commonly used in the literature with discussions regarding their major features. Then, this chapter reviews and discusses the performance measures that are widely used to evaluate and compare EC approaches for DOPs, followed by suggestions for future improvement regarding dynamic test and evaluation environments. Finally, this chapter describes in detail a generalized dynamic benchmark generator (GDBG), which has been recently developed and used in the 2009 and 2012 IEEE Competitions on EC for DOPs.

Chapter 2, contributed by Nguyen *et al.*, summarizes main EC methodologies that have been developed over the years for solving DOPs with discussions on the strength and weakness of each approach and their suitability for different types of DOPs. Current gaps, challenging issues and future directions regarding EC methodologies for DOPs are also presented in this chapter.

In Chapter 3, Rohlfshagen and Yao discuss challenges and perspectives on EC for DOPs regarding several key issues, including different problem definitions that have been proposed, the modelling of DOPs in terms of benchmark suites, and the way the performance of an algorithm is assessed. This chapter critically reviews the work done in each of these aspects, points out many gaps and vagueness in the current research, and identifies some promising research directions for the future of the field.

As well as addressing single-objective DOPs, researchers from the EC community have also investigated dynamic multi-objective optimization problems (DMOPs) in recent years. In the last chapter of Part I (Chapter 4), Raquel and Yao provide a survey of EC for DMOPs with regards to the definition and classification of DMOPS, test problems, performance measures and optimization approaches, and identify gaps, challenges and future directions in the domain of EC for DMOPs.

Part II: Algorithm Design

As mentioned before, many EC methodologies have been developed to address DOPs during the last two decades. Part II of the book includes four chapters on the design of different EC methods for solving DOPs with experimental studies.

Particle swarm optimization (PSO) has been widely applied to solve DOPs due to its efficiency of locating optima. In Chapter 5, Li and Yang review PSO with variant enhancements, e.g., diversity, memory, multi-population, adaptive, and hybrid schemes, for solving DOPs, and discuss the weaknesses and strengths of those approaches. A set of typical PSO approaches to solving DOPs are chosen to experimentally compare their performance on the moving peaks problem. Based on the experimental results and relevant analyses, suggestions are given regarding algorithm design of PSO for DOPs in this chapter.

Memetic algorithms, as a class of hybrid EC methods, have also been studied for solving DOPs in recent years in the literature. Chapter 6, contributed by Wang and Yang, investigates the application of memetic algorithms to solving DOPs. A memetic algorithm that integrates a new adaptive hill climbing method as the local search technique is proposed for solving DOPs. In order to address the convergence problem, an adaptive elitism-based immigrants scheme is introduced into the proposed memetic algorithm. Experiments were conducted to investigate the performance of the proposed memetic algorithm in comparison with some other algorithms. The experimental results have showed the efficiency of the proposed memetic algorithm for solving the tested DOPs.

Hybridizing different enhancement approaches (with proper choices) into EAs has been shown beneficial and is becoming a trend in solving DOPs due to the ability of combining different advantages of different enhancement approaches. In Chapter 7, Alba *et al.* propose a new EA that is augmented by the memory, bi-population, local search, and immigrants schemes to solve the dynamic knapsack problem. The two populations inside the algorithm are used to search in different directions in the search space: the first one takes charge of exploration while the second is responsible for exploitation. According to the experimental results, the proposed algorithm is very competitive in comparison with a few existing EAs taken from the literature for solving the dynamic knapsack problems.

Dynamic constrained optimization problems (DCOPs) are a class of challenging DOPs, where constraints are integrated and may also change over time. DCOPs have recently been investigated by the EC community and are in great need of much more research. In Chapter 8, Nguyen and Yao investigate EC for continuous DCOPs. They first present some studies on the characteristics that can make DCOPs difficult to solve by some existing EAs designed for general DOPs, and then introduce a set of benchmark problems with these characteristics and experimentally test several representative EAs on these problems. The experimental results confirm that DCOPs do have special characteristics that can significantly affect the performance of algorithms. Based on the experimental results and analyses, they suggest a list of potential requirements for an algorithm to solve DCOPs effectively.

Part III: Theoretical Analysis

In comparison with the developments of benchmark and test problems and methodologies on EC for DOPs, theoretical analysis of EC for DOPs has been significantly
lagged behind with very limited results. This is mainly because it is very challenging
and difficult to theoretically analyze EC methods, even for stationary optimization
problems, let along for much more challenging DOPs. Although challenging and
difficult, theoretical analysis is very important for the field of EC for DOPs since
the relative lack of theoretical analysis makes it difficult to fully justify the strengths
and weaknesses of EC methods for DOPs. In recent years, it is great to see that some
researchers have started to address this challenging issue – formally analyzing EC
methods for DOPs. Part III of the book includes four chapters and serves as a review
as well as an introduction to some recent research in this important area.

Chapter 9, contributed by Rohlfshagen *et al.*, provides a review of theoretical
advances in the field of EC for DOPs. In particular, the authors argue the importance
of theoretical results, highlight the challenges faced by theoretical researchers, and
summarise the work that has been done so far in the area. They subsequently identify
relevant directions for future research regarding theoretical analysis of EC for DOPs.

In Chapter 10, Tinós and Yang apply the dynamical systems approach to describe
the conventional genetic algorithm as a discrete dynamical system for solving DOPs.
Based on this dynamical system model, they define some properties and classes of
DOPs and analyze some DOPs used by researchers in the field of EC for DOPs.
The analysis of DOPs via the dynamical systems approach allows explaining some
behaviors of algorithms observed in the results of the experiments conducted in
the chapter and, hence, is important to understand the experimental results and to
analyze the similarity of such problems to other DOPs.

In Chapter 11, Richter takes a different viewpoint of solving DOPs by EC methods, i.e., grounding it on the theoretical framework of dynamic fitness landscapes.
The author defines such dynamic fitness landscapes, discusses their properties, and
studies the analytical tools for measuring topological and dynamical landscape properties. Based on these landscape measures, an approach for drawing conclusion regarding characteristic features of a given optimization problem is obtained, which
may allow us to address the question of how difficult the problem is for an EC
approach, and what type of algorithm is most likely to solve it successfully. The
proposed methodology is further experimentally illustrated using the moving peaks
problem in this chapter.

Chapter 12, contributed by Comsa *et al.*, is devoted to the field of analyzing EC
for DMOPs. The authors briefly review some recent work in this field and present
the analysis of a multi-objective genetic algorithm with an external archive and a
combination of Pareto dominance and aggregated fitness function on dynamic multi-
objective subset sum problems.

Part IV: Applications

In recent years, some researchers from the EC community have started to address real-world DOPs since many real-world optimization problems are DOPs. Part IV of the book consists of five chapters that are devoted to apply EC methods to solve real-world DOPs.

Ant colony optimization (ACO) algorithms, as a class of EC methods, have proved to be powerful methods to address DOPs, especially dynamic travelling salesman problems (DTSPs). In Chapter 13, Mavrovouniotis and Yang investigate ACO algorithms with different immigrants schemes, which help to maintain the diversity of the population via transferring knowledge from previous environments to the pheromone trails, to solve DTSPs with traffic factors. The experimental results based on different DTSP test cases show that the proposed ACO algorithms outperform other peer ACO algorithms and that different immigrants schemes are beneficial on different environmental cases.

Nowadays, with the advancement in wireless communications, more and more mobile ad hoc networks (MANETs) appear in different fields in the real world. For MANETs, one of the most important characteristics is the topology dynamics, i.e., the network topology changes over time due to energy conservation or node mobility. This topology dynamics poses big challenges to solve routing problems, which play an important role in MANETs. In Chapter 14, Cheng and Yang investigate the application of several genetic algorithms with appropriate enhancements to solve two typical dynamic routing problems, i.e., the dynamic shortest path routing problem and the dynamic multicast routing problem, in MANETs. The experimental results show that these specifically designed genetic algorithms can quickly adapt to the network topology changes and produce high quality solutions after each change.

The capacitated arc routing problem (CARP) is a classic combinatorial optimization problem that has many applications in the real world. In Chapter 15, Mei et al. investigate two EC methods, a repair-based tabu search and a memetic algorithm with extended neighborhood search, to solve a new dynamic CARP, where stochastic factors are included in the CARP. The objective of the dynamic CARP is to find a robust solution that shows good performance in uncertain environments. For the dynamic CARP, the authors define a robustness measure and design the corresponding repair operator according to the real-world considerations, which is used in the EC methods. Experiments are conducted based on some benchmark instances of the dynamic CARP generated in this chapter, and the preliminary analysis for the fitness landscape of the dynamic CARP is provided.

In Chapter 16, Peng et al. apply EAs to solve the online path planning (OPP) and dynamic weapon target assignment (WTA) problems for the multiple unmanned aerial combat vehicles anti-ground attack task. A dynamic multi-objective EA with historical Pareto set linkage and prediction, denoted LP-DMOEA, is proposed to solve the OPP problem. In the LP-DMOEA, a Bayesian network and fuzzy logic are used to quantify the bias value to each optimization objective in order to intelligently select an executive solution from the Pareto set. For the dynamic WTA problem, an estimation of distribution algorithm with an environment identification based memory scheme, denoted EI-MEDA, is proposed as the optimizer. The proposed approaches are validated via simulation. The results show that LP-DMOEA and EI-MEDA can efficiently solve the OPP and dynamic WTA problems respectively.

Finally, the last chapter in Part IV (Chapter 17), contributed by Ibrahimov *et al.*, presents detailed insights into a project for transitioning a wine manufacturing company from a mostly spreadsheet driven business with isolated silo-operated planning units into one that makes use of integrated and optimised decision making through the use of modern heuristics. The authors present the modelling of business entities and their silo operation and optimization, and pave the path for a further holistic integration to obtain company-wide globally optimised decisions. They argue that the use of computational intelligence methods, including EC methods, is essential in dealing with dynamic and non-linear constraints and solving today's real-world problems as exemplified by the given wine supply chain.

In summary, this book fulfils the original aims well. The four parts of the book represent a variety of work in the area of EC for DOPs. We hope that the publication of this book will further promote this emerging and important research field.

Shengxiang Yang, De Montfort University, U.K.
Xin Yao, University of Birmingham, U.K.
March 2013

Acknowledgements

We would like to thank Dr. Janusz Kacprzyk for inviting us to edit this book in the Springer book series "Studies in Computational Intelligence". We acknowledge the contributors for their fine work and cooperation during the book preparation and the reviewers for carefully reviewing the chapters for the book. We are grateful to Mr Frank Holzwarth, Mr Holger Schäpe, Dr. Thomas Ditzinger, and Dr. Dieter Merkle, from Springer for their strong support and editorial assistance to the book.

We would also like to thank the Engineering and Physical Sciences Research Council (EPSRC) of U.K. for funding two linked research projects: 1) "Evolutionary Algorithms for Dynamic Optimisation Problems: Design, Analysis and Applications" under Grant numbers EP/E060722/1, EP/E060722/2, and EP/E058884/1; and 2) "Evolutionary Computation for Dynamic Optimisation in Network Environments" under Grant numbers EP/K001310/1 and EP/K001523/1. These two projects contributed greatly to the successful edition of this book.

Contents

Part III: Theoretical Analysis

List of Contributors

Enrique Alba
Departamento de Lenguajes y Ciencias de la Computación, Universidad de Málaga,
E.T.S.I. Informática, Campus de Teatinos, 29071 Málaga, Spain,
e-mail: eat@lcc.uma.es

Hajer Ben-Romdhane
LARODEC Laboratory, Institut Supérieur de Gestion, University of Tunis, 41 Rue
de la Liberté, Le Bardo, Tunisia,
e-mail: hajer1br@hotmail.fr

Juergen Branke
Warwick Business School, University of Warwick, Coventry CV4 7AL, U.K.,
e-mail: juergen.branke@wbs.ac.uk

Hui Cheng
Department of Computer Science and Technology, University of Bedfordshire,
Park Square, Luton LU1 3JU, U.K.,
e-mail: hui.cheng@beds.ac.uk

Iulia Maria Comsa
Department of Computer Science, Babes-Bolyai University, Kogalniceanu 1,
Cluj-Napoca 400084, Romania,
e-mail: iulia.m.comsa@gmail.com

Xiaoguang Gao
School of Electronics and Information, Northwestern Polytechnical University,
Xi'an 710129, China,
e-mail: xggao@nwpu.edu.cn

Crina Grosan
Department of Computer Science, Babes-Bolyai University, Kogalniceanu 1,
Cluj-Napoca 400084, Romania, and Department of Information Systems and
Computing, Brunel University, Uxbridge, Middlesex UB8 3PH, U.K.,
e-mail: crina.grosan@brunel.ac.uk

Maksud Ibrahimov
School of Computer Science, University of Adelaide, South Australia 5005,
Australia,
e-mail: maksud.ibrahimov@adelaide.edu.au

Saoussen Krichen
FSJEG de Jendouba, University of Jendouba, Avenue de l'U.M.A , 8189 Jendouba,
Tunisia,
e-mail: saoussen.krichen@isg.rnu.tn

Per Kristian Lehre
School of Computer Science, University of Nottingham, Nottingham NG8 1BB,
U.K.,
e-mail: perkristian.lehre@nottingham.ac.uk

Changhe Li
School of Computer, China University of Geosciences, 388 Lumo Road, Wuhan
430074, China,
e-mail: changhe.lw@gmail.com

Michalis Mavrovouniotis
Centre for Computational Intelligence (CCI), School of Computer Science and
Informatics, De Montfort University, The Gateway, Leicester LE1 9BH, U.K.,
e-mail: mmavrovouniotis@dmu.ac.uk

Yi Mei
School of Computer Science and Information Technology, RMIT University,
Melbourne VIC 3001, Australia,
e-mail: yi.mei@rmit.edu.au

Zbigniew Michalewicz
School of Computer Science, University of Adelaide, South Australia 5005,
Australia. Institute of Computer Science, Polish Academy of Sciences, ul. Ordona
21, 01-237 Warsaw, Poland, Polish-Japanese Institute of Information Technology,
ul. Koszykowa 86, 02-008 Warsaw, Poland,
e-mail: zbigniew.michalewicz@adelaide.edu.au

Arvind Mohais
SolveIT Software, Pty Ltd., 99 Frome Street, Adelaide, SA 5000 Australia,
e-mail: am@solveitsoftware.com

Trung Thanh Nguyen
School of Engineering, Technology and Maritime Operations, Liverpool John
Moores University, Liverpool L3 3AF, U.K.,
e-mail: T.T.Nguyen@ljmu.ac.uk

Maris Ozols
SolveIT Software, Pty Ltd., 99 Frome Street, Adelaide, SA 5000 Australia,
e-mail: mo@solveitsoftware.com

Xingguang Peng
School of Marine Engineering, Northwestern Polytechnical University, Xi'an
710072, China,
e-mail: pxg0510@gmail.com

Carlo Raquel
Centre of Excellence for Research in Computational Intelligence and Applications
(CERCIA), School of Computer Science, University of Birmingham, Edgbaston,
Birmingham B15 2TT, U.K.,
e-mail: crr954@cs.bham.ac.uk

Hendrik Richter
Department of Measurement Technology and Control Engineering, Faculty of
Electrical Engineering and Information Technology, HTWK Leipzig University of
Applied Sciences, D–04251 Leipzig, Germany,
e-mail: richter@eit.htwk-leipzig.de

Philipp Rohlfshagen
Centre of Excellence for Research in Computational Intelligence and Applications
(CERCIA), School of Computer Science, University of Birmingham, Birmingham
B15 2TT, U.K.,
e-mail: philipp.r@gmail.com

Briseida Sarasola
Departamento de Lenguajes y Ciencias de la Computación, Universidad de Málaga,
E.T.S.I. Informática, Campus de Teatinos, 29071 Málaga, Spain,
e-mail: briseida@lcc.uma.es

Sven Schellenberg
SolveIT Software, Pty Ltd., Level 2, 198 Harbour Esplanade, Docklands, VIC
3008, Australia,
e-mail: ss@solveitsoftware.com

Ke Tang
Nature Inspired Computation and Applications Laboratory (NICAL), School of
Computer Science, University of Science and Technology of China, Hefei 230027,
China,
e-mail: ketang@ustc.edu.cn

Renato Tinós
Department of Computing and Mathematics, FFCLRP, University of São Paulo,
Av. Bandeirantes, 3900, 14040-901, Ribeirão Preto, SP, Brazil,
e-mail: rtinos@ffclrp.usp.br

Hongfeng Wang
College of Information Science and Engineering, Northeastern University,
Shenyang 110004, China,
e-mail: hfwang@mail.neu.edu.cn

Demin Xu
School of Marine Engineering, Northwestern Polytechnical University, Xi'an
710072, China,
e-mail: xudm@nwpu.edu.cn

Shengxiang Yang
Centre for Computational Intelligence (CCI), School of Computer Science and
Informatics, De Montfort University, The Gateway, Leicester LE1 9BH, U.K.,
e-mail: syang@dmu.ac.uk

Xin Yao
Centre of Excellence for Research in Computational Intelligence and Applications
(CERCIA), School of Computer Science, University of Birmingham, Edgbaston,
Birmingham B15 2TT, U.K.,
e-mail: x.yao@cs.bham.ac.uk

Part I
Fundamentals

Chapter 1
Evolutionary Dynamic Optimization:
Test and Evaluation Environments

Shengxiang Yang, Trung Thanh Nguyen, and Changhe Li

Abstract. In the last two decades, dynamic optimization problems (DOPs) have drawn a lot of research studies from the evolutionary computation (EC) community due to the importance in real-world applications. A variety of evolutionary computation approaches have been developed to address DOPs. In parallel with developing new approaches, many benchmark and real-world DOPs have been constructed and used to compare them under different performance measures. In this chapter, we describe the concept of DOPs and review existing dynamic test problems that are commonly used by researchers to investigate their EC approaches in the literature. Some discussions regarding the major features of existing dynamic test environments are presented. Typical dynamic benchmark problems and real-world DOPs are described in detail. We also review the performance measures that are widely used by researchers to evaluate and compare their developed EC approaches for DOPs. Suggestions are also given for potential improvement regarding dynamic test and evaluation environments for the EC community.

1.1 Introduction

In the last two decades, dynamic optimization problems (DOPs) have drawn a lot of research studies from the evolutionary computation (EC) community. Especially, in

Shengxiang Yang
Centre for Computational Intelligence (CCI), School of Computer Science and Informatics,
De Montfort University, The Gateway, Leicester LE1 9BH, U.K.
e-mail: syang@dmu.ac.uk

Trung Thanh Nguyen
School of Engineering, Technology and Maritime Operations,
Liverpool John Moores University, Liverpool L3 3AF, U.K.
e-mail: T.T.Nguyen@ljmu.ac.uk

Changhe Li
School of Computer Science, China University of Geosciences, 388 Lumo Road,
Wuhan 430074, China
e-mail: changhe.lw@gmail.com

S. Yang and X. Yao (Eds.): *Evolutionary Computation for DOPs*, SCI 490, pp. 3–37.
DOI: 10.1007/978-3-642-38416-5_1   © Springer-Verlag Berlin Heidelberg 2013

recent years, there has been a growing interest in studying evolutionary algorithms (EAs) for DOPs due to its importance in real-world applications since many real-world optimization problems are DOPs. The research domain of EC for DOPs can be termed as evolutionary dynamic optimization (EDO). DOPs require EAs to track the trajectory of changing optima in the search space [11, 30]. This poses great challenges to traditional EAs due to the convergence problem: once converged, they can not track the changing optima well. Hence, researchers have developed several approaches into EAs to enhance their performance for DOPs [30, 83], e.g., diversity schemes [19, 27, 80], memory schemes [9, 70, 88], multi-population schemes [13, 55, 82], prediction and anticipation schemes [64], and adaptive schemes [45, 84, 85].

In order to study and compare the developed EA approaches for DOPs, there are two important tasks. One important task is to build up proper dynamic test environments. Over the years, in parallel with developing EA approaches for DOPs, researchers have also developed many dynamic benchmark problems, e.g., the moving peaks benchmark (MPB) problem by Branke [9], the XOR DOP generator by Yang and Yao [77, 87, 88], the generalized dynamic benchmark generator (GDBG) by Li and Yang [36], and modelled a number of real-world DOPs, e.g., dynamic knapsack problems [34, 43, 49], dynamic travelling salesman problems [35, 40], dynamic routing problems in communication networks [16, 17, 81], and dynamic vehicle routing problems in logistics and transportation networks [42, 75]. The other important task is to define proper performance measures to compare different EC approaches for DOPs. Over the years, researchers have developed a number of different performance measures to evaluate the developed EA approaches for DOPs, e.g., the offline error measure [9], the accuracy measure [72], and the best-of-generation measure [18], etc.

In this chapter, we present the concept of DOPs and review existing dynamic test problems commonly used by researchers to investigate their EC approaches in the literature. Some discussions regarding the major features and classification of existing dynamic test environments are presented. Some typical dynamic benchmark problems and real-world DOPs, which cover the binary, real, and combinatorial spaces, are also described in detail. We also review the performance measures that are widely used by researchers to evaluate and compare their developed EC approaches for DOPs. Suggestions are also given for potential improvement regarding dynamic test and evaluation environments for the EC community.

The rest of this chapter is organized as follows. The next section first introduces the concept of DOPs, then historically reviews dynamic test problems in the literature, and finally describes the major features and classification of existing dynamic test problems. Section 1.3 describes in detail some dynamic test problems and generators that are commonly used in the literature, covering the binary space, the real space, and the combinatorial space. Section 1.4 reviews the typical performance measures that are used by researchers to compare and justify their algorithms for DOPs. Section 1.5 presents the GDBG system. Finally, Section 1.6 concludes this chapter with some discussions on the future work on constructing dynamic test and evaluation environments for the EC community.

1.2 DOPs: Concepts, Brief Review, and Classification

1.2.1 Concepts of DOPs

In the literature of EC in dynamic environments, researchers usually define optimization problems that change over time as *dynamic problems* or *time-dependent problems*. In this chapter, we define *DOPs* as *a special class of dynamic problems that are solved online by an optimization algorithm as time goes by*.

It is notable that in many EDO studies, the terms "dynamic problems/time-dependent problems" and "DOPs" are not distinguished or are used interchangeably. In these studies, DOPs are either defined as a sequence of static problems linked up by some dynamic rules [4, 61, 62, 71, 73] or as a problem that has time-dependent parameters in its mathematical expression [5, 7, 20, 76], without explicitly mentioning whether the problems are solved online by an optimization algorithm or not. However, it is necessary to distinguish a DOP from a general time-dependent problem because, no matter how the problem changes, from the perspective of an EA or an optimization algorithm in general, a time-dependent problem is only different from a static problem if it is solved in a dynamic way, i.e., the algorithm needs to take into account changes during the optimization process as time goes by [10, 30, 48]. Hence, only DOPs are relevant to EDO research.

1.2.2 Dynamic Test Problems: Brief Review

In order to compare the performance of the developed GA approaches in dynamic environments, researchers have developed a number of dynamic problem generators. Generally speaking, DOPs are constructed via changing (the parameters of) stationary base problem(s). And, ideally, through proper control, different dynamic environments can be constructed from the stationary base problem(s) regarding the characteristics of the environmental dynamics, such as the frequency, severity, predictability, and cyclicity of environmental changes. Below, we briefly review the dynamic test environments that have been used/developed by researchers to test their EC approaches roughly in the time order.

In the early days, the dynamic test environments were quite simple: the environment is just switched between two or more stationary problems or between two or more states of one problem. For example, Cobb and Grefenstette [19] used a dynamic environment that oscillates between two different fitness landscapes. The dynamic 0-1 knapsack problem where the knapsack capacity oscillates between two or more fixed values has been frequently used in the literature [34, 43, 49]. The dynamic bit-matching problem has also been used by researchers for analyzing the performance of EC approaches in dynamic environments [65].

Later in 1999, several researchers have independently developed several dynamic environment generators by changing a base fitness landscape predefined in the multi-dimensional real space [9, 28, 44, 67]. This base fitness landscape consists of a number of peaks. Each peak can change its own morphology independently, such as the height, slope, and location of the peak. The center of the peak with the highest height is taken as the optimal solution of the landscape. Dynamic problems can be created through changing the parameters of each peak.

More recently, the dynamic 0-1 knapsack problem has been extended to dynamic multi-dimensional knapsack problems in several studies [14, 68]. In [78], Yang proposed a dynamic environment generator based on the concept of problem difficulty and unitation and trap functions. An XOR DOP generator, which can generate dynamic environments from any binary encoded stationary problem based on a bitwise exclusive-or (XOR) operator, has been proposed in [77, 87, 88]. In [57, 58, 60], Richter constructed spatio-temporal fitness landscapes based on coupled map lattices (CML) [15], where such properties as modality, ruggedness, information content, epistasis, dynamic severity, and Lyapunov exponents can be defined. Bosman [7] and Nguyen and Yao [52] investigated the online DOPs that have the time-linkage property, i.e., the current solution found by an optimization algorithm affects the future behaviour of the problem. In [53], Nguyen and Yao investigated dynamic constrained optimization problems where constraints change over time. The authors extended this study to provide a full set of dynamic constraine test problems in [50, 51]. In order to develop a unified approach of constructing dynamic problems across the binary space, the real space, and the combinatorial space, the GDBG system was recently proposed in [36, 37], which can be instantiated to construct dynamic test environments for all the three solution spaces.

In recent years, researchers have also studied a number of real-world DOPs. For example, Li *et al.* [35] studied the dynamic travelling salesman problem where the cities may change their locations. Mavrovouniotis and Yang investigated the dynamic travelling salesman problem where cities may join or leave the topology over time [40] and the traffic may change over time [41]. In [16, 17, 81], Cheng *et al.* studied the dynamic shortest path routing and dynamic multi-cast routing problems in mobile ad hoc networks (MANETs). In [42, 75], dynamic vehicle routing problems in logistics and transportation networks have been investigated by the EC community.

1.2.3 Major Characteristics and Classification of DOPs

As briefly reviewed above, many dynamic test problems have been used in the literature. These dynamic test problems have different characteristics and can be classified into different groups based on the following different criteria:

- Time-linkage: Whether the future behaviour of the problem depends on the current solution found by an algorithm or not.
- Predictability: Whether the generated changes are predictable or not.
- Visibility: Whether the changes are visible to the optimisation algorithm and, if so, whether changes can be detected by using just a few detectors.

- Constrained problem: Whether the problem is constrained or not.
- Number of objectives: Whether the problem is single objective or multiple objectives.
- Type of changes: Detailed explanation of how changes occur in the search space.
- Cyclicity: Whether the changes are cyclic/recurrent in the search space or not.
- Periodicity: Whether the changes are periodical or not in time.
- Factors that change: Changes may involve parameters of objective functions, domain of variables, number of variables, constraints, and other parameters.

The common characteristics of the general-purpose dynamic benchmark problems used in the literature are summarized as follows.

- Most dynamic test problems are non time-linkage problems. There are only a couple of general-purpose time-linkage test problems [7, 52] and some problem-specific time-linkage test problems [7, 8].
- Most of the dynamic test generators/problems in the continuous domain are unconstrained or domain constrained, except the two most recent studies [53, 59]
- In the default settings of most general-purpose dynamic test problems, changes are detectable by using just a few detectors. Exceptions are some problem instances in [19, 67] where only one or some peaks move, and in [53, 59] where the presences of the visibility mask or constraints make only some parts of the landscapes change. Due to their highly configurable property some benchmark generators can be configured to create scenarios where changes are more difficult to detect.
- In most cases, the factors that change are the objective functions. Exceptions are one instance in [36] where the dimension also changes and the problems in [53, 59] where the constraints also change.
- Many generators/problems have unpredictable changes in their default settings. Some of the generators/problems can be configured to allow predictable changes, at least in the frequency and periodicity of changes.
- A majority of benchmark generators/problems have cyclic/recurrent changes.
- Most benchmark generators/problems assume periodical changes, i.e., changes occur every fixed number of generations or fitness evaluations. An exception is the work in [64], which also studies the cases where the changes occur in some time pattern.
- Most generators/problems are single-objective. Only a few studies involve dynamic multi-objective problems, e.g., [23, 31].

In the next section, we describe in details some generators/problems that are commonly used in the domain of EC for DOPs in the real space, binary space, and combinatorial space, respectively.

1.3 Typical Dynamic Test Problems and Generators

1.3.1 Dynamic Test Problems in the Real Space

1.3.1.1 The DF1 Generator

The dynamic problem generator, called the DF1 generator, proposed by Morrison and De Jong [44], is a kind of moving peaks benchmark generators. Within DF1, the base landscape in the D-dimensional real space is defined as:

$$f(\mathbf{x}) = \max_{i=1,...,p} \left[H_i - R_i \times \sqrt{\sum_{j=1}^{D} (x_j - X_{ij})^2} \right] \quad (1.1)$$

where $\mathbf{x} = (x_1, \cdots, x_D)$ is a point in the landscape, p specifies the number of peaks (or cones), and each peak i is independently specified by its height H_i, its slope R_i, and its center $X_i = (X_{i1}, \cdots, X_{iD})$. These peaks are blended together by the *max* function. The fitness at a point on the surface is assigned the maximum height of all optima at that point; the optima with the greatest height at a point is said to be visible at that point.

DF1 creates dynamic problems by changing the features, i.e., the location, height, and slope, of each peak independently. The dynamics are controlled by the Logistics function given by:

$$\Delta_t = A \cdot \Delta_{t-1} \cdot (1 - \Delta_{t-1}) \quad (1.2)$$

where A is a constant value in the range [1.0, 4.0] and Δ_t is used as the step size of changing a particular parameter (i.e., the location, height, or slope) of peaks at iteration t after scaled by a scale factor s in order to reduce step sizes that may be larger than intended for each step.

The logistics function allows a wide range of dynamic performance by a simple change of the value of A, from simple constant step sizes, to step sizes that alternate between two values, to step sizes that rotate through several values, to completely chaotic step sizes. More details on the DF1 generator can be found in [44].

1.3.1.2 The Moving Peaks Benchmark (MPB) Problem

Branke [9] proposed the MPB problem, which has been widely used as dynamic benchmark problems in the literature. Similar to the DF1 generator described above, the MPB problem consists of a multi-dimensional fitness landscape in the real space with a number of peaks, where each peak has three features, i.e., the height, width, and central position. Within the MPB problem, the optima can be changed by changing the three features of each peak independently or in a correlative way.

For the D-dimensional landscape, the MPB problem is defined as follows:

$$F(\mathbf{x},t) = \max_{i=1,...,p} \frac{H_i(t)}{1 + W_i(t) \sum_{j=1}^{D} (x_j(t) - X_{ij}(t))^2}, \quad (1.3)$$

Table 1.1 Default settings for the MPB problem

Parameter	Value
p (the number of peaks)	10
U (change frequency)	5000
height severity	7.0
width severity	1.0
peak shape	cone
basic function	no
s (the shift length)	1.0
D (the number of dimensions)	5
λ (the correlation coefficient)	0
S (the range of allele values)	[0, 100]
H (the range of the height of peaks)	[30.0, 70.0]
W (the range of the width of peaks)	[1, 12]
I (the initial height for all peaks)	50.0

where $W_i(t)$ and $H_i(t)$ are the height and width of peak i at time t, respectively, and $X_{ij}(t)$ is the j-th element of the location of peak i at time t. The p independently specified peaks are blended together by the *max* function. The position of each peak is shifted in a random direction by a vector $\mathbf{v_i}$ of a distance s (s is also called the shift length, which determines the severity of the problem dynamics), and the move of a single peak can be described as follows:

$$\mathbf{v}_i(t) = \frac{s}{|\mathbf{r} + \mathbf{v}_i(t-1)|}((1-\lambda)\mathbf{r} + \lambda \mathbf{v}_i(t-1)), \tag{1.4}$$

where the shift vector $\mathbf{v}_i(t)$ is a linear combination of a random vector $\mathbf{r}$ and the previous shift vector $\mathbf{v}_i(t-1)$ and is normalized to the shift length s. The correlated parameter λ is set to 0, which implies that the peak movements are uncorrelated.

More formally, a change of a single peak can be described as follows:

$$H_i(t) = H_i(t-1) + height\_severity * \sigma \tag{1.5}$$

$$W_i(t) = W_i(t-1) + width\_severity * \sigma \tag{1.6}$$

$$\mathbf{X}_i(t) = \mathbf{X}_i(t)(t-1) + \mathbf{v}_i(t) \tag{1.7}$$

where σ is a normal distributed random number with mean zero and variation of 1.

The default settings for the MPB benchmark typically used in the literature can be found in Table 1.1, which are corresponding to Scenario 2 in [9]. In Table 1.1, the change frequency (U) means that environment changes every U fitness evaluations.

1.3.2   Dynamic Test Problems in the Binary Space

1.3.2.1   The Dynamic Bit-Matching Problem

The dynamic bit-matching problem has been used by researchers for the analysis of the performance of EAs in dynamic environments [21, 65]. For example, Stanhope and Daida [65] analyzed the behaviour of a simple (1+1) EA based on the dynamic bit-matching problem. In the dynamic bit-matching problem, an algorithm needs to find solutions that minimize the Hamming distance to an arbitrary target pattern (i.e., the match string) that may change over time. Given a solution $\mathbf{x} \in \{0,1\}^L$ (L is the length of the binary encoding of a solution for the problem) and the target pattern $\mathbf{a} \in \{0,1\}^L$ at time t, the Hamming distance between them is calculated as:

$$d_{Hamming}(\mathbf{x}(t), \mathbf{a}(t)) = \sum_{i=1}^{i=L} |x_i(t) - a_i(t)| \qquad (1.8)$$

The dynamics of the problem is controlled by two parameters, g and d, which control the number of generations between changes and the degree (Hamming distance) by which the target pattern $\mathbf{a}$ is altered (for each change, d distinct and randomly chosen bits in $\mathbf{a}$ are inverted), respectively. For example, setting $(g,d) = (0,0)$ results in a stationary function whereas setting $(g,d) = (10,5)$ means that every 10 generations, the target pattern $\mathbf{a}$ changes by 5 bits randomly.

1.3.2.2   Dynamic Knapsack Problems (DKP) and Dynamic
Multi-dimensional Knapsack Problems (DMKP)

The knapsack problem [32] is a classical NP-hard combinatorial optimization problem, where the solution space belongs to the binary space. Given a set of items, each of which has a weight and a profit, and a knapsack with a fixed capacity, the problem aims to select items to fill up the knapsack to maximize the total profit while satisfying the capacity constraint of the knapsack. Suppose there are n items, and $\mathbf{w}$, $\mathbf{p}$, and C, denote the weights of items, the profits of items, and the capacity of the knapsack, respectively. Then, the knapsack problem can be defined as follows:

$$Max\, f(\mathbf{x}) = \sum_{i=1}^{n} p_i \cdot x_i \qquad (1.9)$$

$$subject\,to: \quad \sum_{i=1}^{n} w_i \cdot x_i \leq C, \qquad (1.10)$$

where $\mathbf{x} \in \{0,1\}^n$ is a solution, $x_i \in \{0,1\}$ indicates whether item i is included in the subset or not, p_i is the profit of item i, and w_i is the weight of item i.

The above knapsack problem has been frequently used to test the performance of EAs in stationary environments, and its dynamic version has also been used by researchers to test the performance of EAs in dynamic environments [34, 43, 49]. In the dynamic knapsack problem, the system dynamics can be constructed by changing the weights of items, the profits of items, and/or the knapsack capacity over time according to some dynamics, respectively. So, the dynamic knapsack problem can be described as follows:

$$Max\ f(\mathbf{x},t) = \sum_{i=1}^{n} p_i(t) \cdot x_i \qquad (1.11)$$

$$subject\ to: \ \sum_{i=1}^{n} w_i(t) \cdot x_i \leq C(t), \qquad (1.12)$$

where the weight and profit of each item may be bounded in the range of $[l_w, u_w]$, $[l_p, u_p]$, and the capacity of knapsack may be bounded in the range of $[l_c, u_c]$.

Similarly, the static multi-dimensional knapsack problem (MKP) belongs to the class of NP-complete problems, which has a wide range of real-world applications, such as cargo loading, selecting projects to fund, budget management, etc. In the MKP, we have a number of resources (knapsacks), each of which has a capacity, and a set of items, each of which has a profit and consumes some amount of each resource. The aim is to select items to maximize the total profit while satisfying the capacity constraints for all resources.

The DMKP has recently been used to investigate the performance of EAs for DOPs [14, 68]. As in the DKP, in the DMKP, the profit and resource consumption for each item as well as the capacity of each resource may change over time. Let $\mathbf{r}, \mathbf{p},$ $\mathbf{c}$ denote the resource consumptions of items, the profits of items, and the capacities of resources, respectively. Then, the DMKP can be defined as follows:

$$Max\ f(\mathbf{x},t) = \sum_{i=1}^{n} p_i(t) \cdot x_i \qquad (1.13)$$

$$subject\ to: \ \sum_{i=1}^{n} r_{ij}(t) \cdot x_i \leq c_i(t), j = 1, 2, \cdots, m \qquad (1.14)$$

where n is the number of items, m is the number of resources, x_i and p_i are as defined above, $r_{ij}(t)$ denotes the resource consumption of item i for resource j at time t, and $c_i(t)$ is the capacity constraint of resource i at time t. The system dynamics can be constructed by changing the profits of items, resource consumptions of items, and the capacity constraints of resources within certain upper and lower bounds over time according to some dynamics, respectively.

1.3.2.3   The XOR DOP Generator

In [77, 87], an XOR DOP generator that can generate dynamic environments from any binary encoded stationary problem using a bitwise exclusive-or (XOR) operator has been proposed. Given a stationary problem $f(\mathbf{x})$ ($\mathbf{x} \in \{0,1\}^l$ where l is the length of binary encoding), DOPs can be constructed from it as follows. Suppose the environment is changed every τ generations. For each environmental period k, an XORing mask $\mathbf{M}(k)$ is first incrementally generated as follows:

$$\mathbf{M}(k) = \mathbf{M}(k-1) \oplus \mathbf{T}(k), \tag{1.15}$$

where "$\oplus$" is the XOR operator (i.e., $1 \oplus 1 = 0$, $1 \oplus 0 = 1$, $0 \oplus 0 = 0$) and $\mathbf{T}(k)$ is an intermediate binary template randomly created with $\rho \times l$ ($\rho \in [0.0, 1.0]$) ones inside it for environmental period k. Initially, $\mathbf{M}(0)$ is set to a zero vector. Then, the individuals at generation t are evaluated using the following formula:

$$f(\mathbf{x}, t) = f(\mathbf{x} \oplus \mathbf{M}(k)), \tag{1.16}$$

where $k = \lfloor t/\tau \rfloor$ is the environmental period index.

With this XOR DOP generator, the environmental dynamics can be tuned by two parameters: τ controls the speed of environmental changes while ρ controls the severity of changes. The bigger the value of τ, the slower the environment changes. The bigger the value of ρ, the more severe the environment changes.

The aforementioned XOR DOP generator in fact can construct *random dynamic environments* because there is no guarantee that the environment will return to a previous one after certain changes. In order to test the performance of memory based EAs, the XOR generator has been extended to construct *cyclic dynamic environments* in [79] and *cyclic dynamic environments with noise* further in [88].

With the XOR generator, cyclic dynamic environments can be constructed as follows. First, we can generate $2K$ XOR masks $\mathbf{M}(0), \cdots, \mathbf{M}(2K-1)$ as the *base states* in the search space randomly or in a certain patter. Then, the environment can cycle among these base states in a fixed logical ring. Suppose the environment changes every τ generations, then an individual at generation t is evaluated as follows:

$$f(\mathbf{x}, t) = f(\mathbf{x} \oplus \mathbf{M}(I_t)) = f(\mathbf{x} \oplus \mathbf{M}(k\%(2K))), \tag{1.17}$$

where $k = \lfloor t/\tau \rfloor$ is the index of current environmental period and $I_t = k\%(2K)$ is the index of the base state that the environment is in at generation t.

The $2K$ XOR masks can be generated in the following way. First, we construct K binary templates $\mathbf{T}(0), \cdots, \mathbf{T}(K-1)$ that form a random partition of the search space with each template containing $\rho \times l = l/K$ bits of ones[1]. Let $\mathbf{M}(0) = \mathbf{0}$ denote the initial state. Then, the other XOR masks are generated iteratively as follows:

$$\mathbf{M}(i+1) = \mathbf{M}(i) \oplus \mathbf{T}(i\%K), i = 0, \cdots, 2K-1 \tag{1.18}$$

So, the templates $\mathbf{T}(0), \cdots, \mathbf{T}(K-1)$ are first used to create K masks till $\mathbf{M}(K) = \mathbf{1}$ and then orderly reused to construct another K masks till $\mathbf{M}(2K) = \mathbf{M}(0) = \mathbf{0}$. The Hamming distance between two neighbour XOR masks is the same and equals $\rho \times l$. Here, $\rho \in [1/l, 1.0]$ is the distance factor, determining the number of base states.

---

[1] In the partition each template $\mathbf{T}(i)$ ($i = 0, \cdots, K-1$) has randomly but exclusively selected $\rho \times l$ bits set to 1 while other bits set to 0. For example, $\mathbf{T}(0) = 0101$ and $\mathbf{T}(1) = 1010$ form a partition of the 4-bit search space.

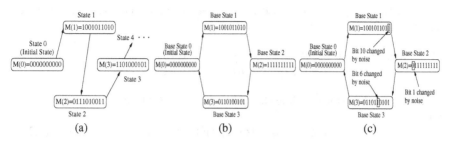

Fig. 1.1 Illustration of three kinds of dynamic environments constructed from a 10-bit encoded function with $\rho = 0.5$: (a) random, (b) cyclic, and (c) cyclic with noise

From the above cyclic environment generator, we can further construct cyclic dynamic environments with noise as below. Each time the environment is about to move to a next base state $\mathbf{M}(i)$, $\mathbf{M}(i)$ is bitwise flipped with a small probability p_n. Figure 1.1 illustrates the construction of random, cyclic, and cyclic with noise dynamic environments respectively from a 10-bit function with $\rho = 0.5$, where the XORing mask is used to represent the environmental state.

This XOR DOP generator has two properties. One is that the distances among the solutions in the search space remains unaltered after an environmental change. The other is that the properties of the fitness landscape are not changed after an environmental change, which facilitates the analysis of the behavior of algorithms. Recently, the XOR DOP generator has been extended to construct dynamic problems in the real space [66]. In [66], two continuous dynamic problem generators were proposed using the linear transformation of individuals. The first generator does the linear transformation by changing the direction of some axes of the search space while the second one uses successive rotations in different planes.

1.3.3   Dynamic Test Problems in the Combinatorial Space

1.3.3.1   The Spatio-Temporal Fitness Landscapes

In [57, 58], Richter constructed spatio-temporal fitness landscapes based on Coupled Map Lattices (CML). The idea of using CML to construct dynamic fitness landscapes is interesting since CML facilitate efficient computing of the fitness landscape and can reveal a broad variety of complex spatio-temporal behavior [15]. In [59], Richter further analyzed and quantified the properties of spatio-temporal fitness landscapes constructed from CML using topological and dynamical landscape measures such as modality, ruggedness, information content, epistasis, dynamic severity, and two types of dynamic complexity measures, Lyapunov exponents and bred vector dimension.

In order to build spatio-temporal landscape based on CML, Richter defined a lattice grid with $I \times J$ cells in 2-dimension. A height at time t is assigned to each cell referred to $h(i,j,t)$, where i and j denote the indices of cells. The change of the height of a cell is as follows:

$$h(i,j,t) = (1-\varepsilon)g(h(i,j,t)) + \frac{\varepsilon}{4}[g(h(i-1,j,t)) \\ +g(h(i+1,j,t))+g(h(i,j-1,t))+g(h(i,j+1,t))] \quad (1.19)$$

where ε is the diffusion coupling strength and $g(h(i,j,t))$ is logistic mapping function by:

$$g(h(i,j,t)) = \alpha h(i,j,t)(1-h(i,j,t)) \quad (1.20)$$

and the period boundary conditions is rendered by:

$$h(I+1,j,t) = h(1,j,t), h(i,J+1,t) = h(i,1,t) \quad (1.21)$$

To convert this integer system into real-valued fitness landscape, scaling factors s_1 and $s_2 \in \mathbb{R}$ are employed for the vertical (i) and horizontal extension (j) so that the search space variable $x = (x_1,x_2)^T$ is obtained by a rounding condition: $(i,j)^T = (\lceil s_1 x_1 \rceil, \lceil s_2 x_2 \rceil)^T$. Finally, the spatio-temporal fitness landscape in 2-D is produced by:

$$f(x,t) = \left\{ \begin{array}{cc} h(\lceil s_1 x_1 \rceil, \lceil s_2 x_2 \rceil, t) \; for & \begin{array}{c} 1 \leq \lceil s_1 x_1 \rceil \leq I \\ 1 \leq \lceil s_2 x_2 \rceil \leq J \end{array} \\ 0 & otherwise \end{array} \right\}, t \geq 0 \quad (1.22)$$

1.3.3.2 Dynamic Travelling Salesman Problems (DTSPs)

TSP is another classical NP-complete combinatorial problem. DTSPs have a wide range of real applications, especially in the optimization of dynamic networks, like network planning and designing, load-balance routing, and traffic management.

In [35], a DTSP is defined as a TSP with a dynamic cost (distance) matrix as:

$$D(t) = \{d_{ij}(t)\}_{n*n} \quad (1.23)$$

where $d_{ij}(t)$ is the cost from city i to city j, n is the number of cities. DTSP can be defined as $f(x,t)$, the objective of DTSP is to find a minimum-cost route containing all cities at time t. It can be described as:

$$f(x,t) = Min(\sum_{i=1}^{n} d_{T_i,T_{i+1}}(t)) \quad (1.24)$$

where $x_i \in 1,2,\cdots,n$ denotes the i-th city in the solution such that $x_{n+1} = x_1$ and, if $i \neq j$, $x_i \neq x_j$.

1.3.3.3 Dynamic Routing Problems in Communication Networks

In [16, 17, 81], Cheng *et al.* studied the dynamic shortest path routing and dynamic multi-cast routing problems in mobile ad hoc networks (MANETs).

The MANETs in [16, 17, 81] were modeled within a fixed geographical graph G_0, which is composed by a set of wireless nodes and a set of communication links connecting two neighbor nodes that fall into the radio transmission range. To simulate the wireless network in dynamic environments in real world, two different kind of models were created, which are the general dynamics model and the worst dynamics model, respectively. In the general dynamics model, periodically or randomly, some nodes are scheduled to sleep or some sleeping nodes are scheduled to wake up due to energy conservation. While, in the worst model, each change is produced manually by removal of a few links on the current best multi-cast tree.

The models can be described by a MANET $G(V, E)$ and a multi-cast communication request from node s to a set of receivers R with a delay upper bound δ. So, the dynamic delay-constrained multi-cast routing problem is to find a series of trees $\{T_i | i \in \{0, 1, \ldots\}\}$ over a series of graphs $\{G_i | i \in \{0, 1, \ldots\}\}$, which satisfy the dylay constraint and have the least tree cost as follows:

$$\max_{r_j \in R} \left\{ \sum_{l \in P_T(s, r_j)} d_l \right\} \leq \delta \tag{1.25}$$

$$C(T_i) = \min_{T \in G_i} \left\{ \sum_{l \in T(V_T, E_T)} c_l \right\} \tag{1.26}$$

where $G_i(V_i, E_i)$ is MANET topology after the i-th change, $R = \{r_0, r_1, \cdots, r_m\}$ is a set of receivers of a multi-cast request, $T_i(V_{T_i}, E_{T_i})$ is a multi-cast tree with nodes V_{T_i} and links E_{T_i}, $P_T(s, r_j)$ is a path from s to r_j on the tree T_i, d_l represents the transmission delay on the communication link l, C_{T_i} is the cost of the tree T_i, and $\delta(P_i)$ is the total transmission delay on the path P_i.

1.3.3.4 Dynamic Vehicle Routing Problems

In [42, 75], dynamic vehicle routing problems have been investigated by the EC community. In [75], a freight transportation planning models was proposed. A standardized container with an extent of roughly $7.5\text{m} \times 2.6\text{m} \times 2.7\text{m}$, termed by a swap body b, was considered as the basic unit of freight. All possible transportation means are referred as F, and all trucks $tr \in F$ can carry $\hat{v}(tr) = 2$ swap bodies at once whereas the capacity limits of trains $z \in F$ are usually between 30 and 60 ($\hat{v}(z) \in [30, 60]$). The schedules of trains, e.g., routes, departure times, and arrival times, are fixed. Freight trucks can take any route on the map, but must perform cyclic tours. The locations where a freight may be collected or delivered are referred to L. Each transportation order has a fixed time window $[\check{t}_s, \hat{t}_s]$ in which it must be collected from its source $l_s \in L$ and a destination location and time window $[\check{t}_d, \hat{t}_d]$ in which it must be delivered to its destination $l_d \in L$. It also has a volume v that is the capacity of a swap body times by an integer. Therefore, a transportation order o can be described by a tuple $o = \langle l_s, l_d, [\check{t}_s, \hat{t}_s], [\check{t}_d, \hat{t}_d], v \rangle$. In the model, all orders that need more than one ($v > 1$) swap body are split into multiple orders where each requires one swap body.

Finally, the planning process becomes a set of R tours, each tour r is described by a tuple $r = \langle l_s, l_d, f, \check{t}, \hat{t}, \underline{b}, \underline{o} \rangle$ where l_s and l_d are the start and destination locations, $\check{t}$ and $\hat{t}$ are the departure and arrival time, $\underline{b} = \{b_1, b_2, \ldots\}$ is a set of swap bodies which are carried by the vehicle $f \in F$ and contains the goods assigned to the orders $\underline{o} = \{o_1, o_2, \ldots\}$.

1.4 Performance Metrics

In addition to develop different dynamic benchmark generators and problems for testing EAs for DOPs, another relevant issue is how to compare different algorithms. Over the years, researchers have developed a number of different performance measures to evaluate the developed EA approaches for DOPs. The widely used performance measures, which can be classified into two main groups: optimality-based and behaviour-based, are reviewed as follows.

1.4.1 Optimality-Based Performance Measures

Optimality-based performance measures are measures that evaluate the ability of algorithms in finding the solutions with the best objective/fitness values (fitness-based measures) or finding the solutions that are closest to the global optimum (distance-based measures). This type of measures is by far the most common in EDO. The measures can be categorised into groups as follow:

1.4.1.1 Best-of-Generation

This measure is calculated as the averages for many runs of the best values at each generation on the same problem. It is usually used in two ways: First, the best value in each generation is plotted against the time axis to create a performance curve. This measure has been used since the early research in [5, 18, 25, 27, 28]. This measure is still one of the most commonly used measures in the literature. The advantage of such performance curves is that they can show the whole picture of how the tested algorithm has performed. However, because the performance curve is not quantitative, it is difficult to compare the final outcome of different algorithms and to see if the difference between two algorithms is statistically significant [47].

To improve the above disadvantage, a variation of the measure is proposed where the best-of-generation values is averaged over all generations [86]. The measure is described below:

$$\overline{F}_{BOG} = \frac{1}{G} \times \sum_{i=1}^{i=G} \left(\frac{1}{N} \times \sum_{j=1}^{j=N} F_{BOG_{ij}} \right) \tag{1.27}$$

where $\overline{F}_{BOG}$ is the mean best-of-generation fitness, G is the number of generations, N is the total number of runs, and $F_{BOG_{ij}}$ is the best-of-generation fitness of generation i of run j of an algorithm on a particular problem. An identical measure to the $\overline{F}_{BOG}$, but with different name, the *collective mean fitness*, was proposed by Morrison [47] at the same time.

Recently the idea of plotting performance curves was adapted in [3] to create two measures: the *area below a curve*, which is calculated as the definite integral of F_{BOG} (or other measures such as F_C or offline error/performance) over the optimisation process; and the *area between curves*, which is the area spanned between the performance curves of two algorithms.

The $\overline{F}_{BOG}$ is one of the most commonly used measures. The advantage of this measure, as mentioned above, is to enable algorithm designers to quantitatively compare the performance of algorithms. The disadvantage of the measure and its variants is that they are not normalised, hence can be biased by the difference of the fitness landscapes at different periods of change. For example, if at a certain period of change the overall fitness values of the landscape is particularly higher than those at other periods of changes, or if an algorithm is able to get particular high fitness value at a certain period of change, the final $\overline{F}_{BOG}$ or F_C might be biased toward the high fitness values in this particular period and hence might not correctly reflect the overall performance of the algorithm. Similarly, if $\overline{F}_{BOG}$ is used averagely to evaluate the performance of algorithms in solving a group of problems, it is also biased toward problems with larger fitness values.

1.4.1.2 Best-Error-Before-Change

Proposed in [67] and named *Accuracy* by the authors, this measure is calculated as the average of the smallest errors (the difference between the optimum value and the value of the best individual) achieved at the end of each change period (right before the moment of change).

$$E_B = \frac{1}{m} \sum_{i=1}^{m} e_B(i) \qquad (1.28)$$

where $e_B(i)$ is the best error just before the ith change happens; m is the number of changes.

This measure is useful in situations where we are interested in the final solution that the algorithm achieved before the change. The measure also makes it possible to compare the final outcome of different algorithms. However, the measure also has three important disadvantages. First, it does not say anything about how the algorithms have done to achieve the current performance. As a result, the measure is not suitable if what users are interested in is the overall performance or behaviours of the algorithms. Second, similar to the best-of-generation measure, this measure is also not normalised and hence can be biased toward periods where the errors are relatively very large. Third, the measure requires that the global optimum value at each change is known.

This measures is adapted as the basis for one of the complementary performance measures in the CEC'09 competition on dynamic optimisation [36].

1.4.1.3 Modified Offline Error and Offline Performance

Proposed in [11] and [12], the *modified offline error* is measured as the average over, at every evaluations, the error of the best solution found since the last change of the environment. This measure is always greater than or equal to zero and would be zero for a perfect performance.

$$E_{MO} = \frac{1}{n} \sum_{j=1}^{n} e_{MO}(j) \qquad (1.29)$$

where n is the number of generations so far, and $e_{MO}(j)$ is the best error since the last change gained by the algorithm at the generation j.

A similar measure, the *modified offline performance*, is also proposed in the same reference to evaluate algorithm performance in case the exact values of the global optima are not known

$$P_{MO} = \frac{1}{n} \sum_{j=1}^{n} F_{MO}(j) \qquad (1.30)$$

where n is the number of generations so far, and $F_{MO}(j)$ is the best performance since the last change gained by the algorithm at the generation j.

With this type of measures, the faster the algorithm to find a good solution, the higher the score. Similar to the $\overline{F}_{BOG}$, the offline error/performance are also useful in evaluating the overall performance of an algorithm and to compare the final outcomes of different algorithms. These measures however have some disadvantages. First, they require that the time a change occurs is known. Second, similar to $\overline{F}_{BOG}$, these measures are also not normalised and hence can be biased under certain circumstances.

In [51][Sect. 5.3.2], the offline error/performance was modified to measure the performance of algorithms in dynamic constrained environments. Specifically, when calculating Eq. (1.29) for dynamic constrained problems, the authors only consider the best errors/fitness values of *feasible* solutions at each generation. If in any generation there is no feasible solution, the measure will take the worst possible value that a feasible solution can have for that particular generation.

Recently based on the modified offline error a new measure named *best known peak error* (BKPE) [6] was proposed to measure the convergence speed of the algorithm in tracking optima. Different to the modified offline error, in BKPE at each generation the error is calculated for each known peak, i.e. it is the difference between the best found solution in the peak and the top of the peak. Then immediately before a change, the error of the best individual on a known peak is added to the total error for the run.

1.4.1.4 Optimisation Accuracy

The *optimisation accuracy* measure (also known as the *relative error*) was initially proposed in [24] and was adopted in [72] for the dynamic case:

$$accuracy_{F,EA}^{(t)} = \frac{F(best_{EA}^{(t)}) - Min_F^{(t)}}{Max_F^{(t)} - Min_F^{(t)}} \tag{1.31}$$

where $best_{EA}^{(t)}$ is the best solution in the population at time t, $Max_F^{(t)} \in \mathbb{M}$ is the best fitness value of the search space and $Min_F^{(t)} \in \mathbb{M}$ is the worst fitness value of the search space. The range of the accuracy measure ranges from 0 to 1, with a value of 1 and 0 represents the best and worst possible values, respectively.

The optimisation accuracy have the same advantages as the $\overline{F}_{BOG}$ and E_{MO} in providing quantitative value and in evaluating the overall performance of algorithms. The measure has an advantage over $\overline{F}_{BOG}$ and E_{MO}: it is independent to fitness rescalings and hence become less biased to those change periods where the difference in fitness becomes particularly large. The measure, however, has a disadvantage: it requires information about the absolute best and worst fitness values in the search space, which might not always be available in practical situations. In addition, as pointed by the author himself [72], the optimisation accuracy measure is only well-defined if the complete search space is not a plateau at any generation t, because otherwise the denominator of Eq. (1.31) at t would be equal to zero.

1.4.1.5 Normalised Scores

Another useful way to avoid the possible biases caused different fitness scales in different change periods and/or different problems is to use the *normalised score* [51], which evaluates the overall performance of an algorithm compared to other peer algorithms in solving a group of problems in a normalised way. The idea is that, given a group of n tested algorithms and m test instances (which could be m different test problems or m change periods of a problem), for each instance j the performance of each algorithm is normalised to the range $(0, 1)$ so that the best algorithm in this instance j will have the score of 1 and the worst algorithm will get the score of 0. The final overall score of each algorithm will be calculated as the average of the normalised scores from each individual instance. According to this calculation, if an algorithm is able to perform best in all tested instances, it will get an overall score of 1. Similarly, if an algorithm performs worst in all tested instances, it will get an overall score of 0.

Given a group of n tested algorithms and m test instances, a formal description of the *normalised score* of the ith algorithm is given in Eq. (1.32):

$$S_{norm}(i) = \frac{1}{m} \sum_{j=1}^{m} \frac{|e_{max}(j) - e(i,j)|}{|e_{max}(j) - e_{min}(j)|}, \forall i = 1 : n. \tag{1.32}$$

where $e(i,j)$ is the modified offline error of algorithm i in test instance j; and $e_{max}(j)$ and $e_{min}(j)$ are the largest and smallest errors among all algorithms in solving instance j. In case the offline errors of the algorithms are not known (because global optima are not know), we can replace them by the offline performance to get exactly the same score. The normalised score S_{norm} can also be calculated based on the best-of-generation values.

The normalised score has two major advantages. First, it is unbiased. The fact that an algorithm might get a very large or very small error on a particular problem or on a particular change period will not bias the overall score as it does when we use the traditional measures. Second, it does not need the knowledge of the global optima or the absolute best and worst fitness values of a problem.

The normalised score, however, also has its own disadvantages: First, S_{norm} is only feasible in case an algorithm is compared to other peer algorithms because the scores are calculated based on the performance of peer algorithms. Second, S_{norm} only shows the relative performance of an algorithm in comparison with other peer algorithms in the corresponding experiment. It cannot be used solely as an absolute score to compare algorithm performance from different experiments. For this purpose, we need to gather the offline errors/offline performance/best-of-generation of the algorithms first, then calculate the normalised score S_{norm} for these values. For example, assume that we have calculated S^A_{norm} for all algorithms in group A, and S^B_{norm} for all algorithms in group B in a separated experiment. If we need to compare the performance of algorithms in group A with algorithms in group B, we cannot compare the S^A_{norm} against S^B_{norm} directly. Instead, we need to gather the $E_{MO}/P_{MO}/F_{BOG}$ of all algorithms from the two groups first, then based on these errors we calculate the normalised scores S^{AB}_{norm} of all algorithms in the two groups.

1.4.1.6   Distance-Based Measures

Although most of the optimality-based measures are fitness-based, some performance measures do rely on the distances from the current solutions to the global optimum to evaluate algorithm performance. In [74], a performance measure, which is calculated as the minimum distance from the individuals in the population to the global optimum, was proposed. In [63], another distance-based measure was introduced. This measure is calculated as the distance from the mass centre of the population to the global optimum.

Euclidean distance-based measures are also commonly used to evaluate the performance of dynamic multi-objective (DMO) optimisation algorithms. In [91] the performance of DMO algorithms are evaluated based on the *generational distance* (GD) [69] between the approximated front (which contains the current best function values) and the Pareto optimal front at the moment just before a change occurs. In [23] two measures, one is based on the minimum Euclidean distance between members of the approximated front and the Pareto front, and the other is based on the minimum Euclidean distance between members of the approximated set and the Pareto set, were proposed. In [29], these two measures were extended using the idea of modified offline-error. In [38], a modified version of the original GD named *reversed GD* was proposed for the dynamic case. In [26], an offline measure named *variable space generational distance* was also proposed and was calculated based on the distance between the approximated set and the Pareto set at each time step.

The advantage of distance-based measures is that they are independent to fitness rescalings and hence are less affected by possible biases caused by the difference in fitness of the landscapes in different change periods. The disadvantages of these measures are that they require knowledge about the exact position of the global optimum, which is not always available in practical situation. In addition, compared to some other measures this type of measures might not always correctly approximate the exact adaptation characteristics of the algorithm under evaluated, as shown in an analysis in [72].

1.4.2 Behaviour-Based Performance Measures

Behaviour-based performance measures are those that evaluate whether EDO algorithms exhibit certain behaviours that are believed to be useful in dynamic environments. Example of such behaviours are maintaining high diversity through out the run; quickly recovering from a drop in performance when a change happens, and limiting the fitness drops when changes happen. These measures are usually used complementarily with optimality-based measures to study the behaviour of algorithms. They can be categorised into the following groups.

1.4.2.1 Diversity

Diversity-based measures, as their name imply, are used to evaluate the ability of algorithms in maintaining diversity to deal with environmental dynamics. There are many diversity-based measures, e.g. *entropy* [43], *Hamming distance* [54, 56, 80], *moment-of-inertia* [46], *peak cover* [11], and *maximum spread* [26] of which Hamming distance-based measures are the most common.

Hamming distance-based measures for diversity have been widely used in static evolutionary optimisation and one of the first EDO research to use this measure for dynamic environments is the study of [54] where the *all possible pair-wise Hamming distance* among all individuals of the population was used as the diversity measure. In [56] the measure was modified so that only the Hamming distances among the best individuals are taken into account.

A different and interesting diversity measure is the *moment-of-inertia* [46], which is inspired from the fact that the moment of inertia of a physical, rotating object can be used to measure how far the mass of the object is distributed from the centroid. Morrison and De Jong [46] applied this idea to measuring the diversity of an EA population. Given a population of P individuals in N-dimensional space, the coordinates $C = (c_1, ..., c_N)$ of the centroid of the population can be computed as follows:

$$c_i = \frac{\sum_{j=1}^{P} x_{ij}}{P} \tag{1.33}$$

where x_{ij} is the ith coordinate of the jth individual and c_i is the ith coordinate of the centroid.

Given the computed centroid above, the moment-of-inertia of the population is calculated as follows:

$$I = \sum_{i=1}^{N} \sum_{j=1}^{P} (x_{ij} - c_i)^2 \tag{1.34}$$

In [46], the authors proved that the moment-of-inertia measure is equal to the pairwise Hamming distance measure. The moment-of-inertia, however, has an advantage over the Hamming distance measure: it is more computationally efficient. The complexity of computing the moment-of-inertia is only linear with the population size P while the complexity of the pair-wise diversity computation is quadratic.

Another interesting, but less common diversity measure is the *peak cover* [11], which counts the number of peaks covered by the algorithms over all peaks. This measure requires full information about the peaks in the landscape and hence is only suitable in academic environments.

Diversity measures are also used in dynamic multi-objective approaches. In [26] the *maximum spread* commonly used in static MO was modified for the dynamic case by calculating the average value of the maximum spread over all generations when time goes by. In [38], the diversity-based *hypervolume* (HV) measure [69] commonly used in static MO was extended to a dynamic measure HVR(t), which is the ratio between the dynamic HV of the approximated front and the Pareto front.

In dynamic constrained environments, a diversity-related measure was also proposed [51][Sect 5.3.2], which counts the percentage of solutions that are infeasible among the solutions selected in each generation. The average score of this measure (over all tested generations) is then compared with the percentage of infeasible areas over the total search area of the landscape. If the considered algorithm is able to treat infeasible diversified individuals and feasible diversified individuals on an equal basis (and hence to maintain diversity effectively), the two percentage values should be equal.

1.4.2.2   Performance Drop after Changes

Some EDO studies also develop measures to evaluate the ability of algorithms in restricting the drop of fitness when a change occurs. Of which, the most representative measures are the measures *stability* [72], *satisficability* and *robustness* [56].

The measure *stability* is evaluated by calculating the difference in the fitness-based *accuracy* measure (see Eq. (1.31)) of the considered algorithm between each two time steps

$$stab_{F,EA}^{(t)} = max\{0, accuracy_{F,EA}^{(t-1)} - accuracy_{F,EA}^{(t)}\} \qquad (1.35)$$

where $accuracy_{F,EA}^{(t)}$ has already been defined in Eq. (1.31).

The *robustness* measure is similar to the measure *stability* in that it also determines how much the fitness of the next generation of the EA can drop, given the current generation's fitness. The measure is calculated as the ratio of the fitness values of the best solutions (or the average fitness of the population) between each two consecutive generations.

The *satisficability* measure focuses on a slightly different aspect. It determines how well the system is in maintaining a certain level of fitness and not dropping below a pre-set threshold. The measure is calculated by counting how many times the algorithm is able to exceed a given threshold in fitness value.

1.4.2.3 Convergence Speed after Changes

Convergence speed after changes, or the ability of the algorithm to recover quickly after a change, is also an aspect that attracts the attention of various studies in EDO. In fact many of the optimality-based measures, such as the offline error/performance, best-of-generation, relative-ratio-of-best-value discussed previously can be used to indirectly evaluate the convergence speed. In addition, in [72], the author also proposed a measure dedicated to evaluating the ability of an adaptive algorithm to react quickly to changes. The measure is named *reactivity* and is defined as follows:

$$
react^{(t)}_{F,A,\varepsilon} = min \left\{ t' - t | t < t' \leq maxgen, t' \in \mathbb{N}, \frac{accuracy^{(t')}_{F,A}}{accuracy^{(t)}_{F,A}} \geq (1 - \varepsilon) \right\} \cup \{maxgen - t\}
\tag{1.36}
$$

where *maxgen* is the number of generations. The *reactivity* measure has a disadvantage: it is only meaningful if there is actually a drop in performance when a change occurs. Otherwise, the value of the measure *reactivity* is always zero and nothing can be said about how well the algorithm reacts to changes. In situations like the dynamic constrained benchmark problems in [53] where the total fitness level of the search space may increase after a change, the measure *reactivity* cannot be used.

To avoid the disadvantage of *reactivity* and to provide more insights on the convergence behaviour of algorithms, recently a pair of measures, the *recovery rate* (RR) and the *absolute recovery rate* (ARR) were proposed [51]. The RR measure is used to analyse *how quick it is for an algorithm to recover from an environmental change and to start converging on a new solution before the next change occurs.*

$$
RR = \frac{1}{m} \sum_{i=1}^{m} \frac{\sum_{j=1}^{p(i)} [f_{best}(i,j) - f_{best}(i,1)]}{p(i) [f_{best}(i,p(i)) - f_{best}(i,1)]}
\tag{1.37}
$$

where $f_{best}(i,j)$ is the fitness value of the best solution since the last change found by the tested algorithm until the jth generation of the change period i, m is the number of changes and $p(i), i = 1 : m$ is the number of generations at each change period i. The RR score would be equal to 1 in the best case where the algorithm is able to recover and converge on a solution immediately after a change, and would be equal to zero in case the algorithm is unable to recover from the drop at all.

The ARR measure is used to analyse *how quick it is for an algorithm to start converging on the global optimum before the next change occurs:*

$$
ARR = \frac{1}{m} \sum_{i=1}^{m} \frac{\sum_{j=1}^{p(i)} [f_{best}(i,j) - f_{best}(i,1)]}{p(i) [f^*(i) - f_{best}(i,1)]}
\tag{1.38}
$$

where $f_{best}(i,j), i, j, m, p(i)$ are the same as in Eq. (1.37) and $f^*(i)$ is the global optimal value of the landscape at the ith change. The ARR score would be equal to 1 in the best case when the algorithm is able to recover and converge on the global optimum immediately after a change, and would be equal to zero in case the algorithm is unable to recover from the change at all. Note that in order to use the measure ARR we need to know the global optimum value at each change period.

Beside the advantage of working even in case there is no drop in performance, the RR and ARR measures can also be used together in an *RR-ARR diagram* (see [51][Fig. 5.2, page 118]) to indicate if an algorithm is able to converge on the global optimum; or if it has suffered from slow convergence or pre-mature convergence.

1.4.2.4   Fitness Degradation over Time

A recent experimental observation [2] showed that in DOPs the performance of an algorithm might degrade over time due to the fact that the algorithm fails to follow the optima after some changes have occurred. To measure this degradation, in [2] a measure named β −*degradation* was proposed. The measure is calculated by firstly using linear regression (over the accuracy values achieved at each change period) to create a regression line, then evaluate the measure as the slope of the regression line. A positive β −*degradation* value might indicate that the algorithm is able to keep track with the moving optima. The measure however does not indicate whether the degradation in performance is really caused by the long-term impact of DOP, or simply by an increase in the difficulty level of the problem after a change. In addition, a positive β −*degradation* value might also not always be an indication that the algorithm is able to keep track with the moving optima. In problems where the total fitness level increases, like in the dynamic constrained benchmark problems in [53] mentioned above, a positive β −*degradation* can be achieved even when the algorithm stays at the same place.

1.4.3   Discussion

There are some open questions about performance measures in EDO. First, it is not clear if optimality is the only goal of real-world DOPs and if existing performance measures really reflect what practitioners would expect from optimisation algorithms. So far, only a few studies, e.g., [51, 56, 90], tried to justify the meaning of the measures by suggesting some possible real-world examples where the measures are applicable. It would be interesting to find the answer for the question of what are the main goals of real-world DOPs, how existing performance measures reflect these goals and from that investigate if it is possible to make the performance measures more specific (if needed) to suit practical requirements. In [51, chapter 3], a first attempt has been be made to find out more about the main optimisation goals of real-world DOPs and the link between existing performance measures and the goals of real-world applications.

Second, as shown in the literature review in this section, many optimality-based measures are not normalised and hence might be biased by fitness rescalings and other disproportionate factors caused by the changing landscapes. The *accuracy* measure [72] is among the few studies that tried to overcome this disadvantage by normalising the fitness values at each change period using a window of the maximum and minimum possible values. This approach, however, requires full knowledge of the maximum and minimum possible values at each change period, which might not be available in practical situations. The normalised score proposed in [51] offers an alternative way to compare the performance of algorithms in a normalised way without using problem-specific knowledge.

Third, although the behaviour-based measures are usually used complementary with the optimality-based measures, it is not clear if the earlier really correlate with the latter. Recent studies [2] have shown that the behaviour-based measure *stability* does not directly relate to the quality of solutions and the results of the behaviour-based measure *reactivity* are "usually insignificant" [1, 2]. It would be interesting to systematically study the relationship between behaviour-based measures and optimality-based measures, and more importantly the relationship between the quality of solutions and the assumptions of the community about the expected behaviours of dynamic optimization algorithms.

1.5   The Generalized Dynamic Benchmark Generator (GDBG)

The dynamic test problems described in Section 1.3 are based on different search spaces. In order to develop a unified approach of constructing dynamic problems across the binary, real, and combinatorial spaces, a GDBG system has recently been proposed in [36, 37]. For the GDBG system, DOPs can be defined as follows:

$$F = f(x, \phi, t) \tag{1.39}$$

where F is the optimization problem, f is the cost function, x is a feasible solution in the solution set $\mathbf{X}$, t is the real-world time, and ϕ is the system control parameter, which determines the solution distribution in the fitness landscape. The objective is to find a global optimal solution x^* such that $f(x^*) \leq f(x) \forall x \in \mathbf{X}$ (without loss of generality, minimization problems are considered).

In the GDBG system, the dynamism results from a deviation of solution distribution from the current environment by tuning the system control parameters. It can be described as follows:

$$\phi(t+1) = \phi(t) \oplus \Delta\phi \tag{1.40}$$

where $\Delta\phi$ is a deviation from the current system control parameters. Then, we can get the new environment at the next moment $t + 1$ as follows:

$$f(x, \phi, t+1) = f(x, \phi(t) \oplus \Delta\phi, t) \tag{1.41}$$

The system control parameters decide the distribution of solutions in the solution space. They may be different from one specific instance to another instance. The GDBG system constructs dynamic environments by changing the values of these system control parameters. There are seven change types of the system control parameters in the GDBG system, which are defined as follows:

- T1 (small step):

$$\Delta\phi = \alpha \cdot \|\phi\| \cdot r \cdot \phi_{severity} \tag{1.42}$$

- T2 (large step):

$$\Delta\phi = \|\phi\| \cdot (\alpha \cdot sign(r) + (\alpha_{max} - \alpha) \cdot r) \cdot \phi_{severity} \qquad (1.43)$$

- T3 (random):

$$\Delta\phi = N(0,1) \cdot \phi_{severity} \qquad (1.44)$$

- T4 (chaotic):

$$\phi(t+1) = A \cdot \phi(t) \cdot (1 - \phi(t)/\|\phi\|) \qquad (1.45)$$

- T5 (recurrent):

$$\phi(t+1) = \phi_{min} + \|\phi\|(\sin(\frac{2\pi}{P}t + \varphi) + 1)/2 \qquad (1.46)$$

- T6 (recurrent with noise):

$$\phi(t+1) = \phi_{min} + \|\phi\|(\sin(\frac{2\pi}{P}t + \varphi) + 1)/2 + N(0,1) \cdot noise_{severity} \qquad (1.47)$$

- T7 (dimensional change):

$$D(t+1) = D(t) + sign \cdot \Delta D, \qquad (1.48)$$

where $\|\phi\|$ is the change range of ϕ, $\phi_{severity} \in (0,1)$ is change severity of ϕ, ϕ_{min} is the minimum value of ϕ, $noisy_{severity} \in (0,1)$ is noisy severity in recurrent with noisy change. $\alpha \in (0,1)$ and $\alpha_{max} \in (0,1)$ are constant values, which are set to 0.02 and 0.1 in the GDBG system. A logistics function is used in the chaotic change type, where A is a positive constant between $(1.0, 4.0)$, if ϕ is a vector, the initial values of the items in ϕ should be different within $\|\phi\|$ in chaotic change. P is the period of recurrent change and recurrent change with noise, φ is the initial phase, r is a random number in $(-1, 1)$, $sign(x)$ returns 1 when x is greater than 0, returns -1 when x is less than 0, otherwise, returns 0. $N(0,1)$ denotes a normally distributed one dimensional random number with mean zero and standard deviation one. For T7, ΔD is a predefined constant, which the default value of is 1. If $D(t) = Max\_D$, $sign = -1$; if $D(t) = Min\_D$, $sign = 1$. Max_D and Min_D are the maximum and minimum number of dimensions. When the number of dimensions deceases by 1, just the last dimension is removed from the fitness landscape, and the fitness landscape of the left dimensions does not change. When the number of dimensions increases by 1, a new dimension with a random value is added into the fitness landscape. Dimensional change *only happens following* the non-dimensional change.

The GDBG system can be instantiated to construct specific DOP instances in the binary space, real space, and combinatorial space, respectively. Both the DF1 [44] and MPB [9] generators have a disadvantage: the challenge per change is unequal for algorithms when the position of a peak bounces back from the search boundary. From the GDBG system, we can construct two different real space DOPs using a rotation method on the peak position to overcome that shortcoming. The two benchmark instances are: dynamic rotation peak benchmark generator (DRPBG) and dynamic composition benchmark generator (DCBG), described as follows.

1.5.1 Dynamic Rotation Peak Benchmark Generator

The DRPBG uses a similar peak-composition structure to those of the MPB [9] and DF1 [44] generators. Given a problem $f(x, \phi, t)$, $\phi = (\mathbf{H}, \mathbf{W}, \mathbf{X})$, where $\mathbf{H}$, $\mathbf{W}$, and $\mathbf{X}$ denote the peak height, width, and position, respectively. The function of $f(x, \phi, t)$ is defined as follows:

$$f(x, \phi, t) = \max_{i=1}^{m}\left(\mathbf{H}_i(t)/(1 + \mathbf{W}_i(t) \cdot \sqrt{\sum_{j=1}^{n}\frac{(x_j - \mathbf{X}_j^i(t))^2}{n}})\right) \tag{1.49}$$

where m is the number of peaks and n is the number of dimensions.
$\mathbf{H}$ and $\mathbf{W}$ change as follows:

$$\mathbf{H}(t+1) = \text{DynamicChanges}(\mathbf{H}(t)) \tag{1.50}$$

$$\mathbf{W}(t+1) = \text{DynamicChanges}(\mathbf{W}(t)) \tag{1.51}$$

where in the height change, *height_severity* should read $\phi\_h_{severity}$ according to Eq. (1.49) and $\|\phi\_h\|$ is the height range. Accordingly, *width_severity* should read $\phi\_w_{severity}$ and $\|\phi\_w\|$ is the width change.

A rotation matrix $R_{ij}(\theta)$ is obtained by rotating the projection of $\vec{x}$ in the plane $i - j$ by an angle θ from the i-th axis to the j-th axis. The peak position $\mathbf{X}$ is changed by the following algorithm:

Step 1:   Randomly select l dimensions (l is an even number) from the n dimensions to compose a vector $r = [r_1, r_2, ..., r_l]$.

Step 2:   For each pair of dimension $r[i]$ and dimension $r[i+1]$, construct a rotation matrix $R_{r[i], r[i+1]}(\theta(t))$, where:

$$\theta(t) = \text{DynamicChanges}(\theta(t-1)). \tag{1.52}$$

Step 3:   A transformation matrix $A(t)$ is obtained by:

$$A(t) = R_{r[1], r[2]}(\theta(t)) \cdot R_{r[3], r[4]}(\theta(t)) \cdots R_{r[l-1], r[l]}(\theta(t)) \tag{1.53}$$

Step 4:   Update the peak position by:

$$\mathbf{X}(t+1) = \mathbf{X}(t) \cdot A(t) \tag{1.54}$$

where the change severity of θ ($\phi\_\theta_{severity}$) is set to l in Eq. (4), the range of θ should read $\|\phi\_\theta\|$, $\|\phi\_\theta\| \in (-\pi, \pi)$. For the value of l, if n is an even number,

$l = n$; otherwise $l = n - 1$. Note that, for recurrent and recurrent with noisy changes, $\|\phi_-\theta\|$ is within $(0, \pi/6)$.

1.5.2 Dynamic Composition Benchmark Generator

The dynamic composition functions, which are extended from the static composition functions devised by Suganthan et al. [39], can be described as follows:

$$F(x,\phi,t) = \sum_{i=1}^{m} (w_i \cdot (f_i'((x - O_i(t) + O_{iold})/\lambda_i \cdot M_i) + H_i(t))) \qquad (1.55)$$

where the system control parameter $\phi = (O, M, H)$, $F(x)$ is the composition function, $f_i(x)$ is i-th basic function used to construct the composition function, m is the number of basic functions, M_i is orthogonal rotation matrix for each $f_i(x)$, $O_i(t)$ is the optimum of the changed $f_i(x)$ caused by rotating the landscape at the time t, O_{iold} is the optimum of the original $f_i(x)$ without any change. O_{iold} is 0 for all the basic functions. The weight value w_i for each $f_i(x)$ is calculated as:

$$w_i = exp(-sqrt(\frac{\sum_{k=1}^{n} (x_k - o_i^k + o_{iold}^k)^2}{2n\sigma_i^2})) \qquad (1.56)$$

$$w_i = \begin{cases} w_i & \text{if } w_i = max(w_i) \\ w_i \cdot (1 - max(w_i)^{10}) & \text{if } w_i \neq max(w_i) \end{cases} \qquad (1.57)$$

$$w_i = w_i / \sum_{i=1}^{m} w_i \qquad (1.58)$$

where σ_i is the converge range factor of $f_i(x)$, whose default value is 1.0, λ_i is the stretch factor for each $f_i(x)$, which is defined as:

$$\lambda_i = \sigma_i \cdot \frac{X_{max} - X_{min}}{x_{max}^i - x_{min}^i} \qquad (1.59)$$

where $[X_{max}, X_{min}]^n$ is the search range of $F(x)$ and $[x_{max}^i, x_{min}^i]^n$ is the search range of $f_i(x)$.

In Eq. (1.55), $f_i'(x) = C \cdot f_i(x)/|f_{max}^i|$, where C is a constant, which is set to 2000, and f_{max}^i is the estimated maximum value of $f_i(x)$, which is estimated as:

$$f_{max}^i = f_i(x_{max} \cdot M_i) \qquad (1.60)$$

In the DCBG, M is initialized using the above transformation matrix construction algorithm and then remains unchanged. The dynamism of the system control parameter H and O are changed as the parameters H and X in dynamic rotation peak benchmark generator. Note that, for both DRPBG and DCBG, chaotic change of peaks locations directly operates on the value of each dimension instead of using rotation matrix due to simulating chaotic systems in real applications.

Table 1.2 Details of the basic benchmark functions

Name	Function	Range
Sphere	$f(x) = \sum_{i=1}^{n} x_i^2$	[-100,100]
Rastrigin	$f(x) = \sum_{i=1}^{n} (x_i^2 - 10\cos(2\pi x_i) + 10)$	[-5,5]
Weierstrass	$f(x) = \sum_{i=1}^{n} (\sum_{k=0}^{k_{max}} [a^k \cos(2\pi b^k (x_i + 0.5))])$ $-n \sum_{k=0}^{k_{max}} [a^k \cos(\pi b^k)], a = 0.5, b = 3, k_{max} = 20$	[-0.5,0.5]
Griewank	$f(x) = \frac{1}{4000} \sum_{i=1}^{n} (x_i)^2 - \prod_{i=1}^{n} \cos(\frac{x_i}{\sqrt{i}}) + 1$	[-100,100]
Ackley	$f(x) = -20\exp(-0.2\sqrt{\frac{1}{n}\sum_{i=1}^{n} x_i^2}) - \exp(\frac{1}{n}\sum_{i=1}^{n}\cos(2\pi x_i)) + 20 + e$	[-32,32]

Table 1.3 Default settings for the GDBG used for CEC 2009 Competition on Dynamic Optimization

Parameter	Value
number of dimensions D	fixed: 10; changing: [5-15]
search range	$x \in [-5,5]^D$
number of functions or peaks p	$p = 10$
change frequency U	$10,000 \times D$ fitness evaluations
number of changes K	$K = 60$
period P	$P = 12$
severity of noisy $noisy_{severity}$	$noisy_{severity} = 0.8$
chaotic constant A	$A = 3.67$
step severity α	$\alpha = 0.04$
maximum of α	$\alpha_{max} = 0.1$
height range	$h \in [10, 100]$
initial height $initial\_height$	$initial\_height = 50$
severity of height change $\phi\_h_{severity}$	$\phi\_h_{severity} = 5.0$
sampling frequency s_f	$s\_f = 100$

Five basic benchmark functions are used in the GDBG system. Table 1.2 shows the details of the five functions.

1.5.3 Dynamic Test Problems for the CEC 2009 Competition

The GDBG was used to construct dynamic test problems for the CEC 2009 Competition on Dynamic Optimization [37], where the following seven different particular functions are defined: the rotation peak function with 10 peaks (F_1 with $p = 10$), the rotation peak function with 50 peaks (F_1 with $p = 50$), the composition of Sphere's functions (F_2), the composition of Rastrigin's functions (F_3), the composition of Griewank's functions (F_4), the composition of Ackley's functions (F_5), and the hybrid composition function (F_6). The detailed settings of each function can be found in [37] and the general parameter settings are given in Table 1.3.

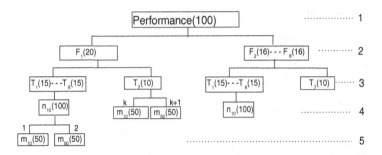

Fig. 1.2 Overall performance measurement

There are 49 test cases in total constructed from the seven test problems in the GDBG benchmark. For an algorithm on each test case, the offline error (e_{off}) and its standard variance (STD) are recorded, which are defined as in [37] as follows:

$$e_{off} = \frac{1}{R * K} \sum_{r=1}^{R} \sum_{k=1}^{K} e_{r,k}^{last} \qquad (1.61)$$

$$STD = \sqrt{\frac{\sum_{r=1}^{R} \sum_{k=1}^{K} (e_{r,k}^{last} - e_{off})^2}{R * K - 1}} \qquad (1.62)$$

where R and K are the total number of runs and the number of environmental changes for each run, respectively, and $e_{r,k}^{last} = |f(\mathbf{x}_{best}(r,k)) - f(\mathbf{x}^*(r,k))|$, where $\mathbf{x}^*(r,k)$ is the global optimum of the k-th environment and $\mathbf{x}_{best}(r,k)$ is the position of the best particle of the last generation of the k-th environment during the r-th run.

The calculation of the overall performance of an algorithm on all 49 test cases is as illustrated in Fig. 1.2, where F_1-F_6 denote the six functions defined in the GDBG benchmark in [37], T_1-T_7 represent the seven change types, n_{10} means that the number of dimensions is ten, and m_{10} and m_{50} denote that the number of peaks is 10 and 50, respectively. Each test case i is assigned a weight w_i and the sum of weights of all the test cases is 1.0. The mark obtained by an algorithm on test case $i \in \{1,\ldots,49\}$ is calculated by:

$$mark_i = \frac{w_i}{R * K} \sum_{r=1}^{R} \sum_{k=1}^{K} \left(r_{rk}^{last} / \left(1 + \frac{1}{S} \sum_{s=1}^{S} (1 - r_{rk}^s) \right) \right) \qquad (1.63)$$

where r_{rk}^{last} is the relative ratio of the best particle fitness of the last generation to the global optimum of the k-th environment, r_{rk}^s is the relative ratio of the best particle's fitness to the global optimum at the s-th sampling during the k-th environment (*initial population should be sampled*), and $S = U/s\_f$ is the total number of samples for each environment. The relative ratio r_{rk}^s is defined by

$$r_{rk}^s = \begin{cases} \frac{f(\mathbf{x}_{best}(r,k,s))}{f(\mathbf{x}^*(r,k))}, & f = F_1 \\ \frac{f(\mathbf{x}^*(r,k))}{f(\mathbf{x}_{best}(r,k,s))}, & f \in \{F_2, F_3, F_4, F_5, F_6\} \end{cases} \tag{1.64}$$

where $\mathbf{x}_{best}(r,k,s)$ is the position of the best particle up to the s-th sample in the k-th environment during the r-th run.

The overall performance of an algorithm on all the test cases is then calculated as follows:

$$performance = 100 \times \sum_{i=1}^{49} mark_i \tag{1.65}$$

1.6   Conclusions and Discussions

Developing proper dynamic test and eveluation environments is an important task in studying EC for DOPs. In this chapter, we present the concept of DOPs and review existing dynamic test problems commonly used by researchers to investigate their EC approaches in the literature. Some discussions regarding the major features and classification of existing dynamic test environments are presented. Some typical dynamic benchmark problems and real-world DOPs, which cover the binary, real, and combinatorial spaces, are also described in detail. We also review the performance measures that are widely used by researchers to evaluate and compare their developed EC approaches for DOPs.

The review has identified the common assumptions of the community about the characteristics of DOPs, which can be summarised as follows:

- *Optimisation goals*: *Optimality is the primary goal or the only goal in a majority of academic EDO studies*, as evidently shown by the large number of optimality-based measures reviewed in Section 1.4.1. Some studies do pay attention to developing other complementary measures (e.g. the behaviour-based measures in Section 1.4.2), but these complementary measures mainly focus on analysing the behaviours of the algorithms rather than checking if the algorithms satisfy users requirements.
- *The time-linkage property*: *Non time-linkage (the algorithm does not influence the future dynamics) is the main focus of current academic EDO research*, as evidently shown by the fact that almost all commonly used general-purpose benchmark problems are non-time-linkage.
- *Constraints*: *Unconstrained or bounded constrained problems are the main focus of academic research*, especially in the continuous domain, as shown by the majority of academic benchmark problems. There is a clear lack of studies on constrained and dynamic constrained problems.
- *Visibility and detectability of changes*: Current EDO methods assume that changes either are known or can be easily detected using a few detectors.

- *Factors that change*: The major aspect that changes in academic problems is the objective function.
- *Reason for tracking*: The main assumption is that the optima (local or global) after change is close to the optima (local or global) before change, as shown in a majority of benchmark problems (although in the Moving peaks [9] and DF1 [44] benchmarks the new global optima are not close to the previous global optima, they are still close to a previous local optima). Due to that, tracking is preferred to restarting.
- *Predictability*: The predictability of changes has increasingly attracted the attention of the community. However, the number of studies in this topic is still relatively small compared to the unpredictable case
- *Periodicity*: The periodicity of changes is a given assumption in many mainstream approaches as *memory* and *prediction*.

The review in this chapter showed that not many of the assumptions above are backed up by evidence from real-world applications. This leads to the question of whether these academic assumptions still hold in real-world DOPs and, if yes, then whether these assumptions are representative in real-world applications and in what type of applications do they hold. So far there is very little reserach aiming at answering these questions. One exception is the recent study in [51, chap. 3] where a large set of recent "real"[2] real-world dynamic optimisation problems has been reviewed to investigate the real characteristics of real-world problems and how they relate to the characteristics of current academic benchmark problems. The reserch in [51] has pointed out that there are certain gaps between current EDO academic research and real-world applications. In future research on EDO, further investigations should be made to close these gaps and accordingly to bring EDO research closer to realistic scenarios.

Acknowledgements. This work was supported by the Engineering and Physical Sciences Research Council (EPSRC) of U.K. under Grant numbers EP/E060722/1, EP/K001310/1, and EP/E058884/1, and partially supported by an UK ORS Award and a studentship from the School of Computer Science, University of Birmingham and an EU-funded project named "Intelligent Transportation for Dynamic Environment (InTraDE)".

References

[1] Alba, E., Saucedo Badia, J., Luque, G.: A study of canonical gas for nsops. In: Doerner, et al. (eds.) Metaheuristics. Operations Research/Computer Science Interfaces Series, vol. 39, pp. 245–260. Springer (2007)

---

[2] Only references that actual use real-world data or solve problems in actual real-world situations were considered. Benchmark problems, even if designed to simulate real-world applications, were not considered unless there is evidence that the data used to create the benchmark were taken from real-world applications.

[2] Alba, E., Sarasola, B.: Measuring fitness degradation in dynamic optimization problems. In: Di Chio, C., et al. (eds.) EvoApplicatons 2010, Part I. LNCS, vol. 6024, pp. 572–581. Springer, Heidelberg (2010)

[3] Alba, E., Sarasola, B.: Abc, a new performance tool for algorithms solving dynamic optimization problems. In: Proc. 2010 IEEE World Congr. on Comput. Intell. (WCCI 2010), pp. 734–740 (2010)

[4] Aragón, V.S., Esquivel, S.C.: An evolutionary algorithm to track changes of optimum value locations in dynamic environments. J. of Comp. Sci. and Tech. 4(3), 127–134 (2004)

[5] Bäck, T.: On the behavior of evolutionary algorithms in dynamic environments. In: Proc. 1998 IEEE Int. Conf. Evol. Comput., pp. 446–451 (1998)

[6] Bird, S., Li, X.: Informative performance metrics for dynamic optimisation problems. In: Proc. 9th Annual Conf. on Genetic and Evol. Comput., pp. 18–25 (2007)

[7] Bosman, P.A.N.: Learning and anticipation in online dynamic optimization. In: Yang, S., Ong, Y.-S., Jin, Y. (eds.) Evolutionary Computation in Dynamic and Uncertain Environments. SCI, vol. 51, pp. 129–152. Springer, Heidelberg (2007)

[8] Bosman, P.A.N., Poutré, H.L.: Learning and anticipation in online dynamic optimization with evolutionary algorithms: the stochastic case. In: Proc. 9th Annual Conf. on Genetic and Evol. Comput., pp. 1165–1172 (2007)

[9] Branke, J.: Memory enhanced evolutionary algorithms for changing optimization problems. In: Proc. 1999 IEEE Congr. Evol. Comput., pp. 1875–1882 (1999)

[10] Branke, J.: Evolutionary approaches to dynamic environments - updated survey. In: Proc. 2001 GECCO Workshop on Evolutionary Algorithms for Dynamic Optimization Problems, pp. 27–30 (2001)

[11] Branke, J.: Evolutionary Optimization in Dynamic Environments. Kluwer Academic Publishers (2002)

[12] Branke, J., Schmeck, H.: Designing evolutionary algorithms for dynamic optimization problems. In: Tsutsui, S., Ghosh, A. (eds.) Theory and Application of Evolutionary Computation: Recent Trends, pp. 239–262. Springer (2003)

[13] Branke, J., Kaußler, T., Schmidth, C., Schmeck, H.: A multi-population approach to dynamic optimization problems. In: Proc. 4th Int. Conf. Adaptive Comput. Des. Manuf., pp. 299–308 (2000)

[14] Branke, J., Orbayı, M., Uyar, Ş.: The role of representations in dynamic knapsack problems. In: Rothlauf, F., et al. (eds.) EvoWorkshops 2006. LNCS, vol. 3907, pp. 764–775. Springer, Heidelberg (2006)

[15] Chazottes, J., Fernandez, B.: Dynamics of Coupled Map Lattices and of Related Spatially Extended Systems. Springer, Heidelberg (2005)

[16] Cheng, H., Yang, S.: Multi-population genetic algorithms with immigrants scheme for dynamic shortest path routing problems in mobile ad hoc networks. In: Di Chio, C., et al. (eds.) EvoApplicatons 2010, Part I. LNCS, vol. 6024, pp. 562–571. Springer, Heidelberg (2010)

[17] Cheng, H., Yang, S.: Genetic algorithms with immigrants schemes for dynamic multicast problems in mobile ad hoc networks. Engg. Appl. of Artif. Intell. 23(5), 806–819 (2010)

[18] Cobb, H.G.: An investigation into the use of hypermutation as an adaptive operator in genetic algorithms having continuouis, time-dependent nonstationary environments. Technical Report AIC-90-001, Naval Research Laboratory, Washington, USA (1990)

[19] Cobb, H.G., Grefenstette, J.J.: Genetic algorithms for tracking changing environments. In: Proc. 5th Int. Conf. Genetic Algorithms, pp. 523–530 (1993)

[20] Cruz, C., Gonzalez, J.R., Pelta, D.A.: Optimization in dynamic environments: A survey on problems, methods and measures. Soft Comput. 15(7), 1427–1448 (2011)

[21] Droste, S.: Analysis of the (1+1) EA for a dynamically changing onemax-variant. In: Proc. 2002 IEEE Congr. Evol. Comput., pp. 55–60 (2002)

[22] Eberhart, R.C., Kennedy, J.: A new optimizer using particle swarm theory. In: Proc. 6th Int. Symp. on Micro Machine and Human Science, pp. 39–43 (1995)

[23] Farina, M., Deb, K., Amato, P.: Dynamic multiobjective optimization problems: test cases, approximations, and applications. IEEE Trans. Evol. Comput. 8(5), 425–442 (2004)

[24] Feng, W., Brune, T., Chan, L., Chowdhury, M., Kuek, C., Li, Y.: Benchmarks for testing evolutionary algorithms. Technical Report, Center for System and Control, University of Glasgow (1997)

[25] Gaspar, A., Collard, P.: From gas to artificial immune systems: Improving adaptation in time dependent optimization. In: Proc. 1999 IEEE Congr. Evol. Comput., vol. 3, pp. 1859–1866 (1999)

[26] Goh, C.K., Tan, K.C.: A competitive-cooperative coevolutionary paradigm for dynamic multiobjective optimization. IEEE Trans. Evol. Comput. 13(1), 103–127 (2009)

[27] Grefenstette, J.J.: Genetic algorithms for changing environments. In: Proc. 2nd Int. Conf. Parallel Problem Solving from Nature, pp. 137–144 (1992)

[28] Grefenstette, J.J.: Evolvability in dynamic fitness landscapes: a genetic algorithm approach. In: Proc. 1999 IEEE Congr. Evol. Comput., vol. 3, pp. 2031–2038 (1999)

[29] Hatzakis, I., Wallace, D.: Dynamic multi-objective optimization with evolutionary algorithms: a forward-looking approach. In: Proc. 8th Annual Conf. on Genetic and Evol. Comput., pp. 1201–1208 (2006)

[30] Jin, Y., Branke, J.: Evolutionary optimization in uncertain environments – a survey. IEEE Trans. Evol. Comput. 9(3), 303–317 (2005)

[31] Jin, Y., Sendhoff, B.: Constructing dynamic optimization test problems using the multi-objective optimization concept. In: Raidl, G.R., et al. (eds.) EvoWorkshops 2004. LNCS, vol. 3005, pp. 525–536. Springer, Heidelberg (2004)

[32] Kellerer, K., Pferschy, U., Pisinger, D.: Knapsack Problems. Springer (2004)

[33] Kennedy, J., Eberhart, R.C.: Particle swarm optimization. In: Proc. 1995 IEEE Int. Conf. on Neural Networks, pp. 1942–1948 (1995)

[34] Lewis, J., Hart, E., Ritchie, G.: A comparison of dominance mechanisms and simple mutation on non-stationary problems. In: Eiben, A.E., Bäck, T., Schoenauer, M., Schwefel, H.-P. (eds.) PPSN 1998. LNCS, vol. 1498, pp. 139–148. Springer, Heidelberg (1998)

[35] Li, C., Yang, M., Kang, L.: A new approach to solving dynamic traveling salesman problems. In: Wang, T.-D., Li, X., Chen, S.-H., Wang, X., Abbass, H.A., Iba, H., Chen, G.-L., Yao, X. (eds.) SEAL 2006. LNCS, vol. 4247, pp. 236–243. Springer, Heidelberg (2006)

[36] Li, C., Yang, S.: A generalized approach to construct benchmark problems for dynamic optimization. In: Li, X., et al. (eds.) SEAL 2008. LNCS, vol. 5361, pp. 391–400. Springer, Heidelberg (2008)

[37] Li, C., Yang, S., Nguyen, T.T., Yu, E.L., Yao, X., Jin, Y., Beyer, H.-G., Suganthan, P.N.: Benchmark generator for CEC 2009 competition on dynamic optimization. Technical Report 2008, Department of Computer Science, University of Leicester, U.K. (2008)

[38] Li, X., Branke, J., Kirley, M.: On performance metrics and particle swarm methods for dynamic multiobjective optimization problems. In: Proc. 2007 IEEE Congr. Evol. Comput., pp. 576–583 (2007)

[39] Liang, J.J., Suganthan, P.N., Deb, K.: Novel composition test functions for numerical global optimization. In: Proc. 2005 IEEE Congr. Evol. Comput., pp. 68–75 (2005)

[40] Mavrovouniotis, M., Yang, S.: A memetic ant colony optimization algorithm for the dynamic traveling salesman problem. Soft Comput. 15(7), 1405–1425 (2011)

[41] Mavrovouniotis, M., Yang, S.: Memory-based immigrants for ant colony optimization in changing environments. In: Di Chio, C., et al. (eds.) EvoApplications 2011, Part I. LNCS, vol. 6624, pp. 324–333. Springer, Heidelberg (2011)

[42] Mei, Y., Tang, K., Yao, X.: Capacitated arc routing problem in uncertain environments. In: Proc. 2010 IEEE Congr. Evol. Comput., pp. 1400–1407 (2010)

[43] Mori, N., Imanishi, S., Kita, H., Nishikawa, Y.: Adaptation to changing environments by means of the memory based thermodynamical genetic algorithm. In: Proc. 1997 Int. Conf. on Genetic Algorithms, pp. 299–306 (1997)

[44] Morrison, R.W., De Jong, K.A.: A test problem generator for non-stationary environments. In: Proc. 1999 IEEE Congr. Evol. Comput., pp. 2047–2053 (1999)

[45] Morrison, R.W., De Jong, K.A.: Triggered hypermutation revisited. In: Proc. 2000 IEEE Congr. Evol. Comput., pp. 1025–1032 (2000)

[46] Morrison, R.W., De Jong, K.A.: Measurement of population diversity. In: Collet, P., Fonlupt, C., Hao, J.-K., Lutton, E., Schoenauer, M. (eds.) EA 2001. LNCS, vol. 2310, pp. 31–41. Springer, Heidelberg (2002)

[47] Morrison, R.W.: Performance measurement in dynamic environments. In: Proc. 2003 GECCO Workshop on Evolutionary Algorithms for Dynamic Optimization Problems, pp. 5–8 (2003)

[48] Morrison, R.W.: Designing Evolutionary Algorithms for Dynamic Environments. Springer, Berlin (2004)

[49] Ng, K.P., Wong, K.C.: A new diploid scheme and dominance change mechanism for non-stationary function optimization. In: Proc. 6th Int. Conf. on Genetic Algorithms, pp. 159–166 (1995)

[50] Nguyen, T.T.: A proposed real-valued dynamic constrained benchmark set. Technical report, School of Computer Science, Univesity of Birmingham (2008)

[51] Nguyen, T.T.: Continuous Dynamic Optimisation Using Evolutionary Algorithms. PhD thesis, School of Computer Science, University of Birmingham (January 2011), http://etheses.bham.ac.uk/1296 and http://www.staff.ljmu.ac.uk/enrtngu1/theses/ phd_thesis_nguyen.pdf

[52] Nguyen, T.T., Yao, X.: Dynamic time-linkage problems revisited. In: Giacobini, M., et al. (eds.) EvoWorkshops 2009. LNCS, vol. 5484, pp. 735–744. Springer, Heidelberg (2009)

[53] Nguyen, T.T., Yao, X.: Benchmarking and solving dynamic constrained problems. In: Proc. 2009 IEEE Congr. Evol. Comput., pp. 690–697 (2009)

[54] Oppacher, F., Wineberg, M.: The shifting balance genetic algorithm: Improving the ga in a dynamic environment. In: Proc. 1999 Genetic and Evol. Comput. Conf., vol. 1, pp. 504–510 (1999)

[55] Parrott, D., Li, X.: Locating and tracking multiple dynamic optima by a particle swarm model using speciation. IEEE Trans. Evol. Comput. 10(4), 440–458 (2006)

[56] Rand, W., Riolo, R.: Measurements for understanding the behavior of the genetic algorithm in dynamic environments: A case study using the shaky ladder hyperplane-defined functions. In: Yang, S., Branke, J. (eds.) GECCO Workshop on Evolutionary Algorithms for Dynamic Optimization (2005)

[57] Richter, H.: Behavior of evolutionary algorithms in chaotically changing fitness land-scapes. In: Yao, X., et al. (eds.) PPSN 2004. LNCS, vol. 3242, pp. 111–120. Springer, Heidelberg (2004)

[58] Richter, H.: Evolutionary optimization in spatio–temporal fitness landscapes. In: Runarsson, T.P., Beyer, H.-G., Burke, E.K., Merelo-Guervós, J.J., Whitley, L.D., Yao, X. (eds.) PPSN 2006. LNCS, vol. 4193, pp. 1–10. Springer, Heidelberg (2006)

[59] Richter, H.: Evolutionary optimization and dynamic fitness landscapes: From reaction-diffusion systems to chaotic CML. In: Zelinka, I., Celikovsky, S., Richter, H., Chen, G. (eds.) Evolutionary Algorithms and Chaotic Systems. SCI, vol. 267, pp. 409–446. Springer, Heidelberg (2010)

[60] Richter, H.: Memory design for constrained dynamic optimization problems. In: Di Chio, C., et al. (eds.) EvoApplicatons 2010, Part I. LNCS, vol. 6024, pp. 552–561. Springer, Heidelberg (2010)

[61] Rohlfshagen, P., Yao, X.: Attributes of dynamic combinatorial optimisation. In: Li, X., et al. (eds.) SEAL 2008. LNCS, vol. 5361, pp. 442–451. Springer, Heidelberg (2008)

[62] Rohlfshagen, P., Yao, X.: Dynamic combinatorial optimisation problems: an analysis of the subset sum problem. Soft Comput. 15(9), 1723–1734 (2011)

[63] Salomon, R., Eggenberger, P.: Adaptation on the evolutionary time scale: A working hypothesis and basic experiments. In: Hao, J.-K., Lutton, E., Ronald, E., Schoenauer, M., Snyers, D. (eds.) AE 1997. LNCS, vol. 1363, pp. 251–262. Springer, Heidelberg (1998)

[64] Simões, A., Costa, E.: Evolutionary algorithms for dynamic environments: Prediction using linear regression and Markov chains. In: Rudolph, G., Jansen, T., Lucas, S., Poloni, C., Beume, N. (eds.) PPSN 2008. LNCS, vol. 5199, pp. 306–315. Springer, Heidelberg (2008)

[65] Stanhope, S.A., Daida, J.M.: Optimal mutation and crossover rates for a genetic algo-rithm operating in a dynamic environment. In: Porto, V.W., Waagen, D. (eds.) EP 1998. LNCS, vol. 1447, pp. 693–702. Springer, Heidelberg (1998)

[66] Tinos, R., Yang, S.: Continuous dynamic problem generators for evolutionary algo-rithms. In: Proc. 2007 IEEE Congr. Evol. Comput., pp. 236–243 (2007)

[67] Trojanowski, K., Michalewicz, Z.: Searching for optima in non-stationary environ-ments. In: Proc. 1999 IEEE Congr. Evol. Comput., vol. 3, pp. 1843–1850 (1999)

[68] Uyar, Ş., Uyar, H.T.: A critical look at dynamic multi-dimensional knapsack problem generation. In: Giacobini, M., et al. (eds.) EvoWorkshops 2009. LNCS, vol. 5484, pp. 762–767. Springer, Heidelberg (2009)

[69] Van Veldhuizen, D.A.: Multiobjective evolutionary algorithms: classifications, analy-ses, and new innovations. PhD thesis, Air Force Institute of Technolog, Wright Patter-son AFB, OH, USA (1999)

[70] Wang, H., Wang, D.-W., Yang, S.: Triggered memory-based swarm optimization in dy-namic environments. In: Giacobini, M. (ed.) EvoWorkshops 2007. LNCS, vol. 4448, pp. 637–646. Springer, Heidelberg (2007)

[71] Weicker, K.: An analysis of dynamic severity and population size. In: Deb, K., Rudolph, G., Lutton, E., Merelo, J.J., Schoenauer, M., Schwefel, H.-P., Yao, X. (eds.) PPSN 2000. LNCS, vol. 1917, pp. 159–168. Springer, Heidelberg (2000)

[72] Weicker, K.: Performance measures for dynamic environments. In: Guervós, J.J.M., Adamidis, P.A., Beyer, H.-G., Fernández-Villacañas, J.-L., Schwefel, H.-P. (eds.) PPSN 2002. LNCS, vol. 2439, pp. 64–73. Springer, Heidelberg (2002)

[73] Weicker, K., Weicker, N.: Dynamic rotation and partial visibility. In: Proc. 2003 IEEE Congr. Evol. Comput., pp. 1125–1131 (2003)

[74] Weicker, K., Weicker, N.: On evolution strategy optimization in dynamic environments. In: Proc. 1999 IEEE Congr. Evol. Comput., vol. 3, pp. 2039–2046 (1999)

[75] Weise, T., Podlich, A., Reinhard, K., Gorldt, C., Geihs, K.: Evolutionary freight transportation planning. In: Giacobini, M., et al. (eds.) EvoWorkshops 2009. LNCS, vol. 5484, pp. 768–777. Springer, Heidelberg (2009)

[76] Woldesenbet, Y.G., Yen, G.G.: Dynamic evolutionary algorithm with variable relocation. IEEE Trans. Evol. Comput. 13(3), 500–513 (2009)

[77] Yang, S.: Non-stationary problem optimization using the primal-dual genetic algorithm. In: Proc. 2003 IEEE Congr. Evol. Comput., pp. 2246–2253 (2003)

[78] Yang, S.: Constructing dynamic test environments for genetic algorithms based on problem difficulty. In: Proc. 2004 IEEE Congr. Evol. Comput., vol. 2, pp. 1262–1269 (2004)

[79] Yang, S.: Memory-enhanced univariate marginal distribution algorithms for dynamic optimization problems. In: Proc. 2005 IEEE Congr. Evol. Comput., vol. 3, pp. 2560–2567 (2005)

[80] Yang, S.: Genetic algorithms with memory and elitism based immigrants in dynamic environments. Evol. Comput. 16(3), 385–416 (2008)

[81] Yang, S., Cheng, H., Wang, F.: Genetic algorithms with immigrants and memory schemes for dynamic shortest path routing problems in mobile ad hoc networks. IEEE Trans. Syst., Man, and Cybern. Part C: Appl. and Rev. 40(1), 52–63 (2010)

[82] Yang, S., Li, C.: A clustering particle swarm optimizer for locating and tracking multiple optima in dynamic environments. IEEE Trans. Evol. Comput. 14(6), 959–974 (2010)

[83] Yang, S., Ong, Y.-S., Jin, Y. (eds.): Evolutionary Computation in Dynamic and Uncertain Environments. Springer (2007)

[84] Yang, S., Richter, H.: Hyper-learning for population-based incremental learning in dynamic environments. In: Proc. 2009 IEEE Congr. Evol. Comput., pp. 682–689 (2009)

[85] Yang, S., Tinos, R.: Hyper-selection in dynamic environments. In: Proc. 2008 IEEE Congr. Evol. Comput., pp. 3185–3192 (2008)

[86] Yang, S., Yao, X.: Dual population-based incremental learning for problem optimization in dynamic environments. In: Proc. 7th Asia Pacific Symp. on Intell. and Evol. Syst., pp. 49–56 (2003)

[87] Yang, S., Yao, X.: Experimental study on population-based incremental learning algorithms for dynamic optimization problems. Soft Comput. 9(11), 815–834 (2005)

[88] Yang, S., Yao, X.: Population-based incremental learning with associative memory for dynamic environments. IEEE Trans. Evol. Comput. 12(5), 542–561 (2008)

[89] Yao, X., Liu, Y.: Fast evolutionary programming. In: Proc. 5th Annual Conf. on Evolutionary Programming, pp. 451–460 (1996)

[90] Yu, X., Jin, Y., Tang, K., Yao, X.: Robust optimization over time – a new perspective on dynamic optimization problems. In: Proc. 2010 IEEE World Congr. Comput. Intell., pp. 3998–4003 (2010)

[91] Zeng, S., Chen, G., Zheng, L., Shi, H., de Garis, H., Ding, L., Kang, L.: A dynamic multi-objective evolutionary algorithm based on an orthogonal design. In: Proc. 2006 IEEE Congr. Evol. Comput., pp. 573–580 (2006)

Chapter 2
Evolutionary Dynamic Optimization: Methodologies

Trung Thanh Nguyen, Shengxiang Yang, Juergen Branke, and Xin Yao

Abstract. In recent years, Evolutionary Dynamic Optimization (EDO) has attracted a lot of research effort and has become one of the most active research areas in evolutionary computation (EC) in terms of the number of activities and publications. This chapter provides a summary of main EDO approaches in solving DOPs. The strength and weakness of each approach and their suitability for different types of DOPs are discussed. Current gaps, challenging issues and future directions regarding EDO methodolgies are also presented.

2.1 Introduction

Many real-world objects are changing over time. For example, people are aging, the climate is changing, the stock market is moving up and down, and so on. As a result, it is important to be able to optimize in a dynamic environment. Changes may affect

Trung Thanh Nguyen
School of Engineering, Technology and Maritime Operations,
Liverpool John Moores University, Liverpool L3 3AF, U.K.
e-mail: T.T.Nguyen@ljmu.ac.uk

Shengxiang Yang
Centre for Computational Intelligence (CCI), School of Computer Science and Informatics,
De Montfort University, The Gateway, Leicester LE1 9BH, U.K.
e-mail: syang@dmu.ac.uk

Juergen Branke
Warwick Business School, University of Warwick, Coventry CV4 7AL, U.K.
e-mail: juergen.branke@wbs.ac.uk

Xin Yao
Centre of Excellence for Research in Computational Intelligence and Applications
(CERCIA), School of Computer Science, University of Birmingham,
Birmingham B15 2TT, U.K.
e-mail: x.yao@cs.bham.ac.uk

S. Yang and X. Yao (Eds.): *Evolutionary Computation for DOPs*, SCI 490, pp. 39–64.
DOI: 10.1007/978-3-642-38416-5_2   © Springer-Verlag Berlin Heidelberg 2013

the objective function, the problem instance, and/or the constraints [16, 124]. Hence, the optimal solution(s) of the problem being considered may change over time.

Formally, a dynamic optimization problem can be defined as follows [71]:

Definition 2.1 (Dynamic optimization problem) *Given a time-dependent problem* f_t, *an optimization algorithm* G *to solve* f_t, *and a given optimization period* $\left[t^{begin}, t^{end}\right]$, f_t *is called a* **dynamic optimization problem** *in the period* $\left[t^{begin}, t^{end}\right]$ *if during* $\left[t^{begin}, t^{end}\right]$ *the underlying fitness landscape that* G *uses to represent* f_t *changes* **and** *G has to react to this change by providing new optimal solutions*[1].

Evolutionary computation (EC) methods are good tools to solve DOPs due to their inspiration from natural systems, which have always been subject to changing environments. The study of applying evolutionary algorithms (EAs) and similar techniques to solving DOPs is termed *evolutionary optimization in dynamic environments* or *evolutionary dynamic optimization* (EDO) in this chapter. Over the last twenty years, a great number of different EDO methodologies have been proposed. The purpose of this chapter is to provide a summary of main EDO approaches in solving DOPs[2]. In-depth discussion of the strength and weakness of each approach will be provided, plus the suitability of each approach for different types of DOPs. Some future research issues and directions regarding EDO will also be presented.

The rest of this chapter is organized as follows. Section 2.2 reviews different approaches that have been developed by researchers to address DOPs. The strength and weakness of different approaches are also discussed there. Section 2.3 presents theoretical development regarding EDO methodologies. Finally, Section 2.4 summarizes the chapter and presents some discussions on current gaps, challenging research issues, and future directions regarding EDO methodologies.

2.2 Optimization Approaches

2.2.1 The Goals of EDO Algorithms

In stationary optimization, in most cases the only goal of optimization algorithms is to find the global optimum as fast as possible. However, in current EDO research where the considered problems are time-varying, the goal of an algorithm turns from finding the global optimum to firstly detecting the changes and secondly tracking the changing optima (local optima or ideally the global optimum) over time. In addition, in case the problem-after-change somehow correlates with the problem-before-change, an optimization algorithm also needs to learn from its previous search

---

[1] A more detailed version of this definition for DOPs was provided in [70, Chapter 4] and [74].

[2] An broader literature review, which is extended from this chapter, covering not only methodologies but also other aspects in EDO, can be found in [71].

experience as much as possible to hopefully advance the search more effectively. Otherwise, the optimization process after each change will simply become the process of solving a different problem starting with the old population/structure.

The following sections will briefly review typical approaches in EDO that have been proposed to satisfy the goals above. We will discuss the strength and weakness of the approaches and their suitability for different types of problems.

2.2.2 Detecting Changes

Many EDO approaches take an explicit action to respond to a change in the environment. This either assumes that changes in the environment are made known to the algorithm, or that the algorithm has to detect the change. If algorithms have to detect changes, they generally follow one of the following approaches: (a) detecting changes by re-evaluating dedicated detectors, or (b) detecting changes based on algorithm behaviors.

2.2.2.1 Detecting Changes by Re-evaluating Solutions

Overview
By far the most common change-detection approach is re-evaluating existing solutions. The algorithm regularly re-evaluates some specific solutions (detectors) to detect changes in their function values and/or feasibility. Detectors can be a part of the population, such as the current best solutions [48, 54, 58, 70], a memory-based sub-population [15, 130], or a feasible sub-population [70, 75]. Detectors can also be maintained separately from the search population. In this case, they can be just a fixed point [25], one or a set of random solutions [26, 82, 98], a regular grid of solutions / set of specifically distributed solutions [65], or a list of found peaks [67, 69].

Strength and Weakenesses
Since using detectors involves additional function evaluations, it might be required to identify an optimal number of detectors to maximize algorithm performance. A majority of existing methods just use one or a small number of detectors. However, in situations where only some parts of the search space change, e.g., see [73, 76, 82] and in a list of real-world problems cited in [70], using only a small number of detectors might not guarantee that changes are detected [82]. Recent attempts have been made to overcome this drawback. In [65, 70, 82], different methods were considered to study the optimal number of detectors depending on the size and complexity of the problem. A theoretical analysis in [65] showed that problem dimensionality is a prominent factor in the success of change detection. This finding was later confirmed by the experiments in [82].

One clear advantage of re-evaluating dedicated detectors is that it allows "robust detection" if a high enough number of detectors is used [82]. Richter [82] also showed that the more difficult the change detection is, the more favorable the approach of re-evaluating dedicated detectors is.

There are some disadvantages in re-evaluating dedicated detectors. First, there is the additional cost due to that detectors have to be re-evaluated at every generation. Second, this approach might not be accurate when used in problems with noisy fitness function because noises may mislead the algorithm to thinking that a change has occured [51]. It may also miss changes if changes did not occur in the region of detectors.

2.2.2.2 Detecting Changes Based on Algorithm Behaviors

Overview
Irregularities in algorithm behaviors can also be used to detect changes. In [31] (and many studies that follow the same idea), changes are detected based on monitoring the drop in the average of best found solutions over a number of generations. In a swarm-based study [51] where the swarm was divided into a tree-based hierarchy of sub-swarms, environmental changes were detected based on observation of changes in the hierarchy itself. In [65], the possibility of detecting changes based on diversity, and the relationship between the diversity of fitness values and the success rate of change detection, were studied. In [82], changes were detected based on statistical hypothesis tests to find the difference between distribution of the populations from two consecutive generations. This technique has been commonly used in environmental change detection in the real-world applications of biomedicine, data mining and image processing, as can be seen from the references cited in [82].

Strength and Weakenesses
The clear advantage of this approach is that it does not require any additional function evaluations. However, because no dedicated detector is used, there is no guarantee that changes are detected [82]. In addition, this approach may cause false positives and hence cause the algorithm to react unnecessarily when no change occurs. Evidence of false positives was found in [51, 76, 82]. Another possible disadvantage is that some methods following this approach might be algorithm-specific, such as the method of monitoring swarm hierarchy in [51].

2.2.3 Introducing Diversity When Changes Occur

2.2.3.1 Overview

In static environments, proper algorithm convergence is required so that the algorithm can focus on finding the best solution. In dynamic environments, however, convergence may result in negative effects. This is because if the dynamic landscape changes in one area and there is no member of the algorithm in this area, the change will become undetected. As a result, the algorithm will not be able to react to the change effectively and hence might fail to track the moving global optimum.

Intuitively one simple solution for this drawback is to increase the diversity of an EA after a change has been detected. This solution is described in the pseudo-code of Algorithm 1.

Algorithm 1 Introducing diversity after detecting a change

---

1. *Initialize:*: Initialize the population
2. *For each generation*

   a. *Evaluate*: Evaluate each member of the population
   b. *Check for changes*: Detect changes by monitoring possible signs of changes, e.g. a reduction in the best fitness values, or re-evaluation of old solutions
   c. *Increase diversity*: If change occurs, increase population's diversity by changing the mutations (sizes or rates) or relocating individuals
   d. *Reproduce*: Reproduce a new population using the adjusted muta-tion/learning/adaptation rate
   e. Return to step 2a

Diversity introduction can be done in many ways, for example, by increasing the current mutation rate as in hyper-mutation [31], by adding randomised individuals [40, 48], by increasing the mutation step size [107, 115], by moving individuals from one sub-population to another [40] or by keeping the sub-populations/individuals away from one another [50, 51].

Some of the first studies following this approach are hyper-mutation [31] and variable local search (VLS) [108, 109]. In his research, Cob [31] proposed an adaptive mutation operator called hyper-mutation whose mutation rate is a multiplication of the normal mutation rate and a hyper-mutation factor. The hyper-mutation is invoked only after a change is detected. In the VLS algorithm, the mutation step size is controlled by a variable local search range. This range is determined by the formula $(2^{BITS} - 1)$ where BITS is a value adjustable during the search [107] or adapted using a learning strategy borrowed from the feature partitioning algorithm by Vavak *et al.* [108].

In [70], hyper-mutation was used to solve dynamic constraint problems. Detectors are placed near the boundary of feasible regions and when the feasibility of these detectors changes, the EA increases its mutation rate to raise the diversity level to track the moving feasible regions. The mutation rate is decreased again once the moving feasible region has been tracked successfully. Riekert and Malan [86] proposed adaptive genetic programming which not only increases mutation, but also reduces elitism and increases crossover probability after a change. The idea of introducing diversity after a change has also been used in dynamic multi-objective optimization (DMO). For example, in a multi-population algorithm for DMO [40], when a change is detected, random individuals and some competitor individuals from other sub-populations are introduced to each sub-population to increase diversity.

Diversity introduction is also used in particle swarm optimization (PSO). Hu and Eberhart [48] introduced a simple mechanism in which a part of the swarm or the whole swarm will be re-diversified using randomization after a change is detected. Janson and Middendorf [51] followed a more sophisticated mechanism where in addition to partial re-diversification, after each change the swarm is divided into several sub-swarms for a certain number of generations. The purpose of this is to prevent the swarm from converging to the old position of the global optimum too quickly. Daneshyari and Yen [34] proposed a cultural-based PSO where after a change, the swarms are re-diversified using a framework of knowledge inspired from the belief space in cultural algorithms. The diversity-introducing approach is still commonly used in many recent EDO algorithms, e.g., [67, 78, 82, 83, 85, 93, 115].

2.2.3.2   Strength and Weakness

In general, methods following the diversity-introducing approach appear to be good in solving problems with continuous changes where changes are small and medium. This is because invoking mutations or distributing individuals around an optimum resembles a type of "local search", which is useful to observe the nearby places of this optimum. Thus, if the optimum does not move very far, it might be tracked [107, 108].

However, this approach has some drawbacks that might make it not so suitable for certain type of problems. First, it is dependent on whether changes are known/easy to detect or not. If a change appears in a place where no individual exists, it will go undetected [65]. Second, it might be difficult to identify the right amount of diversity needed: too small steps will resemble local search while too large steps will result in random search [52]. Third, the approach might not be effective for solving problems with random changes or large changes (changes are severe) because many diversity-introducing methods have their mutation/relocation size restricted to a specific range. The diversity-introducing approach is still commonly used in many recent metaheuristics algorithms, e.g., [67, 70, 78, 82, 83, 85].

2.2.4   Maintaining Diversity during the Search

2.2.4.1   Overview

Another related approach is to maintain population diversity throughout the search process (see Algorithm 2). Methods following this approach do not detect changes

---

Algorithm 2 Maintaining diversity

---

1. *Initialize*: Initialize the population
2. *For each generation*

   a. *Evaluate*: Evaluate each member of the population
   b. *Maintain diversity*: Add a number of new, diversified individuals to the current population, or select more diversified individuals, or explicitly relocate individuals to keep them away from one another.
   c. *Reproduce*: Reproduce a new population
   d. Return to step 2a

---

explicitly. Instead, they rely on their diversity to adaptively cope with the changes. Diversity can be maintained by regularly introducing random individuals (*random immigrants*) to the population [43, 125], by sharing fitness [2], by specifically distributing some sentinel individuals [65], by explicitly keeping individuals from getting close to one another [10, 11], by dedicating one of the objectives for divesity purposes [24], or by combining several of the strategies above [1, 102].

In the random immigrants method [43], in every generation a number of generated random individuals are added to the population to maintain diversity. Experimental results show that the method is more effective in handling dynamics than regular EA [43]. It was also reported that the high diversity level brought by random immigrants also helps in handling constraints [70].

In [65], a different mechanism was proposed in which instead of generating random individuals, the sentinel placement method initializes a number of sentinels which are specifically distributed throughout the search space. Experiments show that this method might get better results than random immigrants and hypermutation in problems with large and chaotic changes [65].

Yang and Yao [125] proposed two other approaches based on the population-based incremental learning (PBIL) algorithm - parallel PBIL (PPBIL2) and dual PBIL (DPBIL). The PBIL algorithm has a probability vector adjusted based on the best found solutions. In PPBIL2, Yang and Yao [125] improved PBIL for DOPs by maintaining two parallel probability vectors: a vector similar to the original one in PBIL and a random initialized probability dedicated to maintain diversity during the search. To improve PBIL2 in dealing with large changes, Yang and Yao [125] proposed the DPBIL where two probability vectors are *dual* with each other, i.e., given the first vector P_1, the second vector P_2 is determined by $P_2[i] = 1 - P_1[i]$ ($i = 1, ..., n$), where n is the number of variables. During the search only P_1 needs to learn from the best generated solution because P2 will change with P_1 automatically. PBIL and DPBIL were also combined with random-immigrants in [126] with better results than the original algorithms.

Another approach to maintain diversity is to reward individuals that are genetically different from their parents [110]. In this approach, in addition to a regular population, the algorithm maintains an additional population where individuals are selected based on their Hamming distance to their parents (to promote diversity) and another population where individuals are selected based on their fitness improvement compared to their parents (to promote exploitation). By observing its own performance in stagnation and population diversity, the algorithm adaptively adjusts the size of the three populations to react to dynamic environments.

Diversity can be maintained in evolution strategies (ESs) by preventing the strategy parameters from converging to zero, e.g. in [53]. The approach of maintaining diversity is also used in PSO to solve dynamic continuous problems. In their charged PSOs [9–11], Blackwell *et al.* applied a *repulsion* mechanism, which is inspired from the atom field, to prevent particles/swarms to get too close to each other. In this mechanism, each swarm is comprised of a nucleus and a cloud of charged particles which are responsible to maintain diversity. There is a repulsion among these particles to keep particles from approaching near to each other. In [34], both the particle selection and replacement mechanisms are modified so that the most diversified particles (in term of Hamming distance) are selected and the particles that have similar positions are replaced. In the compound PSO [59], the degree of particles deviating from their original directions becomes larger when the velocities becomes smaller, and distance information was incorporated as one of the criteria to choose a particle for the update mechanism.

Bui *et al.* [24] used multiple objectives to maintain diversity. The dynamic problem is represented by a problem with two objectives: one original objective and one special objective created to maintain diversity. Similar examples can be found in [1, 102], of which the latter proposed six different types of objectives, including retaining more old solutions; retaining more random solutions; reversing the first objective; keeping a distance from the closest neighbour; keeping a distance from all individuals; and keeping a distance from the best individual. The diversity-maintaining strategy is still the main strategy in many recent approaches, for example, see [6, 9, 10, 29, 35, 39, 42, 50, 70, 125, 126].

2.2.4.2   Strength and Weakness

Methods following the diversity-maintaining approach may be good at solving problems with large changes (e.g. in [70, 73] random immigrants helped significantly improve the performance in dynamic constrained problems where changes are severe due to the presence of disconnected feasible regions), problems with slow changes (as shown in e.g. [2, 125]), and problems with competing peaks (as reported in [27]).

However, the diversity-maintaining approach might suffer from some disadvantages. First, continuously focusing on diversity may slow down, or even distract the optimization process [52]. Second, the approach may become less effective in dealing with small changes where the optima just take a slight move away from their previous places [32].

2.2.5   Memory Approaches

In situations where DOP changes are periodical or recurrent, and hence the optima may return to the regions near their previous locations, it might be useful to re-use previously found solutions to save computational time. The most common way to re-use old solutions in this manner is to maintain memory components in the algorithms. The memory can also play the role as a reserved place for storing old solutions in order to maintain diversity when needed. The memory can be integrated *implicitly* as a redundant representation in the algorithms, or maintained *explicitly* as a separate memory component.

2.2.5.1   Implicit Memory

The most common implicit memory used in EDO algorithms is redundant coding using diploid genomes, e.g., [41, 56, 68, 106, 121]. A diploid EA is usually an algorithm whose chromosomes contain two alleles at each locus. Although most

---

Algorithm 3 Multiploid EA for dynamic optimization

---

1. *Initialize*: Initialize the population and the multiploid representation
2. *For each generation*

   a. *Evaluate:* Evaluate each member of the population
   b. For each individual:
      i. *Detect changes*
      ii. *Adjust the dominance level of each allele* : If there is any change, adjust the dominance to accommodate the current change
      iii. *Select the dominant alleles according to their dominance level*
   c. *Reproduce*: Reproduce a new population using the adjusted mutations
   d. Return to step 2a

---

normal EAs for static problems are haploid, it is believed that diploid, and other multiploid approaches, are suitable for solving dynamic problems [56]. A pseudo code for multiploid approaches in dynamic environments is described in Algorithm 3, where the following three components need to be incorporated: (i) represent the redundant code; (ii)readjust the dominance of alleles; and (iii) check for changes.

The dominance of alleles is usually represented by a table [68, 91] or a mask [33] mapping between genotypes and phenotypes. The dominance then can be changed adaptively among alleles depending on the detection of changes in the landscape.

2.2.5.2 Explicit Memory

Methods that use explicit memory generally follow the steps in Algorithm 4. The memory can be maintained *directly* in the form of previous good solutions [8, 15, 34, 55, 60, 62, 64, 93, 118, 119, 126–128], or it can be maintained *indirectly* in the form of associative information. Various type of associative information can be included, e.g., the environment at the considered time [37, 79]; the list of environmental states

---

Algorithm 4 Using explicit memory

---

1. *Initialize*:

   a. Initialize the population
   b. Initialize the explicit memory

2. *For each generation*

   a. Evaluate each member of the population
   b. Update the memory
   c. Reproduce a new population
   d. Use information from the memory to update the new population
   e. Return to step 2a

---

and state transition probabilities [96]; the successful individuals for certain types of changes[94, 120]; the probability vector that created the best solutions [126]; the distribution statistics information of the population at the considered time [119]; the probability of the occurrence of good solutions in each area of the landscape [84, 85]; or the probability of likely feasible regions [83].

Generally the best found elements (direct or associative) of the current generation will be used to update the memory. These newly found elements will replace some existing elements in the memory, which can be the oldest member [37, 95, 103, 115], the one with the least contributions to the diversity of the population [15, 37, 95, 118, 126], or the one with least contribution to fitness [37]. During the search, usually the best elements in the memory (i.e. the ones that show the best results when being re-evaluated) will be used to replace the worst individuals in the population. Replacement can take place after each generation or after a certain number of generations, or it can be done after each change if the change can be detected.

2.2.5.3   Strength and Weakness

Memory methods are particularly effective for solving problems with periodically changing environments. For example, it was shown that the memory-based versions of EAs and random-immigrant significantly outperform the original algorithms in cyclic dynamic environments [122]. The approach might also be good in slowing down convergence and favour diversity [16, 18].

Memory methods, however, have disadvantages that may require them to be used with some other methods for the best results. First, they might be useful only when optima reappear at their previous locations or if the environment returns to its previous states [15, 16]. Second, they might not be good enough to maintain diversity for the population [15]. Third, the information stored in the memory might become redundant (and obsolete) and consequently may affect the performance of the algorithm. In addition, redundant coding approaches might not be good for cases where the number of oscillating states is large.

2.2.6   Prediction Approaches

2.2.6.1   Overview

In DOPs where changes exhibit regular patterns, it might be helpful to try to learn the patterns from previous search experience, and then try to predict changes in the future. A pseudo code of methods following this approach is shown in Algorithm 5.

One of the common prediction approaches is to predict the movement of the moving optima. Hatzakis and Wallace [45] combined a forecasting technique (autoregressive) with an EA to predict the location of the next optimal solution after a change is detected. The forecasting model (time series model) is created using a sequence of optimum positions found in the past. Experimental results show that if this algorithm can predict the movements of optima correctly, it can work well with

---

Algorithm 5 Prediction approach to solve dynamic problems

---

1. *Initialize phase*:

   a. Initialize the population
   b. Initialize the learning model and training set

2. *Search for optimum solutions and detect changes*
3. *If a change is detected*

   a. Use the current environment state as the input for the learning model
   b. Use the learning model to estimate the type of this current change and/or how the next change should be
   c. Generate new individuals or recall old ones that best match with the estimation
   d. Search for the new optimum using the new population
   e. Update the training set based on the search results

4. Return to step 2

---

very fast changes. A similar approach was proposed in [90] where the movement of optima was predicted using Kalman filters. The predicted information (the next location of the optimum) is incorporated into an EA in three ways: First, the mutation operator is modified by introducing some bias so that individuals' exploration is directed toward the predicted region. Second, the fitness function is modified so that individuals close to the estimated future position are rewarded. Third, some "gifted" individuals are generated at the predicted position, and introduced into the population to guide the search. Experiments on a visual tracking benchmark problem show that the proposed method does improve the tracking of the optimum, both in terms of distance to the real optimum and smoothness of the tracking.

Prediction was also used to determine the locations that individuals should be re-initialized to when a change occurs. In [129] this approach is used to solve two dynamic multi-objective optimization benchmark problems in two ways: First, the solutions in the Pareto set from the previous change periods were used as a time series to predict the next re-initialization locations. Second, to improve the chance of the initial population to cover the new Pareto set, the predicted re-initialization population is perturbed with a Gaussian noise whose variance is estimated based on historical data. Compared with random-initialization, the approach was able to achieve better results on the two tested problems.

Another interesting approach is to predict the time when the next change will occur and which possible environments will appear in the next change [96, 97]. In these works, the authors used two prediction modules to predict two different factors. The first module, which uses either a linear regression [96] or a non-linear regression [97], is used to estimate the generation when the next change will occur. The second module, which uses Markov chain, monitors the transitions of previous environments and based on this data provides estimations of which environment will appear in the next change. Experimental results show that an EA with the proposed predictor is able to perform better than a regular EA in cyclic/periodic environments.

A special class of prediction approaches is dynamic time-linkage optimization [12–14, 72, 74]. Time-linkage problems are problems where the current solutions made by the algorithms can influence the future dynamics. In such problems, it was suggested that the only way to solve the problems effectively is to predict future changes and take into account the possible future outcomes when solving the problems online. Research in [12–14, 72, 74] followed this idea to solve time-linkage problems effectively. Another related study is the anticipation approach [20] in solving dynamic scheduling problems where in addition to finding good solutions, the solver also tries to move the system "into a flexible state" where adaptation to changes can be done more easily. Specifically, because it is observed that in the tested dynamic job-shop scheduling problem, the flexibility of the system can be increased by avoiding early machine idle times, the authors proposed a scheduling approach where in addition to the main optimality objective, solutions with early idle time are penalized. The experimental results show that such an anticipation approach significantly improved the performance of the system.

2.2.6.2   Strength and Weakness

Prediction approaches may become very effective if their predictions are accurate. In this case, the algorithms can detect/track/find the global optima quickly, as shown in [45, 94, 120]. However, prediction/adaptation-based algorithms also have their own disadvantages, mostly due to training errors. These errors might be resulted by the unpredictable nature of the problems. If the changes are stochastic, or history data are misleading, prediction approaches might not get satisfactory results. For example, [72, 74] illustrated a situation where history data are actually inappropriate for the prediction and might even mislead the predictor to get worse results.

Prediction errors might also be due to wrong training data, or lack of training data. As in the case of any learning/predicting/forecasting model, the algorithms may need a large enough set of training data to produce good results. It also means that the prediction can only be started after sufficient training data have been collected, e.g., [12, 13, 96, 97]. In the case of dynamic optimization where there is a need of finding/tracking the optima as quick as possible, this might be a disadvantage.

2.2.7   Self-adaptive Approaches

In certain cases, the self-adaptive mechanisms of EAs and other meta-heuristics can be used effectively to cope with changes. One example is the GA with genetic mutation rate [44], which allows the algorithm to evolve its own mutation strategy parameters during the search process based on the fitness of the population. In this method, the mutation rate is encoded in genes and is influenced by the selection process. The algorithm was tested in both gradual and abrupt dynamic landscapes. The results show that the algorithm has better performance than a conventional GA.

However, it is still not better than hyper-mutation (see section 2.2.3 and [31]). A similar method was proposed by Ursem in his multinational GA (MGA) [104]. Five different parameters (probability for mutation, probability for crossover, selection ratio, mutation variance and distance) are encoded in the genomes of his MGA for adaptation. The adaptation mechanism works well in simple cases where the velocity of moving peaks is constant. However, in cases where the velocity is not constant, the adaptation seems to be not fast enough. These two results show the difficulty of applying adaptive parameter tuning to complex dynamic optimization.

The self-adaptive mechanisms of such EAs as ES or evolutionary programming (EP) were also investigated for using in dynamic optimization. Angeline [3] examined self-adaptive EP (saEP) and showed that the strategy is not effective for all types of tested problems. Bäck [7] showed that the log-normal self-adaptation in ES may perform better than saEP. Experiments pointed out that algorithm implementation and parameter settings have much less influence on ES in dynamic environments than in stationary environments [92] and that ES might be unreliable in rapidly changing environments [114]. Weicker [112] also argued that it is possible that if Gaussian mutation is used in the standard ES, self-adaptation might not be appropriate for dynamic optimization.

Some mathematical analyses on the performance of self-adaptive ES in dynamic environments were proposed. Arnold and Beyer [4] pointed out that the cumulative mutation step-size adaptation of ES can work well on a variant of the sphere model with random dynamics of the optimum. The strategy can realize the optimal mutation step-size for the model. However, in the sphere model with linear dynamics, another research of Arnold and Beyer [5] revealed that the mutation step-size realized by ES is not the optimal one (but the adaptation still ensures that the optimum can be tracked).

2.2.8 Multi-population Approaches

2.2.8.1 Overview

Multi-population approach, which maintains multiple sub-populations concurrently, can be seen as a combination of diversity maintaining/introducing, memory and adaptation. Each sub-population may handle a separate area of the search space. Each of them may also take responsibility for a separate task. For example, some sub-populations may focus on searching for the global optimum while some others may concentrate on tracking any possible changes. These two types of populations then may communicate with each other to bias the search. A typical pseudo-code of the multi-population approach is shown in Algorithm 6.

Methods following the approach of using multiple populations usually need to accomplish two goals: First, they may need to assign different types of tasks to different sub-populations, for example, P_{search} to search and P_{track} to track, so that the search can be done effectively. Second, they need to divide the sub-populations appropriately and make sure that the sub-populations are not overlapped to have the

---

Algorithm 6 Multi-population approach

---

1. *Initialize*:

   a. Initialize the set P_{search} of sub populations finding the global optima
   b. Initialize the set P_{track} of sub populations tracking changes in the landscape

2. *For each generation*:

   a. *Search for optima*: Sub-populations in P_{search} find the global optima
   b. *Track changes*: Sub-populations in P_{track} track any changes
   c. *Maintain diversity*: Re-allocate/split/merge the sub-populations so that they are not overlapped and can cover a larger area of the search space
   d. *Adjust*: Re-adjust each sub-population in P_{search} based on the experience from sub-populations in P_{track}
   e. Reproduce each sub-population
   f. Return to step 2a

---

best diversity and also to avoid the situation where many sub-populations find the same peak.

For the first goal, *assigning different tasks to the sub-populations*, there might be multiple small populations in P_{search} searching for new solutions and there is only one large population in P_{track} to track changing peaks [77], or there might be one large population to search and multiple sub-population for tracking changes [9, 19, 29, 38, 61, 63, 73], or each sub-population can both search for new solutions and track changes [39, 57, 58, 104]. Relating to the goal of assigning the tasks to sub-populations, it should be noted that in dynamic optimization multiple populations are used not only for exploring different parts of the search space, but also for co-evolution [40, 73, 75] or maintaining diversity and balancing between exploitation and exploration [110].

For the second goal, *dividing the sub-populations and making sure that the sub-populations are not overlapping*, there are different approaches, of which the most common is clustering: choosing some solutions in the population as the centres of the future clusters, then defining each sub-population as a hyper-cube or sphere with a given size. All individuals within the range of a hyper-cube/sphere will belong to the corresponding sub-population of that hyper-cube/sphere. For example, the self-organizing scouts (SOS) algorithm [19] keeps the sub-populations from being overlapped by confining each sub-population to a hyper-cube determined by a centre (the most fit individual in the population) and a pre-defined range. If an individual of one sub-population ventures to the area monitored by another sub-population, this individual will simply be discarded and re-initialized (this process is called *exclusion*). The same approach is also used in DE [61, 63] and PSO [10, 58]. For example, in multi-swarm PSO (mPSO) [10], swarms are divided into sub-swarms so that each swarm watches a different peak. In addition, mPSO also maintains a similar mechanism (named anti-convergence) to the P_{search} in SOS so that there is always one free swarm to continue exploring the search space. Another example is the speciation PSO (SPSO) algorithm [58], where each species is a hyper-sphere whose centre is the best-fit individual in the species and each species can be used to track a peak. In recent clustering approaches [57, 115], density-based clustering methods are also used to divide/separate the sub-populations and to allow the algorithms explore different parts of the search landscape.

Other approaches to divide sub-populations are to incorporate some mechanism of penalty/rewarding to keep the sub-populations apart [77], and to estimate the basins of attractions of peaks and use these basins as separate regions for each sub-population [104].

2.2.8.2   Strength and Weakness

Multi-population approaches are thought to have multiple advantages. First, they can maintain enough diversity to adaptively start a new search whenever a new change appears, as illustrated in [17]. Second, they may be able to recall some information from the previous generations thanks to one (or several) population(s) dedicated for retaining old solutions, as shown in their good performance in problems with recurrent changes [15, 104]. Third, they can search/track the moves of multiple optima, as analysed in many existing studies on multi-populations, e.g., [17] and [104]. Finally, they can be very effective for solving problems with competing peaks or multimodal problems. A survey by Moser [66] showed that among 19 surveyed algorithms that are designed to solve the moving peaks benchmark (MPB) [15] with multimodal competing peaks, a majority (15 out of 19) follow the multi-population approach.

There are also disadvantages in using multi-population approaches. First, too many sub-populations may slow down the search. For example, Blackwell and Branke [10] showed that for their multi-swarm PSO algorithm, if the number of sub-populations (swarms) is larger than the number of peaks, the performance of the algorithm decreases. It might also be difficult to identify the appropriate number of sub-populations, as well as the size of each sub-population. Second, the need of calculating the distance/similarity/regional metrics to separate the sub-populations might also affect the performance. Third, in academic research, multi-population approaches have been tested mostly in the continuous domain, and hence more evidence might be needed to confirm their effectiveness on combinatorial problems.

2.3   Theoretical Development of EDO Methodologies

EDO research so far has mainly been empirical. Most theoretical analysis of EDO has just started in recent years with some results. Analysing EAs for DOPs is considered more difficult than analysing EAs for static problems due to the extra dynamics introduced in DOPs. The theoretical studies on EDO methodologies are briefly reviewed as follows.

Initial EDO theoretical works were extensions of the analysis of simple EAs, e.g., the (1+1) EA[3], for static optimization to simple DOPs, e.g., the dynamic bit matching problem [99]. In [99], the authors presented the transition probabilities of the (1+1) EA and showed that even small perturbations in the fitness function could have a significantly negative impact on the performance of the (1+1) EA. Based on this work, Branke and Wang [23] developed an analytical model for a (1, 2)-ES and compared different strategies to handle an environmental change within a generation on the dynamic bit matching problem .

The first hitting time of a (1+1)-ES was analyzed by Droste [36] on the dynamic bit matching problem, where exactly one bit is changed with a given probability p after each function evaluation. It was shown that the expected first hitting time of the (1+1)-ES is polynomial if and only if $p = O(logn/n)$. Arnold and Beyer [4] investigated the tracking behaviour of an $(\mu/\mu, \lambda)$-ES with self-adaptive mutation step-size on a single continuously moving peak. They derived a formula to predict the tracking distance of the population from the target. Jansen and Schellbach [49] presented a rigorous performance analysis of the $(1+\lambda)$-EA on a tracking problem in a two-dimensional lattice and showed that the expected first hitting time strictly increases with the offspring population size (i.e., λ) whereas the expected number of generations to reach the target decreases with λ. In [114], Weicker and Weicker analyzed the behaviour of ESs with several mutation variants on a simple rotating dynamic problem. In [111], Weicker presented a framework for classifying DOPs and used it to analyze how the offspring population size and two special techniques for DOPs affect the tracking probability of a $(1,\lambda)$-ES. Weicker [113] also used Markov models to analyze the tracking behaviour of $(1,\lambda)$-ESs with different mutation operators for a discrete optimization problem with a single moving optimum.

Rohlfshagen et al. [87] analyzed how the magnitude and frequency of changes may affect the performance of the (1+1)-EA on two specially designed pseudo-Boolean functions under the dynamic framework of the XOR DOP generator [118]. They demonstrated two counter-intuitive results, i.e., the algorithm is efficient if the magnitude of change is large and inefficient when the magnitude of change is small, and the algorithm is efficient if the frequency of change is very high and inefficient if the frequency of change is sufficiently low. These results allow us to gain a better understanding of how the dynamics of a function may affect the runtime of an algorithm.

In addition to the above runtime analysis of EDO methodologies, there are also theoretical analysis of dynamic fitness landscape [21, 22, 80, 81, 83, 84, 88, 89, 100, 101]. Readers are referred to [71] for a literature review on research in this area.

---

[3] In a (1+1) EA, there is only one solution maintained in the population. In each iteration, the unique solution acts as the parent to generate an offspring via mutation. If the fitness of the offspring is not worse than the parent, the offspring will replace the parent; otherwise, the parent will survive into the next generation.

2.4 Summary and Future Research Directions

2.4.1 Summary

The review above showed that each EDO approach seems to be suitable only for certain types of DOPs, which conforms to the No Free Lunch theorem [116]. The fact that each approach is likely to be suitable to some particular classes of problems is also the reason why many recent studies try to combine different approaches into one single algorithm to solve the problems better. Overall, multi-population approaches seem to be the most flexible approach to date in the continuous domain.

The review showed that there have been some recent works on the theory behind EDO. These theoretical studies are still quite basic. However, they have made very important first steps toward understanding EDO and will surely act as the basis for further theoretical studies on EDO.

It should be noted that most existing EDO methods were tested and evaluated on academic problems only. This leads to the question of whether these methods would still be effective in real-world DOPs. In the next subsection, we will discuss this question in detail.

2.4.2 The Gaps between Academic Research and Real-World Problems

The lack of a clear link between EDO academic research and real-world scenarios has lead to some criticisms on how realistic current academic problems are. Ursem et al. [105] questioned the importance of current academic benchmarks by stating that "no research has been conducted to thoroughly evaluate how well they reflect characteristic dynamics of real-world problems"; Branke et al. [22] pointed out that "little has been done to characterize and understand the nature of a change in real-world problems"; Rohlfshagen and Yao [88] criticized that "a large amount of effort is directed at an academic problem that may only have little relevance in the real world"; and in [74, 76], it has been showed that there are some classes of real-world problems whose characteristics have not been captured by existing academic research yet. Nguyen and Yao [76] also showed evidence of situations where existing EDO techniques could not solve certain classes of DOPs effectively due to the uncaptured characteristics of DOPs.

Recently, a detailed analysis [70, Chapter 3] of a large set of recent "real"[4] real-world DOPs has been made to investigate the characteristics of real-world problems and how they relate to the characteristics of current academic benchmark problems. This investigation pointed out certain gaps between academic EDO research and real-world DOPs. First, current studies in academic EDO do not cover all types of com-

---

[4] Only references that actually use real-world data or solve problems in actual real-world situations were considered. Benchmark problems, even if designed to simulate real-world applications, were not considered unless there is evidence that the data used to create the benchmark were taken from real-world applications.

mon DOPs yet. There are two types of problems that are very common in real-world situations but received very little attention from the community: dynamic constrained problems and time-linkage problems. Second, although many current EDO academic research works only focus on one major optimization goal: optimality (to find the best fitness value), the study in [70] showed that there might be many other common optimization goals. Third, although most current EDO benchmark problems have only one changing factor, the study in [70] showed that there are also other common types of changing factors: constraints, number of variables, domain ranges, etc.

In summary, the review in [70] showed that besides the characteristics and assumptions commonly used in EDO academic research, real-world DOPs also have other important types of problems and problem characteristics that have not been studied extensively by the EDO community. In order to solve real-world DOPs more effectively, it is necessary to take these characteristics and problem types into account when designing new methodolgoies.

2.4.3   Future Research Directions

As reviewed in this chapter, there have been many studies devoted to EDO methodolgoies and fruitful results have been achieved. However, the research domain of EDO is still relatively young. Much more effort is needed to fully develop and understand the domain of EDO. Some future research directions on EDO methodologies are highlighted and suggested as follows.

First, although a number of EDO approaches have been developed for solving DOPs, new efficient approaches are of great needs. As the review has shown, different methods have different strength and weakness on different DOPs. Hence, it is also worthy to further develop and investigate hybrid methods for DOPs in the future. Here, it is very important to develop adaptive systems that can deal with DOPs of different characteristics. Active adaptability should also be addressed so that future algorithms are able to effectively handle dynamics even without change detection.

Second, as shown earlier in this chapter, most EDO methodologies focus on solving academic problems, where there is no clear link to real-world characteristics. Although there have been some real-world application studies on EDO, e.g., see [28–30, 70, 117] (also see [70, 123] for detailed lists of references), the number of EDO application studies so far is still very limited. One of the important research direction for the EDO community is to consider and model more real-world DOPs, and apply EDO and other meta-heuristic methods to solve them in the future. This will further enhance the applicability and feasibility of EDO in practical situations.

Third, as discussed earlier, theoretical research on EDO is still at the beginning stage. The relative lack of theoretical proof on EDO makes it difficult to evaluate the strength and weakness of EDO algorithms on solving different types of DOPs. As reviewed, the computational complexity analysis of EDO has started with some promising results. However, this area of study needs to be extended significantly to gain more insight as to which DOP is difficult or easy to solve for what types of EDO methods. Here, techniques for analyzing evolutionary optimization on static problems, e.g., drift analysis [46, 47], may be applied or adapted to analyze EDO. It will also be beneficial to analyse the dynamic behaviour of EDO algorithms and how the behaviour satisfies the real-world practitioners' requirements, which are not always the ability to identify/track the global optimum after each change.

Acknowledgements. This work was supported by the Engineering and Physical Sciences Research Council (EPSRC) of UK under Grant numbers EP/E058884/1, EP/K001523/1, EP/E060722/1, and EP/K001310/1, a UK ORS Award, a studentship from the School of Computer Science, University of Birmingham, and an EU-funded project named "Intelligent Transportation for Dynamic Environment (InTraDE)".

References

[1] Abbass, H.A., Deb, K.: Searching under multi-evolutionary pressures. In: Fonseca, C.M., Fleming, P.J., Zitzler, E., Deb, K., Thiele, L. (eds.) EMO 2003. LNCS, vol. 2632, pp. 391–404. Springer, Heidelberg (2003)

[2] Andersen, H.C.: An investigation into genetic algorithms, and the relationship between speciation and the tracking of optima in dynamic functions. Honours thesis, Queensland University of Technology, Brisbane, Australia (1991)

[3] Angeline, P.J.: Tracking extrema in dynamic environments. In: Angeline, P.J., McDonnell, J.R., Reynolds, R.G., Eberhart, R. (eds.) EP 1997. LNCS, vol. 1213, pp. 335–345. Springer, Heidelberg (1997)

[4] Arnold, D.V., Beyer, H.-G.: Random Dynamics Optimum Tracking with Evolution Strategies. In: Guervós, J.J.M., Adamidis, P.A., Beyer, H.-G., Fernández-Villacañas, J.-L., Schwefel, H.-P. (eds.) PPSN 2002. LNCS, vol. 2439, pp. 3–12. Springer, Heidelberg (2002)

[5] Arnold, D.V., Beyer, H.G.: Optimum tracking with evolution strategies. Evol. Comput. 14(3), 291–308 (2006)

[6] Azevedo, C., Araujo, A.: Generalized immigration schemes for dynamic evolutionary multiobjective optimization. In: Proc. 2011 IEEE Congr. Evol. Comput., pp. 2033–2040 (2011)

[7] Bäck, T.: On the behavior of evolutionary algorithms in dynamic environments. In: Proc. 1998 IEEE Int. Conf. on Evol. Comput., pp. 446–451 (1998)

[8] Bendtsen, C.N., Krink, T.: Dynamic memory model for non-stationary optimization. In: Proc. 2002 IEEE Congr. Evol. Comput., pp. 145–150 (2002)

[9] Blackwell, T.: Particle swarm optimization in dynamic environment. In: Yang, S., Ong, Y.S., Jin, Y. (eds.) Evolutionary Computation in Dynamic and Uncertain Environments. SCI, vol. 51, pp. 28–49. Springer, Heidelberg (2007)

[10] Blackwell, T., Branke, J.: Multiswarms, exclusion, and anti-convergence in dynamic environments. IEEE Trans. Evol. Comput. 10(4), 459–472 (2006)

[11] Blackwell, T.M., Bentley, P.J.: Dynamic search with charged swarms. In: Proc. 2002 Genetic and Evol. Comput. Conf., pp. 19–26 (2002)

[12] Bosman, P.A.N.: Learning, anticipation and time-deception in evolutionary online dynamic optimization. In: Yang, S., Branke, J. (eds.) GECCO Workshop on Evolutionary Algorithms for Dynamic Optimization (2005)

[13] Bosman, P.A.N.: Learning and anticipation in online dynamic optimization. In: Yang, S., Ong, Y.S., Jin, Y. (eds.) Evolutionary Computation in Dynamic and Uncertain Environments. SCI, vol. 51, pp. 129–152. Springer, Heidelberg (2007)

[14] Bosman, P.A.N., Poutré, H.L.: Learning and anticipation in online dynamic optimization with evolutionary algorithms: the stochastic case. In: Proc. 2002 Genetic and Evol. Comput. Conf., pp. 1165–1172 (2007)

[15] Branke, J.: Memory enhanced evolutionary algorithms for changing optimization problems. In: Proc. 1999 IEEE Congr. Evol. Comput., vol. 3, pp. 1875–1882 (1999)

[16] Branke, J.: Evolutionary approaches to dynamic environments - updated survey. In: GECCO Workshop on Evolutionary Algorithms for Dynamic Optimization Problems, pp. 27–30 (2001)

[17] Branke, J.: Evolutionary Optimization in Dynamic Environments. Kluwer (2001)

[18] Branke, J.: Evolutionary approaches to dynamic optimization problems – introduction and recent trends. In: Branke, J. (ed.) GECCO Workshop on Evolutionary Algorithms for Dynamic Optimization Problems, pp. 2–4 (2003)

[19] Branke, J., Kaußler, T., Schmidth, C., Schmeck, H.: A multi-population approach to dynamic optimization problems. In: Proc. 4th Int. Conf. Adaptive Comput. Des. Manuf., pp. 299–308 (2000)

[20] Branke, J., Mattfeld, D.: Anticipation and flexibility in dynamic scheduling. Int. J. of Production Research 43(15), 3103–3129 (2005)

[21] Branke, J., Orbayı, M., Uyar, Ş.: The role of representations in dynamic knapsack problems. In: Rothlauf, F., et al. (eds.) EvoWorkshops 2006. LNCS, vol. 3907, pp. 764–775. Springer, Heidelberg (2006)

[22] Branke, J., Salihoglu, E., Uyar, Ş.: Towards an analysis of dynamic environments. In: Proc. 2005 Genetic and Evol. Comput. Conf., pp. 1433–1439 (2005)

[23] Branke, J., Wang, W.: Theoretical analysis of simple evolution strategies in quickly changing environments. In: Proc. 2003 Genetic and Evol. Comput. Conf., pp. 537–548 (2003)

[24] Bui, L., Abbass, H., Branke, J.: Multiobjective optimization for dynamic environments. In: Proc. 2005 IEEE Congr. Evol. Comput., vol. 3, pp. 2349–2356 (2005)

[25] Carlisle, A., Dozier, G.: Adapting particle swarm optimisationto dynamic environments. In: Proc. 2000 Int. Conf. on Artif. Intell., pp. 429–434 (2000)

[26] Carlisle, A., Dozier, G.: Tracking changing extrema with adaptive particle swarm optimizer. In: Proc. 5th World Automation Congr., vol. 13, pp. 265–270 (2002)

[27] Cedeno, W., Vemuri, V.R.: On the use of niching for dynamic landscapes. In: Proc. 1997 IEEE Int. Conf. on Evol. Comput. (1997)

[28] Cheng, H., Yang, S.: Genetic algorithms with immigrants schemes for dynamic multicast problems in mobile ad hoc networks. Eng. Appl. of Artif. Intell. 23(5), 806–819 (2010)

[29] Cheng, H., Yang, S.: Multi-population genetic algorithms with immigrants scheme for dynamic shortest path routing problems in mobile ad hoc networks. In: Di Chio, C., et al. (eds.) EvoApplicatons 2010, Part I. LNCS, vol. 6024, pp. 562–571. Springer, Heidelberg (2010)

[30] Chitty, D.M., Hernandez, M.L.: A hybrid ant colony optimisation technique for dynamic vehicle routing. In: Deb, K., Tari, Z. (eds.) GECCO 2004. LNCS, vol. 3102, pp. 48–59. Springer, Heidelberg (2004)

[31] Cobb, H.G.: An investigation into the use of hypermutation as an adaptive operator in genetic algorithms having continuouis, time-dependent nonstationary environments. Technical Report AIC-90-001, Naval Research Laboratory, Washington, USA (1990)

[32] Cobb, H.G., Grefenstette, J.J.: Genetic algorithms for tracking changing environments. In: Proc. 1993 Int. Conf. on Genetic Algorithms, pp. 523–530 (1993)

[33] Collingwood, E., Corne, D., Ross, P.: Useful diversity via multiploidy. In: Proc. 1996 IEEE Int. Conf. on Evol. Comput., pp. 810–813 (1996)

[34] Daneshyari, M., Yen, G.: Dynamic optimization using cultural based pso. In: Proc. 2011 IEEE Congr. Evol. Comput., pp. 509–516 (2011)

[35] Deb, K., Rao N., U.B., Karthik, S.: Dynamic multi-objective optimization and decision-making using modified NSGA-II: A case study on hydro-thermal power scheduling. In: Obayashi, S., Deb, K., Poloni, C., Hiroyasu, T., Murata, T. (eds.) EMO 2007. LNCS, vol. 4403, pp. 803–817. Springer, Heidelberg (2007)

[36] Droste, S.: Analysis of the (1+1) ea for a dynamically changing onemax-variant. In: Proc. 2002 IEEE Congr. Evol. Comput., pp. 55–60 (2002)

[37] Eggermont, J., Lenaerts, T., Poyhonen, S., Termier, A.: Raising the dead: Extending evolutionary algorithms with a case-based memory. In: Miller, J., Tomassini, M., Lanzi, P.L., Ryan, C., Tetamanzi, A.G.B., Langdon, W.B. (eds.) EuroGP 2001. LNCS, vol. 2038, pp. 280–290. Springer, Heidelberg (2001)

[38] Fernández, J.L., Arcos, J.L.: Adapting particle swarm optimization in dynamic and noisy environments. In: Proc. 2010 IEEE Congr. Evol. Comput., pp. 765–772 (2010)

[39] de França, F.O., Von Zuben, F.J.: A dynamic artificial immune algorithm applied to challenging benchmarking problems. In: Proc. 2009 IEEE Congr. Evol. Comput., pp. 423–430 (2009)

[40] Goh, C.K., Tan, K.C.: A competitive-cooperative coevolutionary paradigm for dynamic multiobjective optimization. IEEE Trans. on Evol. Comput. 13(1), 103–127 (2009)

[41] Goldberg, D.E., Smith, R.E.: Nonstationary function optimization using genetic algorithms with dominance and diploidy. In: Proc. Int. Conf. on Genetic Algorithms, pp. 59–68 (1987)

[42] Gouvêa Jr., M., Araújo, A.: Adaptive evolutionary algorithm based on population dynamics for dynamic environments. In: Proc. 2011 Genetic and Evol. Comput. Conf., pp. 909–916 (2011)

[43] Grefenstette, J.J.: Genetic algorithms for changing environments. In: Proc. 2nd Int. Conf. Parallel Problem Solving from Nature, pp. 137–144 (1992)

[44] Grefenstette, J.J.: Evolvability in dynamic fitness landscapes: A genetic algorithm approach. In: Proc. 1999 IEEE Congr. Evol. Comput., vol. 3, pp. 2031–2038 (1999)

[45] Hatzakis, I., Wallace, D.: Dynamic multi-objective optimization with evolutionary algorithms: a forward-looking approach. In: Proc. 2006 Genetic and Evol. Comput. Conf., pp. 1201–1208 (2006)

[46] He, J., Yao, X.: From an individual to a population: An analysis of the first hitting time of population-based evolutionary algorithms. IEEE Trans. Evol. Comput. 6(5), 495–511 (2002)

[47] He, J., Yao, X.: A study of drift analysis for estimating computation time of evolutionary algorithms. Natural Computing 3(1), 21–35 (2004)

[48] Hu, X., Eberhart, R.: Adaptive particle swarm optimisation: detection and response to dynamic systems. In: Proc. 2002 IEEE Congr. Evol. Comput., pp. 1666–1670 (2002)

[49] Jansen, T., Schellbach, U.: Theoretical analysis of a mutation-based evolutionary algorithm for a tracking problem in lattice. In: Proc. 2005 Genetic and Evol. Comput. Conf., pp. 841–848 (2005)

[50] Janson, S., Middendorf, M.: A hierarchical particle swarm optimizer and its adaptive variant. IEEE Trans. Syst., Man, and Cybern.-Part B: Cybern. 35, 1272–1282 (2005)

[51] Janson, S., Middendorf, M.: A hierarchical particle swarm optimizer for noisy and dynamic environments. Genetic Programming and Evolvable Machines 7(4), 329–354 (2006)

[52] Jin, Y., Branke, J.: Evolutionary optimization in uncertain environments—a survey. IEEE Trans. Evol. Comput. 9(3), 303–317 (2005)

[53] Jin, Y., Sendhoff, B.: Constructing dynamic optimization test problems using the multi-objective optimization concept. In: Raidl, G.R., et al. (eds.) EvoWorkshops 2004. LNCS, vol. 3005, pp. 525–536. Springer, Heidelberg (2004)

[54] Kramer, G.R., Gallagher, J.C.: Improvements to the *CGA enabling online intrinsic in compact EH devices. In: Proc. 2003 NASA DoD Conf. on Evolvable Hardware, pp. 235–231 (2003)

[55] Lepagnot, J., Nakib, A., Oulhadj, H., Siarry, P.: Brain cine mri segmentation based on a multiagent algorithm for dynamic continuous optimization. In: Proc. 2011 IEEE Congr. Evol. Comput., pp. 1695–1702 (2011)

[56] Lewis, J., Hart, E., Ritchie, G.: A comparison of dominance mechanisms and simple mutation on non-stationary problems. In: Eiben, A.E., Bäck, T., Schoenauer, M., Schwefel, H.-P. (eds.) PPSN 1998. LNCS, vol. 1498, pp. 139–148. Springer, Heidelberg (1998)

[57] Li, C., Yang, S.: A clustering particle swarm optimizer for dynamic optimization. In: Proc. 2009 IEEE Congr. Evol. Comput., pp. 439–446 (2009)

[58] Li, X., Branke, J., Blackwell, T.: Particle swarm with speciation and adaptation in a dynamic environment. In: Proc. 2006 Genetic and Evol. Comput. Conf., pp. 51–58 (2006)

[59] Liu, L., Wang, D., Yang, S.: Compound particle swarm optimization in dynamic environments. In: Giacobini, M., et al. (eds.) EvoWorkshops 2008. LNCS, vol. 4974, pp. 616–625. Springer, Heidelberg (2008)

[60] Louis, S.J., Xu, Z.: Genetic algorithms for open shop scheduling and re-scheduling. In: Cohen, M.E., Hudson, D.L. (eds.) Proc. ISCA 11th Int. Conf. on Computers and their Applications, pp. 99–102 (1996)

[61] Lung, R.I., Dumitrescu, D.: A new collaborative evolutionary-swarm optimization technique. In: Proc. 2007 Genetic and Evol. Comput. Conf., pp. 2817–2820 (2007)

[62] Mavrovouniotis, M., Yang, S.: Memory-based immigrants for ant colony optimization in changing environments. In: Di Chio, C., et al. (eds.) EvoApplications 2011, Part I. LNCS, vol. 6624, pp. 324–333. Springer, Heidelberg (2011)

[63] Mendes, R., Mohais, A.: Dynde: a differential evolution for dynamic optimization problems. In: Proc. 2005 IEEE Congr. Evol. Comput., pp. 2808–2815 (2005)

[64] Mori, N., Kita, H., Nishikawa, Y.: Adaptation to a changing environment by means of the feedback thermodynamical genetic algorithm. In: Eiben, A.E., Bäck, T., Schoenauer, M., Schwefel, H.-P. (eds.) PPSN 1998. LNCS, vol. 1498, pp. 149–158. Springer, Heidelberg (1998)

[65] Morrison, R.W.: Designing Evolutionary Algorithms for Dynamic Environments. Springer, Berlin (2004) ISBN 3-540-21231-0

[66] Moser, I.: Review - all currently known publications on approaches which solve the moving peaks problem. Tech. Rep., Swinburne University of Technology, Melbourne, Australia (2007)

[67] Moser, I., Hendtlass, T.: A simple and efficient multi-component algorithm for solving dynamic function optimisation problems. In: Proc. 2007 IEEE Congr. Evol. Comput., pp. 252–259 (2007)

[68] Ng, K.P., Wong, K.C.: A new diploid scheme and dominance change mechanism for non-stationary function optimization. In: Proc. 6th Int. Conf. on Genetic Algorithms, pp. 159–166 (1995)

[69] Nguyen, T.T.: Tracking optima in dynamic environments using evolutionary algorithms - rsmg report 5. Tech. Rep., School of Computer Science, University of Birmingham (2008), http://www.cs.bham.ac.uk/~txn/unpublished/reports/Report_5_Thanh.pdf

[70] Nguyen, T.T.: Continuous Dynamic Optimisation Using Evolutionary Algorithms. Ph.D. thesis, School of Computer Science, University of Birmingham (2011), http://etheses.bham.ac.uk/1296 and http://www.staff.ljmu.ac.uk/enrtngu1/theses/phd_thesis_nguyen.pdf

[71] Nguyen, T.T., Yang, S., Branke, J.: Evolutionary dynamic optimization: A survey of the state of the art. Swarm and Evol. Comput. 6, 1–24 (2012)

[72] Nguyen, T.T., Yang, Z., Bonsall, S.: Dynamic time-linkage problems - the challenges. In: IEEE RIVF Int. Conf. on Computing and Communication Technologies, Research, Innovation, and Vision for the Future, pp. 1–6 (2012)

[73] Nguyen, T.T., Yao, X.: Benchmarking and solving dynamic constrained problems. In: Proc. 2009 IEEE Congr. Evol. Comput., pp. 690–697 (2009)

[74] Nguyen, T.T., Yao, X.: Dynamic time-linkage problems revisited. In: Giacobini, M., et al. (eds.) EvoWorkshops 2009. LNCS, vol. 5484, pp. 735–744. Springer, Heidelberg (2009)

[75] Nguyen, T.T., Yao, X.: Solving dynamic constrained optimisation problems using stochastic ranking and repair methods. IEEE Trans. Evol. Comput. (2010) (submitted), http://www.staff.ljmu.ac.uk/enrtngu1/Papers/Nguyen_Yao_dRepairGA.pdf

[76] Nguyen, T.T., Yao, X.: Continuous dynamic constrained optimisation - the challenges. IEEE Trans. Evol. Comput. 16(6), 769–786 (2012)

[77] Oppacher, F., Wineberg, M.: The Shifting Balance Genetic Algorithm: Improving the GA in a Dynamic Environment. In: Proc. 1999 Genetic and Evol. Comput. Conf., vol. 1, pp. 504–510 (1999)

[78] Parrott, D., Li, X.: Locating and tracking multiple dynamic optima by a particle swarm model using speciation. IEEE Trans. Evol. Comput. 10(4), 440–458 (2006)

[79] Ramsey, C.L., Grefenstette, J.J.: Case-based initialization of genetic algorithms. In: Proc. 5th Int. Conf. on Genetic Algorithms, pp. 84–91 (1993)

[80] Richter, H.: Behavior of evolutionary algorithms in chaotically changing fitness landscapes. In: Yao, X., et al. (eds.) PPSN 2004. LNCS, vol. 3242, pp. 111–120. Springer, Heidelberg (2004)

[81] Richter, H.: Evolutionary optimization in spatio–temporal fitness landscapes. In: Runarsson, T.P., Beyer, H.-G., Burke, E.K., Merelo-Guervós, J.J., Whitley, L.D., Yao, X. (eds.) PPSN 2006. LNCS, vol. 4193, pp. 1–10. Springer, Heidelberg (2006)

[82] Richter, H.: Detecting change in dynamic fitness landscapes. In: Proc. 2009 IEEE Congr. Evol. Comput., pp. 1613–1620 (2009)

[83] Richter, H.: Memory design for constrained dynamic optimization problems. In: Di Chio, C., et al. (eds.) EvoApplicatons 2010, Part I. LNCS, vol. 6024, pp. 552–561. Springer, Heidelberg (2010)

[84] Richter, H., Yang, S.: Memory based on abstraction for dynamic fitness functions. In: Giacobini, M., et al. (eds.) EvoWorkshops 2008. LNCS, vol. 4974, pp. 596–605. Springer, Heidelberg (2008)

[85] Richter, H., Yang, S.: Learning behavior in abstract memory schemes for dynamic optimization problems. Soft Comput. 13(12), 1163–1173 (2009)

[86] Riekert, M., Malan, K.M., Engelbrecht, A.P.: Adaptive genetic programming for dynamic classification problems. In: Proc. 2009 IEEE Congr. Evol. Comput., pp. 674–681 (2009)

[87] Rohlfshagen, P., Lehre, P.K., Yao, X.: Dynamic evolutionary optimisation: An analysis of frequency and magnitude of change. In: Proc. 2009 Genetic and Evol. Comput. Conf., pp. 1713–1720 (2009)

[88] Rohlfshagen, P., Yao, X.: Attributes of dynamic combinatorial optimisation. In: Li, X., et al. (eds.) SEAL 2008. LNCS, vol. 5361, pp. 442–451. Springer, Heidelberg (2008)

[89] Rohlfshagen, P., Yao, X.: On the role of modularity in evolutionary dynamic optimisation. In: Proc. 2010 IEEE Congr. Evol. Comput., pp. 3539–3546 (2010)

[90] Rossi, C., Abderrahim, M., Díaz, J.C.: Tracking moving optima using kalman-based predictions. Evol. Comput. 16(1), 1–30 (2008)

[91] Ryan, C.: The degree of oneness. In: Proc. 1st Online Workshop on Soft Computing, pp. 43–49 (1996)

[92] Salomon, R., Eggenberger, P.: Adaptation on the evolutionary time scale: A working hypothesis and basic experiments. In: Hao, J.-K., Lutton, E., Ronald, E., Schoenauer, M., Snyers, D. (eds.) AE 1997. LNCS, vol. 1363, pp. 251–262. Springer, Heidelberg (1998)

[93] Simões, A., Costa, E.: Memory-based chc algorithms for the dynamic traveling salesman problem. In: Proc. 2011 Genetic and Evol. Comput. Conf., pp. 1037–1044 (2011)

[94] Simões, A., Costa, E.: An immune system-based genetic algorithm to deal with dynamic environments: Diversity and memory. In: Pearson, D.W., Steele, N.C., Albrecht, R. (eds.) Proc. 2003 Int. Conf. on Neural Networks and Genetic Algorithms (ICAN-NGA 2003), pp. 168–174 (2003)

[95] Simões, A., Costa, E.: Improving memory's usage in evolutionary algorithms for changing environments. In: Proc. 2007 IEEE Congr. Evol. Comput., pp. 276–283 (2007)

[96] Simões, A., Costa, E.: Evolutionary algorithms for dynamic environments: Prediction using linear regression and markov chains. In: Rudolph, G., Jansen, T., Lucas, S., Poloni, C., Beume, N. (eds.) PPSN 2008. LNCS, vol. 5199, pp. 306–315. Springer, Heidelberg (2008)

[97] Simões, A., Costa, E.: Improving prediction in evolutionary algorithms for dynamic environments. In: Proc. 2009 Genetic and Evol. Comput. Conf., pp. 875–882 (2009)

[98] Singh, H.K., Isaacs, A., Nguyen, T.T., Ray, T., Yao, X.: Performance of infeasibility driven evolutionary algorithm (IDEA) on constrained dynamic single objective optimization problems. In: Proc. 2009 IEEE Congr. Evol. Comput., pp. 3127–3134 (2009)

[99] Stanhope, S.A., Daida, J.M.: Genetic algorithm fitness dynamics in a changing environment. In: Proc. 1999 IEEE Congr. Evol. Comput., vol. 3, pp. 1851–1858 (1999)

[100] Tinos, R., Yang, S.: Continuous dynamic problem generators for evolutionary algorithms. In: Proc. 2007 IEEE Congr. Evol. Comput., pp. 236–243 (2007)

[101] Tinós, R., Yang, S.: An analysis of the XOR dynamic problem generator based on the dynamical system. In: Schaefer, R., Cotta, C., Kołodziej, J., Rudolph, G. (eds.) PPSN XI, Part I. LNCS, vol. 6238, pp. 274–283. Springer, Heidelberg (2010)

[102] Toffolo, A., Benini, E.: Genetic diversity as an objective in multi-objective evolutionary algorithms. Evol. Comput. 11(2), 151–167 (2003)

[103] Trojanowski, K., Michalewicz, Z.: Searching for optima in non-stationary environments. In: Proc. 1999 IEEE Congr. Evol. Comput., vol. 3, pp. 1843–1850 (1999)

[104] Ursem, R.K.: Multinational GA optimization techniques in dynamic environments. In: Proc. 2000 Genetic and Evol. Comput. Conf., pp. 19–26 (2000)

[105] Ursem, R.K., Krink, T., Jensen, M.T., Michalewicz, Z.: Analysis and modeling of control tasks in dynamic systems. IEEE Trans. Evol. Comput. 6(4), 378–389 (2002)

[106] Uyar, A.S., Harmanci, A.E.: A new population based adaptive domination change mechanism for diploid genetic algorithms in dynamic environments. Soft Comput. 9(11), 803–814 (2005)

[107] Vavak, F., Fogarty, T.C., Jukes, K.: A genetic algorithm with variable range of local search for tracking changing environments. In: Ebeling, W., Rechenberg, I., Voigt, H.-M., Schwefel, H.-P. (eds.) PPSN 1996. LNCS, vol. 1141, pp. 376–385. Springer, Heidelberg (1996)

[108] Vavak, F., Jukes, K., Fogarty, T.C.: Learning the local search range for genetic optimisation in nonstationary environments. In: Proc. 1997 IEEE Int. Conf. on Evol. Comput., pp. 355–360 (1997)

[109] Vavak, F., Jukes, K.A., Fogarty, T.C.: Performance of a genetic algorithm with variable local search range relative to frequency for the environmental changes. In: Proc. 3rd Int. Conf. on Genetic Programming, pp. 602–608 (1998)

[110] Wang, Y., Wineberg, M.: Estimation of evolvability genetic algorithm and dynamic environments. Genetic Programming and Evolvable Machines 7(4), 355–382 (2006)

[111] Weicker, K.: An analysis of dynamic severity and population size. In: Deb, K., Rudolph, G., Lutton, E., Merelo, J.J., Schoenauer, M., Schwefel, H.-P., Yao, X. (eds.) PPSN 2000. LNCS, vol. 1917, pp. 159–168. Springer, Heidelberg (2000)

[112] Weicker, K.: Evolutionary algorithms and dynamic optimization problems. Der Andere Verlag (2003)

[113] Weicker, K.: Analysis of local operators applied to discrete tracking problems. Soft Comput 9(11), 778–792 (2005)

[114] Weicker, K., Weicker, N.: On evolution strategy optimization in dynamic environments. In: Proc. 1999 IEEE Congr. Evol. Comput., vol. 3, pp. 2039–2046 (1999)

[115] Woldesenbet, Y.G., Yen, G.G.: Dynamic evolutionary algorithm with variable relocation. IEEE Trans. Evol. Comput. 13(3), 500–513 (2009)

[116] Wolpert, D.H., Macready, W.G.: No free lunch theorems for optimization. IEEE Trans. Evol. Comput. 1(1), 67–82 (1997)

[117] Xing, L., Rohlfshagen, P., Chen, Y., Yao, X.: A hybrid ant colony optimisation algorithm for the extended capacitated arc routing problem. IEEE Trans. Syst., Man and Cybern., Part B: Cybern. 41(4), 1110–1123 (2011)

[118] Yang, S.: Memory-based immigrants for genetic algorithms in dynamic environments. In: Proc. 2005 Genetic and Evol. Comput. Conf., pp. 1115–1122 (2005)

[119] Yang, S.: Associative memory scheme for genetic algorithms in dynamic environments. In: Rothlauf, F., et al. (eds.) EvoWorkshops 2006. LNCS, vol. 3907, pp. 788–799. Springer, Heidelberg (2006)

[120] Yang, S.: A comparative study of immune system based genetic algorithms in dynamic environments. In: Proc. 2006 Genetic and Evol. Comput. Conf., pp. 1377–1384 (2006)

[121] Yang, S.: On the design of diploid genetic algorithms for problem optimization in dynamic environments. In: Proc. 2006 IEEE Congr. Evol. Comput., pp. 1362–1369 (2006)

[122] Yang, S.: Genetic algorithms with memory- and elitism-based immigrants in dynamic environments. Evol. Comput. 16(3), 385–416 (2008)

[123] Yang, S., Jiang, Y., Nguyen, T.T.: Metaheuristics for dynamic combinatorial optimization problems. IMA J. of Management Mathematics (2012), doi:10.1093/imaman/DPS021

[124] Yang, S., Jin, Y., Ong, Y.S. (eds.): Evolutionary Computation in Dynamic and Uncertain Environments. Springer, Heidelberg (2007)

[125] Yang, S., Yao, X.: Experimental study on population-based incremental learning algorithms for dynamic optimization problems. Soft Comput. 9(11), 815–834 (2005)

[126] Yang, S., Yao, X.: Population-based incremental learning with associative memory for dynamic environments. IEEE Trans. Evol. Comput. 12(5), 542–561 (2008)

[127] Yu, E.L., Suganthan, P.N.: Evolutionary programming with ensemble of explicit memories for dynamic optimization. In: Proc. 2009 IEEE Congr. Evol. Comput., pp. 431–438 (2009)

[128] Zeng, S., Shi, H., Kang, L., Ding, L.: Orthogonal dynamic hill climbing algorithm: ODHC. In: Yang, S., Ong, Y.S., Jin, Y. (eds.) Evolutionary Computation in Dynamic and Uncertain Environments. SCI, vol. 51, pp. 79–105. Springer, Heidelberg (2007)

[129] Zhou, A., Jin, Y., Zhang, Q., Sendhoff, B., Tsang, E.: Prediction-based population re-initialization for evolutionary dynamic multi-objective optimization. In: Obayashi, S., Deb, K., Poloni, C., Hiroyasu, T., Murata, T. (eds.) EMO 2007. LNCS, vol. 4403, pp. 832–846. Springer, Heidelberg (2007)

[130] Zou, X., Wang, M., Zhou, A., Mckay, B.: Evolutionary optimization based on chaotic sequence in dynamic environments. In: Proc. 2004 IEEE Int. Conf. on Networking, Sensing and Control, vol. 2, pp. 1364–1369 (2004)

Chapter 3
Evolutionary Dynamic Optimization: Challenges and Perspectives

Philipp Rohlfshagen and Xin Yao

Abstract. The field of evolutionary dynamic optimization is concerned with the study and application of evolutionary algorithms to dynamic optimization problems. In this chapter we highlight some of the challenges associated with the time-variant nature of these problems. We focus particularly on the different problem definitions that have been proposed, the modelling of dynamic optimization problems in terms of benchmark suites and the way the performance of an algorithm is assessed. Amid significant developments in the last decade, several practitioners have highlighted shortcomings with all of these fundamental issues. In this chapter we review the work done in each of these areas, evaluate the criticism and subsequently identify some perspectives for the future of the field.

3.1 Introduction

The field of *evolutionary dynamic optimization* [5, 7, 8, 14, 21, 32, 52] is concerned with the study and application of evolutionary algorithms (EAs) to dynamic optimization problems (DOPs). The dynamics of these problems poses many challenges not commonly found in stationary optimization. In particular, it is generally understood that the problem's dependency on time requires an algorithm not only to *locate* high quality solutions but also to *track* them as closely as possible. This poses numerous problems to traditional EAs and a wealth of new techniques have been proposed in recent years to improve their performances. Indeed, most efforts have been dedicated to the design of new algorithms. However, many additional challenges need to be addressed as each has a fundamental impact on the developments in evolutionary dynamic optimization. In this chapter we highlight some of these challenges.

Philipp Rohlfshagen · Xin Yao
Centre of Excellence for Research in Computational Intelligence and Applications
(CERCIA), School of Computer Science, University of Birmingham,
Birmingham B15 2TT, U.K.
e-mail: philipp.r@gmail.com, X.Yao@cs.bham.ac.uk

S. Yang and X. Yao (Eds.): *Evolutionary Computation for DOPs*, SCI 490, pp. 65–84.
DOI: 10.1007/978-3-642-38416-5_3   © Springer-Verlag Berlin Heidelberg 2013

We focus in particular on three fundamental inter-dependent challenges, each of which may have numerous sub-challenges, as outlined in the relevant sections of this chapter:

1. Problem definition and general framework for dynamic optimization.
2. Specifications of benchmark problems and experimental designs.
3. Problem requirements, notions of optimality and performance measures.

Challenge 1: one of the most fundamental challenges in evolutionary dynamic optimization is the problem definition itself. Although numerous definitions have been proposed, a general framework that allows practitioners to fully articulate their assumptions about a problem's properties is still missing. This has a profound impact on our understanding of the problem domain and the lack of standards often complicates the generalisation of empirical results from specific scenarios to a wider class of problems; this challenge is discussed in Section 3.2.

Challenge 2: the first challenge is partially addressed by the availability of dynamic benchmark problems, used by many practitioners to test and evaluate their algorithms. However, the design of general benchmarks is particularly challenging in the dynamic domain, especially in the combinatorial case where fitness landscape dynamics depend on the algorithm's representation and variation operators. Several practitioners have questioned whether these benchmark problems are representative of real-world dynamics and whether experimental settings are flexible enough to reflect the requirements of tackling real-world DOPs. We review these issues in Section 3.3.

Challenge 3: the third major challenge considered is closely related to the requirements imposed by real-world problems and concerns the way the performance of an algorithm is assessed. A performance measure used should reflect the algorithm's ability to locate satisfactory solutions given the requirements of the problem: an appropriate notion of optimality thus needs to be defined a priori. Many performance measures have been proposed in the past yet most of them have not been linked explicitly to the requirements of the problem. Section 3.4 highlights these issues.

Clearly, all challenges considered are related to one another. For instance, it is difficult to express a problem's requirements without the proper tools to do so. Likewise, it is impossible to model real-world dynamics without a good understanding of them. These interdependencies will become apparent throughout this chapter. Finally, the chapter is concluded in Section 3.5 where we summarise how these challenges have been addressed to date and subsequently outline some perspectives for future work in evolutionary dynamic optimization.

3.2 Challenge I: Problem Definition

3.2.1 Optimization in Uncertain Environments

An arbitrary problem domain

$$F : X \to Y \tag{3.1}$$

may be modelled as a functional mapping from some search space X to some other domain Y. The *search space* X corresponds to the set of the candidate solutions, each with quality $f(x)$ for each problem instance $f \in F$. The function f is usually referred to as the *objective function* and the n elements x_i are known as the *decision* or *design* variables. The set of all f-values corresponding to all elements in X is denoted $\hat{f}$. The majority of real-world problems have inequality and/or equality constraints as specified by the functions $g : X \rightarrow Y^p$ and $h : X \rightarrow Y^q$ respectively; a solution is considered feasible if $g_i(x) \leq 0$, $i = 1, \ldots, p$ and $h_j(x) = 0$, $j = 1, \ldots, q$ (see [36]).

A function $f \in F$, as defined above, constitutes an *search problem* once a *notion of optimality* has been defined: a *solution concept* is required to distinguish between those points in X that constitute (acceptable) solutions and those that do not [17]. In the majority of cases where one has to content with satisficing (rather than optimising), this notion is defined trivially as those points that minimise or maximise f (see [44], cited in [37]); without loss of generality, we will consider the goal of maximisation for the remainder of this chapter.

A *search algorithm* that implements the specified solution concept may subsequently be used to locate the desired solutions. In the case of maximisation, the best possible solutions is known as the function's *global optimum* x^*, $f(x^*) \geq f(x)$, $\forall x \in X$ which the algorithm should find as quickly as possible. Alternatively, one is interested in the best solution returned by the algorithm given a predetermined amount of time. A common obstacle encountered by the algorithm is the presence of *local optima*, defined as points $x \in X$ such that $f(x^*) \geq f(x)$, $\forall x \in N(x^*)$ where $N(x)$ is the *neighbourhood* of x, determined by the algorithm's variation operators.

Traditionally, the majority of work in evolutionary computation has concentrated on stationary optimization problems. Nevertheless, a significant effort has also been devoted to other domains, particularly those characterised by uncertainty. In [21], Jin and Branke review four distinct types of uncertain environments: noisy optimization problems, robust optimization, approximate optimization and DOPs, the focus of this chapter.

Jin and Branke [21] consider DOPs to be the set of functions that are deterministic at any point in time, but dependent on time $t \in \mathbb{N}$:

$$F : X \times \mathbb{N} \rightarrow Y \qquad (3.2)$$

The definition set forth in Eq. (3.2) is a straightforward extension to Eq. (3.1). This generality of this definition is not surprising considering that in principle any component of f may change over time (i.e., an arbitrary $f_a \in F$ may transition to any $f_b \in F$). However, it is questionable how useful such a general problem definition is: it implies, for instance, that the class of DOPs includes all those problems that do not actually change over time. Indeed, the Handbook of Approximation Algorithms and Metaheuristics [28] states that a general definition of DOPs does not exist (as of 2007). Nevertheless, numerous extended problem definitions have been proposed in the past that attempt to capture additional properties of DOPs; some of these are briefly reviewed next.

3.2.2 Problem Definitions

The fundamental difference between stationary and dynamic optimization problems is, of course, time: in principle, any component of a problem may depend on time, leading to a significant variety of problem dynamics. Several attempts have been made in the past to capture some of these characteristics.

Trojanowski and Michalewicz [48] define a model M of the problem P as $M(P) = (D, F, C)$ where D is the time-variant domain of the decision variables, F the potentially time-variant objective function and C the set of time-variant constraints. The dynamics of the domain may affect either the dimensionality of the problem or the interval of each decision variable. The authors note that this framework accounts for both stationary and dynamic optimization problems.

Weicker [51, 52] proposes a much more detailed problem definition, hoping "to establish a basis for comparison and classification of non-stationary functions as well as for theoretical results on problem hardness and algorithm's power." [51, p 160]. Weicker assumes that each (continuous) dynamic function consists of several static functions, each of which follows some dynamic rule, defined by a sequence of distance preserving coordinate transformations, rotations and fitness rescalings. In other words, the dynamics are defined with respect to the changes in $\hat{f}$ and Weicker comments that these properties cannot easily be applied outside the artificial domain (also see [11]).

Rohlfshagen and Yao [40, 41] restrict their problem definition to those DOPs where only the parameters (coefficients) of the function change. The dynamic equivalence of the stationary problem $f(x, \delta)$ is subsequently defined as $f(x, \delta(t))$, where $\delta(t) \in \Delta$ are the parameters (including constraints) of the function at time t. The dynamics of the problem then correspond to different trajectories through the metric space Δ:

$$\delta(T) \longrightarrow \delta(T+1) \longrightarrow \delta(T+2) \longrightarrow \ldots \qquad (3.3)$$

where T is a period index such that $\delta(T) \neq \delta(T+1), \forall T$. It is assumed that time advances with every call to the objective function and the function changes according to the frequency of change $\tau : \mathbb{N} \times \mathbb{N} \to \mathbb{N}$. The transitions are governed by a time-variant mapping $\mathcal{T} : \Delta \times \mathbb{N} \to \Delta$ that maps from one problem instance to another such that $\delta(T+1) = \mathcal{T}(\delta(T))$.

Nguyen and Yao [35] propose an extended problem definition to account for numerous aspects the authors believe have not been captured by previous problem definitions, particularly *time-linkage* (see Section 3.3.3). The authors propose the *full-description form* of a function: the idea is to model the dynamic function as a parameterised set of stationary functions. It is those external parameters that are subsequently varied: "most common types of changes in dynamic (time-linkage) problems can be represented as changes in the parameter space if we can formulate the problem in a general enough full-description form." [35, p 739].

The majority of problem definitions concentrates primarily on the *type* of dynamics rather than their frequency. A common simplifying assumption is to equate every call to the objective function with one time step. In other words, the algorithm is treated as a discrete clock. The majority of practitioners assume a time-invariant frequency of change and often correlated the time of change with the attributes of the algorithm (see Section 3.3.4).

More recently, robust optimization over time (ROOT) [18, 23, 59] was proposed as a new problem formulation as it seems to capture some characteristics of real-world problems. It considers the cost of changing a solution, albeit implicitly, and prefers the solution that is of high quality over a longer period of time even with some minor environmental changes. A new solution is generated dynamically only when the current solution does not perform well.

3.2.3   Characterisation of Dynamics

Numerous ways have been proposed in the past to characterise and classify a problem's dynamics. Trojanowski and Michalewicz [48] consider cases where the function is either stationary or dynamic, with or without stationary/dynamic constraints. The authors subsequently classify dynamics based on their regularity and distinguish between random changes, non-random and non-predictable changes and non-random and predictable changes. The second classification refers to cases where functional dependencies exist but are too complex to be captured by the algorithm. The third classification allows the algorithm to predict future structural properties and may be divided further into cyclical and non-cyclical changes. Finally, the authors also consider the visibility of change (i.e., whether the algorithm needs to detect changes) and distinguish between continuous and discrete dynamics. In the former, the fitness function differs every time a sample is taken while in the latter, there are periods of stagnation.

In [11], Branke lists classifications proposed by De Jong [25] who distinguishes between alternating problems, problems with changing morphology, drifting landscapes and abrupt and discontinuous problems. Branke himself [8] considers frequency of change, severity of change, predictability of change and cycle length/cycle accuracy. The frequency of change determines the number of function evaluations that may be performed between changes and the severity of change refers to the displacement of the global optimum. The predictability of change implies the existence of exploitable structural properties while the cycle length refers to the reappearance of a previously encountered problem instance in a deterministic setting.

Younes et al. [58] differentiate between dynamically significant and insignificant changes: a change is insignificant if it does not alter the structure of the problem instance to such a degree that affects the behaviour of the algorithm used to tackle said problem. Similarly, Karaman and Etaner-Uyar [26] define an effective environment change as one that affects the selection probabilities of the candidate solutions.

In [50], Ursem et al. propose a new benchmark generator designed to capture more closely the properties of real-world problems. The authors consider the following attributes some of which are closely related to those considered by De Jong [25]: the problem's periodicity describes the temporal correlation between successive values of a variable while the stochasticity of a variable describes the degree of randomness inherent in its dynamics. Drift is present in a variable if the value of the variable has a tendency to change towards one direction only and finally, the authors consider the dependencies amongst the time-variant parameters of the problem.

The multitude of different classifications highlights the variety of properties that DOPs may exhibit, in particular if one considers all types of problems that correspond to Eq. (3.2). The ability to accurately describe a specific class of dynamic problems is thus vital. For instance, Pelta points out that we still lack a clear understanding of what actually constitutes an instance of a DOP (personal communication). In the following we highlight the importance to be able to specify a problem's properties and discuss several attributes a mathematical framework should be able to capture.

3.2.4 Problem Properties, Assumptions and Generalisations

EAs are commonly employed as *black-box* algorithms (see [15]), using as guidance only the f-values returned by the objective function. Clearly, one has to make assumptions regarding the utility of the f-values in the search for the global optimum and almost all search algorithms make the implicit assumption that, at least in expectation, higher f-values, found in the vicinity of the currently sampled search points, will lead to the global optimum. If these assumptions break down, the performance of the algorithm will be degraded. Indeed, it has been shown that over all functions $f : A \to B$ (A and B are finites sets, B totally ordered [16]) closed under permutation, any two algorithms α and β will perform equally, precisely because the expected overall structural properties are removed from this set (No Free Lunch; see [16, 53]). Subsequently, practitioners tend to concentrate on subsets of functions that are "reasonable" [45] in that they are "simple" and "natural" [16]. In this case, the performance of a particular algorithm may be superior to another algorithm because the assumptions that were made during the design of the algorithm are *aligned* with the underlying probability distribution of functions encountered by the algorithm [53, p67].

The No Free Lunch theorem for optimization has also been proven for the case of dynamic optimization problems [53]. It is thus important that any assumptions inherent in algorithm's design are communicated clearly to allow other practitioners to apply the algorithm efficiently. In particular, "when an algorithm is developed for specific problems, the boundaries of its performance should be clear, and we expect estimates of reasonably good performance within and (at least modestly) outside its 'seen' instance distribution." [13, p22]. Clearly articulated assumptions may thus facilitate a more reliable generalisation from benchmark instances to unseen problem instances; this may overcome possible limitations of testing the algorithm on a specific set of functions [13]. As Section 3.2.2 has shown, however, very few tools exist to allow practitioners in evolutionary dynamic optimization to clearly express their assumptions about the attributes of the problems considered. Indeed, as the review of classifications of dynamics in Section 3.2.3 has shown, most concepts are described in high-level terms and structural properties such as the *magnitude of change* are often described in general terms such as "small", "medium" and "large".

Most practitioners make the assumption that the global optimum will, on average, move relatively little and that it may thus be beneficial to restrict the search to the vicinity of the previously found global optimum: given the algorithm has already sampled a previous problem instance by the time a change is encountered, "a natural attempt to speed up optimization after a change would be to somehow use knowledge about the previous search space to advance the search after a change." [21, p 311]. This *distance-based assumption* is probably the most widely considered one in the literature yet it usually relates to properties in the problem's fitness landscape (see Section 3.3.2) and thus depends on the algorithm's chosen representation and variation operators. It is thus necessary to consider dynamics that affect the problem itself and it is vital to gain a better understanding of how changes in the problem lead to different types of dynamics. For instance, there is some empirical evidence show that the distance by which the global optimum is displaced is directly proportional to the fitness distance correlation of the instance [40] which is determined by the chosen algorithm.

3.3 Challenge II: Benchmark Problems

3.3.1 Benchmark Problems

The systematic evaluation and comparison of different algorithms necessitates the use of unified benchmark problems these algorithms may be tested on (see [58]). In order to account for a variety of different dynamics, numerous benchmark problems have been proposed in the past.

The most widely used benchmark problem [14] is MOVING PEAKS, due to Branke [8] ; this problem is conceptually identical to DF1, proposed by Morrison independently around the same time [33] . MOVING PEAKS is based on the continuous domain and corresponds to a retrospective implementation of a dynamic fitness landscape. It is modelled as a "field of cones" [33], where each cone may be controlled individually to model different ranges of dynamics. The n-dimensional function with m peaks is specificed as

$$f(x,t) := \max_{i=1,\ldots,m} \frac{h_i(t)}{1 + w_i(t) \sum_{j=1}^{n} (x_j - p_{ij}(t))^2} \tag{3.4}$$

where p entails the peaks' coordinate positions and w and h describe the width and height of each peak respectively. The initial morphology is randomly generated within the bounds specified by the user and the locations, widths and heights of all peaks are subsequently varied over time. The motivation behind this approach is to create a problem where the majority of changes are smooth and correlated with the occasional drastic change when the height of the peaks change in such a way that any of the local optima surpasses the currently global one [6].

XOR DOP [54] is probably the most widely used benchmark problem for the combinatorial domain and generates a dynamic version of any static pseudo-Boolean problem: given a static fitness function $f(x), x \in \{0,1\}^n$, its dynamic equivalence is simply

$$f(x(t) \oplus m(T)) \tag{3.5}$$

where $\oplus$ is the bit-wise exclusive-or operator. The period index $T = \lceil t/\tau \rceil$ is determined by the update period τ. The binary mask $m(T) \in \{0,1\}^n$ is initially $m(1) = 0$ and is subsequently altered for each period k as follows: $m(T) = m(T-1) \oplus p(T)$ where $p(T) \in \{0,1\}^n$ is a randomly created template for period T that contains exactly $\lfloor \rho n \rfloor$ ones. The value of $\rho \in [0,1]$ thus controls the magnitude of change which is specified as the *Hamming distance* between two binary points, $d(x,y) = \sum_{i=1}^{n} |x_i - y_i|$. The values τ and ρ are usually kept constant throughout the execution of the algorithm. XOR DOP has also been extended to generate cyclical and noisy cyclical environments (see [56, 57]).

There are numerous additional dynamic problems that have been used in the evaluation of new algorithms and a full overview may be found in [14]. The simplest functions considered correspond to the dynamic ONEMAX (which may be made dynamic using XOR DOP) and the moving sphere function, $f(x) = g(|x^* - x|)$, where the dynamics, with magnitude θ and direction v, are modelled as $x^*(t+1) = x^*(t) + \theta v$ (see [42] for details). These functions are frequently employed in theoretical studies. Other noteworthy examples include in the continuous domain a continuous benchmark with dynamic constraints [34], a test suite based on a multi-objective optimization concept [22], and, in the combinatorial case, a deceptive function generator [55], a general purpose combinatorial benchmark based on permutations of the mapping [58], the dynamic 0/1 knapsack problem [19, 39] and the multiple knapsack problem [10]. Finally, Li and Yang [29] propose a generalised benchmark that can be used in both continuous and combinatorial spaces. Similarly, Tinos and Yang [46] propose an extension of XOR DOP to the continuous domain.

3.3.2 Combinatorial Fitness Landscapes

A significant number of studies in evolutionary dynamic optimization have focussed on the continuous domain, particularly making use of MOVING PEAKS, DF1 and numerous other dynamic variants of well-known continuous functions (e.g., dynamic Rastrigin or dynamic Rosenbrock; see [14, p 4]). One of the reasons for this is the role played by the algorithm in *visualising* the problem's dynamics. Given some combinatorial problem $f : X \rightarrow Y$, it is often necessary to transform X into a domain that is suitable for the algorithm. This transformation is achieved using *representations*: a representation is a mapping $f_g : X_g \rightarrow X_p$ that transforms an element $x_g \in X_g$ to an element $x_p \in X_p$; the former is commonly referred to as a *genotype* and the latter as a *phenotype*. The objective function then corresponds to the composite mapping $f = f_p \circ f_g = f_p(f_g(x_g), \delta)$, where $f_p : X_p \times \Delta \rightarrow \mathbb{R}$ (see [43]).

A significant number of different representations have been proposed for a variety of domains and their impact on the algorithm's behaviour is often estimated by means of the *fitness landscape* metaphor, an abstract visualisation of the problem's topological properties. A fitness landscape may be specified by the tuple $\mathcal{L} = (X_g, f_{p \circ g}, d_g)$. The distance metric d_g is determined by the algorithm's variation operators and determines the features of the space X_g, including the presence of local optima. The chosen representation translates those features to the space X_p and since the algorithm's selection mechanism determines the search points visited next, the combined effect of representation and variation operators determines the overall topology of the search space.

The notion of a fitness landscape also applies in the dynamic domain. However, the fitness landscape may now change over time and it follows that the choice of representation and variation operators not only determines the structure of the stationary fitness landscape at a particular point in time but also the transition from one fitness landscape to another. This makes it partiuclarly challenging to develop general benchmark problems for the combinatorial domain. Younes et al [58] discuss this issue and propose to induce dynamics by altering the mapping f_g. The XOR DOP benchmark, on the other hand, overcomes this issue altogether by restricting itself to the pseudo-Boolean domain. Furthermore, as the analysis of XOR DOP has shown, the rotations of the search points preserve all structural properties of the problem (i.e., the underlying function remains stationary) [46, 47].

3.3.3 Real-World Dynamics

The previous section has highlighted the wide variety of problems that have been considered in evolutionary dynamic optimization. Numerous studies have been dedicated to real-world problems yet the majority of work concentrates on artificial benchmark problems: in [14], the authors present an extensive review of the problems considered in evolutionary dynamic optimization. The authors distinguish between *synthetic* dynamic problems and *real world* applications. The former has a total of 144 references listed, whereas the latter lists only 21 references. However, numerous practitioners have criticised these benchmarks as it has not been established how well they represent the requirements of real-world problems.

Morrison and De Jong [33, p 2048] specify several properties they desire from a useful benchmark generator, including an "easily modifiable landscape complexity that is scalable to complexity levels representative of problems found in nature". The benchmark problems reviewed above do allow full control over important aspects of the dynamics (e.g., frequency and magnitude of change). However, they have been criticised by Ursem et al. [50, p 1] as "no research has been conducted to thoroughly evaluate how well they reflect characteristic dynamics of real-world problems.". The same authors point out that these benchmarks "focus on how the landscape changes instead of the underlying dynamics" (p 379) and conclude, based on the analysis of their specific control-problem, meant to resemble a real-world problem, that "the resulting fitness landscapes looked surprisingly different from the landscapes that can be generated with traditional test-case generators" (p 387).

Branke et al. make a similar observation: "although these artificial benchmark problems are quite popular, it is not clear how closely their dynamics resemble the dynamics of real-world problems." [11, p 1434]. Likewise, Cruz et al. [14] point out that it is not clear what real-world attributes should be modelled by the benchmark problems. Such arguments are supported by the ommission of crtitical attributes such as dynamic constraints, as pointed out by [34] in the case of dynamic continuous functions. Furthermore, additional aspects such as *time-linkage* (e.g., [4, 30]) are rarely, if ever, captured.

Indeed, very few studies exists that examine possible dynamics encountered in real-world scenarios and without such real-world data, it remains difficult to judge what classes of dynamics may be of interest. Branke et al. [11, p 1434] point out that "to our knowledge, no one has ever attempted to characterize and measure the dynamism of the fitness landscape of a given dynamic real-world problem." One of the very few studies on this subject matter is by Andrews and Tuson who attempt to investigate the requirements of practitioners of dynamic optimization with the intent to "facilitate the appropriate choice of research directions" [2]. The authors attempt to answer the questions regarding what aspects of the problem change, how they change and what measure of success is of interest to the practitioner. Unfortunately, the study is too small to be representative but the results indicate the importance of dynamic constraints and, more importantly, the relevance of different notions of optimality that depend on the application of interest.

There are additional drawbacks associated with the (desired) simplicity of the most common benchmarks: as each peak in MOVING PEAKS (and indeed DF1) is controlled individually, the complexity of the problem is necessarily limited and most practitioners consider a relatively low number of peaks. Likewise, as XOR DOP does not actually alter the underlying search space (the search points are rotated by some degree but the underlying function is static) it is limited in its applicability [46, 47]. It is useful at this stage to draw some parallels to the stationary domain: MOVING PEAKS, DF1, and XOR DOP are somewhat similar to the NK-fitness landscape [27] which may be used to generate problem instances with different characteristics. However, benchmarks such as this are usually used to study particular attributes of an algorithm, which is subsequently validated on problems representative of real-world scenarios (e.g., OR-library [3] which provides commonly used NP-hard optimization problems). There is a need for an equivalent in the dynamic case and it may be possible to extract relevant information from the few studies that do deal with real-world problems.

3.3.4 Experimental Settings

The testing of algorithms in the dynamic domain is a complex endeavour as the dynamics of a problem allow for a wide range of different scenarios and Cruz et al. [14] point out the need to develop experimentation protocols that allows replication of experiments conducted in empirical studies. In particular, the authors point out that the concept of a *problem instance* is not as clearly defined as in the stationary optimization case, particularly because the dynamics of the problem are insufficiently detailed. The authors also point out the need for new measures that may be used to better understand the algorithm's behaviour during its execution (i.e., beyond simple performance measures) and criticise the lack of non-parametric statistical tests in most empirical studies.

One of the most prominent attributes of DOPs is the magnitude of change and most practitioners consider a range of values when evaluating their algorithm. However, the choice of these values, or indeed the dynamic mechanism itself, is often not motivated. In particular, the dynamics considered are often random and only little or no attention is paid to their long-term characteristics or the attributes of the current configuration of the problem. Furthermore, the frequency of change is often chosen specifically with a particular algorithm in mind: in order to avoid outdated f-values, most practitioners adopt their experimental setups specifically to eradicate this issue. In particular, as changes usually assumed to occur at unknown points in time, it has become a common practice to measure the frequency of change in generations (i.e., in periods of N function evaluations) and to make changes occur at intervals that coincide precisely with the generational cycle of the algorithm. The trend has also been extended to the measurement of an algorithm's performance (see Section 3.4.1).

Branke and Wang [12] point out that this requirement is an oversimplification of the requirements of real-world dynamics: the dynamics of a problem should be treated independently of the algorithm. More specifically, DOPs are usually understood to be real-time optimization processes and, as such, should be assumed to be completely decoupled from the algorithm that searches for solutions (i.e., the problem may not be "paused" to grant the algorithm extra computational resources). The authors subsequently investigate both theoretically and empirically the impact of changes that take place during the evaluation of the algorithm's population [9, 12]. The authors point out that it is possible to ignore such changes by updating the model of the problem only at the end of the generation. However, this (a) assumes that changes are detected externally and that the algorithm has full control over the model and (b) potentially implies that solutions are implemented which have been evaluated before the change (i.e., their real f-value or validity is unknown). The issue of change detection itself has been addressed by Richter [38]) who investigates the efficacy of population-based and sensor-based detection mechanism.

3.4 Challenge III: Notions of Optimality

3.4.1 Performance Measures in Evolutionary Dynamic Optimization

A performance measure (or metric) should indicate an algorithm's degree of success in finding desired solutions (as specified by the solution concept) to a particular search problem; it is thus an indicator of how well the algorithm implements the solution concept. In the case of stationary optimization, practitioners are commonly interested in the time taken to find a solution of satisfactory quality or the best solution quality obtained after some limited amount of time. In dynamic optimization, on the other hand, notions of optimality are more complex: as Trojanowski and Michalewicz point out, "when a problem is non-stationary, the situation is more complex, as it is necessary to measure not the final result (which does not exist in the continuous process of tracking the moving optimum), but rather the search process itself (e.g., its reactions to different types of changes)." [48, p 236]. The solution to a DOP is thus generally considered to be a time-series of search points and although each such point indicates the performance of the algorithm at that particular moment in time, the temporal aspect may be accounted for in numerous different ways. and in [14], the authors distinguish between at least 20 distinct measures, citing a total of 162 references.[1] Below we review the basic concept behind the majority of performance measures.

In stationary optimization, one is usually interested in a single solution but other metrics have also been suggested that measure a *trajectory* of solutions. The *offline* performance measure [24]:

$$\text{OFFLINE} = \frac{1}{t} \sum_{i=1}^{t} \max\{f_1, f_2, \ldots, f_i\} \tag{3.6}$$

The offline performance measure is not suitable for use in the dynamic domain as the values f_i are only valid within some time-dependent context (i.e., once a change has taken place, an f-value may not longer be accurate). The offline performance metric has subsequently been modified to account for changes in the problem such that only the best individual *since* the last change is considered [8]:

$$\text{M\_OFFLINE} = \frac{1}{t} \sum_{i=1}^{t} \max\{f_{\lfloor t/\tau \rfloor_1}, \ldots, f_{\lfloor t/\tau \rfloor_i}\} \tag{3.7}$$

where τ corresponds to a constant period between changes. The modified offline measure thus takes into account the trajectory of the *best* points found so far in between changes and thus requires knowledge regarding the points in time a change occurs and it has thus become a more common practice to adopt a *current-best* metric, rather than a *best-so-far* one [20]; it should be noted that under *elitism*, these measures are identical.

The resulting measure, originally proposed by Morrison, is called the *collective mean fitness* [31, 32] and attempts to capture the *typical* performance of the algorithm across a representative range of dynamics:

$$\text{COLLECTIVE} = \frac{1}{G} \sum_{i=1}^{G} (BoG_i) \tag{3.8}$$

---

[1] This classification is slightly misleading as several of the measures listed are identical or very similar to one another.

where $BoG_i = \max_f(P_i)$ is the best-of-generation (BoG) and G the number of generations. In order to obtain a representative performance from an experimental point of view, the value $\max_f(P)$ is averaged over K trials.

3.4.2 Existence of a Model

Almost all practitioners evaluate their algorithms using an offline approach (i.e., modified offline performance), taking into account only the best solution found so far (or the currently best solution). This compares against the *online* approach where every call to the objective functions requires testing in the real world and hence contributes towards the algorithm's overall performance. The offline approach assumes the existence of a model of the problem [8]: a model of the problem may be used to evaluate candidate solutions at a significantly reduced cost as no testing/implementation in the real world is required. Some authors claim that this approach is a closer reflection of optimality than the online case: Alba and Sarasola [1] report about the online measure that "it is nowadays usually not reported because it provides no information about the best values found, which are in fact the values of interest in real implementations".

However, this reasoning neglects the issue of obtaining an accurate model in the first place and Younes et al. suggest that it may be unreasonable to assume an accurate model of a dynamic real-world process [58]. Mori and Kita [30] point out that it is necessary to account for possible discrepancies between the real world and the model. This is particularly stringent as DOPs are commonly viewed as real-time optimization processes and using an offline evaluation measure necessarily implies potential discrepancies between the time a solution is evaluated on the model and implemented in the real world. Furthermore, in many cases, a model may be difficult to obtain in the first place. In [35], for instance, the authors consider the case where the dynamics affect the actual function form and hence extensive prior knowledge would be required to develop an accurate model of the problem.

3.4.3 Notions of Optimality

The variety of performance measures that have been suggested to measure the performance of EAs in the dynamic domain highlights the additional complexity imposed by a problem's dynamics. However, it also raises an important question: what solutions are we actually interested in? Alternatively, one might ask what actually constitutes a solution to a problem. This is known as a solution concept (see [17]): a binary predicate that identifies elements in X that are solutions from those that are not. This differentiation is usually based upon some number of measurable attributes and the solution concept should ideally provide a way to evaluate individual points in the search space that yields a gradient the algorithm can exploit. In many cases, the notion of optimality is so trivial, it receives no further attention. In more complex domains, however, notions of optimality require a significantly more elaborate formulation [17] and dynamic optimization appears to be one such domain. Here, similarly to multi-objective optimization, one is interested in solutions that consist of multiple search points.

It is therefore impossible to state which performance measures are more meaningful - this depends entirely on the desired solution concept envisioned by the practitioner. It is possible, however, to highlight the intrinsic attributes of each measure to allow practitioners to choose a performance metric that is aligned with their intended solution concept. In particular, there are several aspects that determine an appropriate measure. First and foremost is (a) the goal of the practitioner (e.g., optimum tracking versus robustness) and (b) the availability of a model (i.e., online vs. offline). The goal may then be refined further: robustness, for instance, may refer to genotypic or phenotypic variation (or both). Finally, (c) knowledge regarding the global optimum and the times of change may be exploited in numerous different ways.

The majority of practitioners assesses the performance of an algorithm using one of the performance measures reviewed above. In most cases, this corresponds to a trajectory of f-values, sampled every generation during the algorithm's execution (i.e., Eq. (3.8)). Nevertheless, other measures have been used also: Trojanowski and Michalewicz [49], for instance, only considers a single solution per update period. More generally, the desired solution to the DOPs is specified by points in time where a solution is required, presumably because it is implemented in the real world. We call this the *implementation schedule*:

Definition 3.1 (Implementation Schedule). A set of points in time within $t = 1$ and $t = t_{end}$ that specify when a solution is to be implemented.

The collective mean fitness has led to a rather worrying trend as the candidate solutions of interest corresponds to the best-of-generation and are thus tailored specifically towards the algorithm (i.e., dependent on population size). Furthermore, it is not obvious whether an algorithm developed for one implementation schedule necessarily performs well for another. In other words, the comparative performance of two algorithms may differ depending on the trajectory of solutions considered. This is particularly true if one considers an algorithm-independent schedule of dynamics (which is part of the problem's specifications).

Definition 3.2 (Schedule of Dynamics). A set of points in time within $t = 1$ and $t = t_{end}$ at which the underlying optimization problem changes.

It is therefore vital for practitioners to clearly state their goals and to choose a performance measure accordingly. In other words, choosing a performance measure should be done retrospectively, given the desired solution concept. It is therefore vital to have the means to better express solution concepts as notions such as "tracking the optimum" are insufficient for this purpose.

3.5 Implications, Perspectives and Conclusions

3.5.1 Summary

In this chapter we have presented a critical assessment of the field of evolutionary dynamic optimization and found that the advances in recent years, although significant in many aspects, are somewhat overshadowed by a lack of clarity regarding the requirements imposed by the problems' dynamics. In particular, we looked at three fundamental challenges faced by practitioners in evolutionary dynamic optimization.

The first challenge concerns the problem definition itself: generally speaking, the class of dynamic optimization problems (DOPs) entails all those problems that are deterministic at any moment in time but change as time goes by. Numerous extended problem definitions have been proposed in the past, each with a different degrees of expressional power, yet a commonly accepted framework to fully describe the properties of DOPs is still missing. The extent to which such a framework is actually possible remains to be seen but the need to fully articulate one's assumptions about the structural properties of a DOP is fundamental to the remainder of the field.

The second challenge concerned the benchmark problems used to test, evaluate and compare the new algorithms proposed for the dynamic domain. The most commonly used benchmark, MOVING PEAKS, is modelled as a dynamic multi-modal fitness landscape where individuals peaks change in height, width and position over time; similar approaches for the combinatorial domain have proven difficult given the algorithm-dependency of combinatorial fitness landscapes. Numerous practitioners have questioned whether the dynamics modelled by these benchmarks reflect those found in real-world problems. This is particularly relevant for a field of research where the majority of publications are motivated by the real-world relevance of such problems. Finally, concern has also been expressed with regard to the experimental setups used to test new algorithms on these benchmarks. In particular, the properties of the dynamics often appear to be tailored specifically towards the algorithm to be tested. The same holds for the performance measure used, as demonstrated by challenge 3.

The third challenge looked at the variety of performance measures used in dynamic optimization. Amongst the most popular is the collective mean fitness measure which takes into account the best solution found at every generation of the algorithm's execution. A DOP is usually seen as some real-time online optimization process where the algorithm is required to *locate* and *track* high-quality solutions over time. However, only little has been done to identify the requirements of real-world problems including, for instance, the assumption of a perfect model that allows for offline performance measures. Although this appears trivial in the case of stationary optimization, it requires further consideration in the dynamic case. Finally, clear notions of optimality need to be established a priori and independently of chosen performance measures.

3.5.2   Implications and Perspectives

The recent interest in the dynamic domain sparked the development of numerous new algorithms designed specifically to *track* high quality solutions over time. The success of these algorithms is usually demonstrated by comparative empirical studies that show improvements in performance over traditional EAs. Such studies rely on three fundamental aspects: a particular class of problems that is of interest, a way to model the representative characteristics of these problems and finally, a way to assess the performance of the algorithm. This chapter has highlighted some of the criticism that has been raised with respect to each of these aspects.

One of the most fundamental aspects is to better understand the class of problems itself and to have the means to fully describe instances of each problem. Many benchmark suites exists in the case of stationary optimization (e.g., OR-library [3]) yet an equivalent is still missing for the dynamic domain. Instead, universal benchmarks like MOVING PEAKS are used to simulate a wide array of different dynamics yet these benchmarks have been criticised for their potential lack of realism; this phenomenon appears to extend beyond the problem considered to the experimental setup. A formal framework may allow one to dissect real-world problems and to model their properties accordingly. This appears an essential requirement at this stage to drive forward the field of evolutionary dynamic optimization. Closely related to this issue is the notion of optimality that is used to define the search problem. A wide range of performance measures have been suggested yet it is impossible to differentiate between them as it remains unclear what actually constitutes a desired solution. In particular, it is essential for practitioners to clearly state their goals and assumptions and to design their algorithms accordingly.

The last decade of research in evolutionary dynamic optimization has made significant progress and numerous important results have been obtained. What is required is a way to integrate all these results to make the communal effort somewhat more standardised and streamlined.

3.5.3   Conclusions

The problem's dependency on time introduces many complexities not found in stationary optimization. Most notably, the dynamics introduce a significant variety of possible scenarios and it is vital to (a) identify those cases that are of interest and (b) have the means to describe such cases unambiguously. This is particularly relevant in light of the No Free Lunch theorem for dynamic optimization [53]: if one considers the most abstract of settings, including a set of functions F, a transition function $\mathcal{T} : F \times \mathbb{N} \rightarrow F$ and a frequency of change $\tau : \mathbb{N} \times \mathbb{N} \rightarrow \mathbb{N}$, there are an infinite number of possible "problem instances". The reduction of this to a well-defined set of DOPs with exploitable structural properties remains a key issue in evolutionary dynamic optimization.

Closely related to this issue is the notion of optimality considered and this too should be grounded in real-world requirements. The explicit articulation of solution concepts should facilitate an improved development of new algorithms. In particular, assuming that the dynamics of the problem are non-random, one should aim to reduce the *black-box* uncertainty of future problem instances over time. Currently, this is achieved primarily using samples of the search space taken at different points in time. The next step is to extrapolate from these measurements, using tools from related fields of research, including time-series prediction, optimal control theory, online learning and data mining. This would allow one to gain a better estimate of the problem's future states and emphasises the issue of building appropriate models to deal with the problem's dynamics. This, in turn, allows a more efficient transfer of knowledge from one state of the problem to another. Benchmark problems and experimental settings should subsequently be designed to account for this.

Acknowledgements. This work was partially supported by the Engineering and Physical Sciences Research Council (EPSRC) of UK under Grant EP/E058884/1.

References

[1] Alba, E., Sarasola, B.: Abc, a new performance tool for algorithms solving dynamic optimisation problems. In: Proc. IEEE World Congr. Comput. Intell., pp. 734–740 (2010)

[2] Andrews, M., Tuson, A.: Dynamic optimisation: A practitioner requirements study. In: Proc. 24th Annual Workshop of the UK Planning and Scheduling Special Interest Group (2005)

[3] Beasley, J.E.: Or-library: Distributing test problems by electronic mail. J. of Oper. Res. Society 41(11), 1069–1072 (1990)

[4] Bosman, P.A.N.: Learning, anticipation and time-deception in evolutionary online dynamic optimization. In: Yang, S., Branke, J. (eds.) GECCO Workshop on Evolutionary Algorithms for Dynamic Optimization (2005)

[5] Branke, J.: Evolutionary algorithms for dynamic optimization problems - a survey. Tech. Rep. 387, Insitute AIFB, University of Karlsruhe (1999)

[6] Branke, J.: Memory enhanced evolutionary algorithms for changing optimization problems. In: Proc. 1999 IEEE Congr. Evol. Comput., vol. 3, pp. 1875–1882 (1999)

[7] Branke, J.: Evolutionary approaches to dynamic environments - updated survey. In: GECCO Workshop on Evolutionary Algorithms for Dynamic Optimization Problems, pp. 27–30 (2001)

[8] Branke, J.: Evolutionary Optimization in Dynamic Environments. Kluwer (2002)

[9] Branke, J., Kulzhabayeva, G., Uyar, S.: Addressing change within a generation. Tech. Rep., University of Karlsruhe (2008)

[10] Branke, J., Orbayı, M., Uyar, Ş.: The role of representations in dynamic knapsack problems. In: Rothlauf, F., et al. (eds.) EvoWorkshops 2006. LNCS, vol. 3907, pp. 764–775. Springer, Heidelberg (2006)

[11] Branke, J., Salihoglu, E., Uyar, Ş.: Towards an analysis of dynamic environments. In: Beyer, H.G.G. (ed.) Genetic and Evolutionary Computation Conference, pp. 1433–1439. ACM (2005)

[12] Branke, J., Wang, W.: Theoretical analysis of simple evolution strategies in quickly changing environments. In: Cantú-Paz, E., et al. (eds.) GECCO 2003. LNCS, vol. 2723, pp. 537–548. Springer, Heidelberg (2003)

[13] Corne, D.W., Reynolds, A.P.: Optimisation and generalisation: Footprints in instance space. In: Schaefer, R., Cotta, C., Kołodziej, J., Rudolph, G. (eds.) PPSN XI, Part I. LNCS, vol. 6238, pp. 22–31. Springer, Heidelberg (2010)

[14] Cruz, C., González, J.R., Pelta, D.A.: Optimization in dynamic environments: a survey on problems, methods and measures. Soft Comput. 15(7), 1427–1448 (2011)

[15] Droste, S., Jansen, T., Tinnefeld, K., Wegener, I.: A new framework for the valuation of algorithms for black-box optimization. In: Proc. 7th Int. Workshop Foundations of Genetic Algorithms, pp. 197–214 (2002)

[16] Droste, S., Jansen, T., Wegener, I.: Optimization with randomized search heuristics the (a)nfl theorem, realistic scenarios, and difficult functions. Theoretical Computer Sci. 287 (2002)

[17] Ficici, S.G.: Solution concepts in coevolutionary algorithms. Ph.D. thesis, Brandeis University (2004)

[18] Fu, H., Sendhoff, B., Tang, K., Yao, X.: Characterizing environmental changes in robust optimization over time. In: Proc. 2012 IEEE Congr. Evol. Comput., pp. 551–558 (2012)

[19] Goldberg, D.E., Smith, R.E.: Nonstationary function optimization using genetic algorithms with dominance and diploidy. In: Grefenstette, J.J. (ed.) Proc. Int. Conf. Genetic Algorithms, pp. 59–68. Lawrence Erlbaum Associates (1987)

[20] Grefenstette, J.J.: Evolvability in dynamic fitness landscapes: A genetic algorithm approach. In: Proc. 1999 IEEE Congr. Evol. Comput., vol. 3, pp. 2031–2038 (1999)

[21] Jin, Y., Branke, J.: Evolutionary optimization in uncertain environment - a survey. IEEE Trans. Evol. Comput. 9(3), 303–317 (2005)

[22] Jin, Y., Sendhoff, B.: Constructing dynamic optimization test problems using the multi-objective optimization concept. In: Raidl, G.R., et al. (eds.) EvoWorkshops 2004. LNCS, vol. 3005, pp. 525–536. Springer, Heidelberg (2004)

[23] Jin, Y., Tang, K., Yu, X., Sendhoff, B., Yao, X.: A framework for finding robust optimal solutions over time. Memetic Comput. 5(1), 3–18 (2012)

[24] Jong, K.D.: Analysis of the behavior of a class of genetic adaptive systems. Ph.D. thesis, Department of Computer and Communication Science, University of Michigan (1975)

[25] Jong, K.D.: Evolving in a changing world. In: Raś, Z.W., Skowron, A. (eds.) ISMIS 1999. LNCS, vol. 1609, pp. 512–519. Springer, Heidelberg (1999)

[26] Karaman, A., Uyar, A.S.: A novel change severity detection mechanism for the dynamic 0/1 knapsack problem. In: Proc. 10th Int. Conf. Soft Computing (2004)

[27] Kauffman, S.A.: The Origins of Order. Oxford University Press (1993)

[28] Leguizamon, G., Blum, C., Alba, E.: Handbook of approximation algorithms and meta-heuristics, pp. 24.1–24.X. CRC Press (2007)

[29] Li, C., Yang, S.: A generalized approach to construct benchmark problems for dynamic optimization. In: Li, X., et al. (eds.) SEAL 2008. LNCS, vol. 5361, pp. 391–400. Springer, Heidelberg (2008)

[30] Mori, N., Kita, H.: Genetic algorithms for adaptation to dynamic environments - a survey. In: Proc. 26th Annual Conf. IEEE Industrial Electronics Society, vol. 4, pp. 2947–2952 (2000)

[31] Morrison, R.W.: Performance measurement in dynamic environments. In: Branke, J. (ed.) GECCO Workshop on Evolutionary Algorithms for Dynamic Optimization Problems, pp. 5–8 (2003)

[32] Morrison, R.W.: Designing Evolutionary Algorithms for Dynamic Environments, pp. 3–540. Springer, Berlin (2004) ISBN 3-540-21231-0

[33] Morrison, R.W., DeJong, K.A.: A test problem generator for non-stationary environments. In: Proc. 1999 IEEE Congr. Evol. Comput., vol. 3, pp. 2047–2053 (1999)

[34] Nguyen, T.T., Yao, X.: Benchmarking and solving dynamic constrained problems. In: Proc. 2009 IEEE Congr. Evol. Comput., pp. 690–697. IEEE Press (2009)

[35] Nguyen, T.T., Yao, X.: Dynamic time-linkage problems revisited. In: Giacobini, M., et al. (eds.) EvoWorkshops 2009. LNCS, vol. 5484, pp. 735–744. Springer, Heidelberg (2009)

[36] Papadimitriou, C.H., Steiglitz, K.: Combinatorial Optimization: Algorithms and Complexity. Dover (1998)

[37] Rand, W., Riolo, R.: Measurements for understanding the behavior of the genetic algorithm in dynamic environments: A case study using the shaky ladder hyperplane-defined functions. In: Yang, S., Branke, J. (eds.) GECCO Workshop on Evolutionary Algorithms for Dynamic Optimization (2005)

[38] Richter, H.: Detecting change in dynamic fitness landscapes. In: Proc. 11th IEEE Congr. Evol. Comput., pp. 1613–1620 (2009)

[39] Rohlfshagen, P., Lehre, P.K., Yao, X.: Dynamic evolutionary optimisation: An analysis of frequency and magnitude of change. In: Proc. 2009 Genetic and Evol. Comput. Conf., pp. 1713–1720 (2009)

[40] Rohlfshagen, P., Yao, X.: Attributes of dynamic combinatorial optimisation. In: Li, X., et al. (eds.) SEAL 2008. LNCS, vol. 5361, pp. 442–451. Springer, Heidelberg (2008)

[41] Rohlfshagen, P., Yao, X.: Dynamic combinatorial optimisation problems: An analysis of the subset sum problems. Soft Comput. 15(9), 1723–1734 (2011)

[42] Rossi, C., Barrientos, A., del Cerro, J.: Two adaptive mutation operators for optima tracking in dynamic optimization problems with evolution strategies. In: Proc. 9th Annual Conf. Genetic and Evol. Comput., pp. 697–704 (2007)

[43] Rothlauf, F.: Representations for Genetic and Evolutionary Algorithms, 2nd edn. Springer (2006)

[44] Simon, H.: Models of Man. Wiley, New York (1957)

[45] Thompson, R.K., Wright, A.H.: Additively decomposable fitness functions. Tech. Rep., University of Montana, Computer Science Department (1996)

[46] Tinos, R., Yang, S.: Continuous dynamic problem generators for evolutionary algorithms. In: Proc. 2007 IEEE Congr. Evol. Comput., pp. 236–243 (2007)

[47] Tinós, R., Yang, S.: An analysis of the XOR dynamic problem generator based on the dynamical system. In: Schaefer, R., Cotta, C., Kołodziej, J., Rudolph, G. (eds.) PPSN XI. LNCS, vol. 6238, pp. 274–283. Springer, Heidelberg (2010)

[48] Trojanowski, K., Michalewicz, Z.: Evolutionary algorithms for non-stationary environments. In: Proc. 8th Workshop on Intell. Inform. Syst., pp. 229–240 (1999)

[49] Trojanowski, K., Michalewicz, Z.: Searching for optima in non-stationary environments. In: Proc. 1999 IEEE Congr. Evol. Comput., vol. 3, pp. 1843–1850. IEEE (1999)

[50] Ursem, R.K., Krink, T., Jensen, M.T., Michalewicz, Z.: Analysis and modeling of control tasks in dynamic systems. IEEE Trans. Evol. Comput. 6(4), 378–389 (2002)

[51] Weicker, K.: An analysis od dynamic severity and population size. In: Deb, K., Rudolph, G., Lutton, E., Merelo, J.J., Schoenauer, M., Schwefel, H.-P., Yao, X. (eds.) PPSN 2000. LNCS, vol. 1917, pp. 159–168. Springer, Heidelberg (2000)

[52] Weicker, K.: Evolutionary algorithms and dynamic optimization problems. Der Andere Verlag (2003)

[53] Wolpert, D.H., Macready, W.G.: No free lunch theorems for optimization. IEEE Trans. Evol. Comput. 1(1), 67–82 (1997)

[54] Yang, S.: Non-stationary problem optimization using the primal-dual genetic algorithms. In: Sarker, R., Reynolds, R., Abbass, H., Tan, K.C., McKay, R., Essam, D., Gedeon, T. (eds.) Proc. 2003 IEEE Congr. Evol. Comput., vol. 3, pp. 2246–2253 (2003)

[55] Yang, S.: Constructing dynamic test environments for genetic algorithms based on problem difficulty. In: Proc. 2004 IEEE Congr. Evol. Comput., vol. 2, pp. 1262–1269 (2004)

[56] Yang, S.: Memory-enhanced univariate marginal distribution algorithms for dynamic optimization problems. In: Proc. 2005 IEEE Congr. Evol. Comput., vol. 3, pp. 2560–2567 (2005)

[57] Yang, S., Yao, X.: Population-based incremental learning with associative memory for dynamic environments. IEEE Trans. Evol. Comput. 12(5), 542–561 (2008)

[58] Younes, A., Calamai, P., Basir, O.: Generalized benchmark generation for dynamic combinatorial problems. In: Yang, S., Branke, J. (eds.) GECCO Workshop on Evolutionary Algorithms for Dynamic Optimization (2005)

[59] Yu, X., Jin, Y., Tang, K., Yao, X.: Robust optimization over time – a new perspective on dynamic optimization problems. In: Proc. 2010 IEEE Congr. Evol. Comput., pp. 3998–4003 (2010)

Chapter 4
Dynamic Multi-objective Optimization: A Survey of the State-of-the-Art

Carlo Raquel and Xin Yao

Abstract. Many optimization problems involve multiple objectives, constraints and parameters that change over time. These problems are called dynamic multi-objective optimization problems (DMOPs) and have recently attracted a lot of research. In this chapter, we provide a survey of the state-of-the-art on the field of dynamic multi-objective optimization with regards to the definition and classification of DMOPS, test problems, performance measures and optimization approaches. We provide a comprehensive definition of DMOPs and identify gaps, challenges and future works in dynamic multi-objective optimization.

4.1 Introduction

Multi-objective evolutionary algorithms (MOEAs) have been applied to solve various real-world optimization problem involving two or more objectives which may be in conflict with one another. MOEAs evolve a population of candidate solutions to find a set of optimal solutions in a single run. This set of optimal solution is called Pareto optimal front (POF) while the solutions are called nondominated solutions.

In the past years, there have been significant contributions made on solving multi-objective optimization problems which are static in nature. However, there are many optimization problems that involve objectives, constraints and parameters that could change over time. These problems are called dynamic multi-objective optimization problems (DMOPs). The dynamic multi-objective optimization problem can be generally defined as:

$$\begin{cases} \min_{x^n} f(x,t) = \{f_1(x,t), f_2(x,t), ... f_M(x,t)\} \\ s.t. \, g(x,t) \le 0, \, h(x,t) = 0 \end{cases} \tag{4.1}$$

Carlo Raquel · Xin Yao
Centre of Excellence for Research in Computational Intelligence and Applications
(CERCIA), School of Computer Science, University of Birmingham,
Birmingham B15 2TT, U. K.
e-mail: {crr954,x.yao}@cs.bham.ac.uk

S. Yang and X. Yao (Eds.): *Evolutionary Computation for DOPs*, SCI 490, pp. 85–106.
DOI: 10.1007/978-3-642-38416-5_4   © Springer-Verlag Berlin Heidelberg 2013

where x is the vector of decision variables bounded by the decision space X^n and f is the set of objectives to be minimized with respect to time t. We denote the dynamic Pareto optimal front at time t as PF_t^* which is the set of nondominated solutions with respect to the objective space at t. The dynamic Pareto optimal set, denoted as PS_t^* is the set of nondominated solutions with respect to the decision space at t.

Whenever there are changes in the environment that affect the solutions of the DMOP, the optimization goal becomes tracking the changing PF^*. This poses the challenge to the optimization algorithm to find and converge to the new PF^* before the problem landscape changes again. While fast convergence to the new Pareto front is a desirable capability of an optimization algorithm, it should be able to address the issue of lack of diversity when the problem landscape changes in order to explore the new search space.

The purpose of this chapter is to provide a comprehensive definition of dynamic multi-objective optimization, a review of existing dynamic multi-objective optimization problems and their classifications, performance metrics and optimization techniques used in solving DMOPs.

4.2   Comprehensive Definition of Dynamic Multi-objective Optimization

Most DMOPs are described using a general definition such as Eq. (4.1) which does not explicitly specify the dynamic fitness function, parameters and constraints. In this section, we model the behavior of dynamic multi-objective optimization problem by directly adapting definitions proposed by Nguyen *et al.* in [26]. We add a dynamic driver D_N to model the changing number of objectives and constraints.

Definition 4.1 (Full-description form [26]). Given a finite set of functions $F = \{f_1(x), ..., f_n(x)\}$; a full-description form of F is a tuple

$$\langle f_\gamma(x), \{c_1, ..., c_n\} \rangle \tag{4.2}$$

where $f_\gamma(x)$ is a mathematical expression with its set of parameters $\gamma \in R^m$, $\{c_1, ..., c_n\}$, $c_i \in R^m$ is a set of vectors so that

$$f_\gamma(x) \to f_1(x), \gamma = c_1 \tag{4.3}$$

$$\cdots$$

$$f_\gamma(x) \to f_n(x), \gamma = c_n \tag{4.4}$$

Each function $f_i(x), i = 1 : n \in N^+, i > 1$ is called an instance of the full-description form at $\gamma = c_i$.

Definition 4.2 (Dynamic Driver [26]). Given a tuple $\langle \hat{f}, \gamma_t, t \rangle$ where t is a time variable, $\hat{f}$ is a full-description form of the set of functions $F = \{f_1(x), ..., f_n(x)\}$ with respect to the set of m-element vectors $\{c_1, ...c_n\}, c_i \in R^m$ and $\gamma_t \in R^m$ is an m-element vector containing all m parameters of $\hat{f}$ at the time t; we call a mapping $D(\gamma_t, t) : R^m \times N^+ \rightarrow R^m$ a dynamic driver of $\hat{f}$ if

$$\gamma_{t+1} = D(\gamma_t, t) \in \{c_1, ..., c_n\} \forall t \in N^+ \tag{4.5}$$

and γ_{t+1} is used as the set of parameters of $\hat{f}$ at the time $t+1$.

Definition 4.3 (Time unit [26]). When a time-dependent problem is being solved, a time unit, or a unit for measuring time periods in the problem, represents the time durations needed to complete one function evaluation of that problem. The number of evaluations (or time units) that have been evaluated so far since we started solving the problem is measured by the variable $\tau \in N^+$.

Definition 4.4 (Change step and frequency of change   [26]). When a time-dependent problem is being solved, a change step represents the moment when the problem changes. The number of change steps that have occurred so far in a time-dependent problem is measured by the variable $t \in N^+$. Obviously t is a time-dependent function of τ-the number of evaluations made so far since we started solving the problem; $t(\tau) : N^+ \rightarrow N^+$. Its dynamics is controlled by a problem-specific time-based dynamic driver

$$\begin{cases} D_t(t(\tau), \tau) = t(\tau) + 1 & when\, a\, change\, occurs \\ D_t(t(\tau), \tau) = t(\tau) & otherwise \end{cases} \tag{4.6}$$

Definition 4.5 (Optimization algorithms and dynamic solutions [26]). Given a time-dependent problem $f_{D(\gamma_t)} = \langle \hat{f}, D(\gamma_t) \rangle$ at the change step t and a set P_t of k_t solutions $x_1, ..., x_{k_t} \in S_t \subseteq R^d$ is the search space, an optimization algorithm G to solve $\hat{f}_{D(\gamma_t)}$ can be seen as a mapping $G_t : R^{d \times k_t} \rightarrow R^{d \times k_{t+1}}$ capable of producing a solution set P_{t+1} of k_{t+1} optimized solutions $x_1^{G[t^b, t^e]}$ that we get by applying an algorithm G to solve $\hat{f}_{D(\gamma_t)}$ of a given initial population P_{t^b-1} during the period $[t^b, t^e], t^b \geq 1$ is given by

$$X_{f_t}^{G[t^b t^e]} = \bigcup_{t=t^b}^{t^e} P_t = \bigcup_{t=t^b}^{t^e} G_t(P_{t-1}) \tag{4.7}$$

Definition 4.6 (Time-linkage dynamic driver [26]). Given a tuple $\left\langle \hat{f}, \gamma_t, t, X_{\hat{f}}^{G[1,t]} \right\rangle$ where t is a time variable, $\hat{f}$ is a full-description form of the set of functions $F = \{f_1(x), ..., f_n(x)\}$ with respect to the set of m-element vectors $\{c_1, ...c_n\}, c_i \in R^m$ and $\gamma_t \in R^m$ is an m-element vector containing all m parameters of $\hat{f}$ at the time t; and

$X_{\hat{f}}^{G[1,t]}$ is a set of k d-dimensional solutions achieved by applying an algorithm G to solve f during the period $[1,t]$; we call a mapping $D\left(\gamma_t, X_{\hat{f}}^{G[1,t]}, t\right) : R^m \times R^{d \times k} \to R^m$ a time linkage dynamic driver of $\hat{f}$ if

$$\gamma_{t+1} = D\left(\gamma_t, X_{\hat{f}}^{G[1,t]}\right) \in \{c_1, ..., c_n\} \forall t \in N^+ \tag{4.8}$$

and γ_{t+1} is used as the set of parameters of $\hat{f}$ at time $t+1$.

Definition 4.7 (Dynamic multi-objective optimization problem). Given a tuple $\langle \hat{f}, \hat{C}, D_P, D_D, D_N, D_T, G \rangle$ a dynamic multi-objective optimization problem in the period $\left[1, \tau^{end}\right]$ function evaluations, $\tau^{end} \in N^+$ can be defined as

$$optimise \left\{ \sum_{\tau=1}^{\tau^{end}} \hat{f}_{\gamma\left(t_\tau, X_{\hat{f}}^{G[1,t]}\right)} (x_t) \right\} \tag{4.9}$$

subject to

$$\hat{C}_{\gamma\left(t, X_{\hat{f}}^{G[1,t]}\right)}^{i=1:k \in N^+} (x_t, t_\tau) \leq 0; \ and \ l\left(t_\tau, X_{\hat{f}}^{G[1,t]}\right) \leq x \leq u\left(t_\tau, X_{\hat{f}}^{G[1,t]}\right) \tag{4.10}$$

where

- $\hat{f}_1, .., \hat{f}_m$ are the full-description forms of the m objective functions, $m \geq 2$
- $\hat{C}_1, ..., \hat{C}_k$ are the full-description forms of k dynamic constraints
- D_P is the dynamic driver for parameters in objective and constraint
- D_D is the dynamic driver for parameters for domain constraints
- D_N is the dynamic driver for the changing number of objectives and constraints
- D_T is the dynamic driver for times and frequency of changes
- G is the algorithm used to solve the problem
- $\tau \in \left[1, \tau^{end} \cap N\right]$ is the number of function evaluations done so far
- t_τ or $t(\tau) \in N^+$ is the current change step; $t(\tau)$ is controlled by D_T
- $X_{\hat{f}}^{G[1,t]}$ is the set of solutions achieved by applying algorithm G to solve $\hat{f}$ during $[1,t]$

- $l(t_\tau), u(t_\tau) \in R^n$ are domain constraints; $\begin{cases} l(t_\tau+1) = D_D\left(l(t_\tau), X_{\hat{f}}^{G[1,t]}, t_\tau\right) \\ u(t_\tau+1) = D_D\left(u(t_\tau), X_{\hat{f}}^{G[1,t]}, t_\tau\right) \end{cases}$

4.3 Dynamic Multi-objective Test Problems

Recently, there have been significant contributions made in solving dynamic multi-objective optimization problems (DMOPs) but only few have been devoted to defining and classifying them. In [9], Farina *et al.* classified problems based on the changes on the Pareto optimal solutions (PS_t) at time t and the Pareto optimal front (PF_t) at time t. The following are the four different types of DMOPs that they have proposed:

- Type I: where PS_t^* changes, while PF_t^* remains the same
- Type II: where both PS_t^* and PF_t^* changes
- Type III: where PF_t^* changes while PS_t^* remains the same
- Type IV: where both PS_t^* and PF_t^* remain the same

This classification shows the difficulty of solving DMOPs by describing the combination of changes in the Pareto set and front but it does not take into consideration the other aspects of dynamism such changes in constraint and time-linkage problems, changing number of parameters, objectives or constraints. Also, it does not look into the sources of dynamism in the problem.

In [11], Goh and Tan provided a more general classification based on the spatial and temporal components of the DMOP. The spatial component is divided into physical and non-physical attributes of change. The physical attributes could have a changing PS_t^*, PF_t^*, fitness landscape. The non-physical attributes indicate whether there is a random, trend or periodic change in the physical attributes. The temporal attributes refer to how the dynamic change is triggered such as random, fixed interval, scheduled, conditional or none at all. A DMOP may be described by one or more of these spatial and temporal components. This type of classification provides more details in terms of the given DMOP but have the same limitation as the first classification.

In [29], Tantar *et al.* proposed a component-oriented classification which highlights the sources of dynamism of the optimization problem as well as the state dependency of these components. Let $H(F_\sigma, D, x, t)$ model the behavior of a DMOP having F_σ as the multi-objective support function, D as the vector of time dependent functions, t as the time and σ as the environment derived set of parameters, generally defined as a constant over time. This classification separates the static and dynamic components of the DMOP. However, it does not incorporate dynamic constrained problems and those with changing dimensions of parameter, objectives and constraints. Their proposed classification are the following classification:

- 1st order: Dynamic parameter evolution modeled as $H(F_\sigma, D, x, t) = F_\sigma(D(x, t))$
- 2nd order: Dynamic function evolution modeled as $H(F_\sigma, D, x, t) = D(F_\sigma, x, t))$
- 3rd order: Dynamic state-dependency evolution (state-parameter dependency and state-function dependency) modeled as $H\left(F_\sigma, T^{[t-j,t]}, x, t\right)$ given a transformation function T over states t and $t - j$
- 4th order: Online dynamic evolution where the environment changes over time modeled as $H(F_\sigma, D, x, t) = F_{D(\sigma,t)}(D(x, t))$

While a classification based on the dynamic changes on the Pareto set and front is important, a classification based on the sources of dynamism provides more insights on the difficulty of solving DMOPs. A DMOP can be further described in terms of its spatial or temporal attributes. Given all these classifications together, they all describe different aspects of the dynamic problem. But what is missing is the inclusion of DMOPs with dynamic constraints in the classification. By using Definition 4.7, we can classify DMOPs based on their (1) full description form, (2) four dynamic drivers, and (3) the algorithms used. In this way, we are able to separate the static and dynamic components of the problem. The sources of dynamism are also explicitly specified.

4.3.1 Dynamic Multi-objective Optimization Test Problems

Benchmark problems are important in dynamic multi-objective optimization in order to develop and test effective algorithms. We review general-purpose test problems based on the characteristic of their Pareto set and front as well as their sources of dynamism.

Table 4.1 shows information about each set of test problems. The list is by no means comprehensive as only general test suites were considered. Most of the test problems represent Type I, II and III problems. Only the last set provides two Type IV test problems. All the sets of test problems include problems dealing with dynamic function evolution. *DIMP* and *T* provide test problems that involve dynamic parameter evolution. The set T proposed problems dealing with time linkage. The test problems T3 and T4 were designed so that current solutions depend on decision variables and solutions found in a previous state. There are no set of test function that deals with online dynamic evolution. Tantar *et al.* [29] provided a dynamic MNK-landscape problem as a proof of principle for this class of problems.

All of the test problems are unconstrained or domain constrained although it should be noted that dynamic constrained multi-objective test problems have already been designed such as extending a static constrained optimization problem [36]. In general, the source of dynamism comes mostly from the changing objective function. More test suites dealing with constrained problems, dynamic parameters, time linkage and online dynamic problems are needed to be developed. All the test problems have fixed number of objective functions and constraints. Thus, test problems with changing number of these sources of dynamism needs to be designed as well.

4.4 Performance Measures

In order to evaluate the performance of algorithm solving DMOPs, performance metrics are commonly used. These metrics should be able to assess how well can optimization algorithms track the changing Pareto front as well as assess the quality of the generated solution sets. Performance measures proposed in literature are

Table 4.1 Dynamic Multi-objective Test Problems

Problem	Description	Problem Type	Dynamic Parameter Evolution	Dynamic Function evolution	Dynamic state-dependency evolution	Dynamic constraint evolution	Online dynamic evolution
FDA [9]	This set of test problems consists of static objective functions together with time-dependent functions F which controls the density of Pareto solutions, G controls the shape of PS_t^* and H controls the shape of PF_t^*. Mehnen et al. [23] proposed a generalisation of the FDA.	FDA1 is Type I, FDA2 is Type III, FDA3 is Type II. FDA4 is Type I and FDA5 is Type II	no	yes	no	no	no
dMOP [10]	This set was developed based on the construction guidelines provided by Farina et al. in [9]. The PF_t^* of the test functions changes from convex to concave.	dMOP1 is Type III, dMOP2 is Type II and dMOP3 is Type I with the variable that controls the spread of the solution also changes.	no	yes	no	no	no
DSW [23]	They extended the static multi-objective problems proposed by Schaffer [28] to analyze the specific effect of each dynamic component of a problem. The test functions have a parabolic Pareto front.	DSW1, DSW2 and DSW3 are Type II	no	yes	no	no	no
HE [16]	The set was adapted from ZDT3 test problem [38] to become dynamic. These problems generate a discontinuous Pareto front	HE1 and HE2 are both Type III	no	yes	no	no	no
DIMP [20]	In this set of test functions, each decision variable is given a different rate of change with the exception of x_1 assigned to control the spread of the solutions	DIMP1 and DIMP2 are both Type I	yes	yes	no	no	no
T [17]	This set of test functions incorporates time linkage and dynamic number of parameters and objective functions	T1 is Type IV, T2 is Type III, T3 is Type IV and T4 is Type IV	yes	yes	yes	no	no

mostly designed for single objective dynamic optimization problems. However, there are also performance measures recently proposed for dynamic multi-objective optimization which are often extension of performance measures from single objective dynamic optimization. They can be classified as either measures designed for problem in which the Pareto fronts are known or unknown.

4.4.1 Performance Measures for Problems with Known Pareto Front

In order to measure the convergence in decision and objective spaces, Farina *et al.* [9] proposed the following:

$$c_x(t) = \frac{1}{np} \sum_{j=1}^{np} \min_{i=1:nh} \left\| \frac{PS_i^*(t) - PS_j^{sol}(t)}{R(t) - U(t)} \right\| \qquad (4.11)$$

$$c_f(t) = \frac{1}{np} \sum_{j=1}^{np} \min_{i=1:nh} \left\| PF_i^*(t) - PF_j^A(t) \right\| \qquad (4.12)$$

where nh is the number of solutions of PF_i^* and PS_i^*, np is the number of generated solutions, $R(t)$ is the time-dependent nadir point, and $U(t)$ is the time-dependent utopia point.

Hatzakis and Wallace [15] used these measures to get the time average of the convergence measures at particular time samples.

$$\bar{c}_x(t) = \frac{1}{np\tau} \sum_{t=1}^{\tau} \sum_{j=1}^{np} \min_{i=1:nh} \left\| \frac{PS_i^*(t) - PS_j^A(t)}{R(t) - U(t)} \right\| \qquad (4.13)$$

$$\bar{c}_f(t) = \frac{1}{np\tau} \sum_{t=1}^{\tau} \sum_{j=1}^{np} \min_{i=1:nh} \left\| PF_i^*(t) - PF_j^A(t) \right\| \qquad (4.14)$$

Li *et al.* [21] proposed the measure reversed $GD(t)$ or $rGD(t)$, which is based in the generational distance measure (GD) proposed by Veldhuizen [30]. The GD measures the average distance of the nondominated solutions Q_t found from the known Pareto front and only measures the convergence of an optimization algorithm. They reversed the process by measuring the average distance of the known Pareto front to the generated Pareto front making it possible to measure convergence diversity of the solution found with respect to PF_t^*.

$$rGD(t) = \frac{\sum_{i=1}^{|PF_t^*|} d_i}{|PF_t^*|} \qquad (4.15)$$

where $d_i = min_{k=1}^{|Q_t|} \sqrt{\sum_{j=1}^{M} \left(f_j^{*(i)} - f_j^{(k)} \right)^2}$. In here, $f_j^{(k)}$ is the j-th objective function value of the k-th member of the Q_t. It is called reversed because it iterates over

PF_t^* instead of iterating over Q_t. The term d_i is the Euclidean distance between the nearest solution in Q_t for each of the sampling point in PF_t^*. There are drawbacks in using rGD(t) as a performance measure. First, the Pareto front should be known a priori and it is strongly dependent on the distribution of the sampling points of PF_t^*.

In [21], Li *et al.* also proposed $HVR(t)$ which is based on the hypervolume (HV) metric proposed in [30]. The hypervolume indicator measures the space covered by a set of solutions Q_t in the objective space using a reference point W. Hypervolume is computed as

$$HV = volume\left(\cup_{i=1}^{Q} v_i\right) \tag{4.16}$$

The HV is basically the union of all hypercubes v_i with respect to a reference point W where $i \in Q$. Li *et al.* proposed $HVR(t)$ for dynamic environment as follows:

$$HVR(t) = \frac{HV(Q_t)}{HV(PF_t^*)} \tag{4.17}$$

They suggested that the reference point $W(t)$ be the worst value in each objective dimension in Q_t. $HVR(t)$ measures how well a MOEA tracks the moving optima by how well $HV(Q_t)$ covers $HV(P_t^t)$. This performance measure also requires that the Pareto front is known and it is dependent on the distribution of the sampling points on PF_t^*.

In order to apply these metrics in measuring the performance of a an algorithm, they proposed the collective mean error (CME) for the two metrics.

$$CME_{rGD} = \frac{1}{T}\sum_{t=1}^{T} rGD(t) \tag{4.18}$$

$$CME_{HVR} = \frac{1}{T}\sum_{t=1}^{T} HVR(t), \tag{4.19}$$

where T is the number of iterations of a run. The best value of CME_{rGD} is 0.0 while 1.0 for CME_{HVR}.

Goh and Tan [10] proposed the two measures variable space GD (VD) and maximum spread for dynamic multi-objective environment.

$$VD_{offline} = \frac{1}{\tau}\sum_{t=1}^{\tau} VD \cdot I(t) \tag{4.20}$$

$$MS_{offline} = \frac{1}{\tau}\sum_{t=1}^{\tau} MS \cdot I(t) \tag{4.21}$$

$$I(t) = \begin{cases} 1, & t\%\tau_T \\ 0, & otherwise \end{cases} \tag{4.22}$$

where % is the modulo operator. $VD_{offline}$ measures the distance between PS_t^* and PS_t^A.

$$VD = \frac{1}{n_{PS}} \cdot \left(n_{PS} \sum_{i=1}^{n_{PS}} d_i^2 \right)^{\frac{1}{2}} \quad (4.23)$$

where n_{PS} is $\left\| PS_t^A \right\|$ and d_i is the Euclidean distance of the ith member of the PS_t^A and the closest member of PS_t^*.

In [35], Zhang proposed three performance measures which deals with an algorithm's convergence ability and the diversity of solutions it produces. The first one is the convergent ratio CR which measures the consistency of the algorithm in tracking the Pareto front. It is based on the coverage metric proposed by Zitzler and Thiele [39] defined as

$$C(X,Y) = \frac{|\{ y \in Y \mid \exists x \in X, s.t., x \} \prec y|}{|Y|} \quad (4.24)$$

which computes the coverage ratio of the finite sets X and Y. Based on this, CR is defined as

$$CR = \frac{1}{KT} \sum_{i=1}^{T} \sum_{j=1}^{K-1} \frac{1}{K-j} \sum_{l=j+1}^{K} C(X_{ij}, X_{il}) \quad (4.25)$$

where K is the number of runs done by the algorithm and C is the coverage metric. The comparison made is between the simulation runs. A lower value for CR indicates the consistency of the algorithm in tracking the Pareto front but it does not provide an indication regarding its convergence to the true Pareto front. The second metric is the average density (AD) metric which measures how evenly distributed the generated solutions are for every problem change. AD is defined as

$$AD = \frac{1}{\tau} \sum_{i=1}^{\tau} \frac{1}{d'} \cdot \left(\frac{1}{n_{PF}} \sum_{i=1}^{n_{PF}} (d_i' - d)^2 \right)^{\frac{1}{2}} \quad (4.26)$$

$$d' = \frac{1}{n_{PF}} \sum_{i=1}^{n_{PF}} d_i', \quad (4.27)$$

where n_{PF} is $n_{PF} = \left| PF_t^A \right|$, d_i is the Euclidean distance between the ith member of generated Pareto set and the closes member of the true Pareto front. The third metric is average coverage scope (AS) measures the average coverage width of the generated solutions by the algorithm.

$$AS = \frac{1}{K\tau} \sum_{i=1}^{\tau} \sum_{i=1}^{K} \max_{1 \geq j, k \geq M} \left\{ \left\| f_{t,j}^A - f_{t,k}^A \right\|, f_{t,j}^A, f_{t,k}^A \in PS_t^A \right\} \quad (4.28)$$

where $f_{t,i}^A$ is the i-th objective in PF_t^A.

Tantar *et al.* [29] proposed the optimal subpattern assignment measure (OSPA) metric indexPerformance measure!OSPA which was originally used on the multiobject tracking problem. Consider two Pareto fronts A and B with cardinality m and n respectively. They defined the distance between two points x and y cut off by $c > 0$ as $d^{(c)}(x,y) = min\{c, d(x,y)\}$. To compare the two fronts, we need to determine which solutions from B are in the neighborhood of a given solution from A by computing all the permutations of solutions from B, $\Pi_{|B|}$. They proposed two criteria namely localization and cardinality:

$$M_{loc}(X,Y) := \left(\frac{1}{n} min_{\pi \in \Pi_n} \sum_{i=1}^{m} d^{(c)}\left(x_i, y_{\pi(i)}\right)^p \right)^{\frac{1}{p}} \tag{4.29}$$

$$M_{card}(X,Y) := \left(\frac{c^p(n-m)}{n} \right)^{\frac{1}{p}} \tag{4.30}$$

The localization criteria measures the quality of the coverage of A in relation to B while the cardinality penalty function which is only used when the two sets A and B have different cardinality. The advantage of this metric is that it is able to measure the coverage quality without the need for external reference points. The localization criteria has the disadvantage of computing the entire set of permutations but they have suggested the use of the Hungarian algorithm to estimate the measure in polynomial time.

4.4.2   Performance Measures for Problems with Unknown Pareto Fronts

In [5], Camara et al. proposed a performance measure called $acc_{unk}(t)$ indexPerformance measure!accuracy for dynamic multi-objective optimization problems with unknown Pareto front based on the accuracy measure proposed by Weicker [32] for single-objective dynamic environments computed as

$$accuracy_{F,EA}^{(t)} = \frac{F(best_{EA}^{(t)}) - min_F^{(t)}}{max_F^t - min_F^t}, \tag{4.31}$$

where f is the fitness function, $best_{EA}^{(t)}$ is the best solution in the population at time t while $max_F^{(t)}$ is the maximum fitness value in the search space while $mean_F^{(t)}$ is the minimum fitness value in the search space. Camara et al. adapted this measure as accuracy within a window or offset which provides a measure with the current approximate Pareto front and those with those past and future approximate Pareto front. The measure requires that the windows or phases are detected first and then the lengths for each of the windows will be calculated. Algorithm 1 shows the way on how the length for each window is calculated. The input is a set of N hypervolume values for the approximate Pareto fronts while the output is a set S of the lengths of each of the phases.

They noted that other suitable conditions could replace the condition for the if statement in line 6 of Algorithm 1. Once the windows and their lengths are

---

Algorithm 1 Calculation of lengths [5]

1: **for** $i = 2$ to N **do**
2:     $\triangle HV_i = HV_i - HV_{i-1}$
3: **end for**
4: $length = 1$
5: **for** $i = 2$ to N **do**
6:     **if** $\triangle HV_i \geq |\triangle HV_{i-1} + \triangle HV_{i+1}|$ **then**
7:         $S \leftarrow S \cup length$
8:         $length = 1$
9:     **else**
10:         $length = length + 1$
11:     **end if**
12: **end for**

---

calculated, the accuracy for each approximate front is measured. If the optimization problem is minimization, we compute the accuracy as

$$acc_{unk}^{minimizing}(t) = \frac{HV(Q_t)}{HV_{max}(Q_p)} \tag{4.32}$$

If the problem is maximization, we use the following formula:

$$acc_{unk}^{maximizing}(t) = \frac{HV_{min}(Q_p)}{HV(Q_t)}, \tag{4.33}$$

where $HV(Q_t)$ is the approximate Pareto front at time t in the phase being considered, $HV_{max}(Q_p)$ is the maximum hypervolume of the fronts in the phase being considered while $HV_{min}(Q_p)$ is the minimum. While this performance measure does not require that the Pareto front be known a priori, it was designed as an offline measure and does not take into account the diversity of solutions found.

Camara et al. in [5] adapted Weicker's [32] stability indexPerformance measure!stability and reactivity measures whose computation are dependent on the accuracy measure. A dynamic optimization algorithm is stable if its convergence ability is not severely affected by environment changes.

$$stab(t) = \begin{cases} stab(t) = acc(t) - acc(t-1), & \text{if } stab \geq 0 \\ 1, & otherwise \end{cases} \tag{4.34}$$

The reactivity metric indexPerformance measure!reactivity measures the ability of an optimization algorithm to react quickly to environmental changes and is computed as follows:

$$reac_\varepsilon(t) = min\left\{\left\{t' - t \mid t < t' \leq maxgen, t' \in N, \frac{acc(t')}{acc(t)} \geq (1 - \varepsilon)\right\} \cup \{maxgen - t\}\right\} \tag{4.35}$$

Most performance measures for dynamic optimization are designed for measuring convergence to the Pareto front at particular time samples. The convergence measures are the coverage measures ($\bar{c}_f(t)$), collective mean error of reversed generational distance (CME_{rGD}), collective mean error of hypervolume ratio (CME_{HVR}), variable space generational distance ($VD_{offline}$), coverage ratio (CR), localization criteria (M_{loc}) and the accuracy measure. On the other hand, the metric $\bar{c}_x(t)$ measures the convergence to the Pareto set. The other metrics are for measuring diversity of the generated solutions which includes the maximum spread (MS), average density (AD) and average coverage scope (AS). The other aspects of the performance of the optimization algorithm can be evaluated by using the stability and reactability metrics. These different classes of performance metrics indicate that convergence is not the only optimization goal of dynamic multi-objective optimization. Many performance measures require that the with known Pareto front. There is only one proposed performance measure for problems with unknown Pareto front but with the limitation of being an offline measure. Also, it requires the values of the maximum and minimum hypervolume for each window for each problem change which might not be practical to use in real world problems. Thus, we need to carefully consider which set of performance criteria best suits real-world optimization problems. It is important to know whether there is correlation between the chosen set of metrics such as convergence and diversity in terms of the quality of solution produced.

4.5   Dynamic Multi-objective Optimization Approaches

Most approaches in solving DMOPs focus on tracking the moving optima once change is detected in the problem landscape. Whenever a change is detected, it is often inefficient to restart the optimization process with new population. There are various approaches proposed in literature which are discussed in the next sections.

In dynamic optimization, it is important to maintain diversity in the population in order to improve the process of tracking the moving optima. There are three ways of handling population diversity are: diversity introduction, diversity maintenance and using multiple population. Aside from diversity maintenance technique, prediction-based and memory approaches are also being used to solve DMOPs.

4.5.1   Diversity Introduction

Convergence to the optimal region or solution is ideal when an optimization algorithm is solving a static optimization problem. However, convergence in dynamic optimization could lead to several problems. First, when the optimization algorithm has converged to one area in the problem landscape and the landscape change occurred in another area, it would not be possible to detect this change. Second, the optimization algorithm could find it difficult to find the global optima due to lack of diversity since it has already converge in a particular region in the landscape. One way of overcoming this problem is to introduce diversity whenever the problem changes.

The simplest way of introducing diversity is using random restart. Here, the entire population is randomly re-initialized if a problem change is detected. The problem with this approach is information loss since we cannot exploit the information given by the optimal solution found since they are all re-initialized. Also, optimization algorithms that use random restart requires time to converge again to the optimal solution but if the problem changes are fast, then this approach might not be able to find the optimal solution.

Another approach is hypermutation [6] proposed by Cobb which dramatically increases mutation whenever a problem change is detected. The step size of the increase is fixed which poses the problem of being unadaptive. Vavak *et al.* [31] proposed an adaptive mutation using variable local search (VLS) which only gradually increase mutation when there is no detected improvement. In order to control the size of mutation, they proposed a variable local search range by using the formula $2^{BITS} - 1$ where the value of $BITS$ adjusted during the optimization process. The problem approaches such as hypermutation and VLS is determining the correct mutation size or radius size. Taking too small mutation step size would be similar to random restart while too big step sizes would be like random search [18].

Deb *et al.* [8] extended the Nondominated Sorting Genetic Algorithm - II (NSGA-II) algorithm to handle a dynamic multi-objective problem specifically the dynamic hydrothermal power scheduling problem. Algorithm 2 outlines one iteration of DNSGAII.

Algorithm 2 Dynamic Nondominated Sorting Genetic Algorithm II (DNSGAII)

1: Evaluate all solutions in the child population Q_t
2: Test for changes in the environment
3: Randomly select and re-evaluate 10% of solutions from the parent population P_t
4: **if** there is a change in the environment then re-evaluate the entire parent population P_t
   then
5:    Combine C_t and P_t into R_t
6: **end if**
7: Apply nondominated sorting on R_t
8: Truncate R_t to the size of P_t by performing crowding sort
9: **if** there is a change in the environment then choose a percentage of solutions from R_t
   then
10:    Either replace the chosen solutions with random solutions or mutate the selected solutions
11:    Perform tournament selection
12:    Perform recombination and mutation to generate Q_{t+1}
13: **end if**

One of the modifications in the original algorithm is adding a way to detect problem changes by randomly re-evaluating 10% of the individuals in the population for each generation. Two strategies have been proposed whenever a problem change is detected. The algorithm can be restarted by introducing diversity using random initialization or mutation of a selected number of solutions from the population. The algorithm was tested on a two-objective dynamic problem and applied to the problem of dynamic hydrothermal power scheduling. The weakness of the approach is that it solved the problem as an offline optimization problem which assumes that the problem does not change for a period of time. The approach has only been shown to work on two problems and should be tested on different optimization problems.

In a dynamic multi-objective optimization algorithm using multi-population proposed by Goh and Tan [10], diversity introduction is performed whenever a problem change is detected by introducing random individuals as well as some competitor individuals from other sub-population into the other sub-population.

Diversity introduction approaches depend on the ability of the optimization algorithm to detect problem changes. Also, they may not effectively work when the problem changes are random, severe or fast.

4.5.2 Diversity Maintenance

Instead of introducing diversity only during restart, the diversity maintenance approaches aim to preserve diversity throughout the optimization process. Random immigrants could be used by introducing random individuals into the population only at fixed intervals during the optimization process. Unlike random restart, only a portion of the population are replaced. In [14], Grefenstette has shown empirically that random immigrants can improve the performance of evolutionary algorithms in handling dynamics.

Using diversity as an additional objective function to maintain diversity was proposed by Bui et al. [4]. This means we treat the single objective optimization problems as a multi-objective problem by adding a second objective which optimizing the diversity of the populations.

In [24], Sentinel Placement method is used to maintain diversity. A fixed number of sentinels are initialized with values from the different part of the search space. Since they are not removed, they are able to track changes in the landscape and they can be used in the reproduction process. It has been shown empirically that they perform better than random immigrants and hypermutation in handling severe and chaotic changes in the landscape [24].

Zeng et al. proposed Dynamic Orthogonal MOEA (DOMOEA) [34] to solve dynamic multi-objective optimization problem with continuous decision variables. Orthogonal design method is used to improve the fitness values of the population whenever the problem remains static. It treats every problem change as a new optimization problem. It exploits the past information by using the evolutionary results generated before the problem as the initial population of the new problem instance. It employs a linear crossover operator as its diversity maintenance scheme. One iteration in the algorithm is shown in Algorithm 3.

The drawback of this approach is that it can only be used for solving optimization problems with continuous decision variables. The orthogonal design method is mostly useful when the environment change is slow.

Algorithm 3 Dynamic Multi-Objective Evolutionary Algorithm based on Orthogonal Design (DOMOEA)

1: Evaluate all solutions in the child population Q_t
2: Test for changes in the environment
3: Randomly select and re-evaluate a percentage of solutions from the parent population P_t
4: **if** there is a change in the environment **then**
5:     Re-evaluate the entire parent population P_t and set it as P_{t+1}
6: **end if**
7: Perform the selection operation
8: Combine C_t and P_t into R_t
9: Truncate R_t to the size of P_t by performing crowding sort
10: Perform the crossover operation
11: Randomly select two solutions from R_{t+1}
12: **if** a random number is less than the crossover probability **then**
13:     Perform linear crossover
14: **else**
15:     Perform orthogonal crossover
16: **end if**

Maintaining diversity could perform well when the optimization problem has slow changes since it provides time for the optimization algorithm to converge. It might also be effective when there are severe changes in the problem landscape as there is better diversity. However, maintaining diversity over time could slow down the optimization process and may not be as effective when the problem at hand has only small changes.

4.5.3 Multiple Populations

In using a multiple population approach, parts of the search space is simultaneously explored by the different population in order to detect changes or the emergence of new optimal solutions. For instance, one population can be in charge for the current solutions while the other population would explore different regions. In here, one must decide what task to assign to each subpopulation as well as making sure that the subpopulations would not converge to the same location in the search space.

Goh and Tan [10] proposed Dynamic Competitive-Cooperative Coevolutionary Algorithm (dCOEA) that joins together the competitive and cooperative mechanisms to allow adaptive problem decomposition in solving both static and dynamic multi-objective optimization. Stochastic competitors track the moving optima. It exploits evolutionary results by storing them in a temporal memory. The algorithm is able to adapt to the changing problem landscape.

Greeff and Engelbrecht [13] proposed the use of Vector Evaluated Particle Swarm Optimiser (VEPSO) to solve dynamic multi-objective optimization problems. One swarm is dedicated to solve one objective function. VEPSO detects if a change in the environment occurs by re-evaluating sentry particles, which are a random number of particles in the swarm, before the next iteration. If their fitness values change then a new problem change is detected. If a change is detected, a portion of the the swarm's particle are reinitialized by changing their position and re-evaluating their personal and neighborhood best. The swarm share knowledge with each other through either a ring or random topology. The algorithm was tested on several test functions but was not compared with other algorithms. The authors will do further work on improving the performance of VEPSO by testing on wider range of functions.

Multi-populations have the advantage of being able to track multiple optima, recall previous optima, effective in solving multi-modal problems and can adequately adapt whenever the problem changes. However, there are disadvantages such as the number of sub-population could affect the performance of the optimization algorithm and setting the distance value to prevent the subpopulation could be an issue and as well the issue of managing recurrent optima.

4.5.4   Prediction-Based Approaches

Most approaches in solving DMOPs focus on tracking the moving optima once change is detected in the problem landscape. Whenever a change is detected, it is often inefficient to restart the optimization process with new population. Proposed approaches in tracking the moving optima involves exploiting the past information. Zeng *et al.* [34] proposed the dynamic orthogonal MOEA (DOMOEA) wherein it treats every change in the problem landscape as a new problem instance. However, they only exploited the past information by reintroducing the nondominated solutions found before the change is detected as the new initial population.

In [15], Hatzakis and Wallace proposed the Dynamic Queuing Multi-objective Optimizer (D-QMOO). The algorithm exploited the past information in order to predict the future behavior of the dynamic multi-objective optimization problem. In here, whenever there is a change in the landscape, the location of optimal solution is estimated by using an autoregressive model to predict the location of the optimal solutions. However, it was only tested on only one DMOP problem and its precision of prediction should be improved further.

In [22], Liu proposed Dynamic Multi-objective Evolutionary Algorithm with Core Estimation of Distribution (CDDMEA) that utilizes core estimation of distribution model to predict the Pareto optimal solution of the next environment. The algorithm was tested on two dynamic multi-objective problems and compared to another algorithm. The performance of the two algorithms were compared by visually inspecting the graph of the generated solutions. Also, it made of a U-measure to measure the diversity of the solutions generated by the two algorithms. The author claimed that CDDMEA performs better than the other algorithm. The use of core estimation of distribution model is a promising approach to predict the solutions to the next problem landscape but the experimental results need a more thorough investigation through the use more performance metrics and tested on various dynamic optimization problem.

Prediction-based methods could work very efficiently if their prediction is correct each time. But the problem with these types of approach depends on how well the predictors are trained. Like any learning process, there could be training errors due to the presence of wrong or inadequate training data.

4.5.5 Memory-Based Approaches

Another approach in dynamic optimization is to use memory schemes which implicitly or explicitly store relevant information from the current solution and environment and reusing these information at a later stage. This approach is specially useful when periodic changes occur in the environment.

An implicit memory scheme uses a redundant representation that can store more information than necessary. The most common implicit memory scheme used by evolutionary algorithm is diploidy [12, 25]. Other forms of implicit memory mechanisms were proposed in [7, 33].

In the case of explicit memory, information is stored in a memory bank separate from the population. Specific storage and retrieval techniques need to be defined for managing the stored information. Ramsey and Grefensette [27] used a case-based memory to store good solutions as well as their environment information. After a problem change, individuals stored in the memory with similar environments are used to reinitialize part of the population. Branke [1] proposed a memory mechanism that only stores information about the individuals and not the environment. The best individuals are stored in a finite memory. When it becomes full, a replacement strategy is used to select which individual will be replaced. Different replacement strategies to maintain diversity are discussed in [2]. Yang [33] introduced the associative memory scheme which stores the solution and a distribution estimate as individuals in a memory. After detecting a problem change, all individuals in the memory are evaluated. The distribution of the best individuals is used to sample for reinitializing part of the solution.

An algorithm based on artificial immune system was proposed by Zhang and Qian [36] called Dynamic Constrained Multi-objective Optimization Artificial Immune System (DCMOAIS) to solve dynamic constrained multi-objective optimization (DCMO) problems. It first runs a T-module to detect problem changes and creates the initial population by using information from previous results. After a problem change is detected, the B-module finds the Pareto optimal solution of the current environment. The M-module then stores the generated nondominated solutions into a repository to which the T-module will use to generate the initial population whenever a new environment change is detected. In the experimental results, the DCMOAIS performed well with low-dimensional DCMO problems has shown potential for DCMO problems with many constraints but needs to provide a more in-depth analysis.

Memory-based approaches are very useful in dynamic environments with periodic changes and may even help to prevent fast convergence. However, redundancy of information could become a problem and may not necessarily promote diversity. Aside from periodic changes, the memory might not be reused.

4.6   Summary and Future Works

Dynamic multi-objective optimization is an important field of research due to its promising application to a wide array of real-world problem involving multiple objectives with dynamic nature. In this chapter, we have provided a comprehensive definition of dynamic multi-objective optimization, reviewed dynamic multi-objective optimization classifications, test problems performance measures and optimization approaches.

Several classification of DMOPs have been proposed such as classes based on the characteristic of the Pareto set and front and sources of dynamism. A more detailed definition of DMOP have been provided in order to accommodate the different characteristic of dynamic multi-objective optimization problems such as dynamic constraints, changing number of parameters and objective functions. As we can see from the review, most test problems are focused on dynamic objective function evolution and mostly unconstrained problems. There is a need to develop a set of benchmark functions on dynamic multi-objective time-linkage problems, constrained problems, problems wit changing number of objective functions and constraints, and online dynamic optimization problems.

The performance measures discussed are intended towards evaluating the optimality of generated solutions and the behavior of the optimization algorithms. Most of them depend on the existence of the known Pareto front. The performance measure which work with unknown Pareto front may not be feasible to use in real world problems due to its need to know the maximum and minimum values of the parameters. New performance measures should be developed that are appropriate to certain classes of DMOPs. Since most evaluation of an optimization algorithm involves using a set of performance measures, studying the compatibility or suitability of using these measures together should be undertaken.

Even though there are many dynamic multi-objective optimization algorithms that have been developed, new algorithms should still be designed that could address the other characteristics of DMOP such as those that deal with time linkage problems, dynamic constrained problems and online dynamic problems. Although DMO algorithms have been applied to solve real world problems, they are not many. Research should be carried out in modeling more real world problems with dynamic features. Lastly, there is a lack of theoretical studies on dynamic multi-objective optimization such computational complexity analysis of algorithms. These studies are important in understanding which class of DMOPS are easy or difficult to solve.

Acknowledgements. This work was supported by the Engineering and Physical Sciences Research Council (EPSRC) of U.K. under Grant EP/E058884/1 and Grant EP/K001523/1. Xin Yao is supported by a Royal Society Wolfson Research Merit Award.

References

[1] Branke, J.: Memory enhanced evolutionary algorithms for changing optimization problems. In: Proc. 1999 IEEE Congr. Evol. Comput., pp. 1875–1882 (2005)

[2] Branke, J.: Evolutionary Optimization in Dynamic Environments. Kluwer Academic Publishers (2001)

[3] Branke, J., Kauler, T., Schmidth, C., Schmeck, H.: A multi-population approach to dynamic optimization problems. In: Proc. 4th Int. Conf. Adaptive Comput. Des. Manuf., pp. 299–308 (2000)

[4] Bui, L., Abbass, H., Branke, J.: Multiobjective optimization for dynamic environments. In: Proc. 2005 IEEE Congr. Evol. Comput., pp. 2349–2356 (2005)

[5] Cámara, M., Ortega, J., de Toro, F.: Performance measures for dynamic multi-objective optimization. In: Cabestany, J., Sandoval, F., Prieto, A., Corchado, J.M. (eds.) IWANN 2009, Part I. LNCS, vol. 5517, pp. 760–767. Springer, Heidelberg (2009)

[6] Cobb, H.: An Investigation into the use of hypermutation as an adaptive operator in genetic algorithms having continuous, time-dependent nonstationary environments. Technical Report, Naval Research Laboratory (1990)

[7] Dasgupta, D., Mcgregor, D.: Nonstationary function Optimization using the structured genetic algorithm. In: Proc. 2nd Int. Conf. Parallel Problem Solving from Nature, pp. 145–154 (1992)

[8] Deb, K., Rao N., U.B., Karthik, S.: Dynamic multi-objective optimization and decision-making using modified NSGA-II: A case study on hydro-thermal power scheduling. In: Obayashi, S., Deb, K., Poloni, C., Hiroyasu, T., Murata, T. (eds.) EMO 2007. LNCS, vol. 4403, pp. 803–817. Springer, Heidelberg (2007)

[9] Farina, M., Deb, K., Amato, P.: Dynamic multiobjective optimization problems: test cases, approximations, and applications. IEEE Trans. Evol. Comput. 8(5), 425–442 (2004)

[10] Goh, C., Tan, K.: A competitive-cooperative coevolutionary paradigm for dynamic multiobjective optimization. IEEE Trans. Evol. Comput. 13(1), 103–127 (2009)

[11] Goh, C.-K., Tan, K.C.: Evolutionary Multi-objective Optimization in Uncertain Environments. SCI, vol. 186. Springer, Heidelberg (2009)

[12] Goldberg, D., Smith, R.: Nonstationary function optimization using genetic algorithm with dominance and diploidy. In: Proc. 2nd Int. Conf. Genetic Algorithms and Their Applications, pp. 59–68 (1987)

[13] Greeff, M., Engelbrecht, A.: Solving dynamic multi-objective problems with vector evaluated particle swarm optimisation. In: Proc. 2008 IEEE Congr. Evol. Comput., pp. 2917–2924 (2008)

[14] Grefenstette, J.: Genetic algorithms for changing environments. In: Proc. Int. Conf. Parallel Problem Solving from Nature, pp. 137–144 (1992)

[15] Hatzakis, I., Wallace, D.: Dynamic multi-objective optimization with evolutionary algorithms: a forward-looking approach. In: Proc. 8th Annual Conf. Genetic and Evol. Comput., pp. 1201–1208 (2006)

[16] Helbig, M., Engelbrecht, A.: Archive management for dynamic multi-objective optimisation problems using vector evaluated particle swarm optimisation. In: Proc. 2011 IEEE Congr. Evol. Comput., pp. 2047–2054 (2011)

[17] Huang, L., Suh, I., Abraham, A.: Dynamic multi-objective optimization based on membrane computing for control of time-varying unstable plants. Inf. Sci. 181(11), 2370–2391 (2011)

[18] Jin, Y., Branke, J.: Evolutionary optimization in uncertain environments-a survey. IEEE Trans. Evol. Comput. 9(3), 303–317 (2005)

[19] Jin, Y., Sendhoff, B.: Constructing dynamic optimization test problems using the multi-objective optimization concept. In: Raidl, G.R., et al. (eds.) EvoWorkshops 2004. LNCS, vol. 3005, pp. 525–536. Springer, Heidelberg (2004)

[20] Koo, W., Goh, C., Tan, K.: A predictive gradient strategy for multiobjective evolutionary algorithms in a fast changing environment. Memetic Computing 2, 87–110 (2010)

[21] Li, X., Branke, J., Kirley, M.: On performance metrics and particle swarm methods for dynamic multiobjective optimization problems. In: Proc. 2007 IEEE Congr. Evol. Comput., pp. 576–583 (2007)

[22] Liu, C.: New dynamic multiobjective evolutionary algorithm with core estimation of distribution. In: Proc. 2010 Int. Conf. Electrical and Control Engineering, pp. 1345–1348 (2010)

[23] Mehnen, J., Wagner, T., Rudolph, G.: Evolutionary optimization of dynamic multi-objective test functions. In: Proc. 2nd Italian Workshop on Evol. Comput. (2006)

[24] Morrison, R.: Designing evolutionary algorithms for dynamic environments. Springer (2004)

[25] Ng, K., Wong, K.: A wew diploid scheme and dominance change mechanism for non-stationary function optimization. In: Proc. 6th Int. Conf. Genetic Algorithms, pp. 159–166 (1995)

[26] Nguyen, T.: Continuous dynamic optimisation using evolutionary algorithms. PhD Thesis, University of Birmingham (2011)

[27] Ramsey, C., Grefenstette, J.: Case-based initialization of genetic algorithms. In: Proc. 5th Int. Conf. Genetic Algorithms, pp. 84–91 (1993)

[28] Schaffer, J.: Multiple objective optimization with vector evaluated genetic algorithms. In: Proc. 1st Int. Conf. on Genetic Algorithms, pp. 93–100 (1985)

[29] Tantar, A., Tantar, E., Bouvry, P.: A classification of dynamic multi-objective optimization problems. In: Proc. 13th Annual Conf. Genetic and Evol. Comput., pp. 105–106 (2011)

[30] Van Veldhuizen, D.: Multiobjective evolutionary algorithms: classifications, analyses, and new innovations. PhD Thesis, Air Force Institute of Technology (1999)

[31] Vavak, F., Jukes, K., Fogarty, T.: Adaptive combustion balancing in multiple burner boiler using a genetic algorithm with variable range of local search. In: Proc. 7th Int. Conf. Genetic Algorithms, pp. 719–726 (1997)

[32] Weicker, K.: Performance measures for dynamic environments. In: Guervós, J.J.M., Adamidis, P.A., Beyer, H.-G., Fernández-Villacañas, J.-L., Schwefel, H.-P. (eds.) PPSN 2002. LNCS, vol. 2439, pp. 64–73. Springer, Heidelberg (2002)

[33] Yang, S.: Associative memory scheme for genetic algorithms in dynamic environments. In: Rothlauf, F., et al. (eds.) EvoWorkshops 2006. LNCS, vol. 3907, pp. 788–799. Springer, Heidelberg (2006)

[34] Zeng, S., Chen, G., Zheng, L., Shi, H., de Garis, H., Ding, L., Kang, L.: A dynamic multi-objective evolutionary algorithm based on an orthogonal design. In: Proc. 2006 IEEE Congr. Evol. Comput., pp. 573–580 (2006)

[35] Zhang, Z.: Multiobjective optimization immune algorithm in dynamic environments and its application to greenhouse control. Appl. Soft Comput. 8(2), 959–971 (2008)

[36] Zhang, Z., Qian, S.: Artificial immune system in dynamic environments solving time-varying non-linear constrained multi-objective problems. Soft Comput. 15(7), 1333–1349 (2011)
[37] Zhou, A., Jin, Y., Zhang, Q., Sendhoff, B., Tsang, E.: Prediction-based population re-initialization for evolutionary dynamic multi-objective optimization. In: Obayashi, S., Deb, K., Poloni, C., Hiroyasu, T., Murata, T. (eds.) EMO 2007. LNCS, vol. 4403, pp. 832–846. Springer, Heidelberg (2007)
[38] Zitzler, E., Deb, K., Thiele, L.: Comparison of multiobjective evolutionary algorithms: Empirical results. Evol. Comput. 8(2), 173–195 (2000)
[39] Zitzler, E., Thiele, L.: Multiobjective evolutionary algorithms: a comparative case study and the strength Pareto approach. IEEE Trans. Evol. Comput. 3(4), 257–271 (1999)

Part II
Algorithm Design

Chapter 5
A Comparative Study on Particle Swarm Optimization in Dynamic Environments

Changhe Li and Shengxiang Yang

Abstract. Particle swarm optimization (PSO) has been shown an effective optimization tool since it was proposed. Due to the efficiency of locating optima, it has been widely applied to solve dynamic optimization problems (DOPs) with variant enhancements, e.g., diversity maintaining schemes, memory schemes, multi-population schemes, adaptive schemes, and hybrid schemes. In this chapter, we categorize and review approaches proposed based on PSO for DOPs. Weaknesses and strengths of those approaches are also discussed in this chapter. In order to investigate the performance of those approaches, a set of typical algorithms based on PSO are chosen to compare their performance on the moving peaks problem (MPB). According to the comparison results, suggestions are also given regarding future algorithms development for DOPs.

5.1 Introduction

Over the last two decades, dynamic optimization problems (DOPs) have drawn many researchers from the swarm intelligence (SI) community. Especially, in recent years, there has been a growing interest in studying particle swarm optimization (PSO) for DOPs in terms of the number of publications. In dynamic environments, the aim is not only to locate the global optima but also to track the trajectory of changing optima. Although PSO has been shown an efficient method for stationary optimization problems, like other evolutionary algorithms (EAs), there exist

Changhe Li
School of Computer Science, China University of Geosciences, 388 Lumo Road,
Wuhan 430074, China
e-mail: changhe.lw@gmail.com

Shengxiang Yang
Centre for Computational Intelligence (CCI), School of Computer Science and Informatics,
De Montfort University, The Gateway, Leicester LE1 9BH, U.K.
e-mail: syang@dmu.ac.uk

S. Yang and X. Yao (Eds.): *Evolutionary Computation for DOPs*, SCI 490, pp. 109–136.
DOI: 10.1007/978-3-642-38416-5_5   © Springer-Verlag Berlin Heidelberg 2013

difficulties to apply it to solve DOPs. The difficulties lie in two aspect: outdated memory due to environment dynamism and diversity loss due to convergence [3]. For the first difficulty, it can be easily solved by re-evaluating particles over time. However, it is hard to solve the diversity loss issue due to the difficulty in balancing the exploration and exploitation during the search progress. Hence, to address the diversity loss issue, different kinds of approaches have been proposed to enhance the performance of PSO in dynamic environments, e.g., diversity maintaining schemes [5, 13], multi-population schemes [12, 15, 18, 25, 33, 39, 52], adaptive schemes [14, 17, 26, 32, 33, 45], hybrid schemes [13, 15, 16, 22, 36, 37, 40, 51], and memory schemes [49].

Although the performance of PSO has been improved by using these proposed approaches, there is no comprehensive study of these approaches in dynamic environments with different change properties, e.g., severe changes, hard-to-detect changes, and some complicated changes, including the number of optima change and dimensional change. This chapter reviews approaches proposed based on PSO for DOPs and they are categorized as well. Strengths and weaknesses of these approaches are also discussed. To compare the performance of algorithms, some algorithms, belongingto different categories, are selected and tested on the moving peaks problem (MPB) [6].

The rest of this chapter is organized as follows. The related researches are reviewed in Section 5.2, including the PSO algorithm, different kinds of enhancements for PSO in dynamic environments, e.g., diversity maintaining schemes, multi-population schemes, adaptive schemes, and hybrid schemes. Section 5.4 provides the comparison results of the involved algorithms on the MPB problem. Finally, conclusions and future research on PSO for DOPs are given in Section 5.5.

5.2   PSO in Dynamic Environments

5.2.1   Particle Swarm Optimization

Similar to other EAs, PSO is a population based stochastic optimization technique. A potential solution in the fitness landscape is called a particle in PSO. Each particle i is represented by a position vector $\mathbf{x}_i$ and a velocity vector $\mathbf{v}_i$, which are updated as follows:

$$v_i'^d = \omega v_i^d + \eta_1 r_1 (x_{pbest_i}^d - x_i^d) + \eta_2 r_2 (x_{gbest}^d - x_i^d) \tag{5.1}$$

$$x_i'^d = x_i^d + v_i'^d, \tag{5.2}$$

where $x_i'^d$ and x_i^d represent the current and previous positions in the d-th dimension of particle i, respectively, v'_i and v_i are the current and previous velocity of particle i, respectively, $\mathbf{x}_{pbest_i}$ and $\mathbf{x}_{gbest}$ are the best position found by particle i so far and the best position found by the whole swarm so far, respectively, $\omega \in (0, 1)$ is an inertia weight, which determines how much the previous velocity is preserved; η_1 and η_2

are the acceleration constants, and r_1 and r_2 are random numbers generated in the interval $[0.0, 1.0]$ uniformly.

Ever since PSO was first introduced, several major versions of the PSO algorithm have been developed [43]. There are two main models of the PSO algorithm, called *gbest* (global best) and *lbest* (local best), respectively. The two models differ in the way of defining the neighborhood for each particle. In the *gbest* model, the neighborhood of a particle consists of the particles in the whole swarm, which share information between each other. On the contrary, in the *lbest* model, the neighborhood of a particle is defined by several fixed particles. The two models give different performances on different problems. It was pointed out [23, 43] that the *gbest* model has a faster convergence speed but also has a higher probability of getting stuck in local optima than the *lbest* model. On the contrary, the *lbest* model is less vulnerable to the attraction of local optima, but has a slower convergence speed than the *gbest* model. In order to give a standard form for PSO, Bratton and Kennedy proposed a standard version of PSO in [8]. In the standard PSO, a local ring population topology is used and the experimental results have shown that the *lbest* model is more reliable than the *gbest* model on many test problems.

5.2.2   PSO in Dynamic Environments

Due to the efficiency of PSO in stationary environments as well as its simpleness, PSO has been widely applied to solve DOPs with enhancements in some perspective. Many ideas have been proposed for PSO to suit different kinds of DOPs. Among these approaches, the population diversity is an inevitable issue in dynamic environments regardless what kind of the approach is used. In this section, the major ideas are categorized into different groups in terms of their main characteristics rather than how to solve the population diversity issue, and they are reviewed as follows.

5.2.2.1   Diversity Maintaining Schemes

The basic idea of diversity maintaining schemes is to maintain the population diversity throughout the run by 1) regularly introducing random individuals into population [29]; 2) designing specialized particles to focus on diversity maintaining [3–5, 11, 16]; 3) increasing diversity after a change by employing the idea of mutation from genetic algorithms (GAs) [13, 36, 37, 53, 55].

The scheme of introducing random individuals is also called random immigrants scheme. This scheme has been widely applied in GAs for solving DOPs. Due to the disadvantage of slow search, it is not studied in PSO until a general diversity maintaining scheme was proposed by Li and Yang in [29] for the MPB problem. In the paper, the random immigrants scheme is triggered when the total number of individuals decreases to a certain level, which is determined by the number of peaks in the MPB problem.

For the second idea, a typical example is a charged PSO (mQSO) model proposed by Blackwell *et al.* [3–5], which is inspired from the atom field, to prevent particles to get too closed to each other. In this mechanism, each swarm is comprised of a nucleus and a cloud of charged particles which are responsible to maintain diversity. There is a repulsion among charged particles to keep them from approaching near to each other. Thereafter, an enhanced version of mQSO was proposed by applying two heuristic rules to further enhance the diversity of mQSO in [11]. One of the two rules is to increase the number of quantum particles and decrease the number of trajectory particles when a change occurs. The other rule is to re-initialize or pause the swarms that have bad performance. The quantum concept was also used in [16], where after a change in environment is detected, some particles change their role from standard particles to quantum particles.

Using the mutation idea from GAs is also studied in PSO for maintaining the populations diversity. In [13], a new multi-strategy ensemble PSO (MEPSO) for DOPs was proposed. In MEPSO, all particles are assigned into two groups. Differential mutation operators are applied in one of the two groups of population to maintain the diversity for exploring promising areas. A collaborative evolutionary swarm optimization (CESO) was proposed in [36]. The diversity is maintained by a population that uses the crowding differential evolution (CDE) algorithm [48]. An algorithm based on the similar idea in CESO, named ESCA, was proposed in [37]. An adaptive mutation operator was developed to maintain the population diversity as well in a cooperative dual-swarm PSO (CDPSO) algorithm [53].

A novel adaptive mutation PSO (AMPSO) algorithm based on fuzzy matter-element analysis for generalized dynamic constraints satisfaction (GDCS) was presented to resolve the coupling domain level and knowledge level constraints in [55]. In AMPSO, the mutation mechanics is introduced to mutate inactive particles and particles with the smallest fitness according to mutation probability, which is intended to make the algorithm converge faster and respond better to changes in dynamic environments.

There are also some other specialized diversity maintaining methods for PSO. In [10], both the particle selection and replacement mechanisms are modified so that the most diversified particles (measured by the Hamming distance) are selected and the particles that have close locations are replaced. In the compound PSO [34], the degree of particles deviating from their original directions becomes larger when the velocities becomes smaller, and distance information is incorporated as one of the criteria to choose a particle for the update mechanism.

5.2.2.2 Adaptive Schemes

To cope with changes in dynamic environments, many adaptive methods are integrated into PSO with the aim of balancing diversity and convergence. The adaptive techniques roughly fall into two categorizes: 1) adaptively assigning the neighborhood of each particle [31, 33, 35, 41]; and 2) adaptively tuning parameters for PSO [14, 44].

An early attempt on adaptive PSO for DOPs is the adaptive PSO in [17] where different environment detection and response methods are investigated on the parabolic and Rosenbrock benchmark functions. An adaptive particle swarm for solving the vehicle routing problem with dynamic requests was presented in [26]. An adaptive memory idea was developed to store good solutions, which are adaptively updated throughout the search progress. When a change occurs, particles are re-initialized by random generation in the search space. However, the *pbest* position of a randomly initialized particle will be preserved by the position in the previous environment rather than uniformly replaced by its corresponding location of x if it is better than the randomly generated position of the particle.

To adaptively organize the neighborhood of particles, many PSO variants have been proposed where particles are adaptively grouped into sub-swarms in each iteration according to certain principles based on their fitness and distance. The first work of this kind of research is the speciation based PSO (SPSO) [31, 41] proposed by Parrott and Li. In SPSO, the number and size of swarms is adaptively adjusted by constructing an ordered list of particles, ranked according to their fitness, with spatially close particles joining a particular species. Recently, a similar idea, called PSO-CP, was proposed in [35]. Instead of the "fittest first" principle in SPSO [31, 41], swarms are constructed based on a "worst first" principle, but each composite particle consists of three fixed particles. Interestingly, thereafter, an opposite version of PSO-CP was proposed in [33], where a composite particle (a sub-swarm) is generated based on the "fittest-oriented" principle.

An adaptive PSO algorithm was proposed in [44]. In the proposed algorithm, exclusion radios and inertia weight are adaptively adjusted by the FCM mechanism and a local search scheme on the best swarm is employed to accelerate the search progress. To encourage local search or global search for different swarms, which are ranked decreasingly according to the fitness of the best particles, an adaptive inertia weight is adopted by the swarm ranking and swarm evolutionary progress. As a result, the last swarm, which is the best swarm, will focus on local search by using the smallest inertia weight, while the first swarm (the worst swarm) will use a large value of inertia weight for exploring promising areas globally.

To adapt PSO in dynamic and noisy environments, an improved mQSO [5] algorithm with an evaporation mechanism, denoted mQSOE, was proposed in [14], in which the evaporation mechanism is implemented by reducing the fitness value of the best position found by each particle along time. In the study, the difficulty in applying this idea is that a high evaporation factor produces a fast adaptation but the particles can not reach a good solution in the optimization process. While, a low evaporation factor achieves a fast convergence with high quality solutions, but the adaptation is very low. In order to address this issue, an adaptive evaporation factor was proposed using the particles' velocity and the difference between the fitness of its *pbest* position and the fitness of its current position.

5.2.2.3   Hybrid Schemes

Hybrid schemes are commonly used to enhance the diversity for PSO in dynamic environments, which can be roughly divided into three categories: 1) hybridization of swarm topologies [52]; 2) hybridization with other domain knowledge [16, 22, 51]; 3) hybridization with other meta-heuristic methods [13, 15, 36, 37, 51, 53].

Two different population topologies are used in two sub-swarms in [52], where the two sub-swarms exchange their best particle at checkpoints. One sub-swarm is used for searching the global optimum and the other is responsible for local search and diversity maintaining.

A hybrid model of PSO with cellular automata was proposed in [16] to address DOps. In the proposed model, a population is split by cells of cellular automata embedded in the search space. Each cell of cellular automata has a specified number of particles. The quantum particles introduced in [5] are also applied in order to find new local optima quickly. The hybrid characteristic of a particle is implemented by changing their role from the normal particles to quantum particles for a few iterations after a change happens.

Inspired from the microbial life, the particles in [22] can reproduce infants and the old ones die. By using the quadratic interpolation method, the infants are able to be reproduced by high quality particles. The algorithm is adapted to perform in continuous search spaces by using Euclidian norm to define the neighborhood in the reproduction procedure. The performance of the proposed approach was compared with some other heuristic optimization algorithms. The results indicate a better performance of the proposed algorithm than other tested algorithms in terms of real-time error, offline performance, and offline error.

In [51], a memetic algorithm with a fuzzy cognition local search method was proposed for improving the quality of individuals. In the algorithm, a self-organized random immigrants scheme is also used to enhance the exploration capacity in a local version of PSO with the ring topology.

A multi-strategy ensemble PSO (MEPSO) was proposed in [13]. In MEPSO, all particles are divided into two groups, which use Gaussian local search and differential mutation to exploit local optima and to explore new promising areas, respectively. To enhance the local search ability, a particle has a probability (p) to perform the Gaussian local search and a probability of $1 - p$ to play the role of the conventional particle.

A collaborative evolutionary swarm optimization (CESO) was proposed in [36]. In CESO, two swarms, which use the crowding DE (CDE) [48] and the PSO model, respectively, cooperate with each other by a collaborative mechanism. The swarm using CDE is responsible for preserving diversity while the PSO swarm is used for tracking the global optimum. The competitive results were reported in [36]. Thereafter, a similar algorithm, called evolutionary swarm cooperative algorithm (ESCA), was proposed in [37] based on the collaboration between a PSO algorithm and an EA. In ESCA, three populations using different EAs are used. Two of them follow the rules of CDE [48] to maintain the diversity. The third population uses the rules of PSO. Three types of collaborative mechanisms are also developed to transmit information among the three populations.

Like the ideas in CESO [36] and ESCA [37], a cooperative dual-swarm PSO (CDPSO) was proposed [53] to deal with DOPs. CDPSO adopts a dual-swarm structure to maintain the swarm diversity and track the changing optima. A fractional global best formation technique is employed to construct artificial global bests which are potential to be better. An adaptive mutation operator is also developed to maintain the population diversity as well. A centralized cooperative strategy, which is based on tabu search, was proposed for tracking the changing optima in [15].

5.2.2.4   Multi-population Schemes

Branke *et al.* proposed a self-organizing scouts (SOS) [7] algorithm, which is an early attempt of a new kind of multi-population methods. Although it is based on genetic algorithms (GAs), it is worth introducing the SOS algorithm here as it is a typical example of using multi-population methods. In SOS, the whole population is composed of a parent population that searches through the entire search space and child populations that track local optima. The parent population is regularly analyzed to check the condition for creating child populations, which are split off from the parent population.

Inspired by a forking mechanism, a multi-swarm optimization algorithm was proposed in [50]. Similar to the SOS algorithm [7], in the multi-swarm algorithm, a larger main swarm is continuously responsible for exploring new peaks and a number of smaller child swarms, split off from the main swarm, are used to track the achieved peaks over the whole run.

Inspired by the SOS algorithm [7], a fast multi-swarm optimization (FMSO) algorithm was proposed in [27] to locate and track multiple optima in dynamic environments. In FMSO, a parent swarm is used as a basic swarm to detect the most promising area when the environment changes, and a group of child swarms are used to search the local optimum in their own sub-spaces. Each child swarm has a search radius, and there is no overlap among all child swarms since they exclude from each other. If the distance between two child swarms is less than their radius, then the whole swarm of the worse one is removed. This guarantees that no more than one child swarm will cover a single peak. Another similar idea of hibernation multi-swarm optimization algorithm (HmSO) was introduced in [21], where a child swarm will hibernate if it is not productive anymore and will be waken up if an environmental change has been detected.

Swarms are dynamic and the size of each swarm is small in [19]. The whole population is divided into many small sub-swarms. The sub-swarms will be regrouped frequently by using different regrouping schemes and information is exchanged among the sub-swarms. Some accelerating operators is apllied to improve local search ability. A change also needs to be detected and adjustments will be performed once changes are detected.

An atomic swarm approach has been adapted by Blackwell and Branke [4, 5] to track multiple optima simultaneously with multiple swarms in dynamic environments. In their approach, a charged swarm is used for maintaining the diversity of the swarm, and an exclusion principle ensures that no more than one swarm surrounds a single peak. In the algorithm, called mQSO in [5], anti-convergence is introduced to detect new peaks by sharing information among all sub-swarms. This strategy was experimentally shown to be efficient for the MPB function [6]. Borrowing the idea of exclusion from [4], Mendes and Mohais developed a multi-population DE algorithm (DynDE) [38] to solve the MPB problem. In their approach, a dynamic strategy for the mutation factor F and probability factor CR in DE was introduced. Recently, an enhanced version of mQSO was proposed by applying two heuristic rules to further enhance the diversity of mQSO in [11]. One of the two rules is to increase the number of quantum particles and decrease the number of trajectory particles when a change occurs. The other rule is to re-initialize or pause the swarms that have bad performance. To increase the performance of mCPSO [5], two strategies were proposed in [40], one of which divides each swarm into two groups depending on the quality of the particles for addressing the loss of diversity, and the other control the number of active swarms during the run using a fuzzy rule. In the approach, swarms with a bad behavior and a certain level of convergence are stopped by using an adaptive and fuzzy rule to increase the efficiency of mCPSO.

Parrott and Li developed the SPSO algorithm [31, 41], which dynamically adjusts the number and size of swarms by constructing an ordered list of particles, ranked according to their fitness, with spatially close particles joining a particular species. At each generation, SPSO aims to identify multiple species seeds within a swarm. Once a species seed has been identified, all the particles within its radius are assigned to that same species. Parrott and Li also proposed an improved version with a mechanism to remove redundant duplicate particles in species in [42]. In [1], Bird and Li developed an adaptive niching PSO (ANPSO) algorithm which adaptively determines the radius of a species by using the population statistics. Based on their previous work, Bird and Li introduced another improved version of SPSO using a least square regression (rSPSO) in [2]. Recently, in order to determine niche boundaries, a vector-based PSO [47] algorithm was proposed to locate and maintain niches by using additional vector operations.

A clustering PSO (CPSO) algorithm has recently been proposed for DOPs in [28, 54]. CPSO applies a hierarchical clustering method to divide an initial swarm into sub-swarms that cover different local regions. CPSO was proposed to attempt to solve some challenging issues associated with multi-population methods. For example, how to guide particles to move toward different promising sub-regions and how to determine the radius of sub-swarms. CPSO has shown some promising results in comparison with several state-of-the-art algorithms in [54]. Recently, Li and Yang [29] proposed a general framework for multiple population methods in undetectable dynamic environments based on the clustering method used in in [28, 54]. An algorithm called clustering PSO with restart (CPSOR) was implemented based on PSO technique. The CPSOR algorithm shows an superior performance compared with other algorithms especially in dynamic environments where changes are hard to detect. Although CPSOR is able to solve DOPs without change detection, it is not able to adapt populations with environmental changes.

Recently, Li et al. [30] proposed another PSO algorithm based on the clustering idea, called adaptive clustering PSO (ACPSO), which is able to make populations adaptable with changes, especially for the number of populations needed. In order to find a proper moment to increase the population diversity, a special technique is proposed by monitoring the decreasing ratio of the number of survived sub-populations. In order to deal with DOPs with complicated changes, e.g., changes of the number of peaks, a novel idea is introduced to figure out the proper number of populations that are really needed when a change happens. The idea is to compare the number of survived populations in the current and previous diversity increasing points. If the number of survived populations in the current check point is larger than that in the previous check point, the total number of individuals will be increased; otherwise, the total number of individuals will be decreased. By using these methods, ACPSO is able to adaptively maintain the population diversity in a proper moment with a proper number of individuals.

A cultural framework was introduced in [10] for PSO where it defines five different kinds of knowledge, named situational knowledge, temporal knowledge, domain knowledge, normative knowledge and spatial knowledge. The information is used to detect changes. Once a change is detected, a diversity based repulsion mechanism is applied among particles as well as a migration strategy among swarms. The knowledge also helps in selecting the leading particles at the personal, swarm, and global level.

Recently, a similar algorithm to SPSO [42], called PSO with composite particles (PSO-CP), was proposed in [35]. In PSO-CP, the whole swarm is partitioned into a set of composite particles by a "worst first" principle, which is opposite to the idea in SPSO [42]. The members of each composite particle is fixed by three particles. Inspired by the composite particle phenomenon in physics, the elementary members in each composite particle interact via a velocity-anisotropic reflection scheme to integrate valuable information. The diversity of each composite particle is maintained by a scattering operator and an integral movement strategy is also introduced to promote the swarm diversity. In [33], a novel version of interactions among particles was proposed based on the previous work of composite model [34, 35]. It creates each composite particle by one fitter particle from the swarm and other two particles randomly generated in its neighborhoods. In order to integrate valuable information for searching the changed optima, a scatter factor is introduced into the velocity-anisotropic reflection (VAR) scheme and a "fitness-and-distance" based pioneer particle identification (PPI) method is introduced. In addition, the composite particles interact with other particles in the swarm using an integral movement strategy, which aims to enhance the diversity of the swarm.

In [24], a multi-environmental cooperative model for parallel meta-heuristics was proposed to handle DOPs that consists of different sub-problems or environments. A parallel multi-swarm approach is used to dealing with different environments at the same time by using different algorithms that will exchange information obtained from these environments. The multi-swarm model was tested on a set of dynamic vehicle routing problems.

Note that many algorithms from other schemes can also be classified into this category in terms of the number of swarms that is used, such as algorithms in [14, 33, 35, 45] from adaptive schemes, algorithms in [5, 29, 36, 37, 53] from diversity maintaining schemes, and algorithms in [13, 15, 16, 36, 37, 52] from hybrid schemes. As they have been introduced before, here we do not review these methods again.

5.2.2.5 Memory Schemes

Memory schemes have been rarely studied in PSO as each particle is able to memorize the best position found so far in its corresponding *pbest* position. Wang *et al.* introduced a triggered generator for a memory-based PSO in [49], where the re-initialization of a population will be immediately triggered once a peak has been found. It was reported that the triggered generator can be more efficient in exploring the search space than the simple re-initialization method, especially when the environment does not change frequently.

5.3 Discussions and Suggestions

5.3.1 Issues with Current Schemes

To solve DOPs, one inevitable issue must be considered, which is how to handle the dynamism in changing environments. So far, most PSO variants developed for DOPs use some change detection methods [5, 28, 31, 36, 37, 46, 54]. One common detection method is to monitor the fitness of one or several detectors (i.e., solution(s) in the search space). A change is detected if the fitness of the detector(s) changes. Once a change has been detected, different kinds of schemes can be applied to increase the diversity. However, in order to use these strategies efficiently, a condition must be applied, that is, the environmental changes must be successfully detected. Hence, a general issue of these detection based schemes is that what the efficiency will be if they fail to detect a change, such as in the dynamic environments where only parts of the fitness landscape changes.

5.3.1.1 Issues with Diversity Maintaining Schemes

Maintaining diversity throughout the runtime is a common idea to deal with changes in dynamic environments. This kinds of methods may be good at solving problems with severe changes as high diversity is maintained over changes. However, the issues below should be addressed.

Firstly, the most important issue is when to increase the population diversity. Increasing diversity too often is not effective because the continuous focus on diversity slows down the optimization process as pointed out in [20], especially for the first (i.e., regularly introducing random individuals) and second approach (i.e., designing specialized particles) of maintaining diversity, described in Section 5.2.2.1. Individuals, which are in explorative status, will be eliminated due to new random immigrants. However, the local areas where they are searching are not sufficiently exploited. On the other hand, not frequently increasing the population diversity will lead to lack of diversity for algorithms to locate and track the global optimum in changing environments. The ideal moment to increase diversity is when changes happen. However, the change detection issue comes again.

Secondly, the magnitude of diversity to be introduced should also be considered. Here, the magnitude of diversity is determined by the number of random immigrants to be introduced. DOPs can be taken as a series of stationary problems. Therefore, we can take the magnitude of diversity to be increased in a specific environment as the initial population size for a stationary problem to be solved. It is related to the structure of the current environments, and is changeable as the environment changes. For example, the magnitude will change if the number of peaks changes in the MPB problem [6].

Thirdly, for the second diversity maintaining approach, i.e., designing specialized particles to maintain diversity, the problem is how to design effective rules for those specialized particles to maintain the diversity. In addition, it is inevitable to involve extra computational cost due to the function of specialized individuals.

5.3.1.2   Issues with Adaptive Schemes

The aim of adaptive schemes is to adaptively maintain the population diversity by using meta-heuristics to adjust the behavior of optimization algorithms. If effective meta-heuristics are designed, this kind of approaches will be a good choice for DOPs. However, there are several issues for current adaptive schemes in the literature.

Firstly, in order to use heuristic rules, the aid of change detection is needed for many adaptive methods, e.g., algorithms in [5, 17, 26, 31, 41]. Secondly, it is hard to develop effective heuristic rules to deal with changes. Although different heuristics have been proposed, e.g., the inertial weight in [44] and evaporation factor in [14], they still suffer from some difficulties.

5.3.1.3   Issues with Hybrid Schemes

Although the aim of optimization algorithms is to locate the global optimum, different methods have different characteristics. For example, algorithms based on the idea of mutation are biased to explore the search space. While algorithms based on PSO will prefer to exploit in a local area. Therefore, it is easy to work out the idea of hybridization of methods that have different search behaviors to solve DOPs. Maintaining diversity is a key issue for solving DOPs. Therefore, in hybrid schemes, methods that are able to maintain the population diversity are considered on the one hand. On the other hand, methods that have fast convergence speed are also considered. And hybrid schemes are normally implemented by multiple populations. Hence, the major issues with hybrid schemes are how to choose effective methods that have different search behaviors, and how to organize them for co-operation during the search.

5.3.1.4   Issues with Multi-population Schemes

Multiple population methods can be roughly classified into three categories according to the *way to create multiple populations*. The first category simply uses a certain number of randomly generated populations. Simply, all populations use the same evolutionary strategy with the same number of individuals. However, to enhance the capability of locating and tracking the global optimum, in many algorithms of this category, the populations are assigned into different groups where each group uses different search methods (e.g., ESCA [37], CESO [36], mQSO [5] and algorithms proposed recently [10, 24, 52, 52, 53]) to serve different purposes (e.g., exploring new promising sub-areas or exploiting local optima). The second category of multi-population methods (e.g., SOS [7], NichePSO [9], FMSO [27], HmSO [21], and FPSO [50]) starts from a main population and maintains it to generate subpopulations by splitting off from the main population when some predefined criteria are satisfied (e.g., the best individual in the main population does not improve for a certain number of iterations). The third category of multi-population methods divides a large randomly generated population into several small sub-populations to make them cover different sub-areas in the search space. For example, SPSO [31], PSO-CP [35], CPSO [28, 54], and multi-swarm accelerating PSO (MSA-PSO) [19] algorithms belong to this category.

Among the three categories of multi-population schemes, the third one seems to be the best in terms of the aim of multi-population methods, which is to divide the search space into different sub-areas and to locate and track multiple peaks in parallel. From this point of view, only the third category is able to create multiple populations without overlapping between them.

The major common problem of these multi-population methods is that, although they improve the performance in locating and tracking multiple optima, they also bring in new issues which are difficult to solve, as discussed below.

It is difficult to define a proper number of populations that is needed in a specific environment. This is because the proper number of populations is determined by the number of peaks (optima) in the fitness landscape. Generally speaking, the more peaks distribute in the fitness landscape, the more populations are needed. Many experimental studies [5, 38, 54] have shown that the optimal number of populations is equal to the number of peaks in the fitness landscape in the MPB problem [6] with ten peaks. However, we should assume that the number of peaks in the fitness landscape is unknown while solving a problem even though it is known for some artificially created problems, e.g., the MPB problem.

How to define a proper search radius for a sub-population is also difficult. It is determined by the structure of the local area where the population is distributed. Here, we assume an ideal situation where a population only covers a single peak in the search space. Generally, a peak with a large width needs a population with a large search area, vice versa. However, the difficulty is how to obtain the structure information of the fitness landscape of the problem to be solved. In addition, different populations should have different search areas because the shapes of different peaks are different.

5.3.2 Future Algorithms for DOPs

Although a number of algorithms have been proposed to solve different kinds of DOPs, far more effective algorithms are needed to address the issues of current methods. From the above review and discussions about the research that has been done for DOPs, there are several suggestions for future algorithm development.

Firstly, algorithms should work without the assistance of change detection techniques. Due to the uncertainty characteristic of changes, it is very hard and even impossible to detect or predict changes as a successful detection is not always guaranteed, especially for real-world problems. Therefore, change detection should be avoided for future algorithms.

Secondly, learning from changing environments to guide the search should be addressed. Learning is an important feature of evolutionary computation to adapt population to environmental changes. It is a progress of discovering the structure information of a problem being considered, which should be used to guide the search, especially for the diversity maintaining issues.

Thirdly, the multi-population idea is the most flexible and cab be a versatile tool for solving DOPs. The reason lies in that any kinds of schemes for DOPs can be applied in multi-population methods. In addition, multi-population methods are effective approaches to locating and tracking multiple optima, which is very helpful to track the global optimum.

In summary, adaptive systems without change detection should be encouraged to solve DOPs with different kinds of characteristics.

5.4   Experimental Study

In this section, in order to compare the performance of algorithms based on different schemes, a number of well-known PSO algorithms are chosen to be tested on the MPB problem [6]. The investigated algorithms are mQSO [5] from diversity maintaining schemes, SPSO [42], rSPSO [2], ACPSO [30], and PSO-CP [35] from adaptive schemes, ESCA [37] from hybrid schemes, and mCPSO [5], CPSO [54], CPSOR [29], and HmSO [21] from multi-population schemes. Note that the categories of these involved algorithms are in terms of their major characteristics, some of them use several schemes, e.g., the ESCA [37] algorithm employs the hybrid scheme, multi-population scheme, and memory scheme. Table 5.1 presents the detailed schemes used for each algorithm. It can be seen that multi-population, diversity maintaining, and memory schemes are used by most algorithms.

In order to use exactly the same fitness landscapes across all environmental changes for a fair comparison, all the peer algorithms involved in this paper were carefully implemented and examined according to their origins where they were proposed. All the peer algorithms use the suggested configurations from the papers where they were proposed on the MPB problem. It should be noted that all algorithms achieved a close performance as they were proposed except the PSO-CP [35] algorithm, whose results are much worse than the results reported in [35] and hence are presented but not discussed in this study.

Table 5.1 Schemes used for all the investigated algorithms, where the "✓" symbol indicates that the corresponding scheme is used by an algorithm while the "-" symbol denotes that corresponding scheme is not used

	ACPSO	CPSOR	CPSO	HmSO	PSO-CP	ESCA	rSPSO	SPSO	mCPSO	mQSO
Without change detection	✓	✓	-	-	-	-	-	-	✓	✓
Adaptive scheme	✓	-	-	-	✓	-	✓	✓	-	-
Diversity maintaining scheme	✓	✓	✓	✓	✓	✓	✓	✓	-	✓
Multi-population scheme	✓	✓	✓	✓	✓	✓	✓	✓	✓	✓
Hybrid scheme	-	-	✓	✓	✓	-	-	-	✓	✓
Memory scheme	✓	✓	✓	✓	✓	✓	✓	✓	✓	✓

5.4.1 Experimental Setup

5.4.1.1 The Moving Peaks Benchmark (MPB) Problem

The MPB problem, proposed by Branke [6], has been widely used as a dynamic benchmark problem in the literature. Within the MPB problem, the optima can be varied by three features, i.e., the location, height, and width of peaks. For the D-dimensional landscape, the problem is defined as follows:

$$F(\mathbf{x},t) = \max_{i=1,\ldots,p} \frac{H_i(t)}{1 + W_i(t) \sum_{j=1}^{D} (x_j(t) - X_{ij}(t))^2}, \qquad (5.3)$$

where $W_i(t)$ and $H_i(t)$ are the height and width of peak i at time t, respectively, and $X_{ij}(t)$ is the j-th element of the location of peak i at time t. The p independently specified peaks are blended together by the *max* function. The position of each peak is shifted in a random direction by a vector $\mathbf{v_i}$ of a distance s (s is also called the shift length, which determines the severity of the problem dynamics), and the move of a single peak can be described as follows:

$$\mathbf{v}_i(t) = \frac{s}{|\mathbf{r} + \mathbf{v}_i(t-1)|} ((1-\lambda)\mathbf{r} + \lambda \mathbf{v}_i(t-1)), \qquad (5.4)$$

where the shift vector $\mathbf{v}_i(t)$ is a linear combination of a random vector $\mathbf{r}$ and the previous shift vector $\mathbf{v}_i(t-1)$, and is normalized to the shift length s. If the correlated parameter λ is set to 0, it implies that the peak movements are uncorrelated.

More formally, a change of a single peak can be described as follows:

$$H_i(t) = H_i(t-1) + height\_severity * \sigma \qquad (5.5)$$

$$W_i(t) = W_i(t-1) + width\_severity * \sigma \qquad (5.6)$$

$$\mathbf{X}_i(t) = \mathbf{X}_i(t)(t-1) + \mathbf{v}_i(t) \qquad (5.7)$$

where σ is a normally distributed random number with mean 0 and variation 1.

The default settings and definition of the benchmark problem used in the experiments of this study can be found in Table 5.2, which are the same for all the involved algorithms. It should be noted that different from the traditional MPB

Table 5.2 Default settings for the MPB problem, where the term "change frequency (U)" means that the environment changes every U fitness evaluations, S denotes the range of allele values, and I denotes the initial height for all peaks. The height of peaks is shifted randomly in the range $H = [30, 70]$ and the width of peaks is shifted randomly in the range $W = [1, 12]$.

Parameter	Value
peaks (number of peaks)	10
change frequency (U)	5000
height severity	7.0
width severity	1.0
peak shape	cone
basic function	no
shift length s	1.0
number of dimensions D	5
correlation coefficient λ	0
percentages of changing peaks *cPeaks*	1.0
step of number of peaks' change *sPeaks*	2
S	[0, 100]
H	[30.0, 70.0]
W	[1, 12]
I	50.0

problem, the percentage of changing peaks (*cPeaks*), which is a new feature, is added in the MPB problem in this study. This feature will make the MPB problem harder to solve because many techniques based on change detection may lose their functions. Another new feature added into the MPB problem is that, the number of peaks is allowed to change by a step size of 2 when a change happens if this feature is activated. This feature is designed to test the adaptability of an algorithm as the number of peaks changes. If this feature is enabled, the number of peaks changes by *peaks* = *peaks* + *sign* · *sPeaks*, where *sign* = 1 if *peaks* <= 10 and *sign* = −1 if *peaks* >= 50 and the initial value of *sign* is one. The traditional MPB problem is a special case of the MPB problem used in this study where *cPeaks* = 1.0 and *sPeaks* = 0.

5.4.1.2   Performance Evaluation

The first performance measure used in this study is the offline error, which is defined as follows:

$$\mu = \frac{1}{K} \sum_{k=1}^{K} (h_k - f_k), \tag{5.8}$$

where f_k is the best solution obtained by an algorithm just before the k-th environmental change, h_k is the optimum value of the k-th environment, μ is the average of all differences between h_k and f_k over the environmental changes, and K is the total number of environments, which is set to $K = 100$ in this study. All the results reported are based on the average over 30 independent runs with different random seeds. Note that the same set of random seeds are used for all the peer algorithms over 30 runs for a fair competition.

Another performance measure used in this study is the offline performance, which is defined as follows:

$$per = \frac{1}{K} \sum_{k=1}^{K} (r_k^{last}/(1 + \sum_{s=1}^{S} (1 - r_k^s)/S)), \tag{5.9}$$

where r_k^{last} is the relative value of the best solution to the global optimum after reaching U fitness evaluations for each change; r_k^s is the relative value of the best solution to the global optimum at the s-th sampling during one change, $S = U/s\_f$, s_f is the sample frequency, whose value was set to 100; $r_k^s = (f(x_k^s) + offset)/(f(x_k^*) + offset)$ for maximization problems and $r_k^s = (f(x_k^*) + offset)/(f(x_k^s) + offset)$ for minimization problems, where $offset$ was set to $fabs(f(x_k^*)) + 1$ and is used to ensure that $(f(x_k^*) + offset)$ is greater than 0.

The offline performance in Eq. (5.9) takes two aspects into account. The first aspect is the solution quality, which reflects the capability of locating and tracking the global optimum for an algorithm. In Eq. (5.9), the better a solution is found by an algorithm, the larger the value of r_k^{last} is obtained and hence, the larger per is achieved by the algorithm (the limit of per is 1.0). The second aspect is the converge capability, which indicates how fast an algorithm can converge to the global optimum after a change happens. The faster the convergence speed is, the better the offline performance is. Therefore, the offline performance generally is able to indicate an algorithm's performance in terms of both the solution quality and the convergence speed. However, the offline error is still used in this study because it is the mostly used performance measurement for EAs in dynamic environments.

5.4.2 Effect on Varying the Shit Length

5.4.2.1 Effect on Varying the Shit Length with Fixed Number of Peaks

Table 5.3 presents the comparison results of all the involved algorithms regarding different values of the shift length for the MPB problem. In Table 5.3 (and following tables in this chapter), the best value achieved among all algorithms regarding the offline error and offline performance respectively for each problem instance is highlighted in bond font.

Generally speaking, the difficulty for an algorithm to locate and track a changing optimum will increases as the shift length increases for the MPB problem. The larger the shift length is, the further a peak moves and hence, the harder for an algorithm to re-locate and track the new peak. This trend can be clearly observed from the comparison of the offline errors and the offline performance for most algorithms except ESCA in Table 5.3. For example, the offline error of ACPSO increases from 0.736 to 1.59 as the shift length (s) increases from 1.0 to 6.0, and accordingly, the offline performance drops from 0.978 to 0.933. The motivation of the ESCA algorithm is to use a swarm with sufficient diversity to re-start a new search for the

Table 5.3 Offline error, standard error ($\pm$), and offline performance for different algorithms on the MPB problem with different shift severities, where the default settings for the MPB problem in Table 5.2 except s were used

s		ACPSO	CPSOR	CPSO	HmSO	PSO-CP	ESCA	rSPSO	SPSO	mCPSO	mQSO
	error	0.666	1.57	0.899	2.68	26.2	13.9	0.698	0.807	2.32	**0.601**
0	std	±0.784	±1.23	±1.05	±1.26	±8.62	±7.95	±0.841	±0.972	±1.64	±0.439
	per	0.989	0.963	0.971	0.953	0.571	0.788	0.987	0.985	0.961	**0.989**
	error	0.61	**0.448**	1.35	3.49	28.3	14.1	2.16	2.28	7.57	1.69
1	std	±0.865	±0.626	±1.12	±1.65	±8.02	±7.18	±1.45	±1.49	±3.57	±0.784
	per	**0.98**	0.978	0.949	0.937	0.534	0.778	0.956	0.954	0.874	0.965
	error	1.15	**0.588**	1.31	3.88	26	12.9	2.9	2.81	7.57	1.84
2	std	±1.47	±0.654	±1.15	±1.66	±7.73	±6.32	±1.54	±1.54	±3.82	±0.723
	per	0.963	**0.967**	0.939	0.922	0.569	0.789	0.938	0.939	0.867	0.958
	error	1.14	**0.749**	1.41	4.22	21.1	11.9	3.4	3.24	8	2.06
3	std	±1.39	±0.7	±1.08	±1.74	±4.74	±4.73	±1.37	±1.34	±3.74	±0.775
	per	0.956	**0.958**	0.93	0.913	0.643	0.802	0.925	0.927	0.853	0.951
	error	1.34	**0.905**	1.51	4.67	20.6	11.4	3.9	4.01	9.08	2.3
4	std	±1.59	±0.843	±1.1	±2.28	±4.35	±3.46	±1.33	±1.4	±4.31	±0.786
	per	0.947	**0.948**	0.922	0.903	0.652	0.805	0.913	0.91	0.83	0.944
	error	1.41	**1.07**	1.58	4.71	21.1	11.1	4.39	4.4	9.94	2.65
5	std	±1.53	±0.839	±1.04	±2.13	±4.81	±2.9	±1.44	±1.44	±4.35	±0.872
	per	0.941	**0.941**	0.915	0.899	0.641	0.805	0.901	0.9	0.811	0.935
	error	1.54	**1.25**	1.52	5.41	21	11.3	5.12	5.19	10.8	2.94
6	std	±1.63	±1.04	±1.06	±2.36	±4.48	±2.82	±1.46	±1.52	±4.58	±0.744
	per	**0.935**	0.933	0.911	0.886	0.64	0.799	0.886	0.884	0.793	0.927

global optimum whenever a change is detected. A DOP will be treated as a series of static problems by the ESCA algorithm and hence the moving distance of peaks will not affect its performance too much.

From the results, CPSOR obtains the best results in terms of both the offline error and the offline performance for most cases. The comparison shows that CPSOR has a fast convergence speed to the new global optimum when a change occurs. It is interesting to see that ACPSO achieves better offline performance than that for CPSOR on the cases of $s = 1$ and $s = 6$ although CPSOR has better offline errors than that of ACPSO. mQSO obtains the best results on the case of $s = 0$.

5.4.2.2   Effect on Varying the Shit Length with Variable Number of Peaks

Although CPSOR achieves better results than ACPSO with different shit lengths on the default settings of the MPB problem in Table 5.3, their performance greatly decreases when the number of peaks is allowed to change. In such environments, the total number of peaks in the fitness landscape of the MPB problem will change according to the description in Sect. 5.4.1.1. Table 5.4 shows the comparison of the offline errors and offline performance of all the algorithms on the MPB problem with changing number of peaks.

Table 5.4 Offline errors, standard error ($\pm$) and offline performance for different algorithms on the MPB problem with different shift severities, where part of the peaks were enabled to change on the MPB problem

s		ACPSO	CPSOR	CPSO	HmSO	PSO-CP	ESCA	rSPSO	SPSO	mCPSO	mQSO
	error	2.15	1.91	**1.62**	3.13	22.2	12	2.8	2.75	4.43	2.67
0	std	±3.14	±2.63	±2.52	±3.06	±12.6	±8.97	±3.25	±3.23	±4.49	±3.05
	per	0.961	**0.963**	0.961	0.95	0.63	0.82	0.957	0.957	0.931	0.959
	error	**1.73**	1.89	1.76	3.65	22.6	12.7	3.93	3.98	6.62	3.25
1	std	±2.52	±2.53	±2.5	±3.3	±12.3	±8.97	±3.67	±3.74	±5.06	±3.1
	per	**0.964**	0.96	0.948	0.938	0.622	0.803	0.934	0.933	0.893	0.944
	error	1.98	1.91	**1.88**	4.16	21.8	12	4.43	4.29	7.59	3.42
2	std	±2.73	±2.51	±2.59	±3.76	±11.9	±8.95	±3.96	±3.84	±5.3	±3.19
	per	**0.955**	0.954	0.939	0.923	0.634	0.808	0.921	0.924	0.873	0.938
	error	2.07	2.05	**2**	4.2	21.9	11.6	4.72	4.7	8.18	3.41
3	std	±2.45	±2.69	±2.84	±3.81	±11.5	±8.8	±4.03	±4.08	±5.54	±3.1
	per	**0.951**	0.947	0.932	0.917	0.63	0.811	0.913	0.913	0.859	0.936
	error	2.31	2.18	**2.07**	4.22	21.5	12.1	4.98	4.89	8.87	3.45
4	std	±2.71	±2.74	±2.84	±3.96	±11.6	±9.23	±4.09	±3.99	±5.79	±3.1
	per	**0.943**	0.941	0.926	0.913	0.636	0.8	0.905	0.906	0.843	0.932
	error	2.39	2.33	**2.09**	4.3	21.5	11.7	5.33	5.38	9.03	3.54
5	std	±2.76	±2.87	±2.8	±4.11	±11.5	±8.9	±4.26	±4.25	±5.78	±3.16
	per	**0.937**	0.934	0.922	0.909	0.636	0.801	0.897	0.895	0.837	0.927
	error	2.49	2.33	**2.21**	4.32	20.9	11.9	5.67	5.62	9.44	3.58
6	std	±3.04	±2.87	±2.97	±4.25	±11.1	±8.95	±4.4	±4.35	±5.92	±3.17
	per	**0.931**	0.931	0.917	0.906	0.643	0.796	0.889	0.889	0.826	0.924

In such kind of dynamic environments, an algorithm needs to adaptively adjust the number of sub-populations that are really needed according to the change of the number of peaks. Among the algorithms, ACPSO is the only one that is able to adaptively adjust the number of sub-populations in such environments.

From the results, ACPSO obtains the best performance regarding the offline performance in Table 5.4. With the help of change detection techniques, CPSO achieves the best offline errors on most cases. However, comparing with ACPSO, it can be seen that ACPSO achieves better offline performance than that of CPSOR on all test cases. The performance of the improved algorithm rSPSO is better than that of the SPSO algorithm, and the mQSO algorithm with anti-convergence and repulsion also outperforms the mCPSO algorithm.

5.4.3 Effect on Varying the Number of Peaks

To compare the performance of the involved algorithms regarding locating and tracking the global optimum in the environments with different numbers of peaks, an experimental study was conducted on the MPB problem with different numbers

Table 5.5 Offline errors, standard error (±), and offline performance of different algorithms on the MPB problem with different numbers of peaks and on the MPB problem with changing number of peaks

peaks		ACPSO	CPSOR	CPSO	HmSO	PSO-CPESCA	rSPSO	SPSO	mCPSO	mQSO
	error	0.188	0.245	**1.4e-4**	1.8	61.6	5.15	1.56	1.86	38.4	3.15
1	std	±0.59	±0.253	±5.9e-4	±2.09	±17.1	±3.96	±1.35	±1.45	±14.8	±1.21
	per	**0.984**	0.956	0.95	0.947	0.521	0.869	0.95	0.948	0.365	0.918
	error	1.75	2.4	1.33	**0.768**	31.9	6.79	0.922	0.986	11.6	4.33
2	std	±1.68	±1.54	±0.913	±1.13	±8.53	±2.97	±0.509	±0.564	±7.04	±1.7
	per	0.955	0.92	0.941	**0.974**	0.529	0.864	0.97	0.97	0.776	0.908
	error	**0.722**	0.936	0.97	3.99	36.1	12.4	1.52	1.52	6.46	1.56
5	std	±0.976	±0.788	±0.755	±1.92	±7.04	±6.12	±0.751	±0.769	±3.83	±0.691
	per	**0.978**	0.962	0.953	0.931	0.456	0.799	0.966	0.966	0.884	0.965
	error	0.645	0.803	**0.641**	1.81	24.7	12.7	1.23	1.25	4.39	1.19
7	std	±0.896	±0.628	±0.523	±1.26	±9.27	±6.94	±0.611	±0.569	±2.34	±0.638
	per	**0.98**	0.965	0.958	0.96	0.581	0.795	0.972	0.971	0.922	0.972
	error	0.61	**0.448**	1.35	3.49	28.3	14.1	2.16	2.28	7.57	1.69
10	std	±0.865	±0.626	±1.12	±1.65	±8.02	±7.18	±1.45	±1.49	±3.57	±0.784
	per	**0.98**	0.978	0.949	0.937	0.534	0.778	0.956	0.954	0.874	0.965
	error	1.41	**1.38**	1.45	3.2	24.9	10.6	3.59	3.59	7.87	2.66
20	std	±1.01	±0.983	±0.965	±0.866	±9.25	±5.25	±1.39	±1.33	±3.21	±1
	per	**0.972**	0.97	0.954	0.948	0.598	0.838	0.941	0.94	0.876	0.954
	error	**1.14**	1.32	1.2	3.1	26.3	9.4	3.01	3.1	5.68	2.87
30	std	±0.941	±1.15	±0.788	±0.854	±11.9	±5.85	±1.44	±1.53	±2.91	±1.43
	per	**0.976**	0.97	0.961	0.951	0.592	0.855	0.949	0.948	0.91	0.95
	error	**1.42**	1.76	1.43	3.19	18.8	9.09	3.27	3.38	5.52	2.8
50	std	±0.836	±0.999	±0.747	±0.731	±7.22	±5.89	±1.27	±1.3	±2.86	±1.04
	per	**0.972**	0.964	0.96	0.95	0.689	0.86	0.945	0.943	0.912	0.952
	error	1.62	2.19	**1.42**	3.5	17	9.53	3.48	3.5	5.64	3.39
100	std	±0.919	±1.11	±0.752	±0.953	±6.63	±5.82	±1.45	±1.47	±2.79	±1.53
	per	**0.97**	0.959	0.963	0.946	0.722	0.855	0.943	0.943	0.911	0.944
	error	1.46	2.22	**0.991**	2.89	13.6	7.99	3.62	3.54	5.16	3.32
200	std	±0.893	±1.39	±0.55	±0.772	±5.8	±5.37	±1.74	±1.76	±2.99	±1.92
	per	**0.973**	0.957	0.972	0.956	0.775	0.877	0.941	0.942	0.918	0.944
	error	**1.73**	1.89	1.76	3.65	22.6	12.7	3.93	3.98	6.62	3.25
changing	std	±2.52	±2.53	±2.5	±3.3	±12.3	±8.97	±3.67	±3.74	±5.06	±3.1
	per	**0.964**	0.96	0.948	0.938	0.622	0.803	0.934	0.933	0.893	0.944

of peaks. Table 5.5 presents the comparison results of all the involved algorithms. The bottom row in Table 5.5 is the results for algorithms on the MPB problem where the number of peaks changes.

Generally speaking, the more peaks there are in the fitness landscape, the harder it is for an algorithm to locate and track the global optimum. This trend can be observed in Table 5.5 for all the algorithms when the number of peaks is less than 100. However, when the number of peaks is larger than 50, the offline errors and the offline performance get better for most algorithms. This is because it gets easier for an algorithm to locate and track a peak whose height is similar to the global optimum as more such peaks will appear when the number of peaks gets larger.

For the case of one peak, which is the most simple problem, only CPSO successfully locate and track the global optimum in every change over all the runs. Although CPSO obtains a smaller offline error than that of ACPSO, ACPSO achieves a better offline performance than CPSO. For the remaining cases, ACPSO achieves the best performance regarding both the offlline error and the offline performance when the number of peaks is larger than 20.

For the case where the number of peaks changes, the offline error and the offline performance obtained by ACPSO are 1.73 and 0.964, respectively, which are much better than that obtained by all the other algorithms. The comparison results in Table 5.5 show that the performance of ACPSO is better than all the peer algorithms in terms of locating and tracking changing optima in the environments with a large number of local optima.

HmSO achieves the best results on the case of *peaks* = 2 where the offline error and the offline performance are 0.768 and 0.974, respectively. For the overall performance, except the algorithms based on the clustering method (ACPSO, CPSOR, and CPSO), mQSO and rSPSO achieve better results than the other algorithms.

5.4.4 Effect on Varying the Number of Dimensions

In this section, the peer algorithms are compared on the MPB problem with different number of dimensions. Table 5.6 and Table 5.7 present the results of all the algorithms in different numbers of dimensions on the MPB problem with 10 and 100 peaks, respectively.

Comparing the results of different algorithms in Table 5.6 and Table 5.7, it can be seen that the performance decreases for all the involved algorithms when the number of dimensions increases. Compared with the results of SPSO, the least square regression, which is used in rSPSO to estimate local optima to accelerate the search, does not work when the number of dimensions is larger than 5 in both tables. Among all the algorithms, CPSOR outperforms all the other algorithms on most test cases. It can also be seen that the effect of increasing the number of dimensions to CPSOR and CPSO is the smallest. For example, the offline error of CPSOR increases from 1.6 to 5.77 when the number of dimensions increases from 3 to 20, while the corresponding error increases from 2.16 to 11.6 for mQSO, which has the best performance except CPSOR, CPSO, and ACPSO.

Table 5.6 Offline errors, standard error ($\pm$), and offline performance for different algorithms on the MPB problem with different numbers of dimensions, where the default settings for the MPB problem in Table 5.2 were used

s		ACPSO	CPSOR	CPSO	HmSO	PSO-CP	ESCA	rSPSO	SPSO	mCPSO	mQSO
3	error	0.149	**0.134**	0.411	2.35	13.6	6.51	0.992	1.03	1.65	0.558
	std	$\pm$0.443	$\pm$0.235	$\pm$0.724	$\pm$1.7	$\pm$2.6	$\pm$2.72	$\pm$0.635	$\pm$0.655	$\pm$1.42	$\pm$0.532
	per	0.988	**0.988**	0.976	0.957	0.768	0.895	0.981	0.977	0.965	0.985
5	error	0.61	**0.448**	1.35	3.49	28.3	14.1	2.16	2.28	7.57	1.69
	std	$\pm$0.865	$\pm$0.626	$\pm$1.12	$\pm$1.65	$\pm$8.02	$\pm$7.18	$\pm$1.45	$\pm$1.49	$\pm$3.57	$\pm$0.784
	per	**0.98**	0.978	0.949	0.937	0.534	0.778	0.956	0.954	0.874	0.965
7	error	1.65	**1.3**	2.36	3.54	28.3	11.9	8.22	8.16	8.32	2.54
	std	$\pm$1.18	$\pm$0.779	$\pm$0.908	$\pm$1.02	$\pm$12.8	$\pm$2.47	$\pm$3.33	$\pm$3.19	$\pm$3.26	$\pm$0.703
	per	**0.965**	0.957	0.931	0.935	0.56	0.806	0.862	0.863	0.865	0.952
10	error	4.55	**3.04**	4.3	10.1	40.3	15.9	114	114	19.9	6.53
	std	$\pm$2.44	$\pm$1.7	$\pm$1.8	$\pm$2.39	$\pm$11.2	$\pm$5.31	$\pm$15.7	$\pm$15.7	$\pm$4.83	$\pm$1.31
	per	**0.922**	0.914	0.892	0.839	0.478	0.746	0.0527	0.0527	0.679	0.891
15	error	7.5	**4.12**	6.46	10.8	53.6	35.5	171	171	28.6	8.58
	std	$\pm$2.97	$\pm$3.65	$\pm$2.67	$\pm$1.88	$\pm$22.7	$\pm$14.4	$\pm$17	$\pm$17	$\pm$9.3	$\pm$1.41
	per	0.881	**0.888**	0.862	0.832	0.337	0.473	0.0405	0.0405	0.561	0.867
20	error	11.4	**7.69**	9.22	18.4	88	45.8	222	222	69.4	12.8
	std	$\pm$4.23	$\pm$6.06	$\pm$3.58	$\pm$2.71	$\pm$35.5	$\pm$13.1	$\pm$18.4	$\pm$18.4	$\pm$21	$\pm$1.57
	per	0.822	**0.836**	0.824	0.724	0.229	0.325	0.0299	0.0299	0.233	0.805

Table 5.7 Offline errors, standard error ($\pm$), and offline performance for different algorithms on the MPB problem with different numbers of dimensions, where the default settings for the MPB problem in Table 5.2 except *peaks* = 100 were used

s		ACPSO	CPSOR	CPSO	HmSO	PSO-CP	ESCA	rSPSO	SPSO	mCPSO	mQSO
3	error	**0.576**	1.6	0.578	1.31	8.96	3.78	1.6	1.67	2.61	2.16
	std	$\pm$0.46	$\pm$1.07	$\pm$0.477	$\pm$0.645	$\pm$4.15	$\pm$2.06	$\pm$0.932	$\pm$1.03	$\pm$1.54	$\pm$1.35
	per	**0.987**	0.971	0.983	0.975	0.854	0.939	0.974	0.971	0.956	0.965
5	error	1.62	2.19	**1.42**	3.5	17	9.53	3.48	3.5	5.64	3.39
	std	$\pm$0.919	$\pm$1.11	$\pm$0.752	$\pm$0.953	$\pm$6.63	$\pm$5.82	$\pm$1.45	$\pm$1.47	$\pm$2.79	$\pm$1.53
	per	**0.97**	0.959	0.963	0.946	0.722	0.855	0.943	0.943	0.911	0.944
7	error	2.27	2.73	**1.62**	4.24	21.4	13	9.13	9.13	7.03	4.04
	std	$\pm$1.19	$\pm$1.52	$\pm$0.812	$\pm$1.08	$\pm$8.36	$\pm$6.45	$\pm$4.78	$\pm$4.78	$\pm$3.35	$\pm$1.8
	per	**0.962**	0.948	0.955	0.936	0.653	0.805	0.86	0.86	0.891	0.933
10	error	4.01	2.83	**2.66**	6.27	36.1	16.2	69.2	69.2	12	6.37
	std	$\pm$1.62	$\pm$1.46	$\pm$1.18	$\pm$1.49	$\pm$12.4	$\pm$7.99	$\pm$11.5	$\pm$11.5	$\pm$4.81	$\pm$1.97
	per	0.937	**0.942**	0.933	0.906	0.444	0.753	0.101	0.101	0.819	0.899
15	error	6.23	**3.49**	4.04	8.59	45	25.8	105	105	16.7	8.68
	std	$\pm$2.21	$\pm$2.48	$\pm$1.95	$\pm$2.69	$\pm$16.9	$\pm$8.88	$\pm$10.5	$\pm$10.5	$\pm$5.28	$\pm$2.27
	per	0.905	**0.921**	0.903	0.871	0.376	0.612	0.0329	0.0329	0.749	0.867
20	error	10.1	5.77	**5.75**	16.5	61.8	39.9	131	131	35.6	11.6
	std	$\pm$4.47	$\pm$4.44	$\pm$3.22	$\pm$3.67	$\pm$26.4	$\pm$17.2	$\pm$10.5	$\pm$10.5	$\pm$14.6	$\pm$2.69
	per	0.847	**0.871**	0.869	0.757	0.266	0.445	0.0263	0.0263	0.489	0.826

Table 5.8 Offline errors, standard error ($\pm$) and offline performance for different algorithms on the MPB problem with different change ratios, where the default settings for the MPB problem in Table 5.2 except $s = 2.0$ were used

s		ACPSO	CPSOR	CPSO	HmSO	PSO-CP	ESCA	rSPSO	SPSO	mCPSO	mQSO
	error	2.86	**1.87**	9.19	7	18.8	10.3	3.94	3.94	6.07	2.56
0.1	std	$\pm$3.84	$\pm$2.86	$\pm$7.6	$\pm$5.75	$\pm$8.91	$\pm$7.13	$\pm$3.77	$\pm$3.65	$\pm$4.16	$\pm$3
	per	0.944	0.95	0.846	0.884	0.66	0.832	0.931	0.931	0.896	**0.952**
	error	2.24	**1.29**	6.86	9.03	22.5	14.4	5.9	6.02	7.07	2.69
0.3	std	$\pm$2.7	$\pm$1.65	$\pm$7.66	$\pm$3.74	$\pm$7.62	$\pm$6.41	$\pm$2.91	$\pm$2.91	$\pm$2.97	$\pm$1.49
	per	0.954	**0.961**	0.882	0.855	0.609	0.772	0.904	0.902	0.883	0.951
	error	1.51	**0.896**	2.52	2.54	20.9	8.89	1.53	1.54	3.4	1.1
0.5	std	$\pm$1.35	$\pm$0.854	$\pm$3.25	$\pm$1.68	$\pm$6.26	$\pm$5.05	$\pm$1.21	$\pm$1.14	$\pm$2.05	$\pm$0.656
	per	0.969	0.974	0.948	0.955	0.635	0.86	0.97	0.969	0.941	**0.977**
	error	1.73	**0.936**	1.77	3.16	19.7	10.6	3.01	2.98	5.95	1.95
0.7	std	$\pm$1.54	$\pm$0.84	$\pm$2.05	$\pm$1.73	$\pm$5.32	$\pm$5.96	$\pm$1.43	$\pm$1.48	$\pm$2.45	$\pm$0.795
	per	0.964	**0.97**	0.951	0.944	0.656	0.83	0.946	0.946	0.9	0.962
	error	1.79	**0.434**	1.1	3.76	19.7	12.7	2.57	2.6	6.58	1.97
0.9	std	$\pm$2.04	$\pm$0.535	$\pm$1.44	$\pm$1.91	$\pm$5.55	$\pm$6.75	$\pm$1.31	$\pm$1.29	$\pm$3.05	$\pm$0.925
	per	0.961	**0.978**	0.955	0.934	0.662	0.799	0.951	0.951	0.89	0.962
	error	0.61	**0.448**	1.35	3.49	28.3	14.1	2.16	2.28	7.57	1.69
1	std	$\pm$0.865	$\pm$0.626	$\pm$1.12	$\pm$1.65	$\pm$8.02	$\pm$7.18	$\pm$1.45	$\pm$1.49	$\pm$3.57	$\pm$0.784
	per	**0.98**	0.978	0.949	0.937	0.534	0.778	0.956	0.954	0.874	0.965

5.4.5 Comparison in Hard-to-Detect Environments

So far, all the comparisons are in the environments where changes are easy to detect. This section presents the comparison of all the involved algorithms in the environments where changes are hard to detect. Such kind of environments are simulated by introducing *cPeaks* in the MPB problem (see Table 5.2). Note that the highest peak (i.e., the global optimum) is guaranteed to change in order to test the performance of algorithms in tracking the global optimum in this experimental study. Table 5.8 shows the comparison of all the algorithms on the MPB problem with the default settings ($s = 1.0$ and *peaks* = 10).

Generally speaking, the tracking of the global optimum will get harder for an algorithm as the changing ratio (*cPeaks*) decreases, especially for algorithms that are based on change detection. For example, when *cPeaks* = 1.0 for the MPB problem, all the peaks are allowed to change, which is the default setting. In such case, any point in the fitness landscape is able to detect changes. However, when *cPeaks* is less than 1.0, a successful change detection will not always be guaranteed depending on whether detectors are in the areas where changes happen or not. *cPeaks* can be taken as the possibility of successful detections. Therefore, *cPeaks* = 0.1 is the hardest case to detect changes in this experiments.

From the results in Table 5.8, the performance of all the algorithms on the problem with *cPeaks* < 1.0 is worse than that on the problem with *cPeaks* = 1.0 regarding the offline error and offline performance. Two possible reasons can be explained.

Table 5.9 Offline errors, standard error (±) and offline performance for different algorithms on the MPB problem with different change ratios, where the default settings for the MPB problem in Table 5.2 except $s = 2.0$ were used

s		ACPSO	CPSOR	CPSO	HmSO	PSO-CP	ESCA	rSPSO	SPSO	mCPSO	mQSO
	error	2.81	**2.13**	9.47	7.58	18.4	10.1	4.8	4.78	7.43	2.55
0.1	std	±3.83	±2.91	±7.64	±5.85	±8.45	±7.14	±4.21	±4.24	±4.52	±2.79
	per	0.941	0.941	0.839	0.871	0.665	0.833	0.915	0.915	0.873	**0.951**
	error	2.29	**1.46**	6.72	8.68	22.6	14.5	6.62	6.68	8.23	2.65
0.3	std	±2.78	±1.62	±7.05	±3.68	±7.43	±7.14	±3.05	±3.05	±3.56	±1.52
	per	0.948	**0.953**	0.881	0.858	0.603	0.769	0.891	0.89	0.862	0.95
	error	1.64	**1.1**	2.56	3.7	19.8	9.37	3.24	3.04	5.77	1.47
0.5	std	±1.45	±0.943	±3.44	±2.25	±5.05	±4.57	±1.67	±1.74	±2.92	±0.818
	per	0.964	0.967	0.943	0.935	0.651	0.85	0.942	0.944	0.901	**0.969**
	error	1.81	**1.16**	1.98	4.04	20.4	11.3	4.05	3.99	7.08	2.33
0.7	std	±1.6	±1.01	±2.3	±2.01	±5.38	±5.85	±1.6	±1.66	±2.75	±0.827
	per	0.957	**0.96**	0.941	0.925	0.64	0.818	0.926	0.926	0.879	0.954
	error	1.77	**0.636**	1.11	4.64	20.3	12.5	3.2	3.28	7.77	2.05
0.9	std	±2.05	±0.788	±1.46	±2.54	±4.55	±7.04	±1.64	±1.61	±3.54	±0.875
	per	0.956	**0.967**	0.946	0.913	0.648	0.797	0.937	0.936	0.866	0.957
	error	1.15	**0.588**	1.31	3.88	26	12.9	2.9	2.81	7.57	1.84
1	std	±1.47	±0.654	±1.15	±1.66	±7.73	±6.32	±1.54	±1.54	±3.82	±0.723
	per	0.963	**0.967**	0.939	0.922	0.569	0.789	0.938	0.939	0.867	0.958

Firstly, a successful change detection will not be always guaranteed due to the partially changed fitness landscape and hence algorithms that are based on change detection will not work well in such dynamic environments, such as HmSO and CPSO. Secondly, individuals of the previous environment will be easily attracted by the unchanged peaks rather than the changed peaks. Individuals on the unchanged peaks normally have good fitness than those on the changed peaks. As a result, those individuals on the changed peaks are likely to move to the areas of unchanged peaks that are covered by individuals of the previous environment due to communications between individuals.

For HmSO and CPSO, change detection is a very important factor in their performance. As explained above, the smaller the value of *cPeaks*, the harder it is for the two algorithms to detect changes. From the corresponding results, it can be seen that the offline error and offline performance obtained by the two algorithms get worse as *cPeaks* decreases. For the other algorithms, which do not totally rely on change detection, the effect is not as serious as for HmSO and CPSO.

CPSOR obtains the best performance among all the involved algorithms, and mQSO achieves the second best performance due to the working mechanism in undetectable environments. ACPSO achieves slightly worse results than CPSOR and mQSO but much better results than the other algorithms. Although CPSOR and mQSO achieve the best performance on the MPB problem with default settings, their performances degrade when the shift length (s) is set to 2.0 in Table 5.9 even if they still outperform the other algorithms. However, the performance of ACPSO

Table 5.10 Offline errors, standard error ($\pm$) and offline performance for different algorithms on the MPB problem with different change ratios, where the default settings for the MPB problem in Table 5.2 except *peaks* $= 50$ and $s = 2.0$ were used

s		ACPSO	CPSOR	CPSO	HmSO	PSO-CP	ESCA	rSPSO	SPSO	mCPSO	mQSO
	error	**3.1**	3.32	11.5	7.02	17.3	10.8	6.2	6.12	7.6	4.56
0.1	std	± 2.08	± 2.11	± 4.68	± 2.72	± 4.63	± 3.95	± 1.88	± 1.92	± 2.29	± 2.02
	per	**0.948**	0.94	0.825	0.892	0.708	0.837	0.904	0.905	0.882	0.927
	error	**2.53**	2.54	4.44	4.04	19.4	12.7	5.49	5.5	6.66	4.2
0.3	std	± 1.86	± 1.8	± 4.59	± 2.43	± 8.15	± 6.66	± 2.59	± 2.55	± 3.31	± 2.55
	per	**0.957**	0.955	0.924	0.935	0.67	0.807	0.914	0.914	0.895	0.933
	error	**1.94**	2.44	2.32	3.57	20.4	9.76	5.44	5.21	6.79	4.33
0.5	std	± 1.73	± 2.18	± 2.51	± 2.14	± 9.57	± 6.33	± 3.19	± 3.14	± 3.86	± 3.11
	per	**0.965**	0.954	0.952	0.941	0.665	0.851	0.912	0.915	0.891	0.929
	error	1.68	2.28	**1.48**	3.25	16.4	9.02	3.77	3.76	5.4	2.86
0.7	std	± 1.12	± 1.42	± 1.2	± 1.35	± 6.83	± 5.36	± 1.6	± 1.71	± 2.76	± 1.58
	per	**0.968**	0.956	0.961	0.945	0.724	0.86	0.936	0.936	0.912	0.951
	error	2.13	2.2	**1.9**	4.35	19.4	11.4	3.79	3.79	6.39	3.32
0.9	std	± 1.27	± 1.12	± 1.11	± 1.4	± 7.52	± 6.75	± 1.2	± 1.17	± 3.08	± 1.4
	per	**0.958**	0.953	0.948	0.926	0.676	0.823	0.935	0.934	0.896	0.943
	error	1.71	2.1	**1.51**	3.72	18.5	10.2	3.86	3.72	6.7	3.02
1	std	± 0.953	± 1.09	± 0.741	± 0.918	± 6.4	± 6.09	± 1.44	± 1.51	± 3.31	± 1.21
	per	**0.964**	0.953	0.953	0.935	0.689	0.839	0.931	0.933	0.89	0.946

is just slightly affected when $s = 2.0$ is used in comparison of that of ACPSO in Table 5.8.

The superior performance of ACPSO is not shown until the problem with a large number of peaks was tested. Table 5.10 presents the comparison of all the algorithms on the MPB problem with $s = 2.0$ and *peaks* $= 50$. From the results, it can be seen that the offline error and the offline performance of ACPSO are much better than those of the other algorithms on most cases. The performance of algorithms that are based on the clustering method, including ACPSO, CPSOR, and CPSO, is better than that of all the other multi-populations methods.

The comparison results in this section show that ACPSO is able to effectively track changing optima without change detection on the MPB problem, even in undetectable environments.

5.5 Conclusions

This chapter reviews the state-of-the-art research of PSO for DOPs, and categorizes them into different groups, including diversity maintaining schemes, adaptive schemes, hybrid schemes, multi-population schemes, and memory schemes. The corresponding advantages and shortcomings are also discussed. This chapter also carries out a comprehensive experimental study with 10 selected well-known PSO variants on the MPB problems.

From the related work review and the comparison study in this chapter, we can draw several conclusions and suggestions for future methodology development of PSO for DOPs.

Firstly, diversity maintaining mechanism must be considered in dynamic environments. As discussed above, the population diversity is the most important issue to solve DOPs. Algorithms will not work in dynamic environments without this mechanism.

Secondly, the multi-population method is effective to track and locate a set of optima and is commonly used by most algorithms. Due to its flexibility, it can be used with any other schemes.

Thirdly, the memory scheme is rarely used in PSO as PSO has an implicit memory mechanism where each particle's *pbest* is able to memorize the best position found by a particle.

Fourthly, population adaptation in dynamic environments without change detection is a promising research direction. From the comparison results, we can see that the ACPSO, which belongs to such kind of algorithms, outperforms all the other algorithms in most cases on the MPB problem with different perspectives.

Acknowledgements. This work was supported by the National Natural Science Foundation of China (NSFC) under Grant 61203306, the Engineering and Physical Sciences Research Council (EPSRC) of U.K. under Grant numbers EP/E060722/1 and EP/K001310/1, and partially by the State Key Laboratory of Synthetical Automation for Process Industries, Northeastern University, China.

References

[1] Bird, S., Li, X.: Adaptively choosing niching parameters in a pso. In: 2006 Genetic Evol. Comput. Conf., pp. 3–10 (2006)

[2] Bird, S., Li, X.: Using regression to improve local convergence. In: Proc. 2007 IEEE Congr. Evol. Comput., pp. 592–599 (2007)

[3] Blackwell, T.M.: Particle swarm optimization in dynamic environments. In: Yang, S., Ong, Y.-S., Jin, Y. (eds.) Evolutionary Computation in Dynamic and Uncertain Environments. SCI, vol. 51, pp. 29–49. Springer, Heidelberg (2007)

[4] Blackwell, T., Branke, J.: Multi-swarm optimization in dynamic environments. In: Raidl, G.R., et al. (eds.) EvoWorkshops 2004. LNCS, vol. 3005, pp. 489–500. Springer, Heidelberg (2004)

[5] Blackwell, T.M., Branke, J.: Multiswarms, exclusion, and anti-convergence in dynamic environments. IEEE Trans. Evol. Comput. 10(4), 459–472 (2006)

[6] Branke, J.: Memory enhanced evolutionary algorithms for changing optimization problems. In: Proc. 1999 IEEE Congr. Evol. Comput., vol. 3, pp. 1875–1882 (1999)

[7] Branke, J., Kaußler, T., Schmidth, C., Schmeck, H.: A multi-population approach to dynamic optimization problem. In: Proc. 4th Int. Conf. Adaptive Comput. Des. Manuf., pp. 299–308 (2000)

[8] Bratton, D., Kennedy, J.: Defining a standard for particle swarm optimization. In: Proc. IEEE Swarm Intel. Symp., pp. 120–127 (2007)

 [9] Brits, R., Engelbrecht, A., van den Bergh, F.: A niching particle swarm optimizer. In: Proc. 4th Asia-Pacific Conf. Simulated Evolution and Learning, vol. 2, pp. 692–696 (2002)
[10] Daneshyari, M., Yen, G.: Dynamic optimization using cultural based pso. In: Proc. 2011 IEEE Congr. Evol. Comput., pp. 509–516 (2011)
[11] del Amo, I.G., Pelta, D.A., González, J.R.: Using heuristic rules to enhance a multi-swarm pso for dynamic environments. In: Proc. 2010 IEEE Congr. Evol. Comput., pp. 1–8 (2010)
[12] del Amo, I.G., Pelta, D.A., González, J.R., Novoa, P.: An analysis of particle properties on a multi-swarm PSO for dynamic optimization problems. In: Meseguer, P., Mandow, L., Gasca, R.M. (eds.) CAEPIA 2009. LNCS, vol. 5988, pp. 32–41. Springer, Heidelberg (2010)
[13] Du, W., Li, B.: Multi-strategy ensemble particle swarm optimization for dynamic optimization. Inform. Sci. 178(15), 3096–3109 (2008)
[14] Fernandez-Marquez, J., Arcos, J.: Adapting particle swarm optimization in dynamic and noisy environments. In: Proc. 2010 IEEE Congr. Evol. Comput., pp. 1–8 (2010)
[15] Gonzalez, J.R., Masegosa, A.D., Garcia, I.J.: A cooperative strategy for solving dynamic optimization problems. Memetic Computing 3(1), 3–14 (2011)
[16] Hashemi, A., Meybodi, M.: A multi-role cellular pso for dynamic environments. In: Proc. 14th Int. CSI Computer Conf., pp. 412–417 (2009)
[17] Hu, X., Eberhart, R.: Adaptive particle swarm optimization: detection and response to dynamic systems. In: Proc. 2002 IEEE Congr. Evol. Comput., vol. 2, pp. 1666–1670 (2002)
[18] Janson, S., Middendorf, M.: A hierarchical particle swarm optimizer for noisy and dynamic environments. Genetic Programming and Evolvable Machines 7(4), 329–354 (2006)
[19] Jiang, Y., Huang, W., Chen, L.: Applying multi-swarm accelerating particle swarm optimization to dynamic continuous functions. In: Proc. 2nd Int. Workshop on Knowledge Discovery and Data Mining, pp. 710–713 (2009)
[20] Jin, Y., Branke, J.: Evolutionary optimization in uncertain environments: a survey. IEEE Trans. Evol. Comput. 9(3), 303–317 (2005)
[21] Kamosi, M., Hashemi, A.B., Meybodi, M.R.: A hibernating multi-swarm optimization algorithm for dynamic environments. In: Proc. World Congr. on Nature and Biologically Inspired Computing, NaBIC 2010, pp. 363–369 (2010)
[22] Karimi, J., Nobahari, H., Pourtakdoust, S.: A new hybrid approach for dynamic continuous optimization problems. Appl. Soft Comput. 12(3), 1158–1167 (2012)
[23] Kennedy, J., Eberhart, R.C.: Swarm Intelligence. Morgan Kaufmann Publishers (2001)
[24] Khouadjia, M., Sarasola, B., Alba, E., Jourdan, L., Talbi, E.: Multi-environmental cooperative parallel metaheuristics for solving dynamic optimization problems. In: Proc. 2011 IEEE Int. Symp. Parallel and Distributed Processing Workshops and PhD Forum (IPDPSW), pp. 395–403 (2011)
[25] Khouadjia, M.R., Alba, E., Jourdan, L., Talbi, E.-G.: Multi-swarm optimization for dynamic combinatorial problems: A case study on dynamic vehicle routing problem. In: Dorigo, M., et al. (eds.) ANTS 2010. LNCS, vol. 6234, pp. 227–238. Springer, Heidelberg (2010)
[26] Khouadjia, M.R., Jourdan, L., Talbi, E.G.: Adaptive particle swarm for solving the dynamic vehicle routing problem. In: Proc. 2010 IEEE/ACS Int. Conf. Computer Systems and Applications, pp. 1–8 (2010)

[27] Li, C., Yang, S.: Fast multi-swarm optimization for dynamic optimization problems. In: Proc. 4th Int. Conf. Natural Comput., vol. 7, pp. 624–628 (2008)

[28] Li, C., Yang, S.: A clustering particle swarm optimizer for dynamic optimization. In: Proc. 2009 IEEE Congr. Evol. Comput., pp. 439–446 (2009)

[29] Li, C., Yang, S.: A general framework of multipopulation methods with clustering in undetectable dynamic environments. IEEE Trans. Evol. Comput. 16(4), 556–577 (2012)

[30] Li, C., Yang, S.: Population adaptation in dynamic environments via multi-population methods. Evol. Comput. (2012) (submitted)

[31] Li, X.: Adaptively choosing neighbourhood bests using species in a particle swarm optimizer for multimodal function optimization. In: Deb, K., Tari, Z. (eds.) GECCO 2004. LNCS, vol. 3102, pp. 105–116. Springer, Heidelberg (2004)

[32] Liu, L., Ranjithan, S.R.: An adaptive optimization technique for dynamic environments. Eng. Appl. Artif. Intell. 23(5), 772–779 (2010)

[33] Liu, L., Wang, D., Tang, J.: Composite particle optimization with hyper-reflection scheme in dynamic environments. Appl. Soft Comput. 11(8), 4626–4639 (2011)

[34] Liu, L., Wang, D., Yang, S.: Compound particle swarm optimization in dynamic environments. In: Giacobini, M., et al. (eds.) EvoWorkshops 2008. LNCS, vol. 4974, pp. 616–625. Springer, Heidelberg (2008)

[35] Liu, L., Yang, S., Wang, D.: Particle swarm optimization with composite particles in dynamic environments. IEEE Trans. Syst., Man, & Cybern. Part B: Cybern. 40(6), 1634–1648 (2010)

[36] Lung, R.I., Dumitrescu, D.: A collaborative model for tracking optima in dynamic environments. In: Proc. 2007 IEEE Congr. Evol. Comput., pp. 564–567 (2007)

[37] Lung, R.I., Dumitrescu, D.: Evolutionary swarm cooperative optimization in dynamic environments. Natural Computing 9(1), 83–94 (2010)

[38] Mendes, R., Mohais, A.S.: Dynde: a differential evolution for dynamic optimization problems. In: Proc. 2005 IEEE Congr. Evol. Comput., pp. 2808–2815 (2005)

[39] Nickabadi, A., Ebadzadeh, M.M., Safabakhsh, R.: Evaluating the performance of dnpso in dynamic environments. In: Proc. 2008 IEEE Int. Conf. Syst., Man, & Cybern., pp. 2640–2645 (2008)

[40] Novoa-Hernandez, P., Corona, C.C., Pelta, D.A.: Efficient multi-swarm pso algorithms for dynamic environments. Memetic Computing 3(3), 163–174 (2011)

[41] Parrott, D., Li, X.: A particle swarm model for tracking multiple peaks in a dynamic environment using speciation. In: Proc. 2004 IEEE Congr. Evol. Comput., pp. 98–103 (2004)

[42] Parrott, D., Li, X.: Locating and tracking multiple dynamic optima by a particle swarm model using speciation. IEEE Trans. Evol. Comput. 10(4), 440–458 (2006)

[43] Poli, R., Kennedy, J., Blackwell, T.: Particle swarm optimization: An overview. Swarm Intell. 1(1), 33–58 (2007)

[44] Rezazadeh, I., Meybodi, M.R., Naebi, A.: Adaptive particle swarm optimization algorithm for dynamic environments. In: Tan, Y., Shi, Y., Chai, Y., Wang, G. (eds.) ICSI 2011, Part I. LNCS, vol. 6728, pp. 120–129. Springer, Heidelberg (2011)

[45] Rezazadeh, I., Meybodi, M.R., Naebi, A.: Particle swarm optimization algorithm in dynamic environments: Adapting inertia weight and clustering particles. In: 2011 Fifth UKSim European Symposium on Computer Modeling and Simulation (EMS), pp. 76–82 (2011), doi:10.1109/EMS.2011.62

[46] Richter, H.: Detecting change in dynamic fitness landscapes. In: 2009 Congr. Evol. Comput., pp. 1613–1620 (2009)

[47] Schoeman, I.L., Engelbrecht, A.P.: A novel particle swarm niching technique based on extensive vector operations. Natural Computing 9(3), 683–701 (2009)

[48] Thomsen, R.: Multimodal optimization using crowding-based differential evolution. In: Proc. 2004 IEEE Congr. Evol. Comput., vol. 2, pp. 1382–1389 (2004)

[49] Wang, H., Wang, D., Yang, S.: Triggered memory-based swarm optimization in dynamic environments. In: Giacobini, M. (ed.) EvoWorkshops 2007. LNCS, vol. 4448, pp. 637–646. Springer, Heidelberg (2007)

[50] Wang, H., Wang, N., Wang, D.: Multi-swarm optimization algorithm for dynamic optimization problems using forking. In: Proc. 2008 Chinese Control and Decision Conf., pp. 2415–2419 (2008)

[51] Wang, H., Wang, D., Huang, M.: Memetic algorithms in dynamic environments. Kongzhi Lilun Yu Yingyong/Control Theory and Applications 27(8), 1060–1068 (2010)

[52] Zheng, X., Liu, H.: A different topology multi-swarm pso in dynamic environment. In: Proc. 2009 IEEE Int. Symp. IT in Medicine Education, vol. 1, pp. 790–795 (2009)

[53] Zheng, X., Liu, H.: A cooperative dual-swarm pso for dynamic optimization problems. In: Proc. 7th Int. Conf. Natural Comput., vol. 2, pp. 1131–1135 (2011)

[54] Yang, S., Li, C.: A clustering particle swarm optimizer for locating and tracking multiple optima in dynamic environments. IEEE Trans. Evol. Comput., 959–974 (2010)

[55] Yin, Y., Sun, L.: Generalized dynamic constraint satisfaction based on extension particle swarm optimization algorithm for collaborative simulation. In: Proc. 10th IEEE Int. Conf. Computer-Aided Design and Computer Graphics, pp. 541–544 (2007)

Chapter 6
Memetic Algorithms for Dynamic Optimization Problems

Hongfeng Wang and Shengxiang Yang

Abstract. Dynamic optimization problems challenge traditional evolutionary algorithms seriously since they, once converged, cannot adapt quickly to environmental changes. This chapter investigates the application of memetic algorithms, a class of hybrid evolutionary algorithms, for dynamic optimization problems. An adaptive hill climbing method is proposed as the local search technique in the framework of memetic algorithms, which combines the features of greedy crossover-based hill climbing and steepest mutation-based hill climbing. In order to address the convergence problem, a new immigrants scheme, where the immigrant individuals can be generated from mutating an elite individual adaptively, is also introduced into the proposed memetic algorithm for dynamic optimization problems. Based on a series of dynamic problems generated from several stationary benchmark problems, experiments are carried out to investigate the performance of the proposed memetic algorithm in comparison with some peer algorithms. The experimental results show the efficiency of the proposed memetic algorithm in dynamic environments.

6.1 Introduction

Many real-world optimization problems are dynamic optimization problems (DOPs), where the function landscapes may change over time and, thus, the optimum of these problems may also change over time. DOPs require powerful heuristics that account for the uncertainty present in the real world. Since evolutionary algorithms (EAs)

Hongfeng Wang
College of Information Science and Engineering, Northeastern University,
Shenyang 110004, China
e-mail: hfwang@mail.neu.edu.cn

Shengxiang Yang
Centre for Computational Intelligence (CCI), School of Computer Science and Informatics,
De Montfort University, The Gateway, Leicester LE1 9BH, U.K.
e-mail: syang@dmu.ac.uk

S. Yang and X. Yao (Eds.): *Evolutionary Computation for DOPs*, SCI 490, pp. 137–170.
DOI: 10.1007/978-3-642-38416-5_6   © Springer-Verlag Berlin Heidelberg 2013

draw their inspiration from the principles of natural evolution, which is a stochastic and dynamic process, they also seem to be suitable for DOPs. However, traditional EAs face a serious challenge for DOPs because they cannot adapt well to the changing environment once converged.

In order to address DOPs, many approaches have been developed [46] and can be grouped into five categories: 1) increasing population diversity after a change is detected, such as the adaptive mutation methods [5, 38]; 2) maintaining population diversity throughout the run, such as the immigrants approaches [12, 44]; 3) memory approaches, including implicit [11, 37] and explicit memory [3, 40, 43, 48] methods; 4) multi-population [4, 27] and speciation approaches [30]; 5) prediction methods [2, 24, 31, 32]. A comprehensive survey on EAs applied to dynamic environments can be found in [15, 25, 45]

In recent years, there has been an increasing concern from the evolution computation community on a class of hybrid EAs, called memetic algorithms (MAs), which hybridize local search (LS) methods with EAs to refine the solution quality. So far, MAs have been widely used for solving many optimization problems, such as scheduling problems [14, 20, 21], combinatorial optimization problems [9, 35, 36], multi-objective problems [10, 13, 19] and other applications [49, 50]. However, these problems for which MAs have been applied are mainly stationary problems. MAs have rarely been applied for DOPs [7, 8, 41]. During the running course of general MAs, they may always exhibit very strong exploitation capacity due to executing efficient local refinement on individuals, but they may lose the exploration capacity as a result of the population converging to one optimum, which needs to be avoided in dynamic environments. Therefore, it becomes an interesting research issue to examine the performance of MAs, which are enhanced by suitable diversity methods, for DOPs.

In this chapter, we investigate the application of an MA with an adaptive hill climbing strategy, which combines the features of crossover-based hill climbing and mutation-based hill climbing in both cooperative and competitive fashions, to address DOPs. In order to address the convergence problem, a new immigrants scheme, where three different immigrants schemes are generalized within a uniform framework, is introduced into the proposed MA to improve its performance in dynamic environments. In addition, two different dominated schemes, which are used to keep the balance of extra computation costs between LS and diversity maintaining, are experimentally investigated in the proposed MAs for DOPs.

The rest of this paper is organized as follows. Section 6.2 describes the proposed MA in detail, including the framework of general genetic algorithm (GA)-based MA, the proposed LS operators and immigrants scheme, and discussion on how to determining the computation costs of LS and diversity maintaining. Section 6.3 introduces a series of DOPs generated by a dynamic problem generator from the stationary test suite. Section 6.4 reports the experimental results and relevant analysis. Finally, Section 6.5 concludes this paper with some discussions on relevant future work.

```
Procedure General GA-based MA:
begin
    parameterize();
    t := 0;
    initializePopulation(P(0));
    evaluatePopulation(P(0));
    E(0) := selectForLocalSearch(P(0));
    localSearch(E(0));
    repeat
        P'(t) := selectForReproduction(P(t));
        P''(t) := crossover(P'(t));
        mutate(P''(t));
        evaluatePopulation(P''(t));
        P(t + 1) := selectForSurvival(P(t), P''(t));
        E(t) := selectForLocalSearch(P(t));
        localSearch(E(t));
        t := t + 1;
    until a stop condition is met
end
```

Fig. 6.1 Pseudo-code for a general GA-based MA

6.2   Investigated Algorithms

6.2.1   Framework of GA-Based Memetic Algorithms

The MAs investigated in this paper are a class of GA-based MAs, which can be expressed by the pseudo-code shown in Fig. 6.1. Within these MAs, a population $P(0)$ of pop_size individuals are generated randomly and then evaluated at the initialization step. Then, a set $E(0)$ of individuals are selected from $P(0)$ to be improved by LS. At each subsequent generation, individuals are selected randomly or proportionally from the current population and undergo the uniform crossover operation with a crossover probability. Uniform crossover is the generalization of n-point crossover which creates offspring by deciding, for each bit of one parent, whether to swap the allele of that bit with the corresponding allele of the other parent. After crossover is executed, the bit-wise mutation operator is performed for each newly generated offspring individual, which may change the allele in each locus of an offspring bitwise (0 to 1 and vice versa) with a mutation probability. Then, the pop_size best individuals among all parents and offspring are selected to proceed into the next generation and a set $E(t)$ of individuals, which are selected from the newly generated population, are improved by the LS strategy.

Obviously, the LS procedure in an MA which is illustrated in the above pseudo-code would include two steps: first, to select individuals from the population P to construct the set E, and then to apply LS operation to each selected individual in E

to refine it. In the following section, we will give the relevant schemes on the LS operator used in this chapter, where the empirical experience and theoretical reasoning are both used in the design. Here, the aim that the above-mentioned pseudo-code is introduced is only to provide a sound basis for understanding the general framework of used MAs in this chapter.

It is more noticeable that some diversity maintaining approaches have to be used in order to address the convergence problem when an MA is applied for DOPs, which always consume extra computational cost since LS operators involve fitness evaluations. We will also discuss the relevant solutions to the problems when diversity schemes are applied and how to keep the balance of the extra computational cost between the diversity schemes and the LS operators, in the later sections.

6.2.2 Local Search

In MAs, EA operations are used for rough global exploration and LS operations are used for directive local refinements to ensure sufficient exploitation during the course of evolving the population. In many relevant researches, LS is applied to each newly generated individual, which would consume a huge number of extra fitness evaluations in the search of higher-quality solution. This traditional scheme seems to be too costly and infeasible for MAs in dynamic environments where the environmental changes can occur with the increment of fitness evaluations. Here, only the best fitness individual *elite* in the population, which means that the set E only comprises one member, should be executed the local refinement considering that *elite* would lead the running course of algorithm with a greater degree.

Another primary issue that affects the behavior of LS is which LS operator should be used to improve the quality of an individual. Among many LS methods available in the literature, hill climbing (HC) is a common strategy. The basic idea is to use stochastic iterative HC as the move acceptance criterion of the search (i.e., move the search from the current individual to a candidate individual if the candidate has a better fitness). In the context of GAs, HC methods may be divided into two ways: crossover-based hill climbing and mutation-based hill climbing, depending on whether crossover or mutation is used as the move operator in a local area. Here, we propose two HC methods, a greedy crossover-based HC (GCHC) and a steepest mutation-based HC (SMHC), in this section. They are specially designed for MAs with binary encoding scheme, which are our concern in this chapter. The two HC methods are described as follows.

1) GCHC: In this strategy, the individual (*chr*) selected for local improvement is taken as one parent and another parent is selected from the current population using the roulette wheel selection scheme. Then, a special uniform crossover is executed between these two parent individuals to generate an offspring. The offspring will replace the individual *chr* if it has a better fitness than the latter. This procedure is outlined in Fig. 6.2, where a maximization optimization problem is assumed.

2) SMHC: The steepest mutation means that the chromosome only changes several bits randomly when executing one mutation operation on it. In SMHC, the

```
Procedure GCHC(chr):
begin
    calculate(ξ, pc_ls);
    for i := 1 to num_ls do
        par_chr := selectParentForCrossover(P);
        for j := 1 to n do
            if random() < pc_ls then
                chi_chr[j] := par_chr[j];
            else
                chi_chr[j] := chr[j];
        endfor
        evaluate(chi_chr);
        if f(chi_chr) > f(chr) then chr := chi_chr;
    endfor
end

GCHC's denotations:
    ξ: a population index used to renew the value of pc_ls
    pc_ls: the crossover probability in GCHC
    par_chr: the proportionally selected parent individual
    n: individual length (problem dependent)
    random(): a random number between 0 and 1
    chi_chr: the new individual generated by performing a uniform crossover operation
        between par_chr and chr
    chr: the individual selected for LS
```

Fig. 6.2 Pseudo-code for the GCHC operator

individual (chr) being improved by LS is picked out and several random bits are changed. If the newly mutated individual has a better fitness, it will replace the individual chr. The SMHC strategy is outlined in Fig. 6.3.

From Fig. 6.2 and Fig. 6.3, it can be seen that two important parameters, pc_{ls} in GCHC and nm_{ls} in SMHC, respectively, may affect their performance. In GCHC the smaller the value of pc_{ls}, the more the offspring inherits from the selected individual chr. This means executing one step LS operation in a smaller area around chr. Similar results can be obtained for nm_{ls} in SMHC. When the value of nm_{ls} is larger, SMHC will perform the LS operation within a wider range around chr.

Therefore, the question that remains to be answered here is how to set the two parameters. Generally speaking, the methods of setting strategy parameters in GAs can be classified into three categories [6]: *deterministic mechanism* where the value of the strategy parameter is controlled by some deterministic rules without any feedback from the search, *adaptive mechanism* where there is some form of feedback information from the search process that is used to direct the setting of a strategy

```
Procedure SMHC(chr):
begin
   calculate(ξ, nm_ls);
   for i := 1 to num_ls do
      for j := 1 to n do
         chi_chr[j] := chr[j];
      endfor
      for k := 1 to nm_ls do
         loc := random(1, n);
         chi_chr[loc] := 1 - chi_chr[loc];
      endfor
      evaluate(chi_chr);
      if f(chi_chr) > f(chr) then chr := chi_chr;
   endfor
end;
```

SMHC's denotations:

nm_{ls}: the number of bits mutated in SMHC

chi_chr: the new individual generated by performing a steepest mutation operation upon chr

loc: a random selected location for flipping

$random(1, n)$: a random integer between 1 and n

Other settings are the same as those for GCHC

Fig. 6.3 Pseudo-code for the SMHC operator

parameter, and *self-adaptive mechanism* where the parameter to be adapted is encoded into the chromosomes and undergoes genetic operators.

Two different parameter-setting methods will be discussed for pc_{ls} and nm_{ls} in the later experiments. In the *deterministic* method, both pc_{ls} and nm_{ls} are set to constant values, which means that the LS operation will always be executed in a local area of a certain fixed range. In the *adaptive* method, a population index ξ which can measure the diversity of the population is considered as the feedback information to direct the change of the values of pc_{ls} and nm_{ls}.

Let the normalized Hamming distance between two individuals $\mathbf{x}_i = (x_{i1}, \ldots, x_{in})$ and $\mathbf{x}_j = (x_{j1}, \ldots, x_{jn})$ be defined by:

$$d(\mathbf{x}_i, \mathbf{x}_j) = \frac{\sum_{k=1}^{n} |x_{ik} - x_{jk}|}{n} \tag{6.1}$$

and ξ is calculated by the following formula:

$$\xi = \frac{\sum_{i=1}^{pop\_size} d(\mathbf{x}^*, \mathbf{x}_i)}{pop\_size}, \tag{6.2}$$

where $\mathbf{x}^*$ denotes the best individual achieved so far. Obviously, the index ξ can measure the convergence state of the population via the Hamming distance. When ξ decreases to zero, it means that the population has lost its diversity absolutely. With the definition of ξ, pc_{ls} and nm_{ls} can be calculated as follows:

$$pc_{ls} = min\{\alpha \cdot \xi \cdot (pc_{ls}^{max} - pc_{ls}^{min}) + pc_{ls}^{min}, pc_{ls}^{max}\} \qquad (6.3)$$

$$nm_{ls} = min\{\beta \cdot \xi \cdot (nm_{ls}^{max} - nm_{ls}^{min}) + nm_{ls}^{min}, nm_{ls}^{max}\}, \qquad (6.4)$$

where pc_{ls}^{max} and pc_{ls}^{min} are the preset maximum and minimum value of pc_{ls} respectively (pc_{ls}^{max}=0.6 and pc_{ls}^{min}=0.1 in the later experiments), nm_{ls}^{max} and nm_{ls}^{min} are the preset maximum and minimum value of nm_{ls} respectively (nm_{ls}^{max}=4 and nm_{ls}^{min}=1 in the later experiments), and α and β are the predefined constants to control the decreasing or increasing speed of pc_{ls} and nm_{ls} respectively. From these formulae, it is easy to understand that both GCHC and SMHC exhibit a wide range LS operations in the presence of a high population diversity (i.e., when $\xi \to 1$) as a result of $pc_{ls} \to pc_{ls}^{max}$ and $nm_{ls} \to nm_{ls}^{max}$. This may help algorithms find the optimum (maybe local optimum) more quickly. However, when the population is converging (i.e., when $\xi \to 0$), $pc_{ls} \to pc_{ls}^{min}$ and $nm_{ls} \to nm_{ls}^{min}$, which limits the LS operations in a very small range in order to perform more efficient local improvement for the selected individual *chr*.

6.2.3   Adaptive Learning Mechanism in Multiple LS Operators

It has been reported that multiple LS operators can be employed in an MA framework [22, 33, 34]. This is because each LS operator makes a biased search, which makes a method efficient for some classes of problems but not efficient for others. That is, LS is problem-dependent. Therefore, how to achieve improved LS operators and avoid utilizing inappropriate LS methods becomes a very important issue. In order to address this problem, many researchers have used multiple LS methods in their MAs. In comparison with traditional MAs that use a single LS operator throughout the run, MAs with multiple LS methods can usually obtain a better performance.

The key idea of using multiple LS operators in MAs is to promote the cooperation and competition of different LS operators, enabling them to work together to accomplish the shared optimization goal. Some researchers [16, 26] have suggested that multiple LS operators should be executed simultaneously on those individuals that are selected for local improvements and that a certain learning mechanism should be adopted to give the efficient LS methods greater chances to be chosen in the later stage. However, Neri *et al.* [23] have also proposed a multiple LS based MA with a non-competitive scheme, where different LS methods can be activated during different population evolution periods. Inspired by these researches, a learning mechanism is discussed to hybridizes the GCHC and SMHC methods described in Section 6.2.2 effectively and an adaptive hill climbing (AHC) strategy is introduced in this study.

In AHC, the GCHC and SMHC operators are both allowed to work in the whole LS loop and are selected by probability to execute one step LS operation at every generation when the MA is running. Let p_{gchc} and p_{smhc} denote the probabilities of applying GCHC and SMHC to the individual that is used for a local search respectively, where $p_{gchc}+p_{smhc}=1$. At the start of this strategy, p_{gchc} and p_{smhc} are both set to 0.5, which means giving a fair competition chance to each LS operator. As each LS operator always makes a biased search, the LS operator which produces more improvements should be given a greater selection probability. Here, an adaptive learning approach is used to adjust the value of p_{gchc} and p_{smhc} for each LS operator. Let η denote the improvement degree of the selected individual when one LS operator is used to refine it and η can be calculated by:

$$\eta = \frac{|f_{imp} - f_{ini}|}{f_{ini}}, \qquad (6.5)$$

where f_{imp} is the final fitness of the selected individual chr after applying LS and f_{ini} is its initial fitness before executing LS operation. At each generation, the degree of improvement of each LS operator is calculated when a predefined number (num_{ls}) of iterations is achieved and then p_{gchc} and p_{smhc} are re-calculated to proceed with the next LS operation.

Suppose $\eta_{gchc}(t)$ and $\eta_{smhc}(t)$ respectively denote the total improvement of GCHC and SMHC at iteration t. The LS selection probabilities $p_{gchc}(t+1)$ and $p_{smhc}(t+1)$ at iteration $(t+1)$ can be calculated orderly by the following formulae:

$$p_{gchc}(t+1) = p_{gchc}(t) + \Delta \cdot \eta_{gchc}(t), \qquad (6.6)$$

$$p_{smhc}(t+1) = p_{smhc}(t) + \Delta \cdot \eta_{smhc}(t), \qquad (6.7)$$

$$p_{gchc}(t+1) = \frac{p_{gchc}(t+1)}{p_{gchc}(t+1) + p_{smhc}(t+1)}, \qquad (6.8)$$

$$p_{smhc}(t+1) = 1 - p_{gchc}(t+1), \qquad (6.9)$$

where Δ signifies the relative influence of the degree of the improvement on the selection probability. The AHC operator can be expressed by the pseudo-code in Fig. 6.4.

From the above discussion, the two different HC strategies, GCHC and SMHC, may not only cooperate to improve the quality of individuals, but also compete with each other to achieve a greater selection probability in the running process of AHC. To promote competition between them, the selection probability of LS operators can be re-calculated according to an adaptive learning mechanism where the LS operator with a higher fitness improvement is rewarded with more chance of being chosen for the subsequent individual refinement.

```
Procedure AHC(chr):
begin
    if p_gchc and p_smhc are not initialized then
        set p_gchc := p_smhc := 0.5;
    calculate(ξ, pc_ls, nm_ls);
    set η_gchc = η_smhc := 0;
    for i := 1 to num_ls do
        if random() < p_gchc then // GCHC is selected
            GCHC(chr);
            update(η_gchc);
        else // SMHC is selected
            SMHC(chr);
            update(η_smhc);
    endfor
    recalculate(p_gchc, p_smhc);
end;

AHC's denotations are the same as those for GCHC and SMHC
```

Fig. 6.4 Pseudo-code for the AHC operator

6.2.4   Diversity Maintaining

For stationary optimization problems, where the fitness landscape or objective function does not change during the course of computation, LS operators in MAs are designed for exploiting information in the current population and genetic operators, for example, mutation, are mostly responsible for enhancing the diversity of population in order to make an efficient jump from a local optimum. Generally speaking, the population will converge to a small area in the whole search space as a result of keeping the sufficient exploitation for the global optimum. Therefore, MAs may gradually loose their population diversity during the running. However, in dynamic environments, the fitness landscape may change over time. That is, the current optimum point may become a local optimum and the past local optimum may become a new global optimum point. Considering that a spread-out population can adapt to these changes more easily, it is very important and necessary to maintain sufficient diversity of the population for MAs all the time.

The immigrants scheme is a quite simple and common method to maintain the diversity level of the population through substituting a portion of individuals in the population with the same number of newly-generated immigrants every generation. Obviously, the method of generating new immigrants becomes a very important issue in this strategy. In the original scheme, immigrants are designed to be generated randomly. However, the random immigrants introduced may divert the searching force of an algorithm and hence degrade the performance due to their lower fitness level. Another thing to notice is that random individuals are not helpful for

```
Procedure GenerateImmigrants():
begin
    for i := 1 to num_im do
        P_im[i] := mutation(elite, pm_im);
        evaluate(P_im[i]);
    endfor
    replace the worst num_im individuals in the
        current population with P_im
end

Denotations:
    pm_im: the mutation probability in generating immigrants
    num_im: the number of generated immigrants
    elite: the best fitness individual in the population
    P_im: the generated immigrants set
```

Fig. 6.5 Pseudo-code for the general immigrants scheme

improving the diversity level when being inserted into a spread-out population. In the literature [44], a special immigrants scheme, called elitism-based immigrants, is proposed. In this approach, the elitism individual, which is the best fitness individual in the population, is used as a base to generate a set of immigrant individuals iteratively by a simple mutation. The key idea behind this scheme is that the immigrants can be guided to make a biased detection around the elite. However, it is also noticeable that this bias may take no effect in a converging population since all the individuals in the current population have distributed around its elite.

Inspired by the complementarity mechanism in nature, a primal-dual mapping operator, where two chromosomes with the maximal distance in the solution space are defined as a pair of primal-dual chromosomes, was be proposed and applied for GAs in dynamic environments. Here, a new immigrants scheme, called dual-based immigrants, is proposed with the hybridization the primal-dual mechanism and elitism-based immigrants scheme. The dual-based immigrants are generated by executing a simple mutation to the dual of the elite individual and replace the same number of worst individuals in the current population. It is obvious that the hyper-immigrants scheme can enhance the diversity level with the maximal degree, which is helpful to improve the performance of EAs with a converging population in dynamic environments especially in the significantly changing environments.

It seems very interesting that the above-mentioned three different immigrants schemes, that is, elitism-based immigrants, random immigrants, and dual-based immigrants, can actually be described in a general framework. As shown in Fig. 6.5, all immigrants are considered to be generated from mutating the elite individual with a probability. It is easy to understand that the mutation probability pm_{im} can be used to control the categories of immigrants. The generated immigrants belong to the elitism-based immigrants when the value of pm_{im} is very small (e.g., $pm_{im} \rightarrow 0$), while the dual-based immigrants can be achieved when executing the mutation to *elite* with a large probability (e.g., $pm_{im} \rightarrow 1$). When $pm_{im} = 0.5$, the mutation operation always creates random immigrants in fact.

The following problem to be addressed is how to set the mutation probability pm_{im}. Here, we will utilize the index of population diversity ξ, discussed in Section 6.2.2, to calculate the value of pm_{im} by the following formula.

$$pm_{im} = max\{min\{(pm_{im}^{max} + pm_{im}^{min} * random() - \xi), pm_{im}^{max}\}, pm_{im}^{min}\} \qquad (6.10)$$

where $random()$ is a random number between 0 and 1, pm_{im}^{max} and pm_{im}^{min} are the preset maximum and minimum value of pm_{im}, respectively ($pm_{im}^{max} = 0.95$ and $pm_{im}^{min} = 0.05$ in the later experiments). From this formula, it is easy to see that the elitism-based immigrants are inserted into the population when the population has a low diversity level ($\xi \rightarrow 1$) due to pm_{im} approaches to pm_{im}^{min}, while the dual-based immigrants are generated when the population is converging ($\xi \rightarrow 0$) as a result of $pm_{im} \rightarrow pm_{im}^{max}$.

6.2.5   Balance between Local Search and Diversity Maintaining

Based on the above description, an adaptive hill climbing operator, where two different LS operators are used in a cooperation fashion, is proposed to enhance the exploitation capacity of algorithm and hence accelerate its tracking the optimum, while a hybrid immigrants scheme, where the different category of immigrants are generated according to the convergence status of population, is used to maintain the population diversity of algorithm and then improve its performance in dynamic environments. However, it is noticeable that the extra computation cost, which means the numbers of fitness evaluation in this paper, would be produced during executing LS and generating immigrants (*step_ls* in LS and *num_im* in diversity scheme) every generation.

One main question that follows when both LS operation and diversity maintaining strategy are introduced into an algorithm is how to keep the balance between them via the extra computational cost. As the generation index is used to set the change period of environment in the later experiments, it is necessary to maintain a constant number of evaluations in each generation in order to have fair comparisons among our investigated MAs and other peer EAs. Therefore, the additional number of fitness evaluations per generation, denoted as *num_aepg*, is also fixed, that is, $num_{ls} + num_{im} = num\_aepg$. Based on this formula, two different methods can be considered to calculate the values of num_{ls} and num_{ls}.

The first one is dominated by the effect of LS that the value of num_{ls} is firstly calculated and then num_{im}'s value is achieved by ($num\_aepg - num_{ls}$). Let $num_{vls}(t)$ denote the number of valid LS operations, which means the corresponding index $\eta > 0$ after executing one LS operation, at the t-th generation, $num_{ls}(t+1)$ can be calculated by:

$$num_{ls}(t+1) = \begin{cases} max\{num_{ls}(t) \cdot \lambda_0, num_{ls}^{min}\}, & \text{if } num_{vls}(t) > 0 \\ min\{num_{ls}(t)/\lambda_0, num_{ls}^{max}\}, & \text{else} \end{cases} \qquad (6.11)$$

where λ_0 is a preset constant between 0 and 1, num_{ls}^{max} and num_{ls}^{min} are the preset maximum and minimum value of num_{ls} respectively ($num_{ls}^{max} = 0.75 * num\_aepg$ and $num_{ls}^{min} = 0.25 * num\_aepg$ in the later experiments). It is easy to understand from this formula that the LS operation would be further enhanced at the next generation, that is, the value of num_{ls} is increased, once there exists even one valid LS operation at a step.

The second method is dominated by the diversity level of the population, i.e., the value of num_{im} is firstly calculated and then num_{ls}'s value is achieved by ($num\_aepg - num_{im}$). Let $num_{im}(t)$ denote the number of generated immigrants at the t-th generation, $num_{ls}(t+1)$'s value can be calculated by the following formula.

$$num_{im}(t+1) = \begin{cases} max\{num_{im}(t) \cdot \lambda_1, num_{im}^{min}\}, & \text{if } \xi > \xi_0 \\ min\{num_{im}(t)/\lambda_1, num_{im}^{max}\}, & \text{else} \end{cases} \qquad (6.12)$$

where λ_1 is a preset constant between 0 and 1 to signify the influence degree of the population diversity upon num_{im}, ξ_0 is also a preset constant between 0 and 1, num_{im}^{max} and num_{im}^{min} are the preset maximum and minimum value of num_{im} respectively ($num_{im}^{max} = 0.75 * num\_aepg$ and $num_{im}^{min} = 0.25 * num\_aepg$ in the later experiments). It is easy to see from this formula that the number of generated immigrants would be increased to maintain the diversity level of the population once its converge degree is below a threshold (ξ_0).

Therefore, the framework of our proposed MAs in this chapter, which hybridizes the AHC operator and immigrants scheme for DOPs, can be summarized by the pseudo-code in Fig. 6.6.

6.3 Dynamic Test Environments

In this chapter, a series of dynamic test environments are constructed by a specific dynamic problem generator from a set of well-studied stationary problems.

Four 100-bit binary-coded functions, denoted OneMax, Plateau, RoyalRoad, and Deceptive respectively, are selected as the stationary functions to construct dynamic test environments. Each stationary function consists of 25 copies of 4-bit building blocks and has an optimum value of 100. Each building block for the four functions is a unitation-based function, as shown in Fig. 6.7. The unitation function of a bit string returns the number of ones inside the string. The building block for OneMax is an OneMax subfunction, which aims to maximize the number of ones in a bit string. The building block for Plateau contributes 4 (or 2) to the total fitness if its unitation is 4 (or 3); otherwise, it contributes 0. The building block for RoyalRoad contributes 4 to the total fitness if all its four bits are set to one; otherwise, it contributes 0. The building block for Deceptive is a fully deceptive sub-function. Generally speaking, the four functions have an increasing difficulty for GAs in the order from OneMax to Plateau, RoyalRoad to Deceptive.

Procedure Proposed GA-based MA:
begin
   parameterize();
   $t := 0$;
   initializePopulation($P(0)$);
   evaluatePopulation($P(0)$);
   calculate($\xi, pc_{ls}, nm_{ls}, pm_{im}$);
   set $num_{ls}(0) := num_{im}(0) := num\_aepg/2$;
   $elite :=$ selectForLocalSearch($P(0)$);
   AHC($elite$);
   GenerateImmigrants();
   repeat
     $P'(t) :=$ selectForReproduction($P(t)$);
     $P''(t) :=$ crossover($P'(t)$);
     mutate($P''(t)$);
     evaluatePopulation($P''(t)$);
     $P(t+1) :=$ selectForSurvival($P''(t), P(t)$);
     calculate($\xi, pc_{ls}, nm_{ls}, pm_{im}$);
     calculate($num_{ls}(t+1), num_{im}(t+1)$);
     $elite :=$ selectForLocalSearch($P(t+1)$);
     AHC($elite$);
     GenerateImmigrants();
     $t := t+1$;
   until a stop condition is met
end;

Fig. 6.6 Pseudo-code for the proposed GA-based MAs

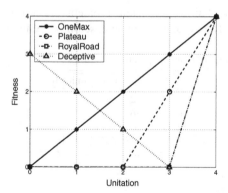

Fig. 6.7 The building blocks for the four stationary functions selected to construct dynamic test problems in this chapter

In [42, 47], an XOR DOP generator was proposed. The XOR DOP generator can generate dynamic environments from any binary-encoded stationary function $f(\mathbf{x})$ ($\mathbf{x} \in \{0,1\}^l$) by a bitwise exclusive-or (XOR) operator. The environment is changed every τ generations. For each environmental period k, an XOR mask $\mathbf{M}(k)$ is incrementally generated as follows:

$$\mathbf{M}(k) = \mathbf{M}(k-1) \oplus \mathbf{T}(k), \qquad (6.13)$$

where "$\oplus$" is the XOR operator (i.e., $1 \oplus 1 = 0$, $1 \oplus 0 = 1$, $0 \oplus 0 = 0$) and $\mathbf{T}(k)$ is an intermediate binary template randomly created with $\rho \times l$ ones for the k-th environmental period. For the first period $k = 1$, $\mathbf{M}(1) = \mathbf{0}$. Then, the population at generation t is evaluated as:

$$f(\mathbf{x},t) = f(\mathbf{x} \oplus \mathbf{M}(k)), \qquad (6.14)$$

where $k = \lceil t/\tau \rceil$ is the environmental index. One advantage of this XOR generator lies in that the speed and severity of environmental changes can be easily tuned. The parameter τ controls the speed of changes while $\rho \in (0.0, 1.0)$ controls the severity of changes. A bigger ρ means more severe changes while a smaller τ means more frequent changes.

The dynamic test environments used in this paper are constructed from the four stationary functions using the aforementioned XOR DOP generator. The change severity ρ parameter is set to 0.1, 0.2, 0.5, and 0.9 respectively in order to examine the performance of algorithms in dynamic environments with different severities: from slight change ($\rho = 0.1$ or 0.2) to moderate variation ($\rho = 0.5$) to intense change ($\rho = 0.9$). The change speed parameter τ is set to 10, 50, and 100 respectively, which means that the environment changes very fast, in the moderate speed, and slowly respectively.

In total, a series of 12 different dynamic problems are constructed from each stationary test problem. The dynamics parameter settings are summarized in Table 6.1.

Table 6.1 The index table for dynamic parameter settings

τ	Environmental Dynamics Index			
10	1	2	3	4
50	5	6	7	8
100	9	10	11	12
$\rho \to$	0.1	0.2	0.5	0.9

6.4 Experimental Study

6.4.1 Experimental Design

In this section, experiments are carried out in order to study the major features of our proposed MAs and to compare their performance with several existing peer algorithms where similar dualism and immigrants methods are also used. The following abbreviations represent the algorithms considered in this paper:

- SGA: Standard GA;
- SGAr: SGA with restart from scratch whenever the environment changes;
- EIGA: GA with the elitism-based immigrants scheme [44];
- RIGA: GA with the random immigrants scheme;
- DIGA: GA with the dual-based immigrants scheme (seen in Section 6.2.4);
- CHMA: MA with the GCHC operator;
- MHMA: MA with the SMHC operator;
- AHMA: MA with the AHC operator;
- EIAHMA: AHMA with the elitism-based immigrants scheme;
- RIAHMA: AHMA with the random immigrants scheme;
- DIAHMA: AHMA with the dual-based immigrants scheme;
- IMAHMA: AHMA with the proposed immigrants scheme in Section 6.2.4;

Some parameters in all algorithms were set as follows. The total number of evaluations per generation was always set to 120 for all algorithms, which means the population size (pop_size) is equal to 120 for SGA and SGAr. In all MAs, EIGA, RIGA and DIGA, pop_size was set to 100 and the additional number of fitness evaluations per generation (num_aepg) was set to 20. The uniform crossover probability pc was set to 0.6 and the bit-wise mutation probability pm was set to 0.01 for all GAs and MAs. The specific parameters in our MAs were set as follows: $\alpha = \beta = 1$, $\Delta = 4$, $\lambda_0 = \lambda_1 = 0.9$, and $\theta_0 = 0.5$. Other parameters in the studied peer algorithms were set the same as their original settings.

For each experiment of an algorithm on a test problem, 20 independent runs were executed with the same set of random seeds. For each run of an algorithm on a DOP, 10 environmental changes were allowed and the best-of-generation fitness was recorded per generation. The overall offline performance of an algorithm is defined as the best-of-generation fitness averaged across the number of total runs and then averaged over the data gathering period, as formulated below:

$$\overline{F}_{BG} = \frac{1}{G} \sum_{i=1}^{G} \left(\frac{1}{N} \sum_{j=1}^{N} F_{BG_{ij}} \right), \tag{6.15}$$

where G is the number of generations (i.e., $G = 10 * \tau$), $N = 20$ is the total number of runs, and $F_{BG_{ij}}$ is the best-of-generation fitness of generation i of run j.

In order to measure the behavior of an algorithm during the course of running, another numeric measure is defined as the best-of-generation fitness averaged across the number of total runs and then averaged from the last change generation τ' to the current generation t. More formally, the running offline performance is defined as:

$$\overline{F}_{BG_t} = \frac{1}{t - \tau'} \sum_{i=\tau'}^{t-\tau'} \left(\frac{1}{N} \sum_{j=1}^{N} F_{BG_{ij}} \right) \tag{6.16}$$

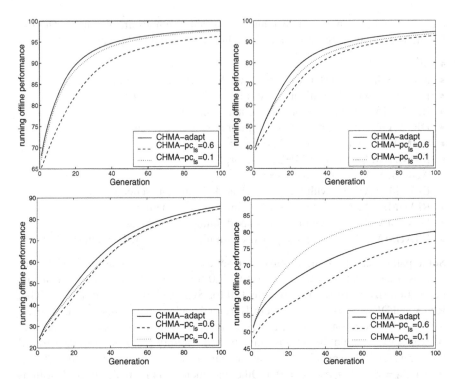

Fig. 6.8 Experimental results with respect to the running offline performance of CHMAs with different pc_{ls} settings on stationary test problems: (a) OneMax, (b) Plateau, (c) RoyalRoad, and (d) Deceptive

6.4.2 Experimental Study on the Effect of LS Operators

In the experimental study on LS operators, we first study the influence of different settings of pc_{ls} in CHMA and nm_{ls} in MHMA, with the aim of determining a robust setting for these two parameters. In particular, we have implemented CHMA just on stationary test problems. Three different settings for pc_{ls} were used: $pc_{ls} = 0.6$ and $pc_{ls} = 0.1$ in the *deterministic* setting and $pc_{ls}^{max} = 0.6$ and $pc_{ls}^{min} = 0.1$ in the *adaptive* setting scheme (see Section 6.2.2). For each run of an algorithm on each problem, the maximum allowable number of generations was set to 100^{1}. The experimental results are shown in Fig. 6.8, where the data were averaged over 20 runs. The results on the Plateau problem are similar to the results on the RoyalRoad problem and are not shown in Fig. 6.8.

From Fig. 6.8, it can be seen that CHMA with *adaptive* pc_{ls} always outperforms CHMAs with the *deterministic* value of pc_{ls} on the OneMax, Plateau and RoyalRoad problems and that a smaller pc_{ls} can help CHMA obtain a better perfor-

---

[1] The number of maximum allowable fitness evaluations is actually 12000 since each algorithm has 120 fitness evaluations per generation.

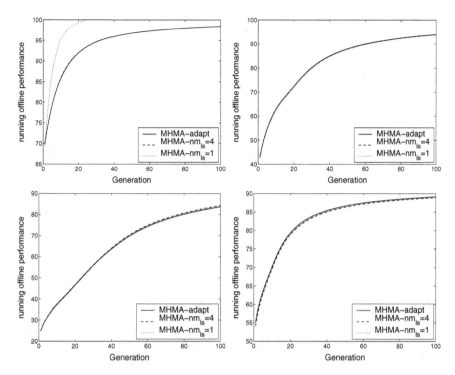

Fig. 6.9 Experimental results with respect to the running offline performance of MHMAs with different nm_{ls} settings on stationary test problems: (a) OneMax, (b) Plateau, (c) RoyalRoad, and (d) Deceptive

mance on the Deceptive problem. Hence, the *adaptive* setting scheme for pc_{ls} will always be used in the following experiments considering that the *deterministic* setting scheme is problem-dependent and the *adaptive* scheme for pc_{ls} always shows a better adaptive capacity on different problems.

Similar experiments were also carried out to test the influence of different settings of nm_{ls} on the performance of MHMA. The value of nm_{ls} was set to 4 and 1 respectively for the *deterministic* scheme and $nm_{ls}^{max} = 4$ and $nm_{ls}^{min} = 1$ in the *adaptive* setting scheme (see Section 6.2.2). The experimental results with respect to the running offline performance are presented in Fig. 6.9.

From Fig. 6.9, it can be observed that the performance curves of the three MHMAs almost overlap together on the Plateau, RoyalRoad and Deceptive problems except that MHMA with $nm_{ls} = 1$ performs better than MHMA with adaptive nm_{ls} and MHMA with $nm_{ls} = 4$ on the OneMax problem. This indicates that adaptively varying the search range of the SMHC operator may not improve the performance of MHMA remarkably. Hence, the value of nm_{ls} will always be set to 1 in the later experiments.

In the following experiments, we investigate the performance of AHMA, MHMA, CHMA and SGA on the stationary test problems in order to examine the validity of

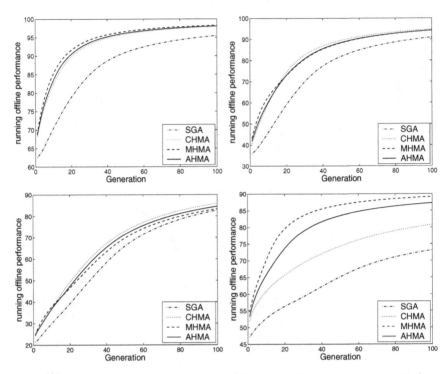

Fig. 6.10 Experimental results with respect to the running offline performance of MAs and SGA on stationary test problems: (a) OneMax, (b) Plateau, (c) RoyalRoad, and (d) Deceptive

our proposed AHC operator. The experimental results with respect to the running offline performance are shown in Fig. 6.10.

From Fig. 6.10, it can be seen that all MAs outperform SGA significantly on all test problems. This shows that proper LS techniques (here AHC in AHMA, SMHC in MHMA, and GCHC in CHMA) can help MAs obtain a much better performance than SGA. Of course, these conclusions have been drawn by many researchers. MHMA exhibits the best performance on the OneMax and Deceptive problems, while CHMA performs best on the Plateau and RoyalRoad problems among the three MAs, which shows that the effect of LS operators is problem-dependent. It can also be seen that AHMA always shows good adaptivity on the four test problems, where AHMA performs better than CHMA on the OneMax and Deceptive problems and better than MHMA on the Plateau and RoyalRoad problems. The results indicate that AHC always does well although it needs to take some time to adjust its LS strategy. Since it is almost impossible for an algorithm to achieve all the characters of a problem in advance, the combination of multiple LS operators within a single MA framework is a good choice for solving optimization problems.

In the above experimental studies, only the elite chromosome is selected for local refinement in order to decrease extra cost due to the fitness evaluations by LS

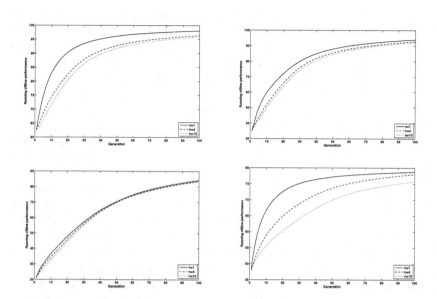

Fig. 6.11 Experimental results with respect to the running offline performance of AHMAs on stationary test problems: (a) OneMax, (b) Plateau, (c) RoyalRoad, and (d) Deceptive

operations. In the final experiments on LS operators, we further investigate the influence of the number of selected individuals for LS upon the performance of AHMA. In order to make a fair comparison, the number of additional fitness evaluations per generation (num_aepg) is always set to 20 and the number of selected individuals (n) is set to 1, 4 and 10, respectively, which means the corresponding iteration number of LS is 20, 5 and 2, respectively. In addition, only the best n individuals in the population would undergo the AHC operation. The experimental results with respect to the running offline performance are shown in Fig. 6.11.

From Fig. 6.11, it can be seen that the performance of AHMAs degrades with the increment of the number of selected individuals for LS, especially on the OneMax, RoyalRoad and Deceptive problems. This means that only executing sufficient local refinement upon the best individual *elite* is a good choice when the extra cost is limited.

6.4.3   Experimental Study on the Effect of Diversity Maintaining Schemes

Immigrants scheme is a common strategy to address the convergence problem of EAs in dynamic environments. In Section 6.2.4, three different immigrants schemes are integrated into a general framework, where different categories of immigrants can be generated by executing the simple mutation on the best fitness individual (*elite*) with different probabilities. In order to investigate the effect of proposed immigrants schemes upon the performance of algorithms in dynamic environments, we firstly carry out the following experiments on AHMAs with different immigrants schemes on DOPs with $\tau=50$ and ρ set to 0.1, 0.2, 0.5 and 0.9, respectively. In all AHMAs, the first method of calculating the values of num_{ls} and num_{im} are used.

The experimental results with respect to the overall offline performance are presented in Table 6.2. The corresponding statistical results of comparing algorithms by the one-tailed t-test with 38 degrees of freedom at a 0.05 level of significance are given in Table 6.3. In Table 6.3, the t-test results regarding Alg. 1–Alg. 2 are shown as "+", "−", or "∼" when Alg. 1 is significantly better than, significantly worse than, or statistically equivalent to Alg. 2, respectively. From Table 6.2 and Table 6.3, several results can be observed and are analyzed below.

First, EIAHMA performs better than RIAHMA and DIAHMA on all dynamic OneMax problems and dynamic Plateau problems when the change severity is small. This is because the immigrant individuals generated by the elitism mechanism can always make a positive guide for the search of the algorithm on the OnmeMax problem. In addition, the similar effect of this immigrants scheme can be obtained when the environmental change is slight, e.g., dynamic Plateau problems with $\rho = 0.1$ and $\rho = 0.2$. It can also be seen that the performance of EIAHMA begins to degrade with the increasing of the value of ρ. When $\rho = 0.5$, EIAHMA performs much worse than RIAHMA on dynamic Plateau, RoyalRoad and Deceptive problems. When the value of ρ increases to 0.9, EIAHMA is always beaten by DIAHMA with a great degree.

Second, RIAHMA exhibits the best performance on most dynamic environments when a random environmental change occurs ($\rho=0.5$). This is easy to understand. When the environment changes with $\rho = 0.5$, almost all building blocks found so far are demolished. Obviously, RIAHMA can adapt to this environmental change more easily since the newly-generated random immigrants always do better than those biased immigrants, that is, the elitism-based immigrants in EIAHMA and the dual-based immigrants in DIAHMA, in re-achieving the demolished building blocks.

Third, DIAHMA performs better than EIAHMA and RIAHMA on dynamic RoyalRoad problems when $\rho = 0.9$ and on all dynamic Deceptive problems. The reason lies in that the dual mechanism may help DIAHMA react to significant environmental changes rapidly and also enable it to escape from the deceptive attractor in the Deceptive problem.

Fourth, IMAHMA always exhibits good performance in most dynamic environments except that it is beaten entirely by EIAHMA on dynamic OneMax problems with the different settings of change severity ρ. In IMAHMA, the probability of generating immigrants (pm_{im}) can be calculated adaptively according to the convergence degree of population. It means IMAHMA can generate different categories of immigrants to adapt well to dynamic environments with different change severities. The experimental results of the good adaptivity of IMAHMA in dynamic environments confirm our expectation of introducing the proposed immigrants scheme in Section 6.2.4 to AHMA for DOPs.

Table 6.2 Experimental results with respect to overall offline performance of AHMAs on dynamic test problems

Dynamics		OneMax Problem			
τ	ρ	IMAHMA	EIAHMA	RIAHMA	DIAHMA
50	0.1	97.11±0.27	97.58±0.19	97.00±0.27	97.08±0.28
50	0.2	95.11±0.37	95.86±0.27	95.03±0.27	95.10±0.24
50	0.5	92.99±0.34	93.54±0.39	93.73±0.32	92.97±0.30
50	0.9	97.18±0.17	97.52±0.31	97.06±0.19	97.09±0.21
Dynamics		Plateau Problem			
τ	ρ	IMAHMA	EIAHMA	RIAHMA	DIAHMA
50	0.1	93.62±0.45	93.76±0.82	93.62±0.42	93.48±0.54
50	0.2	87.53±0.71	87.93±0.98	87.41±0.93	87.41±0.83
50	0.5	81.94±1.00	79.51±1.70	83.65±0.77	81.25±1.11
50	0.9	93.60±0.66	93.48±1.95	93.42±0.60	93.54±0.46
Dynamics		RoyalRoad Problem			
τ	ρ	IMAHMA	EIAHMA	RIAHMA	DIAHMA
50	0.1	85.77±1.19	80.97±3.15	85.64±1.14	85.76±1.13
50	0.2	73.96±1.52	69.76±1.81	73.70±1.82	73.88±1.82
50	0.5	63.83±1.83	55.08±2.20	65.65±1.73	61.55±1.36
50	0.9	85.23±1.02	79.67±4.93	84.93±1.21	85.37±0.94
Dynamics		Deceptive Function			
τ	ρ	IMAHMA	EIAHMA	RIAHMA	DIAHMA
50	0.1	83.44±0.46	76.57±1.42	76.72±1.66	83.99±0.72
50	0.2	81.24±0.47	74.47±0.82	74.14±1.02	82.19±0.54
50	0.5	80.29±0.51	73.50±0.72	74.34±0.62	81.19±0.70
50	0.9	84.07±0.60	77.02±1.45	77.22±1.71	84.06±0.60

Table 6.3 The t-test results of comparing the overall offline performance of AHMAs on dynamic test problems

t-test Result		OneMax				Plateau				RoyalRoad				Deceptive			
$\tau = 50, \rho \Rightarrow$		0.1	0.2	0.5	0.9	0.1	0.2	0.5	0.9	0.1	0.2	0.5	0.9	0.1	0.2	0.5	0.9
IMAHMA – EIAHMA		−	−	−	−	~	~	+	~	+	+	+	+	+	+	+	+
IMAHMA – RIAHMA		~	~	−	+	~	~	−	~	~	~	−	~	+	+	+	+
IMAHMA – DIAHMA		~	~	~	+	~	~	+	~	~	~	+	~	−	−	−	~
EIAHMA – RIAHMA		+	+	~	+	~	+	−	~	−	−	−	−	~	~	−	~
EIAHMA – DIAHMA		+	+	+	+	~	+	−	~	−	−	−	−	−	−	−	−
RIAHMA – DIAHMA		~	~	+	~	~	~	+	~	~	~	+	~	−	−	−	−

Both LS and diversity maintaining scheme require to cost a certain number of additional fitness evaluations. Therefore, it becomes a key problem to balance the two operations considering that the total additional number of evaluations per generation (*num_aepg*) is always constant. In Section 6.2.5, two different schemes are proposed to calculate the corresponding values of additional evaluations consumed

Table 6.4 Experimental results with respect to overall offline performance of IMAHMAs on dynamic test problems

Dynamics		OneMax Problem		
τ	ρ	IMAHMA0	IMAHMA1	IMAHMA2
50	0.1	96.73±0.20	97.13±0.24	96.29±0.24
50	0.2	93.86±0.38	94.86±0.29	92.66±0.45
50	0.5	91.45±0.39	92.89±0.30	90.04±0.52
50	0.9	96.66±0.20	97.04±0.20	96.16±0.24
Dynamics		Plateau Problem		
τ	ρ	IMAHMA0	IMAHMA1	IMAHMA2
50	0.1	92.78±0.62	93.48±0.63	92.39±0.60
50	0.2	86.66±0.86	87.70±0.77	85.44±0.94
50	0.5	80.54±0.96	81.43±1.00	79.55±0.86
50	0.9	92.79±0.54	93.45±0.55	92.15±0.57
Dynamics		RoyalRoad Problem		
τ	ρ	IMAHMA0	IMAHMA1	IMAHMA2
50	0.1	84.99±1.31	85.18±1.76	84.72±1.48
50	0.2	73.74±1.58	74.04±1.62	73.17±1.60
50	0.5	63.71±2.07	64.81±2.14	64.42±1.79
50	0.9	85.14±1.55	85.95±1.34	85.30±1.25
Dynamics		Deceptive Problem		
τ	ρ	IMAHMA0	IMAHMA1	IMAHMA2
50	0.1	83.25±0.57	83.62±0.53	81.26±0.92
50	0.2	79.69±0.82	81.39±0.48	76.51±0.93
50	0.5	77.74±0.61	79.99±0.46	73.98±0.84
50	0.9	82.77±0.87	83.59±0.72	80.88±1.19

by LS (num_{ls}) and immigrants scheme (num_{im}), respectively. In the following experiments, we will examine the effect of proposed schemes upon the performance of IMAHMAs in dynamic environments. For the sake of convenient description of the experiments, IMAHMA0, IMAHMA1 and IMAHMA2 are used to denote IMAHMA with a *deterministic* scheme where both num_{ls} and num_{im} are fixed to $num\_aepg/2$, IMAHMA with the first dominated scheme, and IMAHMA with the second dominated scheme, respectively.

The experimental results respect to the overall offline performance are presented in Table 6.4 and the corresponding statistical results of comparing algorithms by the one-tailed t-test with 38 degrees of freedom at a 0.05 level of significance are given in Table 6.5. From Table 6.4 and Table 6.5, several results can be observed and are analyzed below.

First, IMAHMA1 performs better than IMAHMA0 and IMAHMA2 on all test problems. In IMAHMA1, the value of additional fitness evaluations costed by LS (num_{ls}) is firstly calculated based on the effect of LS. When LS helps to improve the quality of individuals, that is, the number of valid LS (num_{vls}) is larger than zero, num_{ls}'s value would increase in order to execute more sufficient local refinement,

Table 6.5 The t-test results of comparing the overall offline performance of IMAHMAs on dynamic test problems

t-test Result	OneMax				Plateau				RoyalRoad				Deceptive			
$\tau = 50, \rho \Rightarrow$	0.1	0.2	0.5	0.9	0.1	0.2	0.5	0.9	0.1	0.2	0.5	0.9	0.1	0.2	0.5	0.9
AHMA0 − AHMA1	−	−	−	−	−	−	−	−	~	~	~	−	−	−	−	−
AHMA0 − AHMA2	+	+	+	+	+	+	+	+	~	~	~	~	+	+	+	+
AHMA1 − AHMA2	+	+	+	+	+	+	+	+	~	+	~	+	+	+	+	+

while the number of generated immigrants (num_{im}) would decrease as a result of $num_{im} = num\_aepg - num_{ls}$. The experimental results show that this scheme dominated by LS is a good choice to keep balance between LS and diversity maintaining.

Second, IMAHMA2 is always beaten by IMAHMA0 and IMAHMA1 on most dynamic problems. In IMAHMA2, the number of generated immigrants is firstly determined based on the convergence status of population and then the value of num_{ls} is calculated from the formula $num_{ls} = num\_aepg - num_{im}$. When the index ($\xi$) is lower than a threshold, the immigrants' number (num_{im}) would increase in order to improve the diversity level of population. Obviously, the idea behind this scheme is to maintain a spread-out population. However, the experimental results indicate that this simple scheme cannot achieve the original purpose.

Finally, IMAHMA0 exhibits the good performance on the test problems, which validate the necessary of hybridizing LS and diversity maintaining scheme in MAs for DOPs.

6.4.4 Experimental Study on Comparing the Proposed Algorithm with Several Peer GAs on DOPs

In the final experiments, we compare the performance of IMAHMA with several peer GAs on the DOPs constructed in Section 6.3. These peer GAs are SGAr, RIGA, EIGA and DIGA, as described in Section 6.4.1. The experimental results are presented in Table 6.6 to Table 6.9, respectively. The corresponding statistical results are given in Table 6.10. From these tables, several results can be observed and analyzed as follows.

First, IMAHMA always outperforms other peer algorithm on most dynamic problems and underperforms some of these GAs on some dynamic problems when the environment changes slowly, i.e., when $\tau = 50$ or $\tau = 100$. When the environment changes quickly, i.e., when $\tau = 10$, IMAHMA can always locate the optimum (maybe local optimum) more quickly than other GAs because LS has a strong exploitation capacity. This is why IMAHMA performs the best on all dynamic problems with $\tau = 10$. When $\tau = 50$ or $\tau = 100$, IMAHMA performs worse than SGAr on dynamic Plateau and RoyalRoad problems with $\rho = 0.5$. This happens because the random environment always requires algorithms to maintain a sufficient population diversity (see the relevant analysis in Section 6.4.3) and the restart scheme

Table 6.6 Experimental results with respect to overall offline performance of IMAHMA and peer GAs on dynamic OneMax problems

Dynamics		Algorithms				
τ	ρ	IMAHMA	SGAr	EIGA	RIGA	DIGA
10	0.1	91.78±0.60	70.76±0.45	89.50±0.81	87.97±0.66	89.35±0.91
10	0.2	86.38±0.84	70.68±0.36	81.42±0.85	80.58±1.03	81.14±1.09
10	0.5	78.93±0.93	70.66±0.28	71.66±1.73	71.83±0.77	71.52±1.41
10	0.9	91.70±0.42	70.68±0.34	88.32±2.21	87.67±0.84	89.31±0.82
50	0.1	97.11±0.26	91.02±0.14	97.09±0.22	95.06±0.34	97.23±0.25
50	0.2	94.85±0.35	91.10±0.16	94.44±0.45	90.89±0.54	94.50±0.42
50	0.5	92.92±0.34	91.05±0.13	90.05±0.72	89.41±0.32	90.22±0.63
50	0.9	97.11±0.16	91.00±0.31	97.14±0.46	97.08±0.30	97.26±0.24
100	0.1	98.55±0.08	95.50±0.08	98.58±0.06	97.48±0.14	98.59±0.09
100	0.2	97.43±0.13	95.52±0.07	97.21±0.18	95.37±0.24	97.17±0.19
100	0.5	96.46±0.15	95.51±0.08	95.13±0.34	94.69±0.12	95.15±0.33
100	0.9	98.54±0.11	95.52±0.05	98.54±0.25	97.47±0.18	98.60±0.11

Table 6.7 Experimental results with respect to overall offline performance of IMAHMA and peer GAs on dynamic Plateau problems

Dynamics		Algorithms				
τ	ρ	IMAHMA	SGAr	EIGA	RIGA	DIGA
10	0.1	83.03±0.95	47.96±0.60	76.17±2.13	77.53±1.05	76.86±2.72
10	0.2	72.91±1.64	47.77±0.75	63.75±2.02	64.57±1.36	64.31±2.08
10	0.5	57.52±2.02	48.03±0.62	48.12±2.17	49.25±0.87	47.68±1.32
10	0.9	83.46±1.12	47.89±0.72	76.59±4.83	77.79±1.21	78.12±2.74
50	0.1	93.40±0.37	82.99±0.34	92.91±0.83	90.73±0.51	93.01±0.70
50	0.2	87.56±0.96	83.00±0.30	86.45±0.80	82.92±0.83	86.67±1.22
50	0.5	81.42±0.89	82.99±0.40	77.42±1.41	80.37±0.67	78.19±1.18
50	0.9	93.33±0.53	82.81±0.38	92.71±1.94	90.74±0.71	92.93±0.79
100	0.1	96.71±0.32	91.46±0.16	96.24±0.60	95.37±0.33	96.05±0.88
100	0.2	93.32±0.63	91.52±0.17	92.55±0.87	91.13±0.53	92.43±0.93
100	0.5	90.72±0.59	91.51±0.13	86.51±1.19	90.22±0.38	87.18±1.41
100	0.9	96.72±0.31	91.52±0.15	96.07±0.97	95.37±0.36	96.17±0.80

in SGAr can introduce the maximum diversity into the population. The reason why SGAr outperforms IMAHMA only on the Plateau and RoyalRoad problems lies in the intrinsic characteristics of these problems. The OneMax problem is simply unimodal, which is very suitable for a HC search in IMAHMA. Both the Plateau and RoyalRoad problems have higher-order building blocks, which take a HC search much more time to achieve. The Deceptive problem may mislead SGAr's evolution due to the existence of deceptive attractor, which can be escaped from by IMAHMA. IMAHMA is also beaten by DIGA on the dynamic OneMax problems and Deceptive problems with $\rho = 0.5$ and $\tau = 50$ or 100. This is because DIGA can especially fit such an acutely-changing environment and the dual-based immigrants can maintain

Table 6.8 Experimental results with respect to overall offline performance of IMAHMA and peer GAs on dynamic RoyalRoad problems

Dynamics		Algorithms				
τ	ρ	IMAHMA	SGAr	EIGA	RIGA	DIGA
10	0.1	67.27±2.39	32.50±0.87	53.89±4.71	62.30±2.30	53.03±5.37
10	0.2	52.48±2.33	32.52±0.73	43.44±4.53	47.95±1.39	41.75±3.53
10	0.5	38.83±1.53	32.31±0.76	31.67±1.91	33.98±1.48	32.41±1.85
10	0.9	66.09±2.72	32.69±0.73	52.25±4.97	62.29±2.83	54.87±4.79
50	0.1	85.70±1.38	69.32±0.73	76.39±4.13	83.94±1.24	75.84±4.28
50	0.2	74.48±1.10	69.40±1.03	66.33±2.79	72.27±1.54	66.62±3.41
50	0.5	63.40±1.99	69.38±0.83	53.17±2.20	65.58±1.79	53.05±2.44
50	0.9	85.26±1.31	69.10±1.02	77.38±4.31	83.30±1.41	76.28±3.27
100	0.1	91.31±0.98	84.42±0.48	82.56±2.98	90.15±0.99	82.39±4.05
100	0.2	82.64±1.44	84.33±0.43	74.04±2.15	82.13±1.34	73.36±1.96
100	0.5	76.07±2.00	84.39±0.58	62.13±2.17	79.80±1.67	62.55±2.58
100	0.9	91.38±0.79	84.30±0.42	82.99±5.60	90.50±1.10	82.79±3.47

Table 6.9 Experimental results with respect to overall offline performance of IMAHMA and peer GAs on dynamic Deceptive problems

Dynamics		Algorithms				
τ	ρ	IMAHMA	SGAr	EIGA	RIGA	DIGA
10	0.1	72.78±2.04	53.13±0.56	69.15±1.75	64.36±1.55	71.77±1.75
10	0.2	66.05±0.98	52.92±0.50	62.94±1.44	57.88±1.26	63.24±1.38
10	0.5	61.73±0.76	53.18±0.47	56.70±1.01	53.28±0.78	57.55±1.05
10	0.9	73.04±1.54	53.23±0.64	69.18±1.19	64.02±1.19	71.63±1.36
50	0.1	83.91±0.99	66.06±0.57	78.07±1.23	75.47±1.66	84.58±0.49
50	0.2	81.50±0.73	66.02±0.62	75.04±1.10	70.49±1.03	81.79±0.49
50	0.5	80.14±0.60	66.17±0.52	73.77±0.71	68.06±0.82	79.61±0.49
50	0.9	83.64±0.51	66.23±0.70	78.20±1.66	75.28±1.90	84.58±0.46
100	0.1	87.14±0.86	73.85±0.54	79.07±1.09	77.86±1.61	87.21±0.45
100	0.2	86.07±0.55	73.96±0.58	77.45±1.10	75.83±1.14	86.20±0.54
100	0.5	85.95±0.34	73.94±0.63	77.41±0.73	74.89±0.76	85.41±0.40
100	0.9	86.93±0.55	73.90±0.43	78.61±1.91	78.51±1.81	87.04±0.57

a very high fitness level on the OneMax and Deceptive problems. The good performance of IMAHMA over other peer GAs shows that our investigated IMAHMA has a strong robustness and adaptivity in dynamic environments.

Second, on dynamic OneMax and Plateau problems EIGA always outperforms SGAr and RIGA when ρ is set to 0.1 or 0.2, but underperforms them when the value of ρ is set to 0.5 or 0.9. On dynamic RoyalRoad and Deceptive problems, the situations become a little different. EIGA performs better than RIGA on dynamic RoyalRoad problems just when $\tau = 10$ and better than both SGAr and RIGA on all dynamic Deceptive problems. This happens because the elitism-based immigrants scheme can introduce higher fitness individuals, which can adapt better to the current

Table 6.10 The t-test results of comparing the overall offline performance of AHMA and peer EAs on dynamic test problems

t-test Result	OneMax				Plateau				RoyalRoad				Deceptive			
$\tau = 10, \rho \Rightarrow$	0.1	0.2	0.5	0.9	0.1	0.2	0.5	0.9	0.1	0.2	0.5	0.9	0.1	0.2	0.5	0.9
IMAHMA − SGAr	+	+	+	+	+	+	+	+	+	+	+	+	+	+	+	+
IMAHMA − EIGA	+	+	+	+	+	+	+	+	+	+	+	+	+	+	+	+
IMAHMA − RIGA	+	+	+	+	+	+	+	+	+	+	+	+	+	+	+	+
IMAHMA − DIGA	+	+	+	+	+	+	+	+	+	+	+	+	+	+	+	+
SGAr − EIGA	−	−	−	−	−	−	~	−	−	−	~	−	−	−	−	−
SGAr − RIGA	−	−	−	−	−	−	−	−	−	−	−	−	−	−	~	−
SGAr − DIGA	−	−	−	−	−	−	~	−	−	−	~	−	−	−	−	−
EIGA − RIGA	+	+	~	~	−	~	−	~	−	−	−	−	+	+	+	+
EIGA − DIGA	~	~	~	−	~	~	~	~	~	~	~	~	−	~	−	−
RIGA − DIGA	−	~	~	−	~	~	+	~	+	+	+	+	−	−	−	−
$\tau = 50, \rho \Rightarrow$	0.1	0.2	0.5	0.9	0.1	0.2	0.5	0.9	0.1	0.2	0.5	0.9	0.1	0.2	0.5	0.9
IMAHMA − SGAr	+	+	+	+	+	+	−	+	+	+	−	+	+	+	+	+
IMAHMA − EIGA	~	+	+	~	+	+	+	~	+	+	+	+	+	+	+	+
IMAHMA − RIGA	+	+	+	+	+	+	+	+	+	+	−	+	+	+	+	+
IMAHMA − DIGA	~	+	+	−	+	+	+	+	+	+	+	+	−	~	+	−
SGAr − EIGA	−	−	+	−	−	−	+	−	−	+	+	−	−	−	−	−
SGAr − RIGA	−	+	+	−	−	~	+	−	−	−	+	−	−	−	−	−
SGAr − DIGA	−	−	+	−	−	−	+	−	−	+	+	−	−	−	−	−
EIGA − RIGA	+	+	+	+	+	+	−	+	−	−	−	−	+	+	+	+
EIGA − DIGA	~	~	~	~	~	~	−	~	~	~	~	~	−	−	−	−
RIGA − DIGA	−	−	−	−	−	−	−	−	+	+	+	+	−	−	−	−
$\tau = 100, \rho \Rightarrow$	0.1	0.2	0.5	0.9	0.1	0.2	0.5	0.9	0.1	0.2	0.5	0.9	0.1	0.2	0.5	0.9
IMAHMA − SGAr	+	+	+	+	+	+	−	+	+	−	−	+	+	+	+	+
IMAHMA − EIGA	~	+	+	~	+	+	+	+	+	+	+	+	+	+	+	+
IMAHMA − RIGA	+	+	+	+	+	+	+	+	+	~	−	+	+	+	+	+
IMAHMA − DIGA	~	+	+	~	+	+	+	+	+	+	+	+	~	~	+	~
SGAr − EIGA	−	−	+	−	−	−	+	−	+	+	+	~	−	−	−	−
SGAr − RIGA	−	+	+	−	−	+	+	−	−	+	+	−	−	−	−	−
SGAr − DIGA	−	−	+	−	−	−	+	−	+	+	+	+	−	−	−	−
EIGA − RIGA	+	+	+	+	+	+	−	+	−	−	−	−	+	+	+	~
EIGA − DIGA	~	~	~	~	~	~	~	~	~	~	~	~	−	−	−	−
RIGA − DIGA	−	−	−	−	−	−	+	−	+	+	+	+	−	−	−	−

environment, into EIGA's population on dynamic OneMax and Plateau problems when the environment changes slightly, on dynamic RoyalRoad problems when the environment changes quickly, and on all dynamic Deceptive problems due to the intrinsic characteristics of these four kinds of functions.

Third, RIGA always performs better than SGAr when the value of ρ is small on most dynamic test problems. This is because a new environment is close to the previous one when the value of ρ is very small. Therefore, introducing a portion of random individuals into the population as done in RIGA may be more beneficial than re-initializing the whole population as done in SGAr.

Fourth, DIGA exhibits good performance on OneMax, Plateau, and Deceptive problems although DIGA is beaten by SGAr and RIGA on these dynamic problems when $\rho = 0.5$. This also confirms the exception of the dual-based immigrants scheme for GAs in dynamic environments. DIGA performs better than the other GAs on all dynamic Deceptive problems. The reason lies in that the dualism mechanism may help DIGA react to significant environmental changes rapidly and also enable it to escape from the deceptive attractor in the Deceptive function.

Finally, the environmental parameters affect the performance of algorithms. The performance of all algorithms increases when the value of τ increase from 10 to 50 to 100. It is easy to understand. When τ becomes larger, algorithms have more time to find better solutions before the next change. The effect of the changing severity parameter ρ is different. For example, when τ is fixed, the performance curve of IMAHMA always declines when ρ increases from 0.1 to 0.2 to 0.5, but rises when ρ increases from 0.5 to 0.9.

In order to better understand the experimental results, we make a deeper look into the dynamic behavior of these algorithms. The dynamic behavior of different algorithms with respect to the running offline performance is shown in Fig. 6.12 to Fig. 6.15, where τ is set to 50 and ρ is set to 0.1, 0.2, 0.5, and 0.9, respectively. From these figures, it can be easily observed that for the dynamic periods SGAr always performs almost the same as it did for the stationary period (the first 50 generations) and IMAHMA always outperforms other peer GAs for the stationary period on all test problems while their dynamic behaviors are different on different dynamic problems.

On the OneMax and Plateau problems (see Figs. 6.12 and 6.13), the dynamic behavior of IMAHMA for each dynamic period outperforms that for the stationary period when ρ is very small. When $\rho = 0.5$, IMAHMA exhibits almost the same performance in both the stationary and dynamic periods. When the value of ρ increases to 0.9, the performance of IMAHMA in the dynamic period upgrades consistently. This is because that the proposed AHC operator can help IMAHMA achieve the optimum quickly once the population can "find" in the area where the optimum is located. The new optimum may be close to the previous one when the environment changes slightly, while the immigrants scheme in IMAHMA can help the population re-locate to the changing optimum quickly when a significant environmental change occurs. The dynamic behavior of both RIGA and EIGA is affected by the value of ρ. With the increment of dynamic periods, their performance upgrades consistently when $\rho = 0.1$, while their behavior for the dynamic periods underperforms that for the stationary period when $\rho = 0.5$ or 0.9. The behavior of DIGA is very similar to the behavior of RIGA and EIGA when ρ is not very large. However, DIGA exhibits very nice adaptability to a significantly changing environment. Its behavior for the dynamic periods outperforms that for the stationary period when $\rho=0.9$. This

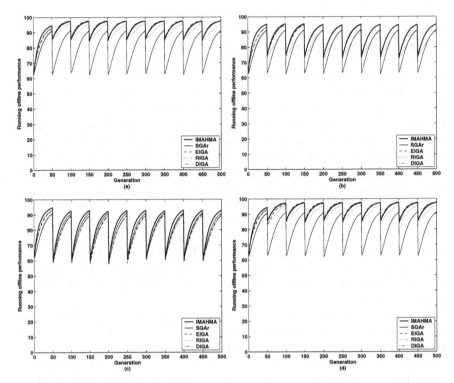

Fig. 6.12 Dynamic behavior of IMAHMA and peer GAs on dynamic OneMax problems with $\tau = 50$ and ρ is set to: (a) $\rho = 0.1$, (b) $\rho = 0.2$, (c) $\rho = 0.5$, and (d) $\rho = 0.9$

happens because the dual-based immigrants scheme can help DIGA re-locate to the changing optimum quickly when a sharp environmental change occurs.

On the RoyalRoad and Deceptive problems (see Fig. 6.14), with the increment of dynamic periods, IMAHMA's performance drops a little when $\rho = 0.5$, while rises when $\rho = 0.1$, 0.2 and 0.9. The reason lies in that IMAHMA does not find the optimum in the stationary period on these two problems. When the environment changes slightly or very significantly, IMAHMA always reruns from the starting points with a higher fitness in the dynamic periods than that in the stationary period, while when $\rho = 0.5$, IMAHMA can only obtain worse starting points in the dynamic periods.

6.5 Conclusions and Future Work

In this chapter, the application of MAs with an adaptive hill climbing strategy for dynamic optimization problems is investigated. In the proposed MA, two local search methods, a greedy crossover-based hill climbing and a steepest mutation-based hill climbing, are used to refine the individual that is selected for local improvements.

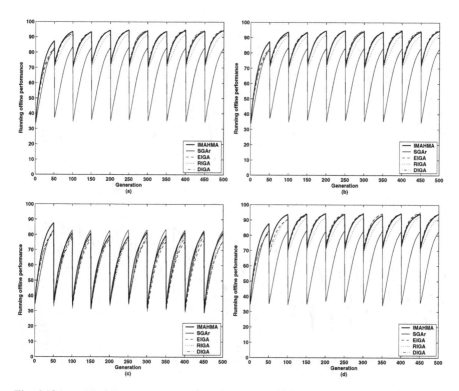

Fig. 6.13 Dynamic behavior of IMAHMA and peer GAs on dynamic Plateau problems with $\tau = 50$ and ρ is set to: (a) $\rho = 0.1$, (b) $\rho = 0.2$, (c) $\rho = 0.5$, and (d) $\rho = 0.9$

A learning mechanism, which gives the more effective LS operator greater chance for the later individual refinement, is introduced in order to execute a robust local search. To maintain a sufficient population diversity for the algorithms to adapt well to the environmental changes, a new immigrants scheme, where the immigrants can be generated from mutating an elite individual with a probability adaptively, is introduced into our proposed MA. In order to keep the balance between LS and diversity maintaining with respect to the extra computation cost, two different dominated schemes are also discussed in this paper.

In order to test the performance of the proposed MA for DOPs, a series of experimental studies have been carried out based on a set of constructed dynamic test problems. From the experimental results, we can draw the following conclusions on the dynamic test problems.

First, MAs enhanced by suitable diversity methods can exhibit a better performance in dynamic environments. For most dynamic test problems, IMAHMA always outperforms other peer GAs.

Second, the immigrants scheme is efficient for improving the performance of MAs in dynamic environments. However, different immigrants schemes have different effects in different dynamic environments. The elitism-based immigrants

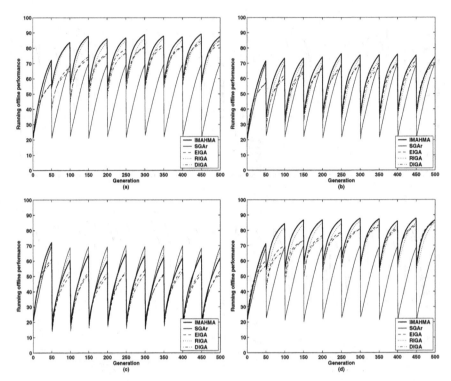

Fig. 6.14 Dynamic behavior of IMAHMA and peer GAs on dynamic RoyalRoad problems with $\tau = 50$ and ρ is set to: (a) $\rho = 0.1$, (b) $\rho = 0.2$, (c) $\rho = 0.5$, and (d) $\rho = 0.9$

scheme is suitable for the slightly-changing environments, the random immigrants scheme performs better when the environmental severity $\rho = 0.5$, and the dualism-based immigrants does better when the environment involves significant changes (i.e., $\rho = 0.9$). The proposed immigrants scheme in Section 6.2.3 is a good choice that generalizes three different immigrants schemes within a common framework.

Third, the effect of LS operators is problem dependent. The AHC strategy can help MAs execute a robust individual refinement since it employs multiple LS operators under the mechanism of cooperation and competition.

Fourth, the difficulty of DOPs depends on the environmental dynamics, including the severity and speed of changes and the difficulty of the base stationary problems. According to our experiments, MAs perform better with the increasing of the frequency of changes, and the effect of the severity of changes is problem dependent.

Generally speaking, the experimental results indicate that the proposed MA, where the adaptive hill climbing operator is used as a local search technique for individual refinement, with adaptive dual mapping and triggered random immigrants schemes, seems a good EA optimizer for DOPs.

For the future work, it is straightforward to introduce other mechanisms, such as memory-based methods [43] and multi-population approaches [30], into MAs

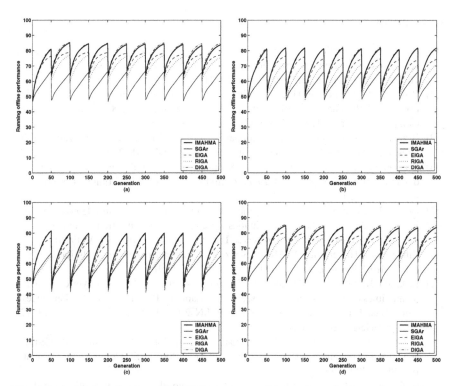

Fig. 6.15 Dynamic behavior of IMAHMA and peer GAs on dynamic Deceptive problems with $\tau = 50$ and ρ is set to: (a) $\rho = 0.1$, (b) $\rho = 0.2$, (c) $\rho = 0.5$, and (d) $\rho = 0.9$

for DOPs. Another interesting research work is to extend the proposed immigrants scheme to other EAs and examine their performance in dynamic environments. In addition, it is also valuable to carry out the sensitivity analysis on the effect of parameters, e.g., α, β, Δ, θ_0, λ_0, and λ_1, on the performance of proposed MAs in the future.

Acknowledgements. The work by Hongfeng Wang was supported by the National Nature Science Foundation of China (NSFC) under Grant 71001018, the National Innovation Research Community Science Foundation of China under Grant 60521003, the China Postdoctoral Science Foundation Under Grant 2012T50266, and the Fundamental Research Funds for the Central Universities Grant N110404019 and Grant N110204005. The work by Shengxiang Yang was supported by the Engineering and Physical Sciences Research Council (EPSRC) of U.K. under Grant EP/K001310/1 and partially by the State Key Laboratory of Synthetical Automation for Process Industries, Northeastern University, China.

References

[1] Baluja, S.: Population-based incremental learning: A method for integrating genetic search based function optimization and competitive learning. Tech. Rep. CMU-CS-94-163, Carnegie Mellon University, USA (1994)

[2] Bosman, P.A.N., Poutré, H.L.: Learning and anticipation in online dynamic optimization with evolutionary algorithms: the stochastic case. In: Proc. 2007 Genetic and Evol. Comput. Conf., pp. 1165–1172 (2007)

[3] Branke, J.: Memory enhanced evolutionary algorithms for changing optimization problems. In: Proc. 1999 IEEE Congr. Evol. Comput., pp. 1875–1882 (1999)

[4] Branke, J., Kaußler, T., Schmidth, C., Schmeck, H.: A multi-population approach to dynamic optimization problems. In: Proc. 4th Int. Conf. Adaptive Comput. Des. Manuf., pp. 299–308 (2000)

[5] Cobb, H.G.: An investigation into the use of hypermutation as an adaptive operator in genetic algorithms having continuous, time-dependent nonstationary environment. Tech. Rep. AIC-90-001, Naval Research Laboratory, Washington, USA (1990)

[6] Eiben, A.E., Hinterding, R., Michalewicz, Z.: Parameter control in evolutionary algorithms. IEEE Trans. Evol. Comput. 3(2), 124–141 (1999)

[7] Eriksson, R., Olsson, B.: On the behavior of evolutionary global-local hybrids with dynamic fitness functions. In: Guervós, J.J.M., Adamidis, P.A., Beyer, H.-G., Fernández-Villacañas, J.-L., Schwefel, H.-P. (eds.) PPSN 2002. LNCS, vol. 2439, pp. 13–22. Springer, Heidelberg (2002)

[8] Eriksson, R., Olsson, B.: On the Performance of Evolutionary Algorithms with life-time adaptation in dynamic fitness landscapes. In: Proc. 2004 IEEE Congr. Evol. Comput., pp. 1293–1300 (2004)

[9] Gallardo, J.E., Cotta, C., Ferndez, A.J.: On the hybridization of memetic algorithms with branch-and-bound techniques. IEEE Trans. Syst., Man, and Cybern.-Part B: Cybern. 37(1), 77–83 (2007)

[10] Goh, C.K., Tan, K.C.: A competitive-cooperation coevolutionary paradigm for dynamic multi-objective optimization. IEEE Trans. Evol. Comput. 13(1), 103–127 (2009)

[11] Goldberg, D.E., Smith, R.E.: Nonstationary function optimization using genetic algorithms with dominance and diploidy. In: Proc. 2nd Int. Conf. on Genetic Algorithms, pp. 59–68 (1987)

[12] Grefenstette, J.J.: Genetic algorithms for changing environments. In: Proc. 2nd Int. Conf. Parallel Problem Solving From Nature, pp. 137–144 (1992)

[13] Hatzakis, I., Wallace, D.: Dynamic multi-objective optimization with evolutionary algorithms: a forward-looking approach. In: Proc. 2006 Genetic and Evol. Comput. Conf., pp. 1201–1208 (2006)

[14] Ishibuchi, H., Yoshida, T., Murata, T.: Balance between genetic search and local search in memetic algorithms for multiobjective permutation flowshop scheduling. IEEE Trans. Evol. Comput. 7(2), 204–223 (2003)

[15] Jin, Y., Branke, J.: Evolutionary optimization in uncertain environments–A survey. IEEE Trans. Evol. Comput. 9(3), 303–317 (2005)

[16] Krasnogor, N., Smith, J.: A tutorial for competent memetic algorithms: model, taxonomy, and design issues. IEEE Trans. Evol. Comput. 9(5), 474–487 (2005)

[17] Lau, T.L., Tsang, E.P.K.: Applying a mutation-based genetic algorithm to processor configuration problems. In: Proc. 8th IEEE Conf. on Tools with Artif. Intell., pp. 17–24 (1996)

[18] Lozano, M., Herrera, F., Krasnogor, N., Molina, D.: Real-coded memetic algorithms with crossover hill-climbing. Evol. Comput. 12(3), 273–302 (2004)

[19] Liu, D., Tan, K.C., Goh, C.K., Ho, W.K.: A multiobjective memetic algorithm based on particle swarm optimization. IEEE Trans. Syst., Man, and Cybern.-Part B: Cybern. 37(1), 42–50 (2007)

[20] Liu, B., Wang, L., Jin, Y.H.: An effective PSO-based memetic algorithm for flow shop scheduling. IEEE Trans. Syst., Man, and Cybern.-Part B: Cybern. 37(1), 18–27 (2007)

[21] Man, S., Liang, Y., Leung, K.S., Lee, K.H., Mok, T.S.K.: A memetic algorithm for multiple-drug cancer chemotherapy schedule optimization. IEEE Trans. Syst., Man, and Cybern.-Part B: Cybern. 37(1), 84–91 (2007)

[22] Neri, F., Toivanen, J., Cascella, G.L., Ong, Y.S.: An adaptive multimeme algorithm for designing HIV multidrug therapies. IEEE/ACM Trans. Comput. Biology and Bioinform. 4(2), 264–278 (2007)

[23] Neri, F., Toivanen, J., Makinen, A.R.E.: An adaptive evolutionary algorithm with intelligent mutation local searchers for designing multidrug therapies for HIV. Applied Intell. 27(3), 219–235 (2007)

[24] Nguyen, T.T., Yao, X.: Dynamic time-linkage problems revisited. In: Giacobini, M., et al. (eds.) EvoWorkshops 2009. LNCS, vol. 5484, pp. 735–744. Springer, Heidelberg (2009)

[25] Nguyen, T.T., Yang, S., Branke, J.: Evolutionary dynamic optimization: A survey of the state of the art. Swarm and Evol. Comput. 6, 1–24 (2012)

[26] Ong, Y.S., Keane, A.J.: Meta-lamarckian learning in memetic algorithms. IEEE Trans. Evol. Comput. 8(2), 99–110 (2004)

[27] Oppacher, F., Wineberg, M.: The shifting balance genetic algorithm: Improving the GA in a dynamic environment. In: Proc. 1999 Genetic and Evol. Comput. Conf., vol. 1, pp. 504–510 (1999)

[28] O'Reilly, U.M., Oppacher, F.: Program search with a hierarchical variable length representation: Genetic programming, simulated annealing and hill climbing. In: Davidor, Y., Männer, R., Schwefel, H.-P. (eds.) PPSN 1994. LNCS, vol. 866, pp. 397–406. Springer, Heidelberg (1994)

[29] O'Reilly, U.M., Oppacher, F.: Hybridized crossover-based search techniques for program discovery. In: Proc. 1995 IEEE Int. Conf. Evol. Comput., pp. 573–578 (1995)

[30] Parrott, D., Li, X.: Locating and tracking multiple dynamic optima by a particle swarm model using speciation. IEEE Trans. Evol. Comput. 10(4), 440–458 (2006)

[31] Simões, A., Costa, E.: Evolutionary algorithms for dynamic environments: Prediction using linear regression and markov chains. In: Rudolph, G., Jansen, T., Lucas, S., Poloni, C., Beume, N. (eds.) PPSN 2008. LNCS, vol. 5199, pp. 306–315. Springer, Heidelberg (2008)

[32] Simões, A., Costa, E.: Improving prediction in evolutionary algorithms for dynamic environments. In: Proc. 2009 Genetic and Evol. Comput. Conf., pp. 875–882 (2009)

[33] Smith, J.E.: Coevolving memetic algorithms: A review and progress report. IEEE Trans. Syst., Man and Cybern.-Part B: Cybern. 37(1), 6–17 (2007)

[34] Talbi, E.G., Bachelet, V.: Cosearch: A parallel cooperative metaheuristic. J. of Mathematical Modelling and Algorithms 5(1), 5–22 (2006)

[35] Tang, J., Lim, M.H., Ong, Y.S.: Diversity-adaptive parallel memetic algorithm for solving large scale combinatorial optimization problems. Soft Comput. 11(10), 957–971 (2007)

[36] Tang, M., Yao, X.: A memetic algorithm for VLSI floor planning. IEEE Trans. Syst., Man and Cybern.-Part B: Cybern. 37(1), 62–69 (2007)

[37] Uyar, A.S., Harmanci, A.E.: A new population based adaptive dominance change mechanism for diploid genetic algorithms in dynamic environments. Soft Comput. 9(11), 803–815 (2005)

[38] Vavak, F., Fogarty, T.C., Jukes, K.: Adaptive combustion balancing in multiple burner boilers using a genetic algorithm with variable range of local search. In: Proc. 7th Int. Conf. on Genetic Algorithms, pp. 719–726 (1996)

[39] Wang, H., Wang, D.: An improved primal-dual genetic algorithm for optimization in dynamic environments. In: King, I., Wang, J., Chan, L.-W., Wang, D. (eds.) ICONIP 2006, Part III. LNCS, vol. 4234, pp. 836–844. Springer, Heidelberg (2006)

[40] Wang, H., Wang, D., Yang, S.: Triggered memory-based swarm optimization in dynamic environments. In: Giacobini, M. (ed.) EvoWorkshops 2007. LNCS, vol. 4448, pp. 637–646. Springer, Heidelberg (2007)

[41] William, E.H., Krasnogor, N., Smith, J.E. (eds.): Recent Advances in Memetic Algorithms. Springer, Heidelberg (2005)

[42] Yang, S.: Non-stationary problem optimization using the primal-dual genetic algorithm. In: Proc. 2003 IEEE Congr. Evol. Comput., vol. 3, pp. 2246–2253 (2003)

[43] Yang, S.: Associative memory scheme for genetic algorithms in dynamic environments. In: Rothlauf, F., et al. (eds.) EvoWorkshops 2006. LNCS, vol. 3907, pp. 788–799. Springer, Heidelberg (2006)

[44] Yang, S.: Genetic algorithms with elitism-based immigrants for changing optimization problems. In: Giacobini, M. (ed.) EvoWorkshops 2007. LNCS, vol. 4448, pp. 627–636. Springer, Heidelberg (2007)

[45] Yang, S., Jiang, Y., Nguyen, T.T.: Metaheuristics for dynamic combinatorial optimization problems. IMA J. of Management Mathematics (2012), doi:10.1093/imaman/DPS021

[46] Yang, S., Ong, Y.S., Jin, Y. (eds.): Evolutionary Computation in Dynamic and Uncertain Environments. Springer, Heidelberg (2007)

[47] Yang, S., Yao, X.: Experimental study on population-based incremental learning algorithms for dynamic optimization problems. Soft Comput. 9(11), 815–834 (2005)

[48] Yang, S., Yao, X.: Population-based incremental learning with associative memory for dynamic environments. IEEE Trans. Evol. Comput. 12(5), 542–561 (2008)

[49] Zhou, Z., Ong, Y.S., Lim, M.H.: Memetic algorithm using multi-surrogates for computationally expensive optimization problems. Soft Comput. 11(9), 873–888 (2007)

[50] Zhu, Z., Ong, Y.S., Dash, M.: Wrapper-filter feature selection algorithm using a memetic framework. IEEE Trans. Syst., Man and Cybern.-Part B: Cybern. 37(1), 70–76 (2007)

Chapter 7
BIPOP: A New Algorithm with Explicit Exploration/Exploitation Control for Dynamic Optimization Problems

Enrique Alba, Hajer Ben-Romdhane, Saoussen Krichen, and Briseida Sarasola

Abstract. Dynamic optimization problems (DOPs) have proven to be a realistic model of dynamic environments where the fitness function, problem parameters, and/or problem constraints are subject to changes. Evolutionary algorithms (EAs) are getting pride of place in solving DOPs due to their ability to match with Nature evolution processes. Several approaches have been presented over the years to enhance the performance of EAs to locate the moving optima in the landscape and avoid premature convergence. We address in this chapter a new bi-population EA augmented by a memory of past solutions and validate it with the dynamic knapsack problem (DKP). We suggest, through the use of two populations, to conduct the search to different directions in the problem space: the first population takes in charge exploring while the second population is responsible for exploiting. Once an environment change is detected, knowledge acquired from the old environment is stored in order to recall it whenever the same state reappears. We illustrate our study by presenting several experiments and compare our results to those of standard algorithms.

Enrique Alba
Departamento de Lenguajes y Ciencias de la Computación, Universidad de Málaga,
E.T.S.I. Informática, Campus de Teatinos, 29071 Málaga, Spain
e-mail: eat@lcc.uma.es

Hajer Ben-Romdhane
LARODEC Laboratory, Institut Supérieur de Gestion, University of Tunis,
41 Rue de la Liberté, Le Bardo, Tunisia
e-mail: hajer1br@hotmail.fr

Saoussen Krichen
FSJEG de Jendouba, University of Jendouba, Avenue de l'U.M.A, 8189 Jendouba, Tunisia
e-mail: saoussen.krichen@isg.rnu.tn

Briseida Sarasola
Departamento de Lenguajes y Ciencias de la Computación, Universidad de Málaga,
E.T.S.I. Informática, Campus de Teatinos, 29071 Málaga, Spain
e-mail: briseida@lcc.uma.es

S. Yang and X. Yao (Eds.): *Evolutionary Computation for DOPs*, SCI 490, pp. 171–191.
DOI: 10.1007/978-3-642-38416-5_7   © Springer-Verlag Berlin Heidelberg 2013

7.1 Introduction

Nowadays, dynamic optimization is getting pride of place in solving real-world problems [27]. Dynamic optimization problems (DOPs) have proved to be a realistic model of dynamic environments where the fitness function, problem parameters, and/or problem constraints are subject to changes. In dynamic domains, decision makers are confronted to uncertainty whether from customers' behavior, the resource availability (budget or material), or external factors influencing their decisions. In network routing problems, for example, network topology and link costs undergo unpredictable and continuous changes due to node/link failure and network dynamics. These changes require the inclusion of new nodes/links or the elimination or the replacement of some existing ones, to remedy the problem and avoid message loss or communication failure. Whatever the problem might be, an optimal solution for the dynamic process should be dispatched over a set of periods [17]. All this is valid also for several other domains, notably industrial (price fluctuations), job shop scheduling (new jobs), TSP (new cities to visit), etc. Researchers have paid a great deal of attention to DOPs to get efficient and accurate solutions for everyday requirements.

To track the moving optima, there exists a widespread appeal to evolutionary algorithms (EAs) as suitable techniques to match with Nature evolution processes [17, 36]. However, the ultimate disadvantage of EAs is their fast convergence to local or wrong optima. To compensate for this problem and thus enhance EAs ability to adapt after changes, many researchers presented modified EAs by means of the following procedures: **1.** increase the diversity once a change occurs (by amplifying the mutation rate [11, 15, 23]), **2.** maintain population diversity throughout the evolutionary process (by introducing new random individuals each generation [9, 15, 22, 33]), **3.** introduce a memory to retrieve pertinent information from previous states (by either implicit or explicit memory [13, 26, 31, 35]), **4.** head the search over different levels via multiple populations [1, 8, 10, 34] and **5.** predict future changes in the landscape (by learning from the past [5, 6, 29, 30]).

The dynamic knapsack problem (DKP) is one of the most important and widely studied DOPs [27], that finds its applications in several domains as Internet advertising, broadcast bandwidth, keyword auctions, etc. A DKP occurs when a decision maker is required to allocate a limited amount of resources (a knapsack) among a given collection of alternatives (or items). As we deal with a dynamic environment, one or more parts of the problem specification may change over time: the fitness function, the values of the parameters or the knapsack capacity. Several EAs were tested via the DKP over the years. Mori *et al.* [22] used the DKP to test their thermodynamical genetic algorithm (GA) in maintaining diversity. A structured GA has been presented in [12]: solutions are structured as multi-layered chromosomes and additional genetic structures are used to store information about useful chromosomes. More recently, Rohlfshagen and Bullinaria [26] proposed a GA based in a relevant genetic phenomenon namely the alternative splicing. Many others EAs were evaluated through DKPs: [13, 14, 16].

In this work, we propose an EA based memory scheme to solve the DKP: the bi-population algorithm (BIPOP). To avoid the premature convergence to local optima we suggest the use of two populations. Each in turn -and via specific search procedures- tries to sweep promising regions in the fitness landscape. This could also strengthen the capacity of the algorithm to keep balance between discovering new regions in the landscape and exploiting the found solutions. Furthermore, we supply BIPOP with a memory in order to quickly recall viable solutions from previous generations if a cyclic problem is faced.

This chapter is organized as follows. Section 7.2 will be devoted to state the problem. We present in Section 7.3 the proposed approach and we draw the corresponding algorithm. Section 7.4 illustrates the proposed approach via numerical experimentations and analyzes the results. Finally, in Section 7.5 we present the concluding remarks.

7.2   Statement of the Problem

In this work we use the DKP to evaluate our algorithmic proposal. In order to better explain our contribution (later, in section 7.3) we give a brief overview of the problem.

Given a set of n items, each with a specific weight w_i as well as an associated value v_i, and a knapsack of capacity c, the DKP is to find the best subset which guarantees the maximization of the knapsack contents without violating the capacity constraint. Besides the standard description of the knapsack problem, the specificity of a DOP is that it takes into consideration potential changes that may affect the problem parameters. One of the most likely scenarios in dynamic domains is the variation of the knapsack capacity along time. In the job shop scheduling for example, machines may break down or wear out. Subsequently, pending jobs will be scheduled over a lesser number of machines. This is reflected in the DKP by a reduction of resources and then the knapsack capacity becomes time-dependent ($c(t)$ with $t = 1, 2, ..., T$). The rest of the problem parameters as well as the fitness function are equally subject to variations in dynamic environments and could be modeled as time-dependent: i.e., $v_i(t)$, $w_i(t)$ and $Z(x, t)$. Formally, the DKP can be stated as follows [22]:

$$\text{Maximize } Z(x, t) = \sum_{i=1}^{n} v_i(t) x_i(t)$$

$$\text{Subject to } \sum_{i=1}^{n} w_i(t) x_i(t) \leq c(t) \quad t = 1, 2, ..., T \quad\quad (7.1)$$

$$x_i(t) = 0 \text{ or } 1; \quad\quad i = 1, 2, ..., n$$

Solutions for this problem are binary-encoded: $[x_1(t)x_2(t)x_3(t)...x_n(t)]$ where $x_i(t) =$ 1 means item i is chosen at t, and 0 means it is discarded.

We suggest to study in this work the DKP where the capacity constraint oscillates between two predefined values $c(1)$ and $c(2)$. This case was introduced by Goldberg and Smith [13] and studied then by several researchers [14, 19]. Our main difference is that we use instances of much larger dimension and difficulty in our work.

7.3   The Proposed Approach: BIPOP-Algorithm

We propose an EA to solve DOPs. To enforce the search towards all possible promising areas in the search space, a multi-population approach is developed. Two independent populations are used in order to sweep the landscape according to two different procedures: the first population takes in charge exploration while the second population is responsible for exploitation. Despite the fact that exploration and exploitation procedures are run independently, the corresponding populations intercommunicate by exchanging elite individuals. Immigrant individuals create further diversity in both populations and could lead the populations to new regions in the search space.

From one period to the next one, the problem can change, and populations must be adjusted. One simple manner to deal with changes is to treat the new environment as a new optimization problem, i.e. to solve the problem from scratch. However, dynamic environments cause small variations on the problem features which usually relate the current problem to precedent scenarios. Thus, solving the problem from scratch once a change is detected might be time and data wasting. In such cases, it will be beneficial to get back information gained in the past in order to push ahead the search after the change. Knowledge induced from previous environments might be memorized and recalled in case of necessity: that is the purpose of using memory schemes. Therefore incorporating a memory in our EA is of significance to quickly adapt to the new environment.

In summary, we are incorporating the most successful techniques in the domain to BIPOP: automatic change detection (the algorithm ignores when changes occur: it detects them), memory management to deal with cyclic problems if this is the case, and immigrants for population diversity. Our multi-population is just a bi-population because we are not speciating or decomposing the problem in any way. Our motivation is this unified vision of the DOP domain, proposing successful existing techniques in BIPOP in an efficient way, and including a new ingredient.

Our main contribution is to make explicit the two forces guiding any search technique: exploration versus exploitation. Most multi-population EAs in the literature (like in [7, 10]), divide the search space into smaller areas and associate each of them with a sub-population that is restricted to search in this specific region. A parent population searches simultaneously for new peaks till it converges to a new area. This area is then never explored in this subpopulation, but assigned to a new spawn to go on searching in it for a solution. Once an environment change is detected, all the child populations are merged together. The working principals of our approach are indeed different from [7, 10], including in our case the use of exactly two populations, fully-connected, and equal-sized: it consists in coordinating concurrently and independently each search force (exploration and exploitation) via two populations while allowing both population to cooperate via immigrants. The specificity of such approach is that is does not restrict any population to search in a limited space but rather immigrant individuals -from one population to the other one- open new areas for both populations. A similar approach has already been adopted in particle swarm optimization to solve DOPs [21, 32] with the main difference in the optimization algorithms being used to evolve each population. Further explanation and a detailed description of the proposed approach will be provided in the subsequent sections. The following lists the notations adopted in this chapter.

j	Index of the current population
t	Period index
g	Generation index
g_c	Exchange generation index
r	Number of random individuals
s	Number of individuals selected from the previous generation
m	Number of individuals retrieved from the memory
G	The total number of generations
Q	The total number of runs
T	The total number of periods
$P_j(g,t)$	Population j at the gth generation and the tth period
M_t	The memory at period t
I', I''	Auxiliary individuals
I	Offspring individual
μ	Population size
$E_j(g,t)$	Set of fitness values of the current population memory
$ctr_j(t)$	Counter of population improvements at period t
$h_j(t)$	The threshold value of population j at the tth period
p_c	Crossover probability
p_m	Mutation probability

7.3.1   Working Principles of BIPOP

BIPOP algorithm (see Algorithm 1) is an EA for uncertain environments, attempting to find good solutions in different scenarios. It encompasses both memory storage and active operators to quickly and precisely deal with changing extrema, that are (in addition) automatically detected.

The algorithm works according to the following way: It starts by generating the two initial populations randomly and saves elite individuals of both populations in the memory as well as initial values of the improvement criteria. Then, the optimization process begins. At the beginning of each generation, stored individuals

---

Algorithm 1 BIPOP Algorithm

$j := 0$
$t = g := 0$
begin
    $\langle P_j(g,t), P_{1-j}(g,t)\rangle := $ **RandomIndividuals**(μ) /* *Populations initialization* */
    $M_t := $ **BestOf**$(P_j(g,t), P_{1-j}(g,t))$ /* *Initialize the memory from the current populations* */
    InitializeImprovCriteria$(h_j(t), ctr_j(t))$
    repeat
        $E_j(g,t) := $ **Evaluate**(M_t)
        if $E_j(g,t) \neq E_j(g-1,t)$ /* *Environmental change detected* */ **then**

            for $j=0$ **to** 1 **do**
                AppendToMemory$(M_t, $ **BestOf**$(P_j(g-1,t)))$
                $P_j(g,t+1) := $ **Adapt**$(P_j(g-1,t))$
                InitializeImprovCriteria$(h_j(t), ctr_j(t))$

            Update(g_c)
            $t++$
        else
            $P_j(g,t) := $ Call **ReproductionAlgo**(j,t)
            if $g = g_c$ **then**

                /* *Copy one individual from population 1 into population 2, and conversely* */
                Replace **WorstOf**$(P_j(g,t))$ by **BestOf**$(P_{1-j}(g-1,t))$

            if *PopImproved*$(P_j(g,t), P_j(g-1,t))$ **then**

                AppendToMemory$(M_t, $ **BestOf**$(P_j(g,t)))$
                UpdateImprovCriteria$(h_j(t), ctr_j(t))$

            $j := 1-j$ /* *Shift from RA to TA and conversely* */
            $g++$

    until *Stopping criteria*;

---

are reassessed in order to detect any change in the problem specifications. If it is the case, the algorithm keeps in the memory fundamental information about the ex-environment and the best individual from the current populations in order to reuse it if a similar environment reappears. Here it is important to notice that only one population is evolved -in turns- at each generation. Besides, we suggest the use of a different approach for each population with the intention to make balance between exploring new regions and exploiting promising ones. Thereby, after a change the new environment parameters are used and the optimization continues. The same population is then adapted to the new environment and the improvement criteria are re-initialized. Otherwise (no change is detected), the population in question is evolved via its specific algorithm: the first population ($j = 0$) is evolved via an

---

Algorithm 2 Reproduction Algorithm

Input : $P_j(g-1,t)$
Output: $P_j(g,t)$
begin

    $P_j(g,t):=$ **Evaluate**$(P_j(g-1,t))$
    if *running-alg=RA* **then**

        $\langle I', I'' \rangle:=$ **Tournament**$(n\text{-}ind, j)$ /* *Select n-ind=2 individuals from population j (j=0 or 1)* */
        $I:=$ **Cross**(I', I'')
        $I:=$ **Mutate**(I)

    else

        /* TA */
        $I':=$ **Tournament**$(n\text{-}ind, j)$ /* *Select n-ind=1 individual for the reproduction* */
        $I:=$ **LocalSearch**(I')

    Replace the **WorstOf**$(P_j(g,t))$ by I

---

exploration algorithm, while the second population ($j = 1$) is evolved using an exploitation algorithm. Here Algorithm 1 calls Algorithm 2 which performs the reproduction and returns the updated population.

BIPOP organizes in the next step the immigrant exchange phase between the two populations. This is done by allowing the copy of the best individual of each population into the other one, at a specific generation g_c and once per period. The inclusion of an individual -issued from a different reproduction approach- into a given population could improve it and shift the search towards more interesting areas. At the end of the generation, the algorithm verifies if the current population is improved in some way (i.g., by calculating the average fitness). In such a case, useful information about the population is kept by saving individuals and updating the improvement criteria. Finally, population j gives way to the other population $(1 - j)$ and a new generation starts.

A. Exploration Algorithm (RA)

The idea behind the RA is to explore the search space, searching new possibilities of solutions. GAs are typical in exploring the landscape and ideal for gathering new solutions, as they search from a set of solutions (population) and not from a single point in the space. However, they do not always insure high solution quality [4].

B. Exploitation Algorithm (TA)

TA consists in searching in the surrounding area of the maintained solutions, in order to improve the population. Unlike the RA, the TA is making use of the best parts of past solutions to enhance the population and try to reach the global optimum. For that purpose, we utilize a local search algorithm (LS). LS approaches perform a complementary search to that of the exploration approach in order to enhance (in general) solutions included in the population.

7.3.2 Construction of BIPOP

We now explain how to construct BIPOP algorithm, from generating basic populations, through the individual reproduction, till the replacement of the old ones.

A. Generation of Initial Populations

Like any EA, BIPOP starts by generating its two basic populations: $P_1(0,0)$ and $P_2(0,0)$. Initial populations are generated randomly by using a uniform distribution. Once an individual is constructed, validation is needed before including it on the current population. In our case, we suggest that an individual is valid if and only if its overall weight fits the capacity constraint. If not, the current individual is repaired by inverting the value of one -or more- randomly chosen bit (from 1 to 0, as the constraint can only be violated by excess of weight in the knapsack).

B. Adapt the Population at the Beginning of Periods

In dynamic processes, an environment change notifies the beginning of a new period: a period is the time interval spent between two environmental changes. Populations need to be re-started before resuming the optimization. We start by verifying if the current state was already visited (by checking information stored about the old environments). If it is the case, the m best stored individuals -corresponding to a similar period- are retrieved. Besides, s individuals are kept from the last generation and r individuals are randomly generated. All these individuals form the new population of size μ.

C. Individual Reproduction

Population evolution is based on the reproduction of individuals. During a given period, current populations are required to evolve from one generation to the next one by enhancing individuals quality. At every generation, a single individual is created to supersede an old one in its same population. Here we must distinguish between the two populations reproduction. For the first population, we suggest to use the standard evolutionary operators: selection, recombination, mutation and replacement. As for the second population, this is problem dependent. In our case (DKP), we adopt a LS for the reproduction. Our LS reproduces the selected individual according to the following way: the individual in question is submitted to a series of bit-inversion independently. As individuals are encoded through the concatenation of n bits, this results in n offsprings. The individual with the higher fitness replaces the worst one in the current population.

7.3.3   Functions Utilized in the Algorithms

The rest of functions used in algorithms work as follows:

- **RandomIndividuals**(k): Generates k random individuals following a uniform distribution U(0,1)
- **BestOf**($P_j(g,t)$), **WorstOf**($P_j(g,t)$): Returns the best/worst individual -in terms of fitness- of the given population
- **Evaluate**(M), **Evaluate**(P): Re-evaluates memory individuals and population individuals respectively
- **AppendToMemory**(M_t, Ind): Saves the individual into the memory
- **Update**(g_c): Updates the g_c parameter, which indicates the next exchange date between the two populations
- **InitializeImprovCriteria**($h_j(t)$, $ctr_j(t)$): Initializes the decision parameters of the population improvement: $h_j(t)$= Average fitness of the initial population and $ctr_j(t) = 0$
- **UpdateImprovCriteria**($h_j(t)$, $ctr_j(t)$): Set $h_j(t)$= Average fitness of the current population and increment $ctr_j(t)$
- **PopImproved**($P_j(g,t)$, $P_j(g-1,t)$): Verifies if population j has improved since the last generation. In this work, we use the average population fitness to decide whether there has been an improvement in the population (or not). A population has improved if the average fitness of $P_j(g,t)$ is greater then the average fitness of $P_j(g-1,t)$ (assuming maximization without loss of generality).
- **Retrieve**(m, M_t): Retrieves m individuals stored in the indicated memory

7.4   Computational Experiments

This section reports on the computational behavior of BIPOP according to several performance measures for dynamic environments. The experimental settings as well as the obtained results are presented in what follows.

7.4.1   Experimental Framework

BIPOP is implemented in Java on an Intel Centrino Duo processor with 2GB of RAM. The algorithm is run on four DKP instances of different sizes: 17, 100, 1000 and 5000-item. We note that, to the best of our knowledge, there is no available data set for the capacity-varying DKP except the (too small) one of Goldberg and Smith [13]. Consequently, we are adapting well-known static instances (see Table 7.1 for references) for which we created dynamism by oscillating the knapsack capacity between two fixed values. The first capacity value is the one defined in the original problem, and we set the second to 50% of the sum of items' weight. Thereby, the problem becomes time-dependent and two temporal optima are considered. We study a fast changing DKP with changes every 10 evaluations and allow different number of environmental changes for each instance: 30 changes in the case of the two smaller instances (size 17 and 100), 150 changes for the instance

Table 7.1 Utilized data set references

Problem size	Reference
17	Goldberg and Smith [13]
100	Liu and Liu [20]
1000	Pisinger [25]
5000	Pisinger [25]

Table 7.2 Set of configuration parameters for BIPOP

Parameter	Specification
Population size	100 individuals
Selection	Binary tournament
Crossover	Two point crossover ($p_c = 1$)
Mutation	Two-bit flip ($p_m = 2/n$)
Replacement	The worst in the current population

with size 1000, and 500 changes in the case of the biggest instance (size 5000). The algorithms were allowed to run longer times on the bigger instances in order to be sufficiently exposed to the problem dynamics.

In the experiments, 30 independent runs were executed. Table 7.2 reports the parameterization of the algorithm.

7.4.2 Analysis

In order to measure the effectiveness and the performance of BIPOP, we run our algorithm as well as four other algorithms over the same instances of DKP. The use of these algorithms is of great importance in positioning our algorithms with regard to a standard algorithm for dynamic environments (GA), a non-standard algorithm adapted -in our case- to solve the DKP (LS), a stochastic optimization algorithm (particle swarm optimization, PSO), and a sanity check algorithm (Random Search, RS). More specifically, we intend to control BIPOP behavior when faced to the performance of GAs in browsing the search space, the efficiency of LS in exploitation, the high adaptability of PSO, and the random shifting in the landscape of RS.

The GA we use here is just a standard one, executing in sequence the evolutionary operators (a two-point crossover and a two-bit flip mutation) once per generation to create a new individual, which replaces the worst individual in the current population. As for the LS, it implements an approach that evolves a set of individuals by applying a series of bit inversions to the selected individual, equal to its number of bits (we use a binary encoding). The offsprings are then evaluated and the best one is retained and introduced in the population. In the same sense, we use the basic binary version of PSO, proposed in [18], which adjusts the solutions positions based on its best history and the position of the best particle in the swarm. RS generates a random solution at each time step, and its results should be bad compared to a guided algorithm; we include it here as a sanity check to show that our technique is having some intelligence in it versus a blind search. These four algorithms are parameterized the same way as BIPOP (see Table 7.2).

A. Average $\underline{Best}\ \underline{Of}\ \underline{G}eneration\ (\overline{BOG})$

The $\overline{BOG}$ is the most commonly used measure to test algorithms in dynamic environments. The strength of this measure is that it quantifies the behavior of the algorithm during the whole optimization without any computational difficulty. It is computed by double-averaging the best fitness ever encountered $(f(BestGeneration_g))$ over all G generations, and over Q runs. The higher the $\overline{BOG}$, the better the algorithm performs. The $\overline{BOG}$ is given by:

$$\overline{BOG} = \frac{1}{Q}\frac{1}{G} \cdot \sum_{q=1}^{Q} \sum_{g=1}^{G} f(BestGeneration_g) \tag{7.2}$$

Table 7.3 $\overline{BOG}$ values for BIPOP, GA, LS and RS

Problem Size	Evals	BIPOP	GA	LS	PSO	RS
17	300	**77.9**	67.5	71.1	66.9	66.8
100	300	**21 910.6**	18 920.9	19 480.4	18 216.3	18 732.6
1000	1500	**351 425.1**	271 954.7	274 666.5	272 347.5	272 597.9
5000	5000	**1 584 746.9**	1 306 435.2	1 305 623.9	1 303 652.2	1 302 972.1

The results of the $\overline{BOG}$ can be seen in Table 7.3. Results show that BIPOP outperforms all considered algorithms for all problem sizes. As to the rest of algorithms, LS seems to be the second best algorithm, although it gets near to the poor performance of PSO and RS with the biggest instance.

B. Off-line Performance

We now measure the average best-so-far during a given period. It provides a monotonically increasing value that indicates how rapidly an algorithm achieves high performance. This measure is also among the most reported in the literature as it is one the few measures that do not require the optimal fitness to be known. Mathematically, it is given by averaging the best fitness during a given period $(f(BestPeriod_t))$ over the total number of periods T (see Eq. (7.3)).

$$Offline = \frac{1}{T} \cdot \sum_{t=1}^{T} f(BestPeriod_t) \tag{7.3}$$

Table 7.4 Comparison of *offline* values for BIPOP, GA, LS, PSO and RS

Problem Size	Periods	BIPOP	GA	LS	PSO	RS
17	30	**78.2**	68.1	72.1	58.7	66.8
100	30	**22 058.9**	19 063.2	19 975	16 026.4	18 732.6
1000	150	**352 286.5**	271 301.1	276 745.2	272 439.3	272 600.8
5000	500	**1 585 777.2**	1 308 333.2	1 307 786.8	1 303 950.7	1 302 974.8

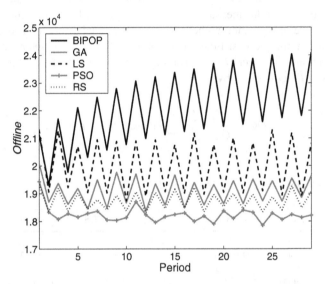

Fig. 7.1 Average *offline* fitness along the time-period obtained via BIPOP, GA, LS, PSO, and RS ($n=100$)

Offline values (Table 7.4) obtained by means of the different algorithms indicate that BIPOP behaves better than the rest of algorithms. Fig. 7.1 shows the final offline performance performed by each algorithm at each period. The highest curve corresponds to BIPOP, which indicates that it performs the best running fitness, and LS comes in a second place.

C. Statistical Significance Test: ANOVA Test

We apply first a Kolmogorov-Smirnov test to check whether the results follow a normal distribution or not. If so, an ANOVA test is done, otherwise we perform a Kruskal-Wallis test. We consider a confidence level of 95%.

The statistical significance tests were performed using the offline performance as base measure. Table 7.5 shows that BIPOP is significantly better than all the other algorithms in (almost) all instances, i.e. all the four algorithms obtain lower fitness values than BIPOP.

Table 7.5 ANOVA test results

	n	RS	PSO	LS	GA
BIPOP	17	▲	▲	—	▲
	100	▲	▲	▲	▲
	1000	▲	▲	▲	▲
	5000	▲	▲	▲	▲
GA	17	—	—	—	
	100	▲	▲	—	
	1000	—	—	▽	
	5000	▲	▲	▲	
LS	17	▲	—		
	100	▲	▲		
	1000	▲	▲		
	5000	▲	▲		
PSO	17	—			
	100	▽			
	1000	—			
	5000	▲			

D. Accuracy

Although the accuracy was originally a metric for static problems, it has been adopted later as a meaningful measurement for DOPs. It represents an outcome measure of the population best solution with regards to the known Maximum fitness and Minimum fitness. Hence, this measure can not be used if we ignore the value of the global optima and the worst fitness value. The major advantage of the accuracy over the measures used so far ($\overline{BOG}$ and *offline*) is that it derives normalized values, which can avoid any potential bias in the result. The accuracy of a given generation g is stated as:

$$Accuracy_g = \frac{f(BestGeneration_g) - Min}{Max - Min} \qquad (7.4)$$

In our case, *Max* corresponds to the global optimum that we generate by an exact algorithm (described in [24]), and *Min* is set to 0. Accuracy values are in the interval [0,1]: the closer to 1, the higher the accuracy. Values achieved by the four algorithms (Table 7.6) are decreasing every time we move to a larger instance size. Despite that BIPOP fulfils the best accuracy level for all problem sizes, scalability is an interesting side effect of our BIPOP due to our economic efficient selection of techniques.

Fig. 7.2 reports the detailed behavior of the accuracy during the whole search process for the problem of size 100. Over the first 30 evaluations, BIPOP and LS provide close performances and overtop the rest of the algorithms. Moving towards the right, BIPOP starts to outstrip, and the difference becomes increasingly important. This improvement is due to the learning about the already visited areas (i.e.,

Table 7.6 Accuracy values for each algorithm

Problem Size	Evals	BIPOP	GA	LS	PSO	RS
17	300	**9.8e-01**	8.5e-01	8.9e-01	8.4e-01	8.4e-01
100	300	**8.7e-01**	7.6e-01	7.6e-01	7.2e-01	7.5e-01
1000	1500	**7.8e-01**	6.1e-01	6.1e-01	6.1e-01	6.1e-01
5000	5000	**7.1e-01**	5.9e-01	5.9e-01	5.9e-01	5.8e-01

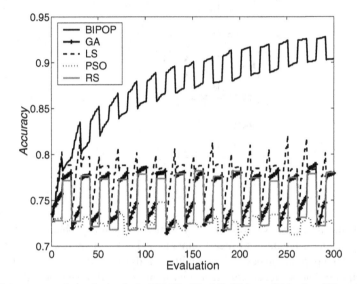

Fig. 7.2 Accuracy values obtained via BIPOP, GA, LS, PSO, and RS after 300 evaluations ($n = 100$)

the memory mechanism is paying off here). LS keeps achieving higher values in comparison with GA, PSO, and RS, which are permanently missing the new optima appearing during the different periods.

E. Stability

The stability is accuracy-dependent: it measures how quick an algorithm is able to adapt after changes. An algorithm is said to be stable if its stability is close to 0 (see Eq. (7.5)).

$$Stability_g = \max\{0, Accuracy_g - Accuracy_{g-1}\} \tag{7.5}$$

The derived results, shown in Table 7.7, are generally satisfying: very close to 0. Nevertheless, PSO showed to be the more stable than the rest of algorithms in all cases, whereas LS is the more unstable. These results indicates also that the stability behavior is independent from the ability of the algorithm to provide solutions of

Table 7.7 Comparison of algorithm behavior according to the stability metric

Problem Size	Evals	BIPOP	GA	LS	PSO	RS
17	300	1.2e-03	1.6e-03	6.8e-03	**7.9e-04**	9.1e-04
100	300	2.1e-03	2.8e-03	4.8e-03	**5.7e-04**	2.7e-03
1000	1500	3.8e-03	3.9e-03	4.2e-03	**7.7e-04**	3.8e-03
5000	5000	2.8e-03	4.2e-03	4.2e-03	**8.1e-04**	4.1e-03

good quality. As to BIPOP, it obtains the second best stability and this implies we have a performing and a rapid algorithm in reacting to changes.

F. Fitness Ratio (FR)

The *FR* is a measure reflecting the quality of the BIPOP results regarding the optimal solution f^* (see Eq. (7.6)). We denote by $f(solution_{kg})$ the fitness of individual k at the gth generation.

$$FR_g = \frac{1}{f^*} \cdot \frac{\sum_{k=1}^{\mu} f(solution_{kg})}{\mu} \qquad (7.6)$$

Table 7.8 *FR* for each algorithm for different problem sizes and number of evaluations

Problem Size	Evals	BIPOP	GA	LS	PSO	RS
17	300	**6.1e-01**	5.5e-01	5.6e-01	5.3e-01	5.4e-01
100	300	**6.4e-01**	6.1e-01	6.1e-01	5.8e-01	5.9e-01
1000	1500	**5.8e-01**	5.6e-01	5.6e-01	5.6e-01	5.6e-01
5000	5000	**5.8e-01**	5.6e-01	5.7e-01	5.7e-01	5.6e-01

Table 7.8 as well as Fig. 7.3 report the obtained results by the different algorithms (Fig. 7.3 shows only the first 150 evaluations for the sake of clarity). We mention the outperforming of BIPOP algorithm in comparison with the rest of algorithms. GA and LS seem to have alike behaviors, although LS gets better values in several periods.

G. Fitness Degradation Measure: $\beta_{degradation}$

An important issue will be to examine the ability of a given algorithm to maintain its performance to produce good solutions as the optimization process advances: $\beta_{degradation}$ measures how much an algorithm degrades in term of solution quality over time [3].

$$y = \beta_{degradation} \, \bar{x} + \varepsilon \qquad (7.7)$$

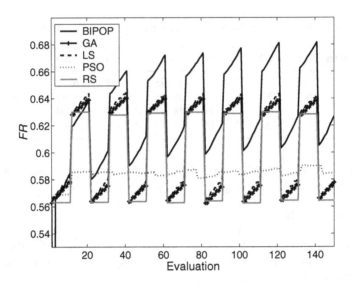

Fig. 7.3 Fitness Ratio behavior along 300 evaluations ($n = 100$)

Table 7.9 Comparison of the $\beta_{degradation}$ of each algorithm for different number of periods

Problem Size	Periods	BIPOP	GA	LS	PSO	RS
17	300	**1.1e-03**	-5.7e-04	-4.2e-04	-3.7e-04	3.3e-04
100	300	**3.4e-03**	1.7e-04	-2.1e-04	-4.2e-04	1.9e-04
1000	1500	**1.7e-03**	-1.6e-05	-1.4e-05	1.1e-08	-1.2e-05
5000	5000	**4.0e-04**	-1.4e-06	-7.8e-07	-4.6e-07	-9.2e-07

$$x_t = \frac{1}{Q} \cdot \sum_{q=1}^{Q} f(BestPeriod_{qt}) \qquad (7.8)$$

We denote, in Eq. (7.7), y an approximation to the overall accuracy, $\bar{x}$ a vector of size T, and $\beta_{degradation}$ the slope of the regression line. In Eq. (7.8), x_t denotes the accuracy of the best solution found in period t averaged over all independent Q runs. If the $\beta_{degradation}$ reveals a positive value, this means the algorithm is well performing and still providing good solutions after running for long: the bigger the improvement in solution quality, the higher the slope value and conversely.

Results of the β slope are presented in Table 7.9. BIPOP provides positive slopes for all instances, which implies a non-degradation in the solution quality. Solutions of RS lose quality with the large problem sizes, while GA, LS and PSO are poor in terms of continuous performance. The algorithms regression lines are drawn in Fig. 7.4. We can see that BIPOP is having the highest slope, GA and RS just show timid increasing slopes, and LS and PSO lines are decreasing which explains the negative $\beta_{degradation}$ values.

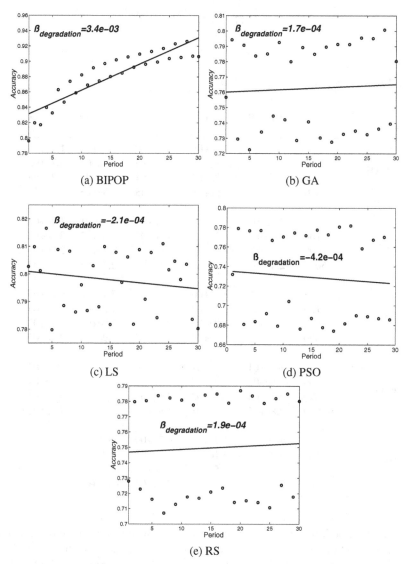

Fig. 7.4 Linear regression after 30 periods for BIPOP, GA, LS, PSO, and RS

H. Area Between Curves (ABC)

The ABC for a pair of algorithms A_1 and A_2 is defined as the area spanned between the performance curves of the two algorithms. Let $p_A(x)$ be the function which determines a certain property values achieved by algorithm A in each generation (BOG in our case). Mathematically, it is modeled as the integral of the difference of p_{A_1} and p_{A_2} in the interval $[1, G]$ [2].

$$ABC_p^{A_1,A_2} = \frac{1}{G} \cdot \int_1^G p_{A_1}(x) - p_{A_2}(x)dx \qquad (7.9)$$

Table 7.10 Comparisons of $ABC_{BOG}^{A_1,A_2}$ values

A_1 \ A_2	Evals	BIPOP	GA	LS	PSO	RS
BIPOP	17		9.3e+00	4.8e+00	1.0e+01	1.1e+01
	100	×	3.4e+03	2.5e+03	3.6e+03	3.7e+03
	1000		7.5e+04	7.4e+04	7.7e+04	7.7e+04
	5000		2.8e+05	2.8e+05	2.8e+05	2.8e+05
GA	17	-9.3e+00		-4.5e+00	-8.9e-01	1.7e+00
	100	-3.4e+03	×	-9.0e+02	2.6e+02	3.1e+02
	1000	-7.5e+05		-6.0e+02	2.4e+03	2.5e+03
	5000	-2.8e+05		8.1e+02	2.7e+03	3.5e+03
LS	17	-4.8e+00	4.5e+00		5.4e+00	6.2e+00
	100	-2.5e+03	9.0e+02	×	1.2e+03	1.2e+03
	1000	-7.4e+04	6.0e+02		3.0e+03	3.1e+03
	5000	-2.8e+05	-8.1e+02		1.9e+03	2.6e+03
PSO	17	-1.0e+01	-8.9e-01	-5.4e+00		8.0e-01
	100	-3.6e+03	-2.6e+02	-1.2e+03	×	4.7e+01
	1000	-7.7e+04	-2.4e+03	-3.0e+03		4.5e+01
	5000	-2.8e+05	-2.7e+03	-1.9e+03		7.4e+02
RS	17	-1.1e+01	-1.7e+00	-6.2e+00	-8.0e-01	
	100	-3.7e+03	-3.1e+02	-1.2e+03	-4.7e+01	×
	1000	-7.7e+04	-2.5e+03	-3.1e+03	-4.5e+01	
	5000	-2.8e+05	-3.5e+03	-2.6e+03	-7.4e+02	

ABC can take both positive and negative values. ABC gets a positive result if the curve obtained by A_1 according to property p has higher values than the curve of A_2, while a negative result indicates that values acquired by A_2 are mostly higher than that provided by A_2.

The ABC measure is applied to each pair of algorithms according to the $\overline{BOG}$ property ($p = \overline{BOG}$). The first group of rows in the Table 7.10 is clearly identifiable because of the positive cell values. This indicates that BIPOP surpasses all the other algorithms. Besides, values corresponding to BIPOP are the larger, which imply a high distance between it and the rest of algorithms in term of performance. LS comes in second place, with positive ABCs when confronted to GA, PSO and RS, whereas RS is the worst, with its entirely negative rows.

7.5   Conclusions

We proposed a bi-population evolutionary algorithm augmented with immigrants, local search, and a memory scheme and evaluated it for the DKP. Our approach tries, over a number of evaluations, to produce fit solutions that survive despite the environmental changes which affect the problem parameters. Two populations are subject to evolution, each one according to its specific algorithm, in order to produce better solutions while making the use of the stored solutions for two purposes: detect and recall.

Experiments were carried out for the proposed algorithm as well as four other algorithms over the same problem instances in order to compare their performances. We measured several metrics to quantify the distance between the algorithms in terms of performance and in terms of ability to maintain good behavior throughout the optimization process. Experimental results always showed that BIPOP is the best in producing better solutions. Besides, it outperforms all the other algorithms according to the applied measures, what is remarkable.

We also detected the good behavior of LS, since it is an efficient operation to quickly go for the new optimum. This, however, was not as good as BIPOP itself, that combined this quick exploitation of existing solutions with a new exploration after the change to avoid pursuing local optima. In any case, this also settles the importance of trajectory-based algorithms in DOPS, not just LS but other algorithms like Variable Neighborhood Search (VNS) could be valuable tools in this domain, and researchers should pay more attention to this because of its good reported results [28].

One aspect that we would like to explore in the future is to adapt our BIPOP to solve different DOPs, and to compare our results to those of well-known DOP algorithms.

Acknowledgements. Authors acknowledge funds from the Spanish Ministry of Economy and Competence, and European FEDER under contract TIN2011-28194 (roadME project, publicly available in URL http://roadme.lcc.uma.es).

References

[1] Alba, E.: Parallel Metaheuristics. John Wiley & Sons, Inc. (2005)

[2] Alba, E., Sarasola, B.: Abc, a new performance tool for algorithms solving dynamic optimization problems. In: Proc. 2010 IEEE Congr. Evol. Comput., pp. 1–7 (2010)

[3] Alba, E., Sarasola, B.: Measuring fitness degradation in dynamic optimization problems. In: Di Chio, C., et al. (eds.) EvoApplicatons 2010, Part I. LNCS, vol. 6024, pp. 572–581. Springer, Heidelberg (2010)

[4] Areibi, S.: Effective exploration & exploitation of the solution space via memetic algorithms for the circuit partition problem. In: Recent Advances in Memetic Algorithms. STUDFUZZ, vol. 166, pp. 161–182. Springer, Heidelberg (2005)

[5] Bosman, P.A.N.: Learning, anticipation and time-deception in evolutionary online dynamic optimization. In: Proc. 2005 Workshops on Genetic and Evol. Comput., pp. 39–47. ACM (2005)

[6] Bosman, P.A.N.: Learning and anticipation in online dynamic optimization. In: Evolutionary Computation in Dynamic and Uncertain Environments. SCI, vol. 51, pp. 129–152. Springer, Heidelberg (2007)

[7] Branke, J., Kau, T., Schmidt, l., Schmeck, H.: A multi-population approach to dynamic optimization problems. In: Proc. 4th Int. Conf. Adaptive Comput. Des. Manuf., pp. 299–308 (2000)

[8] Branke, J., Schmeck, H.: Designing evolutionary algorithms for dynamic optimization problems. In: Ghosh, A., Tsutsui, S. (eds.) Advances in Evolutionary Computing: Theory and Applications. Natural Computing Series, pp. 239–262. Springer-Verlag New York, Inc. (2003)

[9] Cedeno, W., Vemuri, V.: On the use of niching for dynamic landscapes. In: Proc. 1997 IEEE Int. Conf. Evol. Comput., pp. 361–366 (1997)

[10] Cheng, H., Yang, S.: Multi-population genetic algorithms with immigrants scheme for dynamic shortest path routing problems in mobile ad hoc networks. In: Di Chio, C., et al. (eds.) EvoApplicatons 2010, Part I. LNCS, vol. 6024, pp. 562–571. Springer, Heidelberg (2010)

[11] Cobb, H.G., Grefenstette, J.J.: Genetic algorithms for tracking changing environments. In: Proc. 5th Int. Conf. Genetic Algorithms, pp. 523–530 (1993)

[12] Dasgupta, D., Mcgregor, D.R.: Nonstationary function optimization using the structured genetic algorithm. In: Proc. 2nd Int. Conf. Parallel Problem Solving From Nature, pp. 145–154 (1992)

[13] Goldberg, D.E., Smith, R.E.: Nonstationary function optimization using genetic algorithms with dominance and diploidy. In: Proc. 2nd Int. Conf. Genetic Algorithms, pp. 59–68 (1987)

[14] Greene, F.: A method for utilizing diploid/dominance in genetic search. In: Proc. 1st IEEE Conf. Evol. Comput., pp. 439–444. IEEE Press (1994)

[15] Grefenstette, J.: Genetic algorithms for changing environments. In: Proc. 2nd Int. Conf. Parallel Problem Solving from Nature, pp. 137–144 (1992)

[16] Hadad, B., Eick, C.: Supporting polyploidy in genetic algorithms using dominance vectors. In: Angeline, P.J., McDonnell, J.R., Reynolds, R.G., Eberhart, R. (eds.) EP 1997. LNCS, vol. 1213, pp. 223–234. Springer, Heidelberg (1997)

[17] Jin, Y., Branke, J.: Evolutionary optimization in uncertain environments–a survey. IEEE Trans. Evol. Comput. 9, 303–317 (2005)

[18] Kennedy, J., Eberhart, R.: A discrete binary version of the particle swarm algorithm. In: Proc. 1997 IEEE Int. Conf. Syst., Man, and Cybern., vol. 5, pp. 4104–4108 (1997)

[19] Lewis, J., Hart, E., Ritchie, G.: A comparison of dominance mechanisms and simple mutation on non-stationary problems. In: Eiben, A.E., Bäck, T., Schoenauer, M., Schwefel, H.-P. (eds.) PPSN 1998. LNCS, vol. 1498, pp. 139–148. Springer, Heidelberg (1998)

[20] Liu, Y., Liu, C.: A schema-guiding evolutionary algorithm for 0-1 knapsack problem. In: Proc. 2009 Int. Assoc. of Computer Science and Information Technology - Spring Conf., pp. 160–164. IEEE Computer Society (2009)

[21] Lung, R., Dumitrescu, D.: Evolutionary swarm cooperative optimization in dynamic environments. Natural Computing 9(1), 83–94 (2010)

[22] Mori, N., Kita, H., Nishikawa, Y.: Adaptation to a changing environment by means of the thermodynamical genetic algorithm. In: Ebeling, W., Rechenberg, I., Voigt, H.-M., Schwefel, H.-P. (eds.) PPSN 1996. LNCS, vol. 1141, pp. 513–522. Springer, Heidelberg (1996)

[23] Morrison, R., De Jong, K.: Triggered hypermutation revisited. In: Proc. 2000 IEEE Congr. Evol. Comput., vol. 2, pp. 1025–1032 (2000)

[24] Pisinger, D.: A minimal algorithm for the 0-1 knapsack problem. Oper. Res. 45(5), 758–767 (1997)

[25] Pisinger, D.: Core problems in knapsack algorithms. Oper. Res. 47, 570–575 (1999)

[26] Rohlfshagen, P., Bullinaria, J.: Alternative splicing in evolutionary computation: Adaptation in dynamic environments. In: Proc. 2006 IEEE Congr. Evol. Comput., pp. 2277–2284 (2006)

[27] Rohlfshagen, P., Yao, X.: The dynamic knapsack problem revisited: A new benchmark problem for dynamic combinatorial optimisation. In: Giacobini, M., et al. (eds.) EvoWorkshops 2009. LNCS, vol. 5484, pp. 745–754. Springer, Heidelberg (2009)

[28] Sarasola, B., Khouadjia, M.R., Alba, E., Jourdan, L., Talbi, E.-G.: Flexible variable neighborhood search in dynamic vehicle routing. In: Di Chio, C., et al. (eds.) EvoApplications 2011, Part I. LNCS, vol. 6624, pp. 344–353. Springer, Heidelberg (2011)

[29] Simões, A., Costa, E.: Evolutionary algorithms for dynamic environments: Prediction using linear regression and markov chains. In: Rudolph, G., Jansen, T., Lucas, S., Poloni, C., Beume, N. (eds.) PPSN 2008. LNCS, vol. 5199, pp. 306–315. Springer, Heidelberg (2008)

[30] Simões, A., Costa, E.: Improving prediction in evolutionary algorithms for dynamic environments. In: Proc. 11th Annual Conf. Genetic and Evol. Comput., pp. 875–882. ACM (2009)

[31] Wang, H., Wang, D., Yang, S.: Triggered memory-based swarm optimization in dynamic environments. In: Giacobini, M. (ed.) EvoWorkshops 2007. LNCS, vol. 4448, pp. 637–646. Springer, Heidelberg (2007)

[32] Xiangwei, Z., Hong, L.: A cooperative dual-swarm pso for dynamic optimization problems. In: Proc. 7th Int. Conf. Natural Computation, vol. 2, pp. 1131–1135 (2011)

[33] Yang, S.: Genetic algorithms with memory-and elitism-based immigrants in dynamic environments. Evol. Comput. 16, 385–416 (2008)

[34] Yang, S., Li, C.: A clustering particle swarm optimizer for locating and tracking multiple optima in dynamic environments. IEEE Trans. Evol. Comput. 14(6), 959–974 (2010)

[35] Yang, S., Yao, X.: Population-based incremental learning with associative memory for dynamic environments. IEEE Trans. Evol. Comput. 12(5), 542–561 (2008)

[36] Younes, A., Basir, O., Calamai, P.: Adaptive control of genetic parameters for dynamic combinatorial problems. In: Doerner, K.F., Gendreau, M., Greistorfer, P., Gutjahr, W., Hartl, R.F., Reimann, M. (eds.) Metaheuristics. Operations Research/Computer Science Interfaces Series, vol. 39, pp. 205–223. Springer US (2007)

Chapter 8
Evolutionary Optimization on Continuous Dynamic Constrained Problems – An Analysis

Trung Thanh Nguyen and Xin Yao

Abstract. Many real-world dynamic problems have constraints, and in certain cases not only the objective function changes over time, but also the constraints. However, there is little research on whether current algorithms work well on continuous dynamic constrained optimization problems (DCOPs). This chapter investigates this issue. The chapter will present some studies on the characteristics that can make DCOPs difficult to solve by some existing dynamic optimization (DO) algorithms. We will then introduce a set of benchmark problems with these characteristics and test several representative DO strategies on these problems. The results confirm that DCOPs do have special characteristics that can significantly affect algorithm performance. Based on the analyses of the results, a list of potential requirements that an algorithm should meet to solve DCOPs effectively will be proposed.

8.1 Introduction

This chapter attempts to investigate the characteristics, difficulty and solutions of a very common class of problem - dynamic constrained optimization problems (DCOPs). DCOPs are constrained optimization problems that have two properties: (a) the objective functions, the constraints, or both, may change over time, and (b) the changes are taken into account in the optimization process. It is believed that a majority of real-world dynamic problems are DCOPs. However, there are few

Trung Thanh Nguyen
School of Engineering, Technology and Maritime Operations,
Liverpool John Moores University, Liverpool L3 3AF, U.K.
e-mail: T.T.Nguyen@ljmu.ac.uk

Xin Yao
Centre of Excellence for Research in Computational Intelligence and Applications
(CERCIA), School of Computer Science, University of Birmingham,
Birmingham B15 2TT, U.K.
e-mail: X.Yao@cs.bham.ac.uk

S. Yang and X. Yao (Eds.): *Evolutionary Computation for DOPs*, SCI 490, pp. 193–217.
DOI: 10.1007/978-3-642-38416-5_8 　© Springer-Verlag Berlin Heidelberg 2013

studies on continuous dynamic constrained optimization. Existing studies in continuous dynamic optimization only focus on the unconstrained or domain constraint dynamic cases (which in this chapter both are regarded as "unconstrained" problems).

This lack of attention to DCOPs in the continuous domain raises some important research questions: What are the essential characteristics of these types of problems? How well would existing dynamic optimization strategies perform in dynamic constrained environments if most of them are designed for and tested in unconstrained dynamic problems only? Why do they work well or not? How can one evaluate if an algorithm works well or not? And finally, what are the requirements for a "good" algorithm that effectively solves these types of problems?

As a large number of real-world applications are DCOPs, finding the answers to the questions above is essential to have better understanding about the practical issues of DCOPs and to solve this class of problem more effectively. Note that this chapter only investigates the impact of DCOPs on dynamic optimization strategies. For a study on the impact of DCOPs on constraint handling strategies, readers are referred to [23].

The chapter is organized as follows. Section 8.2 discusses the special characteristics from real-world DCOPs and discuss how the characteristics make DCOPs different from unconstrained dynamic optimization problems (DOPs). Section 8.3 reviews related literature about continuous benchmark problems, identifies the gaps between them and real-world problems and proposes a new set of DCO benchmark problems. Section 8.4 discusses the possibility of solving DCOPs using some representative DO strategies. Experimental analyses about the strengths and weaknesses, and the effect of the mentioned characteristics on each strategy will be reported. Based on the experimental results, a list of requirements that algorithms should meet to solve DCOPs effectively is proposed. Finally, Section 8.5 concludes the chapter and identifies future directions.

8.2 Characteristics of Real-World Dynamic Constrained Problems

Constraints make real-world DCOPs very different from the unconstrained or domain constraint problems considered in academic research. In real-world DCOPs the objective function and constraint functions can be combined in three different types: (a) both the objective function and the constraints are dynamic [2, 27, 35]; (b) only the objective function is dynamic while the constraints are static [3, 31, 33]; and (c) the objective function is static and the constraints are dynamic [8, 12, 15]. In all three types, the presence of infeasible areas can affect how the global optimum moves, or appears each change. This leads to some special characteristics which are not found in the unconstrained cases and fixed constrained cases.

First, constraint dynamics can lead to changes in the shape/percentage/structure of the feasible/infeasible areas. Second, objective function dynamics might cause the global optima to switch from one disconnected feasible region to another on problems with disconnected feasible regions, which are very common in real-world constrained problems, especially the scheduling problems [1, 13, 34]. Third, in problems with fixed objective functions and dynamic constraints, the changing infeasible areas might expose new, better global optima without changing the existing optima. One example is the Dynamic 0-1 Knapsack Problem: significantly increasing the capacity of the knapsack can create a new global optimum without changing the existing optimum.

In addition to the three special characteristics above, DCOPs might also have the common characteristics of constrained problems such as global optima in the boundaries of feasible regions, global optima in search boundary, and multiple disconnected feasible regions. These characteristics are widely regarded as being common in real-world applications.

8.3   A Real-Valued Benchmark to Simulate DCOPs Characteristics

8.3.1   Related Literature

In the continuous domain, there is no existing continuous benchmark that fully reflects the characteristics of DCOPs listed in Section 8.2. Among existing continuous benchmarks, there are only few recent studies that are related to dynamic constraints. The first study was [14] in which two simple unimodal constrained problems were proposed. These problems take the time variable t as their only time-dependant parameter and hence the dynamic was created by the increase over time of t. These problems have some important disadvantages which prevent them from being used to capture/simulate the mentioned properties of DCOPs: they only capture a simple linear change. In addition, the two problems do not reflect common situations like dynamic objective + fixed constraints or fixed objective + dynamic constraints and other common properites of DCOPs.

The second study was [29]. In that research, a dynamic constrained benchmark problem was proposed by combining an existing "field of cones on a zero plane" dynamic fitness function with four dynamic norm-based constraints with the square/diamond/sphere-like shapes (see Fig. 2 in [29]). Although the framework used to generate this benchmark problem is highly configurable, the current single benchmark problem generated by the framework in [29] was designed for a different purpose and hence does not simulate the properties mentioned in Section 8.2. For example, the benchmark problem might not be able to simulate common properties of DCOPs such as optima in boundary; disconnected feasible regions; and moving constraints exposing optima in a controllable way. In addition, there is only one single type of benchmark problem and hence it might be difficult to use the problem to evaluate the performance of algorithms under different situations.

The third study was [39]. Based on an existing static test problem [9], the authors assigned six pre-defined values (scalars or matrices) to the coefficients of this static functions to represent six different time steps. Because only six values were given to each coefficient, the dynamic of the problems were defined only up to six time steps. This prevents users from testing the problem in the long run. In addition, there appears to be no specified rule for the dynamics, making it difficult to simulate any dynamic rules from real-world applications. Besides this limitation, the problem also does not reflect common situations such as dynamic objective + fixed constraints, fixed objective + dynamic constraints. It is also unclear if other common properites of DCOPs can be simulated.

The lack of benchmark problems for DCOPs makes it difficult to (a) evaluate how well existing DO algorithms would work on DCOPs, and (b) design new algorithms specialising in DCOPs. Given that a majority of recent real-world DOPs are DCOPs [20], this can be considered an important gap in DO research.

This gap motivates the authors to develop general-purpose benchmark problems to capture the special characteristics of DCOPs. Some initial results involving five benchmark problems were reported in an earlier study [21]. This framework then was extended in [23] to develop full sets of benchmark problems, which are able to capture all characteristics mentioned in the previous section. Two sets of benchmark problems, one with multimodal, scalable objective functions and one with unimodal objective functions, were developed. In this chapter we will describe the benchmark set with unimodal objective functions (many problems in the set still have multiple optima due to the constraints) in detail. Detailed descriptions of the multimodal, scalable set can be found in a technical report [19].

8.3.2 Generating Dynamic Constrained Benchmark Problems

One useful way to create dynamic benchmark problems is to combine existing static benchmark problems with the dynamic rules found in dynamic constrained applications. This can be done by applying the dynamic rules to the parameters of the static problems, as described below.

Given a static function $f_P(x)$ with a set of parameters $P = \{p_1, ... p_k\}$, one can always generalise $f_P(x)$ to its dynamic version $f_{P_t}(x,t)$ by replacing each static parameter $p_i \in P$ with a time-dependent expression $p_i(t)$. The dynamic of the dynamic problem then depends on how $p_i(t)$ varies over time. One can use any type of dynamic rule to represent $p_i(t)$, and hence can create any type of dynamic problem. Details of the concept and a mathematical framework for the idea is described in [19]. Some additional information is provided in [22] (Section 3).

8.3.3 A Dynamic Constrained Benchmark Set

A set of 18 benchmark problems named G24[1] was introduced using the new procedure described in the previous subsection. The general form for each problem in the G24 set is as follows:

---

[1] This benchmark set was named after a static function originally from in [10]. This static function was named G24 in the "CEC06 competition on constrained real-parameter optimization". The static G24 function was adapted to create the $f^{(1)}$, $g^{(1)}$ and $g^{(2)}$ function forms of this DCOP G24 benchmark set.

Table 8.1 The objective function form and set of constraint function forms for each problem

Benchmark problem	objective function
G24_8a & G24_8b	$f(x) = f^{(2)}$
All other problems	$f(x) = f^{(1)}$

Benchmark problem	Set G of constraints
G24_u; G24_uf; G24_2u; G24_8a	$G = \{\emptyset\}$
G24_6a	$G = \left\{ g^{(3)}, g^{(6)} \right\}$
G24_6b	$G = \left\{ g^{(3)} \right\}$
G24_6c	$G = \left\{ g^{(3)}, g^{(4)} \right\}$
G24_6d	$G = \left\{ g^{(5)}, g^{(6)} \right\}$
All other problems	$G = \left\{ g^{(1)}, g^{(2)} \right\}$

$$\text{minimise } f(\mathbf{x})$$
$$\text{subject to } g_i(\mathbf{x}) \leq 0, \, g_i(\mathbf{x}) \in G, \, i = 1,..,n$$

where the objective function $f(\mathbf{x})$ can be one of the function forms set out in Eq. (8.1-8.2), each constraint $g_i(\mathbf{x})$ can be one of the function forms given in Eq. (8.3-8.8), and G is the set of n constraint functions for that particular benchmark problem. The detailed descriptions of $f(\mathbf{x})$ and $g_i(\mathbf{x})$ for each problem are described in Table 8.1.

Eqs. (8.1-8.2) describe the general function forms for the objective functions in the G24 set. Of these function forms, $f^{(2)}$ is used to design the objective function for G24_8a and G24_8b, and $f^{(1)}$ is used to design the objective functions for all other problems. $f^{(1)}$ is modified from a static function proposed in [10] and $f^{(2)}$ is a newly designed function.

$$f^{(1)} = -(X_1 + X_2) \tag{8.1}$$

$$f^{(2)} = -3\exp\left(-\sqrt{\sqrt{(X_1)^2 + (X_2)^2}}\right) \tag{8.2}$$

where $X_i = X_i(x_i, t) = p_i(t)(x_i + q_i(t)); 0 \leq x_1 \leq 3; 0 \leq x_2 \leq 4$ with $p_i(t)$ and $q_i(t)$ $(i = 1, 2)$ as the dynamic parameters, which determine how the dynamic objective function of each benchmark problem changes over time.

Eqs. (8.3-8.8) describe the general function forms for the constraint functions in the G24 set. Of these function forms, $g^{(1)}$ and $g^{(2)}$ were modified from two static functions proposed in [10] and $g^{(3)}, g^{(4)}$, $g^{(5)}$ and $g^{(6)}$ are newly designed functions.

$$g^{(1)} = -2Y_1^4 + 8Y_1^3 - 8Y_1^2 + Y_2 - 2 \tag{8.3}$$

$$g^{(2)} = -4Y_1^4 + 32Y_1^3 - 88Y_1^2 + 96Y_1 + Y_2 - 36 \tag{8.4}$$

$$g^{(3)} = 2Y_1 + 3Y_2 - 9 \tag{8.5}$$

$$g^{(4)} = \begin{cases} -1 \text{ if } (0 \le Y_1 \le 1)\text{or}(2 \le Y_1 \le 3) \\ 1 \text{ otherwise} \end{cases} \tag{8.6}$$

$$g^{(5)} = \begin{cases} -1 \text{ if } (0 \le Y_1 \le 0.5)\text{or}(2 \le Y_1 \le 2.5) \\ 1 \text{ otherwise} \end{cases} \tag{8.7}$$

$$g^{(6)} = \begin{cases} -1 \text{ if } [(0 \le Y_1 \le 1)\text{ and }(2 \le Y_2 \le 3)] \\ \quad \text{or } (2 \le Y_1 \le 3) \\ 1 \text{ otherwise} \end{cases} \tag{8.8}$$

where $Y_i = Y_i(x,t) = r_i(t)(x + s_i(t)); 0 \le x_1 \le 3; 0 \le x_2 \le 4$ with $r_i(t)$ and $s_i(t)$ $(i = 1,2)$ as the dynamic parameters, which determine how the constraint functions of each benchmark problem change over time.

Each benchmark problem may have a different mathematical expression for $p_i(t)$, $q_i(t)$, $r_i(t)$ and $s_i(t)$. Note that although many benchmark problems share the same general function form in Eqs. (8.3-8.8), their individual expressions for $p_i(t)$ and $q_i(t)$ make their actual dynamic objective functions very different. Similarly, the individual expressions for $r_i(t)$ and $s_i(t)$ make each actual dynamic constraint functions very different although they may share the same function form. The individual expressions of $p_i(t)$, $q_i(t)$, $r_i(t)$, and $s_i(t)$ for each benchmark function are described in Table 8.2.

Two guidelines were used to design the test problems: (a) problems should simulate the common properties of DCOPs as mentioned in Section 8.2 and (b) there should always be a pair of problems for each characteristic. The two problems in each pair should be almost identical except that one has a particular characteristic (e.g. fixed constraints) and the other does not. By comparing the performance of an algorithm on the two problems in the pair, it is possible to analyse whether the considered characteristic has any effect on the tested algorithm and to what extent that effect is significant.

Based on the two guidelines above, 18 different test problems were created in [23] (Table 8.2). Each test problem is able to capture one or several of the mentioned characteristics of DCOPs, as shown in Table 8.3. In addition, the problems and their relationships are carefully designed so that they can be arranged in 21 pairs (Table 8.4), of which each pair is a different test case to test a single characteristic of DCOPs (the two problems in each pair are almost identical except that one has a special characteristic and the other does not).

Table 8.2 Dynamic parameters for all test problems in the benchmark set G24. Each dynamic parameter is a time-dependant rule/function which governs the way the problems change (reproduced with permission from [23])

Prob	Parameter settings
G24_u	$p_1(t) = \sin\left(k\pi t + \frac{\pi}{2}\right); p_2(t) = 1; q_i(t) = 0$
G24_1	$p_2(t) = r_i(t) = 1;\ q_i(t) = s_i(t) = 0$ $p_1(t) = \sin\left(k\pi t + \frac{\pi}{2}\right)$
G24_f	$p_i(t) = r_i(t) = 1;\ q_i(t) = s_i(t) = 0$
G24_uf	$p_i(t) = 1;\ q_i(t) = 1$
G24_2	if $(t \bmod 2 = 0)$ $\begin{cases} p_1(t)=\sin\left(\frac{k\pi t}{2}+\frac{\pi}{2}\right) \\ p_2(t)=\begin{cases} p_2(t-1) \text{ if } t>0 \\ p_2(0)=0 \text{ if } t=0 \end{cases} \end{cases}$ if $(t \bmod 2 \neq 0)$ $\begin{cases} p_1(t)=\sin\left(\frac{k\pi t}{2}+\frac{\pi}{2}\right) \\ p_2(t)=\sin\left(\frac{k\pi(t-1)}{2}+\frac{\pi}{2}\right) \end{cases}$ $q_i(t) = s_i(t) = 0;\ r_i(t) = 1$
G24_2u	if $(t \bmod 2 = 0)$ $\begin{cases} p_1(t)=\sin\left(\frac{k\pi t}{2}+\frac{\pi}{2}\right) \\ p_2(t)=\begin{cases} p_2(t-1) \text{ if } t>0 \\ p_2(0)=0 \text{ if } t=0 \end{cases} \end{cases}$ if $(t \bmod 2 \neq 0)$ $\begin{cases} p_1(t)=\sin\left(\frac{k\pi t}{2}+\frac{\pi}{2}\right) \\ p_2(t)=\sin\left(\frac{k\pi(t-1)}{2}+\frac{\pi}{2}\right) \end{cases}$ $q_i(t) = 0$
G24_3	$p_i(t) = r_i(t) = 1; q_i(t) = s_1(t) = 0$ $s_2(t) = 2 - t.\frac{x_2\max - x_2\min}{S}$
G24_3b	$p_1(t) = \sin\left(k\pi t + \frac{\pi}{2}\right);\ p_2(t) = 1$ $q_i(t) = s_1(t) = 0;\ r_i(t) = 1;$ $s_2(t) = 2 - t.\frac{x_2\max - x_2\min}{S}$
G24_3f	$p_i(t) = r_i(t) = 1; q_i(t) = s_1(t) = 0; s_2(t) = 2$
G24_4	$p_2(t) = r_i(t) = 1;\ q_i(t) = s_1(t) = 0$ $p_1(t) = \sin\left(k\pi t + \frac{\pi}{2}\right); s_2(t) = t.\frac{x_2\max - x_2\min}{S}$
G24_5	if $(t \bmod 2 = 0)$ $\begin{cases} p_1(t)=\sin\left(\frac{k\pi t}{2}+\frac{\pi}{2}\right) \\ p_2(t)=\begin{cases} p_2(t-1) \text{ if } t>0 \\ p_2(0) \text{ if } t=0 \end{cases} \end{cases}$ if $(t \bmod 2 \neq 0)$ $\begin{cases} p_1(t)=\sin\left(\frac{k\pi t}{2}+\frac{\pi}{2}\right) \\ p_2(t)=\sin\left(\frac{k\pi(t-1)}{2}+\frac{\pi}{2}\right) \end{cases}$ $q_i(t) = s_1(t) = 0;\ r_i(t) = 1;$ $s_2(t) = t.\frac{x_2\max - x_2\min}{S}$
G24_6a/b/c/d	$p_1(t) = \sin\left(\pi t + \frac{\pi}{2}\right); p_2(t) = 1;$ $q_i(t) = s_i(t) = 0; r_i(t) = 1$
G24_7	$p_i(t) = r_i(t) = 1;\ q_i(t) = s_1(t) = 0;$ $s_2(t) = t.\frac{x_2\max - x_2\min}{S}$
G24_8a	$p_i(t) = -1; q_1(t) = -(c_1 + r_a.\cos(k\pi t))$ $q_2(t) = -(c_2 + r_a.\sin(k\pi t));$
G24_8b	$p_i(t) = -1; q_1(t) = -(c_1 + r_a.\cos(k\pi t))$ $q_2(t) = -(c_2 + r_a.\sin(k\pi t)); r_i(t) = 1;\ s_i(t) = 0$
k	k determines the severity of function changes. $k = 1 \sim$ large; $k = 0.5 \sim$ medium; $k = 0.25 \sim$ small
S	S determines the severity of constraint changes $S = 10 \sim$ large; $S = 20 \sim$ medium; $S = 50 \sim$ small
c_1, c_2, r_a (G24_8a/b only)	$c_1 = 1.470561702; c_2 = 3.442094786232;$ $r_a = 0.858958496.$
i	i is the variable index, $i = 1, 2$

Table 8.3 Properties of each test problem in the G24 benchmark set (reproduced with permission from [23])

Problem	ObjFunc	Constr	DFR	SwO	bNAO	OICB	OISB	Path
G24_u	Dynamic	NoC	1	No	No	No	Yes	N/A
G24_1	Dynamic	Fixed	2	Yes	No	Yes	No	N/A
G24_f	Fixed	Fixed	2	No	No	Yes	No	N/A
G24_uf	Fixed	NoC	1	No	No	No	Yes	N/A
G24_2*	Dynamic	Fixed	2	Yes	No	Yes&No	Yes&No	N/A
G24_2u	Dynamic	NoC	1	No	No	No	Yes	N/A
G24_3	Fixed	Dynamic	2-3	No	Yes	Yes	No	N/A
G24_3b	Dynamic	Dynamic	2-3	Yes	No	Yes	No	N/A
G24_3f	Fixed	Fixed	1	No	No	Yes	No	N/A
G24_4	Dynamic	Dynamic	2-3	Yes	No	Yes	No	N/A
G24_5*	Dynamic	Dynamic	2-3	Yes	No	Yes&No	Yes&No	N/A
G24_6a	Dynamic	Fixed	2	Yes	No	No	Yes	Hard
G24_6b	Dynamic	NoC	1	No	No	No	Yes	N/A
G24_6c	Dynamic	Fixed	2	Yes	No	No	Yes	Easy
G24_6d	Dynamic	Fixed	2	Yes	No	No	Yes	Hard
G24_7	Fixed	Dynamic	2	No	No	Yes	No	N/A
G24_8a	Dynamic	NoC	1	No	No	No	No	N/A
G24_8b	Dynamic	Fixed	2	Yes	No	Yes	No	N/A

DFR	number of Disconnected Feasible Regions
SwO	Switched global Optimum between disconnected regions
bNAO	better Newly Appear Optimum without changing existing ones
OICB	global Optimum is In the Constraint Boundary
OISB	global Optimum is In the Search Boundary
Yes&No	It means OICB/OISB is true at some other changes and false at some others
Path	Indicate if it is easy or difficult to use mutation to travel between feasible regions
Dynamic	The function is dynamic
Fixed	There is no change
NoC	There is no constraint
*	In some change periods, the landscape either is a plateau or contains infinite number of optima and all optima (including the existing optimum) lie in a line parallel to one of the axes

8.4 Challenges to Solve DCOPs

8.4.1 Analysing the Performance of Some Common Dynamic Optimization Strategies in Solving DCOPs

The purpose of this section is to discuss whether the DO strategies commonly used in existing literature can be applied directly to solving DCOPs. We also report our analyses in [23] of whether the special characteristics of DCOPs might have any effect on the performance of these strategies and why. The results of the analysis will also provide insight in understanding how to design suitable algorithms for solving DCOPs.

Table 8.4 The 21 test cases (pairs) to be used in this chapter (reproduced with permission from [23])

	Static problems: Unconstrained vs Fixed constraints		
1	G24_uf (fF, noC)	vs	G24_f (fF, fC)
	Fixed objectives vs Dynamic objectives		
2	G24_uf (fF, noC)	vs	G24_u (dF, noC)
3	G24_f (fF, fC, OICB)	vs	G24_1 (dF, fC, OICB)
4	G24_f (fF, fC, OICB)	vs	G24_2 (dF, fC, ONICB)
	Dynamic objectives: Unconstrained vs Fixed constraints		
5	G24_u (dF, noC)	vs	G24_1 (dF, fC, OICB)
6	G24_2u (dF, noC)	vs	G24_2 (dF, fC, ONICB)
	Fixed constraints vs Dynamic constraints		
7	G24_1 (dF, fC, OICB)	vs	G24_4 (dF, dC, OICB)
8	G24_2 (dF, fC, ONICB)	vs	G24_5 (dF, dC, ONICB)
9	G24_f (fF, fC)	vs	G24_7 (fF, dC, NNAO)
10	G24_3f (fF, fC)	vs	G24_3 (fF, dC, NAO)
	No constraint vs Dynamic constraints		
11	G24_u (dF, noC)	vs	G24_4 (dF, dC, OICB)
12	G24_2u (dF, noC)	vs	G24_5 (dF, dC, ONICB)
13	G24_uf (fF, noC)	vs	G24_7 (fF, dC)
	Moving constraints expose better optima vs not expose optima		
14	G24_3f (fF, fC)	vs	G24_3 (fF, dC, NAO)
15	G24_3 (fF, dC, NAO)	vs	G24_3b (dF, dC, NAO)
	Connected feasible regions vs Disconnected feasible regions		
16	G24_6b (1R)	vs	G24_6a (2DR, hard)
17	G24_6b (1R)	vs	G24_6d (2DR, hard)
18	G24_6c (2DR, easy)	vs	G24_6d (2DR, hard)
	Optima in constraint boundary vs Optima NOT in constr boundary		
19	G24_1 (dF, fC, OICB)	vs	G24_2 (dF, fC, ONICB)
20	G24_4 (dF, dC, OICB)	vs	G24_5 (dF, dC, ONICB)
21	G24_8b (dF, fC, OICB)	vs	G24_8a (dF, noC, ONISB)

dF	dynamic objective func	fF	fixed objective function
dC	dynamic constraints	fC	fixed constraints
OICB	optima in constraint bound	ONICB	opt. not in constraint bound
OISB	optima in search bound	ONISB	optima not in search bound
NAO	better newly appear optima	NNAO	No better newly appear opt
2DR	2 Disconn. feasible regions	1R	One single feasible region
Easy	easy for mutation to travel between disconn. regions	Hard	less easy to travel among regions
noC	unconstrained problem	SwO	Switched optimum between disconnected regions

The strategies being considered are (1) introducing diversity, (2) maintaining diversity and (3) tracking the previous optima. These three are among the four most commonly used strategies (the other strategy is memory-based) to solve DOPs. The diversity-introducing strategy was proposed based on the assumption that by the time a change occurs in the environment, an evolutionary algorithm (EA) might have already converged to a specific area and hence would lose its ability to deal with changes in other areas of the search space. Consequently, it is necessary to increase the diversity level in the population, either by increasing the mutation rate or re-initialising/re-locating the individuals. This strategy was introduced years ago [7] but is still extensively used [18, 26].

The diversity-introducing strategy requires that changes must be visible to the algorithm. To avoid this disadvantage, the diversity-maintaining strategy was introduced so that population diversity can be maintained without explicitly detecting changes [11]. This strategy is still the main strategy in many recent approaches [5, 38].

The third strategy, tracking-previous-optima, is used where the optima might only slightly change. The region surrounding the current optima is monitored to detect changes and "track" the movement of these optima. Similar to the two strategies above, the tracking strategy has also been used for years [7] and it has always been one of the main strategies for solving DOPs. Recently this strategy has been combined with the diversity maintaining/introducing strategy to achieve better performance. Typical examples are the multi-population/multi-swarm approaches, where multiple sub-populations are used to maintain diversity and each sub-population/sub-swarm focuses on tracking one single optimum [5, 6].

8.4.2 Chosen Algorithms and Experimental Settings

8.4.2.1 Chosen Algorithms

Two commonly used algorithms: *triggered hyper-mutation GA* (HyperM [7]) and *random-immigrant GA* (RIGA [11]) were chosen to evaluate the performance of the three strategies mentioned above in DCOPs. HyperM is basically a simple GA with an adaptive mechanism to switch from a low mutation rate (standard-mutation-rate) to a high mutation rate (hyper-mutation-rate, to increase diversity) and vice versa depending on whether or not there is a degradation of the best solution in the population. It represents the "introducing diversity" and "tracking previous optima" strategies in DO.

RIGA is another derivative of a basic GA. After the normal mutation step, a fraction of the population is replaced with randomly generated individuals. This fraction is determined by a random-immigrant-rate (also named replacement rate). By continuously replacing a part of the population with random solutions, the algorithm is able to maintain diversity throughout the search process to cope with dynamics. RIGA represents the "maintaining diversity" strategy in DO.

One reason to choose these algorithms for the test is that their strategies are still commonly used in most current state-of-the-art DO algorithms. Another reason is the strategies in these algorithms are very simple and straightforward, making it easy to test and analyse their behaviour. In addition, because these two algorithms are very well studied, using them would help in comparing new experimental data with existing results. Finally, because both algorithms are developed from a basic GA (actually the only difference between HyperM/RIGA and a basic GA is the

Table 8.5 Test settings for all algorithms used in the chapter

All the	Pop size (pop_size)	5, 15, 25 (medium), 50, 100
algorithms	Elitism	Elitism & non-elitism if applicable
(exceptions	Selection method	Non-linear ranking as in [16]
below)	Mutation method	Uniform, $P = 0.15$.
	Crossover method	Arithmetic, $P = 0.1$.
HyperM	Triggered mutate	Uniform, $P = 0.5$ as in [7].
RIGA	Rand-immig. rate	$P = 0.3$ as in [11].
Benchmark	Number of runs	50
problem	Number of changes	$5/k$ (see below)
settings	Change frequency	250, 500, 1000 (med), 2000, 4000 evaluations
	ObjFunc severity k	0.25 (small), 0.5 (med), 1.0 (large)
	Constr. severity S	10 (small), 20 (medium), 50 (large)

mutation strategy), it would be easier to compare/analyse their performance. The performance of HyperM and RIGA was also compared with a basic GA to see if they work well on the tested problems.

8.4.2.2   Parameter Settings

Table 8.5 shows the detailed parameter settings for HyperM, RIGA and GA. All algorithms use real-valued representations. The algorithms were tested on 18 benchmark problems described in Section 8.3. To create a fair testing environment, the algorithms were tested in a wide range of dynamic settings (different values of population size, severity of change and frequency of change) with five levels: *small, medium small, medium, medium large, large.*

The evolutionary parameters of all tested algorithms were set to similar values or the best known values if possible. The base mutation rate of the algorithms is 0.15, which is the average value of the best mutation rates commonly used for GA-based algorithms in various existing studies on continuous DO, which are 0.1 ([28, 30]) and 0.2 ([4, 6]). For HyperM and RIGA, the best *hyper-mutation-rate* and *random-immigrant-rate* parameter values observed in the original papers [7, 11] were used. The same implementations as described in [7] and [11] were used to reproduce these two algorithms. A crossover rate of 0.1 was chosen for all algorithms because, according to the analysis in [25], this value was one of the few settings where all tested algorithms perform well on this benchmark set.

A further study of the effect of different values of the base mutation rates, hyper-mutation rates, random-immigrant rates and crossover rates on algorithm performance was also carried out. Detailed experimental results and discussion for this analysis can be found in [25] where it was found that the overall behaviours of the algorithms are not different from those using the default/best known settings, except for the followings: (i) When the base mutation rate is very low (≤ 0.01), the performance of GA and HyperM drop significantly; (ii) generally to work well in the tested DCOPs, algorithms need to use high base mutation rates. The range of best mutation rates is 0.3-0.8. (iii) Algorithms like RIGA and HyperM also need high random-immigrant/hyper-mutation rates to solve DCOPs. The best results are usually achieved with the rates of 0.6-0.8; (iv) The suitable range of crossover rate is 0.1-1.0.

8.4.2.3   Constraint Handling

It is necessary to integrate existing DO algorithms with a CH mechanism to use these algorithms for solving DCOPs. That CH mechanism should not interfere with the original DO strategies so that it is possible to correctly evaluate whether the original DO strategies would still be effective in solving DCOPs. To satisfy this requirement, the penalty function approach in [17] was chosen because it is the simplest way to apply existing unconstrained DO algorithms directly to solving DCOPs without changing the algorithms. Also this penalty method can be effective in solving difficult numerical problems without requiring users to choose any penalty factor or other parameter [17].

8.4.2.4   Performance Measures

For measuring the performance of the algorithms in this particular experiment, an existing measure: the *offline error* [6] was modified. The measure is calculated as the average over, at every evaluation, the error of the best solution found since the last change of the environment.

Because the measure above is designed for unconstrained environments, it is necessary to modify it to evaluate algorithm performance in constrained environments: At every generation, instead of considering the best errors/fitness values of *any* solutions regardless of feasibility as implemented in the original measure, only the best fitness values / best errors of *feasible* solutions at each generation are considered. If in any generation there is no feasible solution, the measure takes the *worst possible value* that a feasible solution can have for that particular generation. This measure is called the *modified offline error for DCOPs*, or *offline error* for short.

$$E_{MO} = \frac{1}{num\_of\_gen} \sum_{j=1}^{num\_of\_gen} e_{MO}(j) \qquad (8.9)$$

where $e_{MO}(j)$ is the best *feasible* error since the last change at the generation j.

Five new measures were also proposed to analyse why a particular algorithm might work well on a particular problem. The first two measures are the *recovery rate* (RR) and the *absolute recovery rate* (ARR) to analyse the convergence behaviour of algorithms in dynamic environments. The RR measure is used to analyse *how quickly an algorithm recovers from an environmental change and starts converging to a new solution before the next change occurs*. The new solution is not necessarily the global optimum.

$$RR = \frac{1}{m} \sum_{i=1}^{m} \frac{\sum_{j=1}^{p(i)} [f_{best}(i,j) - f_{best}(i,1)]}{p(i)[f_{best}(i,p(i)) - f_{best}(i,1)]} \tag{8.10}$$

where $f_{best}(i,j)$ is the fitness value of the best feasible solution since the last change found by the tested algorithm until the jth generation of the change period i, m is the number of changes and $p(i), i = 1 : m$ is the number of generations at each change period i. The RR score would be 1 in the best case where the algorithm is able to recover and converge to a solution immediately after a change, and would be close to zero in case the algorithm is unable to recover from the change at all [2].

The RR measure only indicates if the considered algorithm converges to a solution and if it converges quickly. It does not indicate whether the converged solution is the global optimum. For example, RR can still be 1 if the algorithm does nothing but keep re-evaluating the same solution. Because of that, another measure is needed: the *absolute recovery rate* (ARR). This measure is very similar to the RR but is used to analyse *how quick it is for an algorithm to start converging to the global optimum before the next change occurs*:

$$ARR = \frac{1}{m} \sum_{i=1}^{m} \frac{\sum_{j=1}^{p(i)} [f_{best}(i,j) - f_{best}(i,1)]}{p(i)[f^*(i) - f_{best}(i,1)]} \tag{8.11}$$

where $f_{best}(i,j), i, j, m, p(i)$ are the same as in Eq. 8.10 and $f^*(i)$ is the global optimal value of the search space at the ith change. The ARR score would be 1 in the best case when the algorithm is able to recover and converge to the global optimum immediately after a change, and would be zero in case the algorithm is unable to recover from the change at all. Note that the score of ARR should always be less than or equal to that of RR. In the ideal case (converged to global optimum), ARR should be equal to RR[3].

The RR and ARR measures can be used together to indicate if an algorithm is able to converge to the global optimum within the given time frame between changes and if so how quickly it takes to converge. The *RR-ARR diagram* in Fig. 8.1 shows some analysis guidelines.

A third measure, *percentage of selected infeasible individuals*, is proposed to analyse algorithm ability to balance exploiting feasible regions and exploring infeasible regions in DCOPs. This measure finds the percent of infeasible individuals selected for the next generation. The average (over all tested generations) is then compared with the percentage of infeasible areas in the search space. If the considered algorithm is able to accept *infeasible* diversified individuals in the same way as it accepts *feasible* diversified individuals (and hence to maintain diversity effectively), the two percentage values should be equal.

To analyse the behaviour of algorithms using triggered-mutation mechanisms such as HyperM, a fourth measure: *triggered-time count*, which counts the number

---

[2] Note that RR will never be equal to zero because there is at least one generation where $f_{best}(i,j) = f_{best}(i,p(i))$.

[3] Note that to use the measure ARR it is necessary to know the global optimum value at each change period.

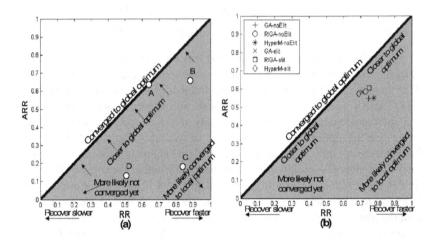

Fig. 8.1 Diagram (a) provides a guideline for analysing the convergence behaviour/recovery speed of an algorithm given its RR/ARR scores. These scores can be represented as the x and y coordinations of a point on the diagonal thick line or inside the shaded area. The position of the point represents the behaviour of the corresponding algorithm. The closer the point is to the right, the faster the algorithm was in recovering and re-converging, and vice versa. In addition, if the point lies on the thick diagonal line (where $RR = ARR$) like point A, the algorithm has been able to recover from the change and converged to the new global optimum. Otherwise, if the point lies inside the shaded area, the algorithm either has converged to a local solution (e.g. point C); or has not been converged yet (e.g. point D - recover slowly; and point B - recover quickly). **Diagram (b)** shows the mapping of the RR/ARR scores of GA, RIGA, and HyperM to the RR-ARR diagram. (Reproduced with permission from [23]).

of times the hyper-mutation-rate is triggered by the algorithm, and a fifth measure: *detected-change count*, which counts the number of triggers actually associated with a change, are also proposed. For HyperM, triggers associated with a change are those that are invoked by the algorithm within v generations after a change, where v is the maximum number of generations (five in this implementation) needed for HyperM to detect a drop in performance. These two measures indicate how many times an algorithm triggers its hyper-mutation; whether each trigger time corresponds to a new change; and if there is any change that goes undetected during the search process.

Note that all of the measures used here are specifically designed for dynamic problems. This creates a problem for the experiments in this chapter because in the G24 benchmark set there are not only dynamic problems but also stationary problems. To overcome this issue, in this study stationary problems are considered a special type of dynamic problem which still have "changes" with the same change frequency as other dynamic problems. However, in stationary problems the "changes" do not alter the search space.

8.4.3   Experimental Results and Analyses

The full *offline-error* results of the tested algorithms on all 18 benchmark problems for all test scenarios are presented in the tables in [24]. These data were further analysed from different perspectives to achieve a better understanding of how existing DO strategies work in DCOPs and how each characteristic of DCOPs would affect the performance of existing DO algorithms. First of all, the average performance of the tested algorithms on each major group of problems under different parameter settings and dynamic ranges were summarised to have an overall picture of algorithm behaviour on different types of problems (see Fig. 8.2). Then the effect of each problem characteristic on each algorithm was analysed in 21 test cases (each case is a pair of almost identical problems, one with a particular characteristic and one without) as shown in Table 8.4 of Section 8.3 (see test results in Figs. 8.4 and 8.5). For each particular algorithm, some further analyses were also carried out using the five newly proposed measures mentioned above. Details of these analyses will be described in the next subsections. Only the summarised results are presented in Fig. 8.2 with different settings (small / medium / large). For other detailed figures and tables, the results will only be presented in the default settings (all parameters and dynamic range are set to medium). For detailed results in other settings, readers are referred to [24].

The experimental results show some interesting findings.

8.4.3.1   The Impact of Different Dynamic Ranges on Algorithm Performance

The summarised results in groups of problems (Fig. 8.2) show that (i) generally the behaviour of algorithms and their relative strengths/weaknesses in comparison with other algorithms still remain roughly the same when the dynamic settings change; and (ii) as expected in most cases algorithms' performance decrease when the conditions become more difficult (magnitude of change becomes larger; change frequency becomes higher; population size becomes much smaller). Among the variations in dynamic settings, it seems that the variations in frequency of change affect algorithms' performance the most, followed by variations in magnitude of changes and in population size.

8.4.3.2   The Effect of Elitism on Algorithm Performance

The summarised results in groups of problems (Fig. 8.2) and the pair-wise comparisons in Fig. 8.4 and Fig. 8.5 reveal an interesting effect of elitism on both unconstrained and constrained dynamic cases: the elitism versions of GA/RIGA/HyperM perform better than their non-elitism counterparts in most tested problems. The reason for this effect (with evidence shown in the next paragraph) is that elitism helps algorithms with diversity-maintaining strategies to converge faster. This effect is independent of the combined CH techniques.

Two measures proposed in Section 8.4.2.4: *recovery rate* (RR) and *absolute recovery rate* (ARR) were used to study the inefficiency of GA/RIGA/HyperM in

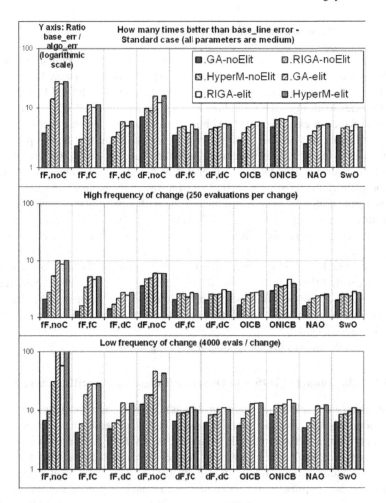

Fig. 8.2 Algorithm performance in groups of problem (part 1 - see Fig. 8.3 for part 2). Performance (vertical axis in logarithmic scale) is evaluated by calculating the ratio between the *base line* (worst error among all scenarios) and the error of each algorithm in each problem to see how many times their performance is better (smaller) than the *base line*. Explanations for abbreviations can be found in Table 8.4.

the non-elitism case. The scores of the algorithms on these measures are given in Fig. 8.1b. The figure shows that none of the algorithms are close to the optimum line, meaning there are problems/ change periods where the algorithms were unable to converge to the global optimum. In addition, for RIGA, its elitism version is closer to the top-right corner while its non-elitism version is closer to the bottom-left corner, meaning that non-elitism makes RIGA converge slower/less accurately. Finally, for GA/HyperM, their elitism versions are closer to the global optimum while their non-elitism versions are closer to the bottom-right corner, meaning that the

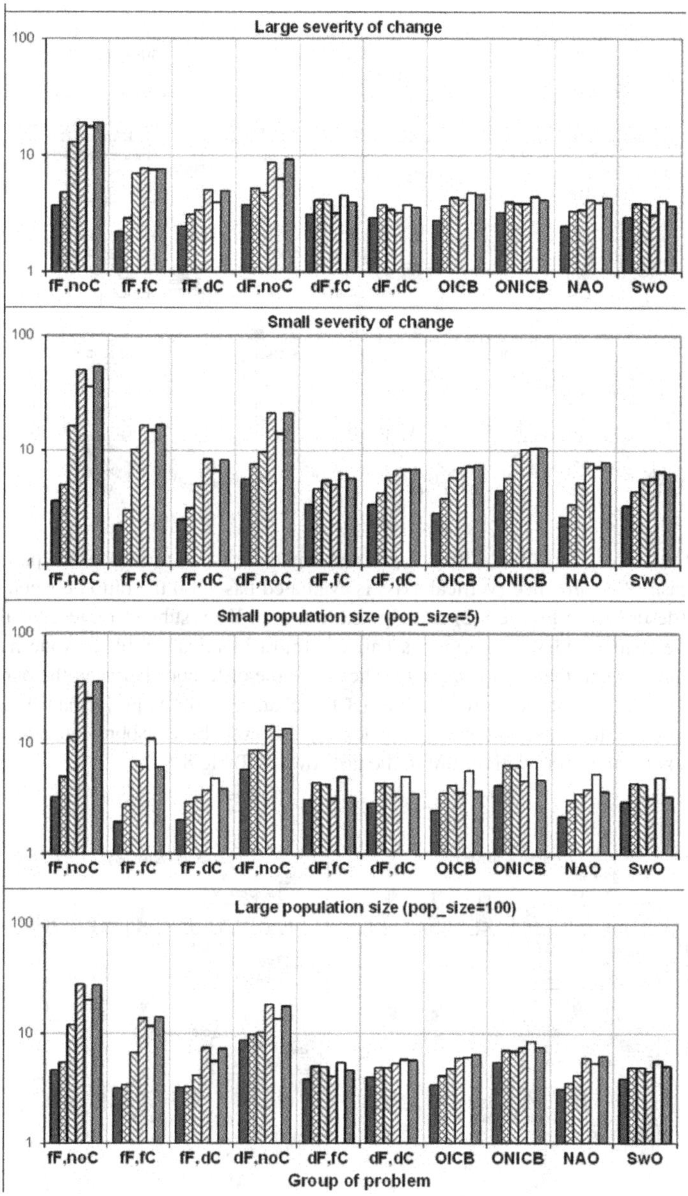

Fig. 8.3 Algorithm performance in groups of problem (part 2 - see Fig. 8.2 for part 1 and explanation)

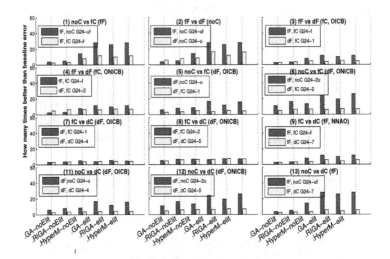

Fig. 8.4 The effect of twelve different problem characteristics on algorithm performance (medium case). Performance (vertical axis) is evaluated based on the ratio between the base line error (described in Figure 8.2) and algorithm errors. Each subplot represents algorithm performance (pair of adjacent bars) in a pair of almost identical problems (one has a special characteristic and the other does not). The larger the difference between the bar heights, the greater the impact of the corresponding DCOP characteristic on performance. Subplots' title represent the test case numbers (in brackets) followed by an abbreviated description. Explanations for the abbreviations are in the last rows of Table 8.4.

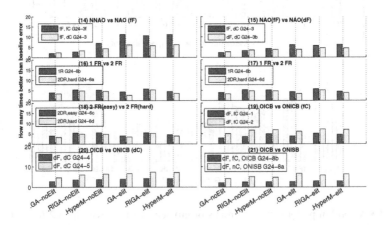

Fig. 8.5 The effect of the other eight different problem properties on algorithm performance (medium case). Instructions to read this figure can be found in Figure 8.4.

non-elitism versions of GA/HyperM are more suceptible to premature convergence. The results hence show that the high diversity maintained by the random-immigrant rate in RIGA and the high mutation rate in GA/HyperM comes with a trade-off: the convergence speed is affected. In such a situation, elitism can be used to speed up the convergence process. Elite members can guide the population to exploit the good regions faster while still maintaining diversity.

8.4.3.3   Effect of Infeasible Areas on Maintaining/Introducing Diversity

Another interesting observation is that the presence of constraints makes the performance of diversity-maintaining/introducing strategies less effective when used in combination with the tested penalty functions. This behaviour can be seen in Fig. 8.2 where the performance of all algorithms in the unconstrained dynamic case (dF+noC) is significantly better than their performance in all dynamic constrained cases (dF+fC, fF+dC, dF+dC). This behaviour can also be seen in the more accurate pair-wise comparisons in Fig. 8.4 and Fig. 8.5: for each pair of problems in which one has constraints and the other does not, GA, RIGA and HyperM always perform worse on the problem with constraints (see pairs 1, 5, 6, 11, 12, 13 in Fig. 8.4 and pair 21 in Fig. 8.5).

The reason for this inefficiency is the use of tested penalty functions prevents diversity-maintaining/introducing mechanisms from working effectively. In solving unconstrained dynamic problems, all diversified individuals generated by the diversity maintaining/introducing strategies are useful because they contribute to either (1) detecting newly appearing optima or (2) finding the new place of the moving optima. In DCOPs, however, there are two difficulties that prevent diversified individuals that are infeasible from being useful in existing DO strategies. One difficulty is many diversified but infeasible individuals might not be selected for the next generation population because they are penalised with lower fitness values by the penalty functions. Consequently, these diversified individuals cannot be used for maintaining diversity unless they are re-introduced again in the next generation. To demonstrate this drawback, the previously proposed measure *percentage of selected infeasible individuals* was used. As can be seen in Table 8.6, in the elitism case the percentage of infeasible solutions in the population (23 - 37.6%) is much smaller than the percentage of infeasible areas over the total search space (60.8%). This means only a few of the diversified, infeasible solutions are retained and hence the algorithms are not able to maintain diversity in the infeasible regions[4].

The second difficulty is that, even if a diversified but infeasible individual is selected for the next generation, it might no longer have its true fitness value. Consequently, environmental changes might not be accurately detected or tracked.

---

[4] Non-elitism algorithms are able to retain more infeasible individuals, of which some might be diversified solutions. However, as shown in Subsection 8.4.3.2, in the non-elitism case this higher percentage of infeasible individuals comes with a trade-off of slower/less accurate convergence, which leads to the generally poorer performance.

Table 8.6 Average *percentage of selected infeasible individuals* over 18 problems. The last row shows the average *percentage of infeasible areas.* (Reproduced with permission from [23])

Algorithms	Percent of infeasible solutions
.GA-elit	23.0%
.RIGA-elit	37.6%
.HyperM-elit	26.4%
.GA-noElit	46.3%
.RIGA-noElit	49.1%
.HyperM-noElit	45.3%
Percentage of infeasible areas	**60.8%**

8.4.3.4 Effect of Switching Global Optima (between Disconnected Feasible Regions) on Strategies That Use Penalty functions

The results show existing DO methods become less effective when they are used in combination with the tested penalty functions to solve a special class of DCOPs: problems with disconnected feasible regions where the global optimum switches from one region to another whenever a change occurs. In addition, the more separated the disconnected regions are, the more difficult it is for algorithms using penalty functions to solve.

The reason for this difficulty is it is necessary to have a path through the infeasible areas that separate the disconnected regions to track the moving optimum. This path might not be available if penalty functions are used because penalties make it unlikely infeasible that individuals are accepted. Obviously the larger the infeasible areas between disconnected regions, the harder it is to establish the path using penalty methods.

Three test cases (pairs of almost identical problems) 16, 17, 18 in Table 8.4 were used to verify the statement above. In all three test cases the objective functions are the same and the global optimum switches between two locations whenever a change occurs. However, each case represents a different dynamic situation. Case 16 tests the situation where in one problem of the pair (G24_6b) there is a feasible path connecting the two locations and in the other problem (G24_6a) the path is infeasible, i.e., there is an infeasible area separating two feasible regions. Case 17 is the same as case 16 except that the infeasible area separating two feasible regions has a different shape. Case 18 tests a different situation where in one problem (G24_6c) the infeasible area separating the two feasible regions is small whereas in the other problem (G24_6d) this infeasible area is large.

The experimental results in these three test cases (pairs 16, 17, 18 in Fig. 8.5) confirm the hypotheses stated in the beginning of this subsection. In cases 16 and 17, the performance of the tested algorithms did decrease when the path between the two regions is infeasible. In case 18, the larger the infeasible area separating the two regions, the worse the performance of the tested algorithms.

Table 8.7 The *triggered-time count* scores and the *detected-change count* scores of HyperM in a pair of problems with moving constraints exposing new optima after 11 changes. (Reproduced with permission from [23])

Algorithms	G24_3 (NAO+fF)				G24_3b (NAO+dF)			
	Trigger Count		Detected Change Count		Trigger count		Detected Change Count	
	Value	stdDev	Value	stdDev	Value	stdDev	Value	stdDev
HyperM-noElit	188.70	8.40	1.74	0.78	199.83	5.88	11.00	0.00
HyperM-elit	0.00	0.00	0.00	0.00	30.43	0.57	11.00	0.00

NAO - Newly Appearing Optimum
fF / dF - fixed / dynamic objective Function

8.4.3.5   Effect of Moving Infeasible Areas on Strategies That Track the Previous Optima

Algorithms relying on tracking previous global optimum such as HyperM might become less effective when the moving constraints expose new, better optima without changing the existing optima. The reason is HyperM cannot detect changes in such DCOPs and hence might not be able to trigger its hyper-mutation rate. With the currently chosen base mutation of 0.15, HyperM is still able to produce good results because the mutation is high enough for the algorithm to maintain diversity. However, in a previous study [21], when a much smaller base mutation rate was used, HyperM becomes significantly worse compared to other algorithms in solving problems like G24_3.

To illustrate this drawback, the newly proposed measures *triggered-time count* and *detected-change count* were used to analyse how the triggered-hypermutation mechanism works on problem G24_3. As can be seen in Table 8.7, HyperM either was not able to trigger its hyper-mutation rate to deal with changes (elitism case, *triggered-time count*=0 & *detected-change count*=0) or was not able to trigger its hyper-mutation rate correctly when a change occurs (non-elitism case, *triggered-time count*~188.7 & *detected-change count*~1.74). It is worth noting in the non-elitism case, most of the trigger times are caused by the selection process because in non-elitism selection the best solution in the population is not always selected for the next generation.

Table 8.7 also shows that in problem G24_3b, which is almost identical to G24_3 except it has its existing optima changed, HyperM was able to detect changes and hence trigger its hyper-mutation timely whenever a change occurs. It shows HyperM only becomes less effective where environmental changes do not change the value of existing optima.

8.4.4   Suggestions to Improve Current Dynamic Optimization Strategies in Solving DCOPs

The experimental results suggest some directions for addressing the drawbacks listed in the previous subsections:

(i) Based on the observation that elitism is useful for diversity-maintaining strategies in solving DCOPs, it might be useful to develop algorithms that support both elitism and diversity maintaining mechanisms.

(ii) Given that methods like HyperM are not able to detect changes because they mainly use change detectors (the best solution in case of HyperM) in the feasible regions, it might be useful to use change detectors and search in both regions and infeasible regions.

(iii) Because experimental results show that tracking the existing optima might not be effective in certain cases of DCOPs, it might be useful to track the moving feasible regions instead. Because after a change in DCOPs the global optimum always either moves along with the feasible areas or appears in a new feasible area, an algorithm able to track feasible areas would have higher chance of tracking the actual global optimum.

Recent experimental results have shown that the directions above could be helpful for improving the performance of DO algorithms in solving DCOPs. The use of elitism was shown to have positive effects in [25, 37], detecting and/or searching in infeasible areas helped improve performance in [25, 29, 32], and tracking feasible areas gave superior results in [21, 25].

8.5   Conclusion and Future Research

In this chapter we have reviewed some important and not well studied characteristics of DCOPs that might cause significant challenges to existing DO strategies. Although these characteristics are common in real-world applications, in the continuous domain they have not been considered in most existing DO studies and they have not been captured in most existing continuous DO benchmark problems.

A set of dynamic constrained benchmark problems for simulating the characteristics of DCOPs, together with eight performance measures, have been discussed to help close this gap.

Using the benchmark problems and measures, we discussed detailed experimental analyses to investigate the strengths and weaknesses of existing DO strategies (GA/RIGA/HyperM) in solving DCOPs. The experimental analyses reveal some interesting findings about the ability of existing algorithms in solving DCOPs. These findings can be categorised as follows.

First, three findings about the performance of existing DO strategies in DCOPs have been identified: (a) the use of elitism might have a positive impact on the performance of existing diversity-maintaining strategies but might have a negative impact on the performance of diversity-introducing strategies if they are not used with diversity-maintaining strategies; (b) the presence of infeasible areas has a negative impact on the performance of diversity-introducing/maintaining strategies; and (c) the presence of switching optima (between disconnected regions) has a negative impact on the performance of DO strategies if they are combined with penalty functions.

Second, based on the findings about the strengths and weaknesses of some existing DO strategies, a list of possible requirements that DO algorithms should meet to solve DCOPs effectively have been suggested. This list of requirements can be used as a guideline to design new algorithms to solve DCOPs in future research.

The results and discussions in this chapter raise some open questions for future research. One direction is to develop new algorithms specialised in solving DCOPs based on our suggested list of requirements. We also plan to apply the results achieved in this chapter to real-world applications, especially to dynamic environments such as container terminals where there is the need to provide dynamic optimization solutions for such problems as dynamic scheduling of automatic-guided vehicles, dynamic allocation of quay-side and stack-side cranes, and dynamic stacking of containers.

Acknowledgements. This work was partially supported by the Engineering and Physical Sciences Research Council (EPSRC) of UK under Grant EP/E058884/1, a UK ORS Award, a studentship from the School of Computer Science, University of Birmingham and an EU-funded project named "Intelligent Transportation for Dynamic Environment (InTraDE)". The programs in this chapter were developed from the source code provided by Williams [36].

References

[1] Aickelin, U., Dowsland, K.: Exploiting problem structure in a genetic algorithm approach to a nurse rostering problem. J. of Sched. 3, 139–153 (2000)

[2] Andrews, M., Tuson, A.L.: Dynamic optimisation: A practitioner requirements study. In: Proc. 24th Annual Workshop of the UK Planning and Scheduling Special Interest Group (2005)

[3] Araujo, L., Merelo, J.J.: A genetic algorithm for dynamic modelling and prediction of activity in document streams. In: Proc. 9th Annual Conf. Genetic and Evol. Comput., pp. 1896–1903 (2007)

[4] Ayvaz, D., Topcuoglu, H., Gurgen, F.: A comparative study of evolutionary optimization techniques in dynamic environments. In: Proc. 8th Annual Conf. Genetic and Evol. Comput., pp. 1397–1398 (2006)

[5] Blackwell, T., Branke, J.: Multiswarms, exclusion, and anti-convergence in dynamic environments. IEEE Trans. Evol. Comput. 10(4), 459–472 (2006)

[6] Branke, J.: Evolutionary Optimization in Dynamic Environments. Kluwer (2001)

[7] Cobb, H.G.: An investigation into the use of hypermutation as an adaptive operator in genetic algorithms having continuois, time-dependent nonstationary environments. Tech. Rep. AIC-90-001, Naval Research Laboratory, Washington, USA (1990)

[8] Deb, K., Rao N., U.B., Karthik, S.: Dynamic multi-objective optimization and decision-making using modified NSGA-II: A case study on hydro-thermal power scheduling. In: Obayashi, S., Deb, K., Poloni, C., Hiroyasu, T., Murata, T. (eds.) EMO 2007. LNCS, vol. 4403, pp. 803–817. Springer, Heidelberg (2007)

[9] Deb, K., Saha, A.: Finding multiple solutions for multimodal optimization problems using a multi-objective evolutionary approach. In: Proc. 12th Annual Conf. Genetic and Evol. Comput., pp. 447–454 (2010)

[10] Floudas, C., Pardalos, P., Adjiman, C., Esposito, W., Gumus, Z., Harding, S., Klepeis, J., Meyer, C., Schweiger, C.: Handbook of Test Problems in Local and Global Optimization. In: Noncovex Optimization and Its Applications, vol. 33. Kluwer Academic Publishers (1999)

[11] Grefenstette, J.J.: Genetic algorithms for changing environments. In: Proc. 2nd Int. Conf. Parallel Problem Solving from Nature, pp. 137–144 (1992)

[12] Ioannou, P., Chassiakos, A., Jula, H., Unglaub, R.: Dynamic optimization of cargo movement by trucks in metropolitan areas with adjacent ports. Tech. Rep., METRANS Transportation Center, University of Southern California, Los Angeles, CA 90089, USA (2002), http://www.metrans.org/research/final/00-15_Final.htm

[13] Kim, H.: Target exploration for disconnected feasible regions in enterprise-driven multi-level product design. American Institute of Aeronautics and Astronautics Journal 44(1), 67–77 (2006)

[14] Liu, C.A.: New dynamic constrained optimization pso algorithm. In: Proc. 4th Int. Conf. Natural Comput., pp. 650–653 (2008)

[15] Mertens, K., Holvoet, T., Berbers, Y.: The DynCOAA algorithm for dynamic constraint optimization problems. In: Proc. 5th Int. Joint Conf. Autonomous Agents and Multiagent Syst., pp. 1421–1423 (2006)

[16] Michalewicz, Z.: The second version of Genocop III: a system which handles also nonlinear constraints, http://www.cs.adelaide.edu.au/zbyszek/EvolSyst/gcopIII10.tar.Z (accessed February 2009)

[17] Morales, K.A., Quezada, C.: A universal eclectic genetic algorithm for constrained optimization. In: Proc. 6th Europ. Congr. Intell. & Soft Comput., pp. 518–522 (1998)

[18] Moser, I., Hendtlass, T.: A simple and efficient multi-component algorithm for solving dynamic function optimisation problems. In: Proc. 2007 IEEE Congr. Evol. Comput., pp. 252–259 (2007)

[19] Nguyen, T.T.: A proposed real-valued dynamic constrained benchmark set. Tech. Rep., School of Computer Science, Univesity of Birmingham (2008), http://www.staff.ljmu.ac.uk/enrtngu1/Papers/DCOPbenchmark.pdf

[20] Nguyen, T.T.: Continuous Dynamic Optimisation Using Evolutionary Algorithms. Ph.D. thesis, School of Computer Science, University of Birmingham (2011), http://etheses.bham.ac.uk/1296 and http://www.staff.ljmu.ac.uk/enrtngu1/theses/phdthesisnguyen.pdf

[21] Nguyen, T.T., Yao, X.: Benchmarking and solving dynamic constrained problems. In: Proc. 2009 IEEE Congr. Evol. Comput., pp. 690–697 (2009)

[22] Nguyen, T.T., Yao, X.: Dynamic time-linkage problems revisited. In: Giacobini, M., et al. (eds.) EvoWorkshops 2009. LNCS, vol. 5484, pp. 735–744. Springer, Heidelberg (2009)

[23] Nguyen, T.T., Yao, X.: Continuous dynamic constrained optimisation - the challenges. IEEE Trans. Evol. Comput. 166, 769–786 (2012)

[24] Nguyen, T.T., Yao, X.: Detailed experimental results of GA, RIGA, HyperM and GA+Repair on the G24 set of benchmark problems. Tech. Rep., School of Computer Science, University of Birmingham (2010), http://www.staff.ljmu.ac.uk/enrtngu1/Papers/DCOPfulldata.pdf

[25] Nguyen, T.T., Yao, X.: Solving dynamic constrained optimisation problems using stochastic ranking and repair methods. IEEE Trans. Evol. Comput. (2010) (submitted), http://www.staff.ljmu.ac.uk/enrtngu1/Papers/ NguyenYaodRepairGA.pdf

[26] Parrott, D., Li, X.: Locating and tracking multiple dynamic optima by a particle swarm model using speciation. IEEE Trans. Evol. Comput. 10(4), 440–458 (2006)

[27] Prata, D.M., Lima, E.L., Pinto, J.C.: Simultaneous data reconciliation and parameter estimation in bulk polypropylene polymerizations in real time. Macromolecular Symposia 243(1), 91–103 (2006)

[28] Richter, H.: Detecting change in dynamic fitness landscapes. In: Proc. 2009 IEEE Congr. Evol. Comput., pp. 1613–1620 (2009)

[29] Richter, H.: Memory design for constrained dynamic optimization problems. In: Di Chio, C., et al. (eds.) EvoApplicatons 2010, Part I. LNCS, vol. 6024, pp. 552–561. Springer, Heidelberg (2010)

[30] Richter, H., Yang, S.: Learning behavior in abstract memory schemes for dynamic optimization problems. Soft Comput. 13(12), 1163–1173 (2009)

[31] Rocha, M., Neves, J., Veloso, A.: Evolutionary algorithms for static and dynamic optimization of fed-batch fermentation processes. In: Ribeiro, B., et al. (eds.) Adaptive and Natural Computing Algorithms, pp. 288–291. Springer (2005)

[32] Singh, H.K., Isaacs, A., Nguyen, T.T., Ray, T., Yao, X.: Performance of infeasibility driven evolutionary algorithm (IDEA) on constrained dynamic single objective optimization problems. In: Proc. 2009 IEEE Congr. Evol. Comput., pp. 3127–3134 (2009)

[33] Tawdross, P., Lakshmanan, S.K., Konig, A.: Intrinsic evolution of predictable behavior evolvable hardware in dynamic environment. In: Proc. 6th Int. Conf. Hybrid Intell. Syst., p. 60 (2006)

[34] Thompson, J.M., Dowsland, K.A.: A robust simulated annealing based examination timetabling system. Comput. Oper. Res. 25(7-8), 637–648 (1998)

[35] Wang, Y., Wineberg, M.: Estimation of evolvability genetic algorithm and dynamic environments. Genetic Programming and Evolvable Machines 7(4), 355–382 (2006)

[36] Williams, K.P.: Simple genetic algorithm (SGA) source code (in C), http://www.kenwilliams.org.uk/code/ga2.c (accessed December 2008)

[37] Yang, S.: Genetic algorithms with memory- and elitism-based immigrants in dynamic environments. Evol. Comput. 16(3), 385–416 (2008)

[38] Yang, S., Yao, X.: Experimental study on population-based incremental learning algorithms for dynamic optimization problems. Soft Comput. 9(11), 815–834 (2005)

[39] Zhang, Z., Liao, M., Wang, L.: Multi-objective immune genetic algorithm solving dynamic single-objective multimodal constrained optimization. In: Proc. 8th Int. Conf. Natural Comput., pp. 864–868 (2012)

Part III
Theoretical Analysis

Chapter 9
Theoretical Advances in Evolutionary Dynamic Optimization

Philipp Rohlfshagen, Per Kristian Lehre, and Xin Yao

Abstract. The field of evolutionary dynamic optimization is concerned with the study and application of evolutionary algorithms to dynamic optimization problems: a significant number of new algorithms have been proposed in recent years that are designed specifically to overcome the limitations faced by traditional algorithms in the dynamic domain. Subsequently, a wealth of empirical studies have been published that evaluate the performance of these algorithms on a variety of benchmark problems. However, very few theoretical results have been obtained during this time. This relative lack of theoretical findings makes it difficult to fully assess the strengths and weaknesses of the individual algorithms. In this chapter we provide a review of theoretical advances in evolutionary dynamic optimization. In particular, we argue the importance of theoretical results, highlight the challenges faced by theoreticians and summarise the work that has been done to date. We subsequently identify relevant directions for future research.

9.1 Introduction

The field of *evolutionary dynamic optimization* is concerned with the study and application of evolutionary algorithms (EAs) to the class of dynamic optimization problems (DOPs): the dependency on time of such problems poses many new challenges to the design of EAs as pointed out by numerous monographs published in early 2000 [5, 35, 56]. This raised noticeable interest in evolutionary dynamic

Philipp Rohlfshagen · Xin Yao
Centre of Excellence for Research in Computational Intelligence and Applications
(CERCIA), School of Computer Science, University of Birmingham,
Birmingham B15 2TT, U.K.
e-mail: philipp.r@gmail.com, X.Yao@cs.bham.ac.uk

Per Kristian Lehre
School of Computer Science, University of Nottingham, Nottingham NG8 1BB, U.K.
e-mail: perkristian.lehre@nottingham.ac.uk

S. Yang and X. Yao (Eds.): *Evolutionary Computation for DOPs*, SCI 490, pp. 221–240.
DOI: 10.1007/978-3-642-38416-5_9   ⓒ Springer-Verlag Berlin Heidelberg 2013

optimization and a significant number of nature-inspired techniques have subsequently been proposed to address the potential shortcomings of traditional EAs in the dynamic domain. The majority of techniques, many of which are reviewed throughout this book, employ additional features, such as the preservation of population diversity, in order to efficiently *track* high quality solutions over time.

The wealth of techniques developed is well documented by the numerous reviews that have been published in the last decade, particularly in 1999/2001 [3, 4], 2005 [26] and 2011 [8]. These reviews not only highlight the significant developments in terms of new algorithms but also the simultaneous lack of theoretical results and although the degree of theoretical results in evolutionary computation is generally overshadowed by the sheer quantity of empirical results, this discrepancy is even more apparent in the case of DOPs: the only review to mention theoretical results is by Jin and Branke [26], limited to a total of four references.

This relative lack of theoretical results, caused primarily by the added difficulty of having to account for the problem's dynamics, makes it difficult to fully assess the strengths and weaknesses of the individual algorithms. Furthermore, the lack of a clearly defined framework has made it difficult for practitioners to fully express their assumptions and to generalise from specific test scenarios to a wider class of problem dynamics. In order to draw attention to this issues, this chapter provides a self-contained overview of theoretical advances in evolutionary dynamic optimization: we argue the importance of theoretical results, highlight the challenges faced by theoreticians and summarise the work that has been done to date. Finally, we subsequently identify relevant directions for future research.

The remainder of this chapter is structured as follows: in section 9.2, we provide a brief overview of evolutionary algorithms and optimization, particularly in uncertain environments. In section 9.3 we then lay the foundation for the review of theoretical results, including an introduction to runtime analysis in the dynamic domain. The review of previous work is found in section 9.4 and finally, the chapter is concluded in section 9.5 where we summarise the reviewed work, assess its implications and outline some prospects for future work.

9.2 Evolutionary Dynamic Optimization

9.2.1 Optimization Problems

An optimization problem $f : X \to Y$ is a mapping, also known as the *objective function*, from a *search space* X to the domain Y (e.g., $\mathbb{R}$); the value $f(x) \in Y$, $x \in X$ indicates the quality of x and the elements x_i are usually referred to as *design* or *decision variables*. The dimensionality of the problem is $|x| = n$ and the set of all f-values, corresponding to all elements in X is denoted as $\hat{f}$. The goal of an optimization algorithm is usually to find the *global optimum* $x^* \in X$ such that $f(x^*) \geq f(x)$, $\forall x \in X$ in as little time as possible. An obstacle faced by the algorithm in doing so are *local optima*, defined as points $x \in X$ such that $f(x) \geq f(z)$, $\forall z \in N(x)$ where $N(x)$ is the *neighbourhood* of x, determined by the algorithm's variation operators. It should be noted that, without loss of generality, we assume that functions are to be maximised. Furthermore, we only consider *combinatorial* optimization problems (i.e., those with a discrete search space).

9.2.2   Optimization in Uncertain Environments

Traditionally, the majority of work in evolutionary computation has concentrated on deterministic stationary optimization problems; nevertheless, a significant effort has also been devoted to problems characterised by uncertainty. In [26], Jin and Branke review four distinct types of uncertain environments as outlined next.

Noisy optimization problems are characterised by an objective function that is subject to noise. This implies that every time a point $x \in X$ is evaluated, the value $f(x)$ varies according to some additive noise z that follows some (usually normal) distribution:

$$f(x) := \int_{-\infty}^{\infty} [f(x) + z] \, p(z) \, dz = f(x), \ z \sim N(0, \sigma^2) \tag{9.1}$$

Algorithms should subsequently work on the *expected* f-values of the search points.

In *robust* optimization, it is the decision variables x_i, $i = 1, \ldots, n$ that are subject to minor perturbations λ *after* the value $f(x)$ has been determined (e.g., manufacturing variances):

$$f(x) := \int_{-\infty}^{\infty} [f(x + \lambda)] \, p(\lambda) \, d\lambda \tag{9.2}$$

Desired solutions are those whose f-values vary within acceptable margins given minor alterations to the solution's decision variables.

The third class of problems considered by Jin and Branke [26] is *approximate optimization* (also known as *surrogate-assisted* optimization) where the objective function is too expensive to be queried continuously. A (meta-) model, which produces approximate f-values with error $e(x)$, is used instead and the algorithm only calls the original objective function intermittently:

$$f(x) := \begin{cases} f(x) & \text{if original objective function is used;} \\ f(x) + e(x) & \text{if meta-model is used.} \end{cases} \tag{9.3}$$

Here it is vital for the algorithm to determine a reasonable trade-off between the accuracy of f-values obtained by the meta-model and the computational cost required to do so.

Finally, the fourth type of uncertain optimization problems corresponds to *dynamic optimization problems* which are *deterministic* at any moment in time but may change over time. The class of DOPs is difficult to define as, in principle, any component of f may change over time and the Handbook of Approximation Algorithms and Metaheuristics [30] states that a general definition of DOPs does not exist. Jin and Branke [26] deliberately keep the problem definition as general as possible:

$$f(x) := f(x, t) \tag{9.4}$$

The dynamics of the problem correspond to the mapping $\mathcal{T} : \mathcal{F} \times \mathbb{N} \to \mathcal{F}$ such that $f(T + 1) = \mathcal{T}(f(T))$. We assume that time t advances with every call to the objective function such that $T\tau \le t < (T + 1)\tau$ where $\tau \ge 1$ is the frequency of change. It follows that T is a period index for each problem instance encountered.

The majority of practitioners attempt to design algorithms that are able to *track* high quality solution as closely as possible over time. In particular, a single solution is considered insufficient and instead, the algorithm should return a *trajectory* of solutions over time. One of the main motivations driving the development of new algorithms is the *transfer of knowledge* from one problem instance encountered to the next [5]: practitioners commonly assume that successive problem instances encountered by the algorithms are correlated to one another, allowing the algorithm to outperform a random restart by making use of the search points found so far.

9.2.3 Evolutionary Algorithms

The field of evolutionary computation provides a variety of nature-inspire metaheuristics that have been utilised successfully to obtain high quality solutions to a variety of NP-hard optimization problems. In this chapter we concentrate exclusively on those algorithms that are understood to be evolutionary algorithms (EAs): EAs are population-based global search algorithms inspired loosely by the general principles of evolutionary systems. Roughly speaking, EAs attempt to obtain solutions of increasing quality by means of *selection, crossover* and *mutation*: selection favours those individuals in the algorithm's population (the multiset P) that represent solutions of higher quality (exploitation) whereas crossover and mutation, the algorithm's *variation operators*, generate offspring from those individuals to advance the search (exploration).

Two simple algorithms that have been analysed theoretically in the dynamic domain are the $(1 + \lambda)$−EA and the $(1 + 1)$−EA. The former maintains at any moment in time a single parent that produces λ offspring by means of mutation. The next generation is chosen from the set of all individuals (i.e., the offspring *and* the parent). The pseudo-code for this algorithm is shown in Algorithm 1. A special (and simpler) variant of this algorithm is the (1+1) EA where the offspring population is limited to a single individual, akin of stochastic local search. Both algorithms rely solely on their mutation operator which alters elements in x with some probability p_m.

9.3 Theoretical Foundation

9.3.1 Introduction to Runtime Analysis

The theoretical foundations for EAs are less well developed than for classical algorithms which are often accompanied with rigorously proven guarantees on the quality of their solutions and bounds on the worst-case cost of obtaining them. In contrast, EAs have traditionally been evaluated empirically on selected problem

---

Algorithm 1 $(1+\lambda)$-EA.

---

set $t=1$
initialise $x(0)$ uniformly at random

while terminate = **false do**
   for $i := 1$ to λ **do**
      $x^i(t) := x(t)$
      Alter each position of $x^i(t)$ with probability p_m.
   end for
   $x^{best}(t) := x^i(t) \mid f(x^i(t)) \geq f(x^j(t)),\ j = 1, \ldots, \lambda$
   if $f(x^{best}(t)) \geq f(x(t))$ **then**
      $x(t+1) := x^{best}(t)$
   end if
   $t := t+1$
end while

---

instances, and strong guarantees about their performance are often not available. This lack of formal performance guarantees is partly because EAs are hard to analyse. In particular, they are designed to simulate some aspect of nature without regard as to whether they can be studied formally or not. In contrast, classical algorithms are often designed specifically with runtime bounds in mind.

Nevertheless, significant progress has been made during the last decade in the runtime analysis of EAs [1, 41, 42]. The *black-box* scenario is the common theoretical framework in most of these studies [17]: the algorithm is assumed to be oblivious to the structure of the function that is to be optimised (i.e., auxiliary information like gradients is not available). Information about the function can only be gained by querying for f-value of search points the algorithm chooses. However, it is assumed that the algorithm knows the class of functions $\mathcal{F}$ (problem) from which the function (problem instance) is taken. As a consequence of the No Free Lunch theorems [16, 61], it is necessary to assume that the function class $\mathcal{F}$ has some structure which can be exploited by the algorithm as otherwise it is impossible to distinguish the (average case) performance of different algorithms (see section 9.3.3).

The expected runtime of an algorithm A on a function f is the expected number of times the algorithm evaluates the objective function before the global optimum is found for the first time (the algorithm's *hitting time*). The expectation is with respect to the random choices made by the algorithm and the expected runtime on $\mathcal{F}$ is the maximum of the expected runtimes over all $f \in \mathcal{F}$. In addition to allow a precise definition of the runtime of particular algorithms, it is also possible to define the complexity of function classes, the so-called black-box complexity [17, 32]. This is the minimum expected runtime on the problem class among all black-box algorithms.

Initial runtime studies were concerned with simple EAs like the (1+1) EA on artificial pseudo-boolean functions [13, 15, 55], highlighting how different components of evolutionary algorithms impact their runtime (e.g., the crossover operator [25, 51], population size [21, 59], diversity mechanisms [19], and selection pressure [31]). This effort also considered other types of meta-heuristics (e.g., ant colony optimization [38, 40], particle swarm optimization [52, 60], and estimation of distribution algorithms [7]) as well as new problem settings (e.g., multi-objective [20, 21, 37] and continuous [23] optimization). A significant amount of work has been directed towards studying classical combinatorial optimization problems, like maximum matching [22], sorting [47], Eulerian cycles [9], minimum spanning trees [39], and more generally matroid optimization problems [44]. For NP-hard problems, the focus has been on interesting sub-classes such as for the vertex cover problem [43], on the algorithm's approximation quality [18, 58], and on fixed-parameter tractability [28, 29]. There has also been some work trying to estimate the runtime on problems close to industrial applications, in particular in software engineering [33, 34].

9.3.2   Runtime Analysis for Dynamic Functions

The rigorous analysis of an algorithm's runtime can be very challenging and it has been common practice in the past to consider relatively simple algorithms (e.g., (1+1) EA) and problems (e.g., ONEMAX; see section 9.3.4). However, in recent years, significant progress has been made and the use of new techniques (e.g., drift analysis) allowed theoreticians to obtains proofs for significantly more complex scenarios. However, runtime analysis in the dynamic domain is further complicated by the dynamics of the function which have to be taken into account in addition to the dynamics of the algorithm. The fundamental impact of this added complexity is illustrated by the need to define a new *notion of optimality* for DOPs as traditional measures are often no longer applicable. In particular, the majority of practitioners is interested in multiple solutions across the life cycle of the problem and hence one has to evaluate and quantify the notion by which the quality of an algorithm is to be judged.

In the stationary case, the goal of an optimization algorithm is usually to find the global optimum in as few steps (number of calls to the objective functions) as possible for any number of inputs n. In his work, Droste [10, 11] translates this notion of optimality directly to the dynamic case: the author considers the expected first hitting time of the (1+1) EA in the continuously changing ONEMAX problem. The expected first hitting time in this case is more accurately referred to as *expected temporal first hitting time* as we are interested in the time the *current* global optimum is first found (which, in turn, may be lost as soon as a change occurs). However, the goal in the dynamic case is usually understood to be the *tracking* of the global optimum over time [5]. In other words, the algorithm has to *repeatedly* locate the global optimum, prompting Droste to mention additional measures that may be taken into account such as the degree to which a found optimum is lost or the average distance to the nearest optimum over time [11, p 56]. The notion of distance to the optimum was also considered by Jansen and Schellbach [24] who quantified this concept via the time until the distance between the sequence obtained by the algorithm and the target point is larger than kd_{max}, assuming that the initial distance is less than d_{max}. Informally, this is the time until the algorithm has "lost" the target point.

In [46], the authors take the concept of dynamic runtime a step further and consider the time required by the algorithm to relocate the global optimum once it has been lost due to update of the function; this measure is called the *second hitting time*, a specific case of the more general *expected i^{th} hitting time*. Considering multiple hitting times, it is natural to take into account the duration with which the algorithm resides at the optimum, a measure called the *séjour time*. These concepts may be formalised as follows.

Definition 9.1 (Dynamic Runtime [46]). Given a search space X and a dynamic fitness function $f : X \times \mathbb{N}_0 \to \mathbb{R}$, let $x(t), t \geq 0$, be the current search point at iteration t of optimization algorithm A on dynamic fitness function f. Then the *hitting times* T_j and the *séjour times* S_i of algorithm A on function f are defined as

$$T_i := \min_t \{t \geq 0 \mid \forall y \in X, f(x(Q_i + t), Q_i + t) \geq f(y, Q_i + t)\},$$

$$S_i := \min_t \{t \geq 0 \mid \exists y \in X, f(x(Z_i + t), Z_i + t) < f(y, Z_i + t)\},$$

where $i \geq 1$ and

$$Q_k := \begin{cases} 0 & \text{if } k = 1, \\ \sum_{l=1}^{k-1}(T_l + S_l) & \text{otherwise} \end{cases}$$

$$Z_k := \begin{cases} T_1 & \text{if } k = 1, \\ T_1 + \sum_{l=1}^{k-1}(S_l + T_{l+1}) & \text{otherwise} \end{cases}$$

This definition of dynamic runtime accounts for the common notion of optimality in the dynamic domain which is concerned with the algorithm's ability to *locate* and *track* the global optimum over time. To be considered efficient in the dynamic domain, it is necessary that the algorithm locates the optimum within reasonable (i.e., polynomial) time and that the second and subsequent hitting times should not be larger than the first: the first hitting time usually assumes a uniformly random distribution from which the initial search points are selected. In the dynamic case, however, once a change takes place, the algorithm has already spent τ steps optimising the function and is thus at a non-random point (e.g., the previous global optimum) when faced with a new instance of the problem. If this is not the case, a restart strategy should be favoured over continuous tracking of the optimum.

9.3.3 No Free Lunches in the Dynamic Domain

Informally, the No Free Lunch theorem for optimization (NFL; [61]) states that any two black-box algorithms a_1 and a_2 perform, on average, identically across the set of all possible functions $\mathcal{F} = Y^X$. Similarly, in the case of time-variant functions, the average performance of any two algorithms is identical across the set of all possible dynamics $\mathcal{T} : \mathcal{F} \times \mathbb{N} \to \mathcal{F}$. The following summarises these results.

Assuming optimization algorithm a only maps to points *not* previously visited, the algorithm corresponds to the mapping $a : d \in D \to \{x \mid x \notin d^x\}$ where

D is the set of all possible samples and d^x the set of unique points sampled so far. Over m iterations, the time-ordered distinct visited points correspond to $d_m \equiv \{(d_m^x(1), d_m^y(1)), \ldots, (d_m^x(m), d_m^y(m))\}$ and it is important to note that this sample contains *all* points sampled by the algorithm, not just those accepted. The performance of an algorithm a iterated m times on function f corresponds to the likelihood that a particular sample d_m^y has been obtained: $P(d_m^y \mid f, m, a)$. The NFL theorem then states that

$$\sum_f P(d_m^y \mid f, m, a_1) = \sum_f P(d_m^y \mid f, m, a_2) \tag{9.5}$$

It follows that for any performance measure $\Phi(d_m^y)$ based on the samples d_m^y, the average performance over all functions is independent of a. Wolpert and McReady [61] extend this analysis to the class of time-variant functions, highlighting two particularly interesting issues:

1. How the dynamic functions are defined.
2. How the performance of an algorithm is measured.

The authors consider the case where the algorithm starts with function f_1 and with each subsequent iteration of the algorithm, the function is transformed to a new function by the bijective mapping $\mathcal{T} : \mathcal{F} \times \mathbb{N} \to \mathcal{F}$. The authors note that an algorithm's performance in the dynamic domain is not trivially defined and propose two measures: in the first scheme, the y-value corresponding to a particular point x is determined by the function at the time the point was evaluated. In the second case, the y-values correspond to the values obtained for each x sampled according to the *final* function encountered.

Subsequently, a similar result to equation 9.5 may then be obtained. In this case, the average is over all possible dynamics $\mathcal{T}$ (rather than functions f):

$$\sum_T P(d_m^y \mid f_1, T, m, a_1) = \sum_T P(d_m^y \mid f_1, T, m, a_2) \tag{9.6}$$

9.3.4 Benchmark Problems

Numerous dynamic benchmark problems have been proposed in the past, allowing practitioners to test, evaluate and compare their algorithms. These benchmarks include tools to generate a wide range of dynamics (e.g., MOVING PEAKS[5] and DF1[35]) and dynamic variants of well-known NP-hard stationary optimization problems (e.g., Travelling Salesman Problem or Scheduling Problems). Naturally, theoreticians have concentrated on simpler problems with well-defined dynamics. The XOR DOP [62, 63] benchmark may be used to impose artificial dynamics on any pseudo-Boolean optimization problem. It is a generalisation of the dynamic pattern match problem that was used in the first attempt to analyse the runtime of an EA in the dynamic domain. Finally, a third problem that has been considered by theoreticians is a simple tracking problem in a lattice. These problems are reviewed below.

9.3.4.1 Dynamic Match Function

Standhope and Daida [49, 50] propose a simple dynamic function for an initial analysis of the behaviour of the (1+1) EA. The function is a generalisation of the well-known ONEMAX problem to the dynamic domain. The f-values in the ONEMAX problem simply correspond to the number of ones found in the solution x:

$$\text{ONEMAX}(x) = \sum_{i=1}^{n} x_i \qquad (9.7)$$

In the case of the dynamic match function, the algorithm must reduce the Hamming distance $d(x,z) = \sum_{i=1}^{n} |x_i - z_i|$ to an arbitrary target pattern (match string) σ_τ that may change over time:

$$f(a, \sigma_\tau) = \sum_{i=1}^{n} \neg(a_i \oplus \sigma_{\tau i}) \qquad (9.8)$$

where $\neg$ is the logical not operator and $\oplus$ logical xor. The dynamics of σ_τ are controlled by two parameters, g and d which control the number of generations between changes and the degree (Hamming distance) by which the target pattern is altered (d distinct and randomly chosen bit in σ are inverted). The values $(0,0)$ result in a stationary function whereas the values $(2,5)$ would imply that every 2 generations, the target changes by 5 bits.

9.3.4.2 The XOR DOP Problem

XOR DOP [62, 63] is the only widely accepted benchmark problem in dynamic optimization for the combinatorial domain and generates a dynamic version of any static pseudo-Boolean problem. It is a generalisation of the dynamic match function and imposes dynamics on *any* stationary pseudo-boolean function $f : \mathbb{B} \to \mathbb{R}$ by means of a bit-wise *exclusive-or* operation that is applied to each search point $x \in \{0,1\}^n$ prior to each function evaluation. The dynamic equivalent of any stationary function is simply

$$f(x(t) \oplus m(T)) \qquad (9.9)$$

where $\oplus$ is the *xor* operator. The vector $m(T) \in \{0,1\}^n$, which initially is equivalent to 0^n, is a binary mask, generated by $m(T) = m(T-1) \oplus p(T)$ where $p(T) \in \{0,1\}^n$ is a randomly created template that contains exactly $\lfloor \rho n \rfloor$ ones. The value of $\rho \in [0,1]$ thus controls the magnitude of change which is specified as the Hamming distance between two binary points. It follows that ρn is the actual number of bits inverted. The period index $T = \lceil t/\tau \rceil$ is determined by the duration $\tau > 0$ between changes.

XOR DOP was analysed by Tinós and Yang [53]: if we assume that the transformation of each encoding $x(t)$ by $m(T)$ yields a vector $z(t) = x(t) \oplus m(T)$, then it is possible to rewrite this expression as $z^n(t) = A(T)x^n(t)$ where $x \in \{0,1\}^n$ is normalised to $x^n(t) \in \{-1,1\}^n$ and where $A(T)$ is a linear transformation:

$$A(\pi) = \begin{bmatrix} A_1(\pi) & 0 & \cdots & 0 \\ 0 & A_2(\pi) & \cdots & 0 \\ & & \ddots & \\ 0 & 0 & \cdots & A_n(T) \end{bmatrix}$$

where

$$A_i(\pi) = \begin{cases} 1 & \text{if } m_i(T) = 0 \\ -1 & \text{if } m_i(T) = 1 \end{cases}$$

for $i = 1, 2, \ldots, n$. It follows that XOR DOP does not alter the underlying function but instead *rotates* each search point x prior to each function evaluation.

This analysis was extended in [54] using a dynamical system analysis. In particular, the authors showed that XOR DOP corresponds to a DOP with permutations: the class of DOPs with permutation are those with dynamics that permute the assignment between elements in X and those in $\hat{f}$. The authors subsequently showed that the xor operator may be replaced with a single mutational step performed whenever the function is meant to change. In other words, given two dynamic processes, one governed by XOR DOP, the other by an initial mutational step, "if both evolutionary processes have the same initial population and parameters, and the fitness function in the first change cycle for the first process is equal to the fitness function in the second process, then the evolution of the population in the two processes is identical, i.e., the two evolutionary processes are equivalent.".

9.3.4.3 Tracking Problem

Weicker [57] considered a simple *tracking problem* modelled on the integer lattice as a sequence of target points $a_1, a_2, \ldots \in \mathbb{Z}^2$ together with a time-variant objective function $f : \mathbb{Z}^2 \times \mathbb{N} \to \mathbb{R}$ which is to be minimised. The first argument to the objective function is a point in the lattice, and the second argument is the time parameter. The function is defined for all x and t as $f(x,t) := \|x - a_t\|$, where $\|\cdot\|$ is the ℓ_1-norm. Essentially, $f(x,t)$ is the Manhattan-distance between the point x and the current target point at time t.

The sequence $a_1, a_2, \ldots$, representing a moving target, is unknown to the algorithm, and can be deterministic, stochastic, or chosen by an adversary. The only assumption made about the sequence is that for some parameter $d_{\max}$, $\|a_t - a_{t+1}\| \leq d_{\max}$ holds for all $t \geq 0$ (i.e., the speed of the target point is no more than $d_{\max}$). The special case $d_{\max} = 0$ corresponds a static optimization problem.

Informally, the objective of an algorithm in the tracking problem is to obtain a sequence of search points $x_1, x_2, \ldots \in \mathbb{Z}^2$ that are close to the sequence of target points. Various aspects of this informal objective have been formalised. First, the algorithm needs to obtain a search point that is within acceptable distance to the target point, then it must track the moving target. Jansen and Schellbach [24] considered the *first hitting-time* $T_{\max_d}(n)$, defined as the number of function evaluations until a search point x_t has been obtained for which $\|x_t - a_t\| \leq d_{\max}$, assuming that the algorithm is provided with an initial search point x_1 for which $\|x_1 - a_1\| = n$.

9.4 Runtime Analysis for Dynamic Functions

9.4.1 First Hitting Times for Pattern Match

The (1+1) EA has been analysed on several variants of the dynamic ONEMAX (match function), extending the work carried out previously on its stationary counterpart (e.g., [2, 14, 36]). The dynamic variants of the ONEMAX differ in their transitions from one instance to the next, a property which may drastically affect the runtime of the algorithm.[1]

The first consideration of a non-empirical analysis is due to Stanhope and Daida [50] who consider the (1+1) EA with simplified mutation operator on the dynamic pattern match function with intergenerational updates. The mutation operator considered inverts exactly r bit, chosen uniformly at random; the mutation rate r remains constant throughout the algorithm execution. The authors first consider the case $(0, 0)$ (i.e., a stationary function) and then generalise to the dynamic case (d, g). The authors use a hypergeometric random variable to describe the probability distribution over neighbouring search points that may be generated by the mutation operator. The dynamics of the pattern matching function are modelled as an additional mutational step that inverts d bits every g generations (c.f., analysis of XOR DOP in section 9.3.4). The authors subsequently derive a distribution function on the fitness if a selected individual. The transition probabilities are validated empirically by a comparison to Monte-Carlo generated fitness distributions and amongst other things, Stanhope and Daida showed that even small perturbations in the fitness function could have a significantly negative impact on the performance of the (1+1) EA.

In a sequence of papers [11, 12], Droste considered the dynamic ONEMAX problem. The initial model considered the target sequence modified by a single, uniformly chosen, bit-flip with probability p in each iteration. The goal of the study was to determine values of p for which the (1+1) EA has polynomial expected first hitting time. Since the parameter setting $p = 0$ corresponds to the static ONEMAX-problem, for which the (1+1) EA has expected first hitting time $O(n \log n)$ [15], it is clear that exponential first hitting time can only occur for strictly larger values. It should be noted that the (1+1) EA considered by Droste [10] was adapted specifically to the dynamic domain by calling the objective function twice during each iteration to prevent the use of outdated f-values.

Droste found that the first hitting time remains polynomial as long as $p = O(\log n / n)$. At this rate, the target sequence is modified $O(\log^2(n))$ times in expectation during a time interval of $n \log n$ iterations. This rate turned out to be critical, as the expected first hitting time becomes exponential for $p = \omega(\log(n)/n)$.

The dynamic ONEMAX-model considered by Droste can be generalised. Instead of only flipping one bit with a certain probability, one can define a random operator $\mathcal{M}$ that acts on the target sequence in each iteration. One such natural operator, is

---

[1] As pointed out in [49], the pattern match function is equivalent to the application of XOR DOP to the ONEMAX function. Furthermore, the runtime analysis may be simplified if the dynamics are viewed as an additional mutation operator that acts directly on x (depending of magnitude and frequency of change; see section 9.3.4.2).

to flip each bit position in the target sequence with some probability p' in each iteration. While the second model leads to a more involved analysis, the results are essentially the same in the two models. Note first that by setting $p' := p/n$, the expected number of bit-flips to the target sequence per time step is the same in the both models. Droste found that the expected first hitting time remains polynomial as long as $p' = O(\log(n)/n^2)$, whereas the expected first hitting time becomes exponential as soon as $p'' = \omega(\log(n)/n^2)$.

9.4.2 Analysis of Frequency and Magnitude of Change

In [46], the authors look at the two most prominent attributes of most DOPs, the *magnitude of change* and the *frequency of change*. The former is generally regarded as the *relatedness* of two successive problem instances, $f(T)$ and $f(T+1)$ and a common assumption is that smaller magnitudes of change are easier to adapt to, primarily by "transferring knowledge from the past" [26, p 311]. The authors attempt to shed light on the question whether this is always the case or whether examples exist where a large magnitude of change may make it easier for the algorithm to relocate the global optimum.

For the magnitude of change, a specially designed function called MAGNITUDE is proposed: informally, this bi-modal function features a local optimum (LOCAL) surrounded by a valley of low f-values (TRAP). Beyond the valley is a region (ZERO) that leads a path that leads to the global optimum (GLOBAL). This stationary function is subsequently made dynamic using the XOR DOP framework to yield a dynamic MAGNITUDE function. The authors subsequently found that for (1+1) EA on MAGNITUDE with an update time $\tau \geq n^2 \log n$, and a magnitude of change θ, the second hitting T_2 satisfies

1. For small magnitudes of change, i.e. when $1 \leq \theta \leq q - cn$,

$$\mathbf{E}[T_2] = e^{\Omega(n)}$$

2. For large magnitudes of change, i.e. when $3q \leq \theta \leq n$,

$$\mathbf{Pr}[T_2 \leq n^2 \log n] = 1 - e^{-\Omega(n)}.$$

The proof idea for the runtime of the (1+1) EA follows directly from the function definition and is based on two concepts: the behaviour of the algorithm during each update period (i.e., in time of stagnation) and the impact of the dynamics on the algorithm, given the algorithm is either at LOCAL or GLOBAL. It is assumed that the time between changes is sufficiently long for the algorithm to reach one of the two optima with high probability; depending on the magnitude of change, different behaviours emerge. The initial search point may be in TRAP or ZERO. The probability to be on PATH is exceedingly small. If the algorithm starts in TRAP, it will be led away from the other regions of the search space towards the point LOCAL. Subsequently, if a small change occurs, the rotated search point will still be in the region TRAP and is hence attracted again to the local optimum. If, on the other hand, the initial search point is not in TRAP, the algorithm is led to the beginning of the path which leads directly to GLOBAL. The situation at GLOBAL is similar to the one at LOCAL. If the magnitude of change is small, the search point will be rotated into the TRAP region. If the magnitude of change is large, on the other hand, the search point will *jump* across the trap into ZERO or PATH. Similarly, if the algorithm is at LOCAL and a large change takes place, the search point is rotated beyond the boundary of the TRAP region.

The authors also looked at the frequency of change and similarly to the result above, showed that a high frequency of change may allow the algorithm to locate the function's global optimum whereas a low frequency of change does not. More specifically, it is shown that the dynamic optimization problem called BALANCE is hard for the (1+1) EA at low frequencies, and easy at high frequencies. Informally, the function is defined as follows: the algorithm is drawn towards the global optimum along the z-axis while the dynamics "tilt" the plane along the x-axis, elevating the f-values at different parts across the y-axis. This potentially draws the algorithm towards a trap-region that allows the algorithm to only get within a specific distance to the global optimum. If the algorithm is not trapped, on the other hand, the global optimum may be found by incremental improvements to the search points sampled.

The function is also made dynamic using the XOR DOP framework but a specially designed mask m is used to alter the search points in specific ways:

$$m(T) := \begin{cases} 0^{n/2} \cdot 0^{n/2} & \text{if } T \bmod 2 = 0, \text{ and} \\ 0^{n/2} \cdot 1^{n/2} & \text{otherwise.} \end{cases}$$

Hence, only the suffix of the point x is affected, and the magnitude of change is equivalent to $n/2$.

Theorem 9.1 ([46]). *The expected first hitting time of (1+1) EA on* BALANCE *with update time τ is*

$$\mathbf{E}[T] = \begin{cases} n^{\Omega(\sqrt{n})} & \text{if } \tau > 40n, \text{ and} \\ O(n^2) & \text{if } \tau = 2. \end{cases}$$

The idea for the proof shows that the algorithm will balance along the centre of the vertical axis when the frequency of change is high, while the algorithm is likely to fall into one of the trap regions when the frequency of change is sufficiently low. This can be proved by analysing the horizontal and vertical *drift*. Informally, the drift of a search point is the distance the search point moves per iteration. The horizontal drift corresponds to the change in number of leading 1-bits in the prefix, and the vertical drift corresponds to the change in number of 1-bits in the suffix. As long as the trap region has not been reached, the position along the vertical axis can be changed by flipping any of at least $n/16$ bits, and no other bits. In contrast, in order to reduce the distance to the optimum along the horizontal axis, it is necessary to flip the single left-most 0-bit, an event that happens with much lower probability. Therefore, the vertical drift is much larger than the horizontal drift. If the frequency of change is sufficiently low, then the current search point will have enough time to reach one of the trap regions before the optimum is found. On the other hand, if the frequency of change is sufficiently high, then the search point will not have time to reach the trap region during one period. In the following period, the vertical drift will be in the opposite direction, and the vertical displacement of the search point is off-set. These informal ideas can be turned into a rigorous analysis using the simplified drift theorem.

9.4.3   Tracking the Optimum in a LATTICE

Jansen and Schellbach [24] analysed the performance of the $(1 + \lambda)$ EA on the tracking problem described in section 9.3.4.3. An offspring is generated by adding to the parent K vectors sampled uniformly at random with replacement from the set $\{(\pm 1, 0), (0, \pm 1)\}$, where K is a Poisson distributed random variable with parameter 1. Hence, the offspring has expected distance 1 from the parent. The analysis assumes that the time steps of the objective function is synchronised with the generation counter of the EA. Hence, in each generation $t \geq 1$, the algorithm evaluates the distance between λ offspring and the current target point a_t. Intuitively, a larger population size should be beneficial for the EA within this scenario, because each generation provides more information about the position of the current target point.

For the special case of $d_{\max} = 0$, (i.e., a static optimization problem), they obtained the following asymptotically tight bound on the first hitting time.

Theorem 9.2 ([24])

$$\mathbf{E}\left[T_{\lambda,0}(n)\right] = \Theta\left(\lambda \cdot \left(1 + \frac{n \cdot \log\log\lambda}{\log\lambda}\right)\right).$$

As the algorithm makes λ function evaluations per generation, and needs to overcome a distance of n, the result informally means that the speed of the algorithm is on the order of $\Theta(\log\lambda / \log\log\lambda)$ per generation. Increasing the population size λ decreases the expected number of generations needed to reach the (static) target point.

The potential difficulty of the tracking problem increases with the parameter $d_{\max}$. In the worst case, the target point moves in the opposite direction of the current search point. Intuitively, if $d_{\max}$ is significantly lower than the speed $\Theta(\log\lambda / \log\log\lambda)$ of the algorithm, then one would expect the algorithm to be able to reach the target point. The following theorem confirms this intuition.

Theorem 9.3 ([24]). *Let* $b := 4/e$, $n' := n - d_{max}$, $\tilde{c} > 1$, *and* $s := \left\lfloor \frac{\log_b \lambda}{2\tilde{c} \log_b \log_b \lambda} \right\rfloor$. *For* $d_{max} \leq (2/3 - o(1/\lambda))s$, *it holds*

$$\mathbf{Pr}\left[T_{\lambda,dmax}(n) = O\left(\lambda(1 + \frac{n'\log\log\lambda}{\log\lambda})\right)\right] = 1 - 2^{-\Omega(n'/s)}.$$

Once the algorithm is within distance $d_{\max}$ of the moving target point, it is intuitive that the algorithm does not loose track of the target point. The following theorem shows that the expected time until the target point is lost grows exponentially with the population size.

Theorem 9.4 ([24]). *Let s be as in Theorem 9.3. For $d_{max} \leq (2/3)s$, and any integer $k \geq 2$, the expected number of generations until the $(1 + \lambda)$ EA has a distance to the target of at least kd_{max} after having a distance of at most d_{max} is bounded below by $e^{\Omega(k\sqrt{\lambda})}$.*

In contrast, when $d_{\max}$ is significantly higher than $\Theta(\log\lambda/\log\log\lambda)$, and the target point moves in an adverserial way, it is to be expected that the tracking problem becomes hard for the $(1+\lambda)$ EA. Jansen and Schellbach [24] provide some theoretically motivated arguments that support this view.

9.5   Conclusions

9.5.1   Summary and Implications

The number of contributions in evolutionary dynamic optimization has risen dramatically in recent years and a wealth of novel evolutionary algorithms (EAs) have been suggested that attempt to *track* the global optimum of some dynamic function over time. However, theoretical results on the expected runtimes of these algorithms are almost non-existent and almost all findings are based exclusively on empirical data. This imbalance may make it difficult to validate an algorithm's performance and to identify a broader class of functions the algorithm may work well on. Furthermore, the lack of theoretical results may lead to incorrect empirical validation of common assumptions about the dynamic domain. In this chapter, we reviewed previous theoretical results in an attempt to highlight some of the gaps in our understanding of evolutionary dynamic optimization. These results may be summarised as follows.

The results by Droste [11, 12] showed how the rate of change plays a crucial role in the algorithm's (temporal) first hitting time. In a more general sense, this is a first step in understanding how subtle differences in the problem's dynamics make the problem tractable or not. This result extended the earlier study by Stanhope and Daida [50] that even small perturbations in the fitness function could have a significantly negative impact on the performance of the (1+1) EA. The results by Rohlfshagen et al. [46] have shown that it is possible to show examples where common assumptions (i.e., that a larger magnitude of change / higher frequency of change makes a DOP harder) break down. This has important ramifications regarding the treatment of such problems and how one generalise empirical results from specific test cases to more general classes of DOPs. Similarly, Chen et al. [6] proved that adaptive and self-adaptive mutations may not perform as well as one might have thought in a dynamic environment. A fixed and non-adaptive scheme can sometime be just as good as any adaptive schemes in a dynamic environment. Finally, Jansen and Schellbach [24] is the only work to consider an offspring population. Within the framework considered, the authors showed that increasing the algorithm's population size decreases the expected number of generations needed to reach the target point and that the expected time until the target point is lost grows exponentially with the population size.

These theoretical studies have some important implications. First of all, they highlight the difficulty in defining the problem itself, including ways to unambiguously describe some of its properties such as the magnitude of change. Furthermore, the performance of an algorithm may be measured in numerous different ways and there are subtle differences between each approach; it is important to better understand these differences as they clearly have an impact on how one evaluates a particular algorithm on a particular problem.

9.5.2 Future Work

As the review in section 9.4 has shown, the scope of existing theoretical results is limited. Nevertheless, the progress to date is essential to further developments in the field and build the basis for future work. We believe the following constitutes important directions for future theoretical work in evolutionary dynamic optimization especially with regard to the significant advances made recently in the runtime analysis of EAs in the stationary domains.

1. **Framework and problem complexity.** A theoretical framework is required that allows practitioners and theoreticians to unambiguously describe different instances of DOPs. This framework would subsequently allow for the classification of different types of DOPs and may subsequently facilitate an analysis of problem complexity. In particular, currently it is not possible to identify and distinguish between those types of DOPs that are easier to solve than stationary problems and those that are harder; empirical evidence from biology seems to suggest that certain types of dynamics allows for faster rates of adaptation (e.g., [27]).

2. **Notion of optimality.** As the review has highlighted, numerous different notions of optimality/runtime may be applied in the dynamic domain and their relationship remains to be established. A general definition of runtime analysis in the dynamic domain should also be grounded in practical requirements and hence be able to account for trajectory-based performance measures as used in most practical applications.

3. **Populations.** One of the main motivations behind the application of EAs to DOPs is their use of populations. In particular, it is thought that the sampling of multiple search points simultaneously allows for better rates of adaptation in the new environments. The work by Jansen and Schellbach [24] provides an initial analysis of the role played by populations yet further examples are required where populations are provably beneficial.

4. **Diversity.** Diversity is considered one of the key issues that determines the performance of an EA on a particular DOP and the majority of algorithms developed aim to maintain high levels of diversity throughout the algorithm's execution. However, it is clear that not all types of diversity are equally useful and

hence a better understanding is required to identify mechanisms that are able to produce useful levels of diversity given a particular DOP.

5. **Crossover.** Related to the issues of populations and diversity comes a better understanding of crossover operators; so far, only EAs with mutation have been considered yet almost all population-based algorithms developed for DOPs also employ crossover operators. Nevertheless, there is little evidence that substantiates the impact of crossover on the algorithm's performance. There is evidence from biology that the evolution of sexual reproduction (i.e., crossover) is directly linked to uncertainty in the environment (e.g., [45, 48]).

6. **Beyond EAs and toy problems.** There are many additional population-based algorithms, such as ant colony optimization and particle swarm optimization, that have already been considered from a practical point of view. Furthermore, it is important to extend the theoretical treatment from simple artificial problems to simple dynamics variants of well-known NP hard problems such as the travelling salesman problem.

Acknowledgements. This work was partially supported by the Engineering and Physical Sciences Research Council (EPSRC) of UK under Grant EP/E058884/1.

References

[1] Auger, A., Doerr, B. (eds.): Theory of randomized search heuristics. World Scientific Publishing (2011)

[2] Back, T.: Optimal mutation rates in genetic search. In: Forrest, S. (ed.) Proc. 5th Int. Conf. Genetic Algorithms, pp. 2–8. Morgan Kaufmann (1993)

[3] Branke, J.: Evolutionary algorithms for dynamic optimization problems - a survey. Tech. Rep. 387, Insitute AIFB, University of Karlsruhe (1999)

[4] Branke, J.: Evolutionary approaches to dynamic environments - updated survey. In: GECCO Workshop on Evolutionary Algorithms for Dynamic Optimization Problems, pp. 27–30 (2001)

[5] Branke, J.: Evolutionary Optimization in Dynamic Environments. Kluwer (2002)

[6] Chen, T., Chen, Y., Tang, K., Chen, G., Yao, X.: The impact of mutation rate on the computation time of evolutionary dynamic optimization. arXiv preprint arXiv:1106.0566 (2011)

[7] Chen, T., Lehre, P.K., Tang, K., Yao, X.: When is an estimation of distribution algorithm better than an evolutionary algorithm? In: Proc. 2009 IEEE Congr. Evol. Comput., pp. 1470–1477. IEEE (2009)

[8] Cruz, C., González, J.R., Pelta, D.A.: Optimization in dynamic environments: a survey on problems, methods and measures. Soft Comput. 15(7), 1427–1448 (2011)

[9] Doerr, B., Klein, C., Storch, T.: Faster evolutionary algorithms by superior graph representation. In: Proc. 1st IEEE Symp. Foundations of Comput. Intell., pp. 245–250 (2007)

[10] Droste, S.: Analysis of the (1+1) ea for a dynamically changing objective function. Tech. rep., Universität Dortmund (2001)

[11] Droste, S.: Analysis of the (1+1) ea for a dynamically changing onemax-variant. In: Proc. 2002 IEEE Congr. Evol. Comput., pp. 55–60 (2002)

[12] Droste, S.: Analysis of the (1+1) ea for a dynamically bitwise changing onemax. In: Cantú-Paz, E., et al. (eds.) GECCO 2003. LNCS, vol. 2723, pp. 909–921. Springer, Heidelberg (2003)

[13] Droste, S., Jansen, T., Wegener, I.: On the optimization of unimodal functions with the (1 + 1) evolutionary algorithm. In: Eiben, A.E., Bäck, T., Schoenauer, M., Schwefel, H.-P. (eds.) PPSN 1998. LNCS, vol. 1498, pp. 13–22. Springer, Heidelberg (1998)

[14] Droste, S., Jansen, T., Wegener, I.: A rigorous complexity analysis of the (1+1) evolutionary algorithm for linear functions with boolean inputs. Evol. Comput. 6(2), 185–196 (1998)

[15] Droste, S., Jansen, T., Wegener, I.: On the analysis of the (1+1) Evolutionary Algorithm. Theoretical Computer Science 276, 51–81 (2002)

[16] Droste, S., Jansen, T., Wegener, I.: Optimization with randomized search heuristics the (a)nfl theorem, realistic scenarios, and difficult functions. Theoretical Computer Science 287 (2002)

[17] Droste, S., Jansen, T., Wegener, I.: Upper and lower bounds for randomized search heuristics in black-box optimization. Electronic Colloquium on Computational Complexity (ECCC) 48, 2003 (2004)

[18] Friedrich, T., Hebbinghaus, N., Neumann, F., He, J., Witt, C.: Approximating covering problems by randomized search heuristics using multi-objective models. In: Proc. 9th Annual Conf. Genetic and Evol. Comput., pp. 797–804 (2007)

[19] Friedrich, T., Oliveto, P.S., Sudholt, D., Witt, C.: Analysis of diversity-preserving mechanisms for global exploration. Evol. Comput. 17(4), 455–476 (2009)

[20] Giel, O.: Zur analyse von randomisierten suchheuristiken und online-heuristiken. Ph.D. thesis, Universität Dortmund (2005)

[21] Giel, O., Lehre, P.K.: On the effect of populations in evolutionary multi-objective optimisation. Evol. Comput. 18(3), 335–356 (2010)

[22] Giel, O., Wegener, I.: Evolutionary algorithms and the maximum matching problem. In: Alt, H., Habib, M. (eds.) STACS 2003. LNCS, vol. 2607, pp. 415–426. Springer, Heidelberg (2003)

[23] Jägersküpper, J.: Probabilistic analysis of evolution strategies using isotropic mutations. Ph.D. thesis, Universität Dortmund (2006)

[24] Jansen, T., Schellbach, U.: Theoretical analysis of a mutation-based evolutionary algorithm for a tracking problem in lattice. In: Beyer, H.G.G. (ed.) Proc. 2005 Genetic and Evol. Comput. Conf., pp. 841–848. ACM (2005)

[25] Jansen, T., Wegener, I.: Real royal road functions–where crossover provably is essential. Discrete Applied Mathematics 149(1-3), 111–125 (2005)

[26] Jin, Y., Branke, J.: Evolutionary optimization in uncertain environment - a survey. IEEE Trans. Evol. Comput. 9(3), 303–317 (2005)

[27] Kashtan, N., Noor, E., Alon, U.: Varying environments can speed up evolution. PNAS 104(34), 13,711–13,716 (2007)

[28] Kratsch, S., Lehre, P.K., Neumann, F., Oliveto, P.S.: Fixed parameter evolutionary algorithms and maximum leaf spanning trees: A matter of mutation. In: Schaefer, R., Cotta, C., Kołodziej, J., Rudolph, G. (eds.) PPSN XI. LNCS, vol. 6238, pp. 204–213. Springer, Heidelberg (2010)

[29] Kratsch, S., Neumann, F.: Fixed-parameter evolutionary algorithms and the vertex cover problem. In: Proc. 2009 Genetic and Evol. Comput. Conf., pp. 293–300 (2009)

[30] Leguizamon, G., Blum, C., Alba, E.: Handbook of approximation algorithms and metaheuristics, pp. 24.1–24.X. CRC Press (2007)

[31] Lehre, P.K.: Negative drift in populations. In: Schaefer, R., Cotta, C., Kołodziej, J., Rudolph, G. (eds.) PPSN XI. LNCS, vol. 6238, pp. 244–253. Springer, Heidelberg (2010)

[32] Lehre, P.K., Witt, C.: Black-box search by unbiased variation. In: Proc. 12th Annual Conf. Genetic and Evol. Comput., pp. 1441–1448. ACM, New York (2010)

[33] Lehre, P.K., Yao, X.: Runtime analysis of search heuristics on software engineering problems. Frontiers of Computer Science in China 3(1), 64–72 (2009)

[34] Lehre, P.K., Yao, X.: Runtime analysis of the (1+1) EA on computing unique input output sequences. Inform. Sci. (2011)

[35] Morrison, R.W.: Designing Evolutionary Algorithms for Dynamic Environments. Springer, Berlin (2004) ISBN 3-540-21231-0

[36] Muhlenbein, H.: How genetic algorithms really work i: Mutation and hillclimbing. In: Manner, R., Manderick, B. (eds.) Proc. 2nd Int. Conf. Parallel Problem Solving from Nature, pp. 15–25 (1992)

[37] Neumann, F.: Combinatorial optimization and the analysis of randomized search heuristics. Ph.D. Thesis, Christian-Albrechts-Universität zu Kiel (2006)

[38] Neumann, F., Sudholt, D., Witt, C.: A few ants are enough: Aco with iteration-best update. In: Proc. 12th Annual Conf. Genetic and Evol. Comput., pp. 63–70. ACM (2010)

[39] Neumann, F., Wegener, I.: Randomized local search, evolutionary algorithms, and the minimum spanning tree problem. Theoretical Computer Science 378(1), 32–40 (2007)

[40] Neumann, F., Witt, C.: Runtime analysis of a simple ant colony optimization algorithm. In: Asano, T. (ed.) ISAAC 2006. LNCS, vol. 4288, pp. 618–627. Springer, Heidelberg (2006)

[41] Neumann, F., Witt, C.: Bioinspired Computation in Combinatorial Optimization. Natural Computation Series. Springer (2010)

[42] Oliveto, P.S., He, J., Yao, X.: Time complexity of evolutionary algorithms for combinatorial optimization: A decade of results. Int. J. of Automation and Computing 4(1), 100–106 (2007)

[43] Oliveto, P.S., He, J., Yao, X.: Analysis of population-based evolutionary algorithms for the vertex cover problem. In: Proc. 2008 IEEE World Congr. Comput. Intell., pp. 1563–1570 (2008)

[44] Reichel, J., Skutella, M.: Evolutionary algorithms and matroid optimization problems. Algorithmica 57(1), 187–206 (2010)

[45] Ridley, M.: The Red Queen: Sex and the Evolution of Human Nature. Penguin Books Ltd. (1993)

[46] Rohlfshagen, P., Lehre, P.K., Yao, X.: Dynamic evolutionary optimisation: An analysis of frequency and magnitude of change. In: Proc. 2009 Genetic and Evol. Comput. Conf., pp. 1713–1720 (2009)

[47] Scharnow, J., Tinnefeld, K., Wegener, I.: Fitness landscapes based on sorting and shortest paths problems. In: Guervós, J.J.M., Adamidis, P.A., Beyer, H.-G., Fernández-Villacañas, J.-L., Schwefel, H.-P. (eds.) PPSN 2002. LNCS, vol. 2439, pp. 54–63. Springer, Heidelberg (2002)

[48] Smith, J.M.: The Evolution of Sex. Cambridge University Press, Cambridge (1978)

[49] Stanhope, S.A., Daida, J.M.: Optimal mutation and crossover rates for a genetic algorithm operating in a dynamic environment. In: Porto, V.W., Waagen, D. (eds.) EP 1998. LNCS, vol. 1447, pp. 693–702. Springer, Heidelberg (1998)

[50] Stanhope, S.A., Daida, J.M. (1+1) genetic algorithm fitness dynamics in a changing environments. In: Proc. 1999 IEEE Congr. Evol. Comput., vol. 3, pp. 1851–1858 (1999)

[51] Storch, T., Wegener, I.: Real royal road functions for constant population size. Theoretical Computer Science 320(1), 123–134 (2004)

[52] Sudholt, D., Witt, C.: Runtime analysis of binary pso. In: GECCO 2008: Proc. 10th Annual Conf. Genetic and Evol. Comput., pp. 135–142. ACM, New York (2008)

[53] Tinos, R., Yang, S.: Continuous dynamic problem generators for evolutionary algorithms. In: Proc. 2007 IEEE Congr. Evol. Comput., pp. 236–243 (2007)

[54] Tinós, R., Yang, S.: An analysis of the XOR dynamic problem generator based on the dynamical system. In: Schaefer, R., Cotta, C., Kołodziej, J., Rudolph, G. (eds.) PPSN XI. LNCS, vol. 6238, pp. 274–283. Springer, Heidelberg (2010)

[55] Wegener, I., Witt, C.: On the analysis of a simple evolutionary algorithm on quadratic pseudo-boolean functions. Journal of Discrete Algorithms 3(1), 61–78 (2005)

[56] Weicker, K.: Evolutionary algorithms and dynamic optimization problems. Der Andere Verlag (2003)

[57] Weicker, K.: Analysis of local operators applied to discrete tracking problems. Soft Comput. 9(11), 778–792 (2005)

[58] Witt, C.: Worst-case and average-case approximations by simple randomized search heuristics. In: Diekert, V., Durand, B. (eds.) STACS 2005. LNCS, vol. 3404, pp. 44–56. Springer, Heidelberg (2005)

[59] Witt, C.: Population size versus runtime of a simple evolutionary algorithm. Theoretical Computer Science 403(1), 104–120 (2008)

[60] Witt, C.: Why standard particle swarm optimisers elude a theoretical runtime analysis. In: Proc. 10th Int. Workshop Foundations of Genetic Algorithms, pp. 13–20. ACM, New York (2009)

[61] Wolpert, D.H., Macready, W.G.: No free lunch theorems for optimization. IEEE Trans. Evol. Comput. 1(1) (1997)

[62] Yang, S.: Non-stationary problem optimization using the primal-dual genetic algorithms. In: Sarker, R., Reynolds, R., Abbass, H., Tan, K.C., McKay, R., Essam, D., Gedeon, T. (eds.) Proc. 2003 IEEE Congr. Evol. Comput., vol. 3, pp. 2246–2253 (2003)

[63] Yang, S.: Memory-based immigrants for genetic algorithms in dynamic environments. In: Proc. 2005 Genetic and Evol. Comput. Conf., pp. 1115–1122. ACM (2005)

Chapter 10
Analyzing Evolutionary Algorithms for Dynamic Optimization Problems Based on the Dynamical Systems Approach

Renato Tinós and Shengxiang Yang

Abstract. The study of evolutionary algorithms for dynamic optimization problems (DOPs) has attracted a rapidly growing interest in recent years. However, few work has addressed the theory in this domain. In this chapter, we use the exact model (or dynamical systems approach) to describe the standard genetic algorithm as a discrete dynamical system for DOPs. Based on this dynamical system model, we define some properties and classes of DOPs and analyze some DOPs used by researchers in the dynamic evolutionary optimization area. The analysis of DOPs via the dynamical systems approach allows explaining some behaviors observed in experimental results. The theoretical analysis of the properties of well-known DOPs is important to understand the results obtained in experiments and to analyze the similarity of such problems to other DOPs.

10.1 Introduction

A significant increasing number of scientific papers on evolutionary algorithms (EAs) applied to dynamic optimization problems (DOPs) have appeared in recent years [4, 11, 25]. The growing interest in studying DOPs is due to its importance to real world applications of EAs, where, often, new solutions should be found in a short time after a change in the problem [2]. Most papers in EAs for DOPs experimentally investigate algorithms for DOPs, and very few investigate the theory

Renato Tinós
Department of Computing and Mathematics, FFCLRP, University of São Paulo,
Av. Bandeirantes, 3900, 14040-901, Ribeirão Preto, SP, Brazil
e-mail: rtinos@ffclrp.usp.br

Shengxiang Yang
Centre for Computational Intelligence (CCI), School of Computer Science and Informatics,
De Montfort University, The Gateway, Leicester LE1 9BH, U.K.
e-mail: syang@dmu.ac.uk

S. Yang and X. Yao (Eds.): *Evolutionary Computation for DOPs*, SCI 490, pp. 241–267.
DOI: 10.1007/978-3-642-38416-5_10   © Springer-Verlag Berlin Heidelberg 2013

behind DOPs [1, 6, 14, 16–18]. Most work on analyzing EAs for DOPs usually extends the analysis of EAs for stationary optimization to simple DOPs.

In [18], the standard genetic algorithm (GA) with mutation and selection is investigated on DOPs with regular changes (see Section 10.3) based on the dynamical system approach (or exact model) of the GA [22]. Despite demanding a large number of equations to track all possible solutions represented by the individuals of the GA, the use of the exact model is very useful as it allows a complete description of the population dynamics [13].

In this chapter, we use the dynamical system approach to define some properties and some classes of DOPs. Then, we analyze DOPs generated by: the XOR DOP Generator [23, 27], the dynamic environment generator based on problem difficulty [24], and three variations of the dynamic 0-1 knapsack problem [15]. The analysis of the properties of a DOP is relevant in order to understand the results obtained in the experiments with EAs and to analyze the similarity of such problems to other problems. In this way, the analysis of DOPs via the dynamical system approach can help researchers to investigate whether DOPs widely used in the evolutionary optimization area are appropriated to represent a given real-world DOP and to identify classes of DOPs where a given algorithm should present good performance.

The rest of this chapter is organized as follows. The exact model for the GA in stationary environments is briefly presented in Section 10.2. Some concepts of DOPs are discussed and formally described based on the dynamical system approach in Section 10.3. Section 10.4 analyses five different examples of DOPs using the dynamical system approach. Finally, the conclusions with discussions on relevant future work are presented in Section 10.5.

10.2   Exact Model of the GA in Stationary Environments

The description of the standard GA as a discrete dynamical system [13], approach known as exact model and proposed by Vose [22], is an attempt to understand the behavior of the GA according to the analysis of its population dynamics. For this purpose, the rule for changing the population probabilities in each generation are defined, and the properties of the resulting discrete dynamical system, like its fixed points, are analyzed. In a GA with binary codification, an individual of a population codifies a possible solution $\mathbf{x} \in \{0,1\}^l$. In the exact model, all possible solutions are represented in a discrete space χ, where each possible solution is enumerated as $\{0,1,\ldots,n-1\}$ and $n = 2^l$. A population is then defined by a n-dimensional vector, where each element defines the proportion of each possible solution in the population, i.e.,

$$\mathbf{p} = \frac{\mathbf{v}}{N}, \tag{10.1}$$

where the k-th element of $\mathbf{v}$ indicates the number of copies of the k-th possible solution in the population with size N. As the sum of the elements of $\mathbf{p}$ is equal to 1, population vectors can be described as members of a simplex Λ, i.e.,

$$\Lambda = \left\{ \mathbf{p} \in \mathbb{R}^n : p_k \geq 0, \text{ for } k = 0, 1, \ldots, n-1 \text{ and } \sum_{k=0}^{n-1} p_k = 1 \right\}, \tag{10.2}$$

where p_k is the k-th element of the population vector $\mathbf{p}$ and the vertices of the simplex represent populations with copies of only one solution. In this way, the population evolution can be described as a trajectory in the simplex and population vectors can be used to describe the probability distribution of the individuals in the search space. Thus, a generational operator $\mathcal{G} : \Lambda \to \Lambda$ can be defined as follows:

$$\mathcal{G}(\mathbf{p}) = \mathbf{p}', \tag{10.3}$$

where $\mathbf{p}'$ is the probability distribution sampled to generate the next population after $\mathbf{p}$, i.e., $\mathcal{G}(\mathbf{p})$ is the expected next population [22].

Each time the generational operator is applied corresponds to one generation. In this way, the vector $\mathcal{G}(\mathbf{p})$ describes the average over all possible populations in the next generation with variance inversely proportional to the population size N. In the limit $N \to \infty$ (infinite population case), the variance goes to zero, and the trajectory of the population in the simplex can be deterministically described. Thus, we can define the evolutionary process of a GA considering its genotypical dynamics for the general case where the generational operator can change in each generation t.

Definition 10.1 (GA). The GA, in the infinite population case, is a discrete dynamical system defined by the successive application of the rule:

$$\mathbf{p}(t) = \mathcal{G}\big(\mathbf{p}(t-1), t\big), \tag{10.4}$$

where $\mathcal{G}(., t) : \Lambda \times \mathbb{N}^+ \to \Lambda$ is the generational operator (map), $\mathbf{p}(t)$ is the expected population at generation $t \in \mathbb{N}^+$, and $\mathbf{p}(0)$ is the initial population vector.

For the stationary case, the generational operator is $\mathcal{G}(., t) = \mathcal{G}(.)$ for all $t \geq 1$, and the trajectory of the population is given by $\mathbf{p}(0), \mathcal{G}(\mathbf{p}(0)), \mathcal{G}^2(\mathbf{p}(0)), \ldots$.

In this way, in generation t for the stationary case, the expected population vector for the infinite population model is given by:

$$\mathbf{p}(t) = \mathcal{G}^t\big(\mathbf{p}(0)\big). \tag{10.5}$$

Let us consider now the standard GA with mutation and proportional selection. Then, the generational operator can be represented by:

$$\mathcal{G} = \mathcal{U} \circ \mathcal{F}, \tag{10.6}$$

where the proportional selection operator is given by:

$$\mathcal{F}(\mathbf{p}) = \frac{F\,\mathbf{p}}{\mathbf{f}^\mathsf{T}\mathbf{p}}, \tag{10.7}$$

f is the vector with the fitness of each solution x_i in χ, and $F = \mathrm{diag}(\mathbf{f})$ is a diagonal matrix generated by **f**. The fitness vector **f** gives information about the structure of the search space. The mutation operator is given by:

$$\mathcal{U}(\mathbf{p}) = U\mathbf{p}, \tag{10.8}$$

where U is the mutation matrix. Each element U_{ij} indicates the probability of generating the i-th element of χ by mutating the j-th element of χ. By Eqs. 10.6-10.8, the generational operator can be written as:

$$\mathcal{G}(\mathbf{p}) = \frac{UF\,\mathbf{p}}{\mathbf{f}^{\mathrm{T}}\mathbf{p}}. \tag{10.9}$$

The analysis of Eq. (10.9) can provide insights in understanding the behavior of the stationary GA. In next section, this analysis will be extended to the dynamic case. For the stationary case, the fixed points of $\mathcal{G}$, i.e., points where $\mathcal{G}(\mathbf{y}) = \mathbf{y}$, are given by the eigenvectors of UF. For each eigenvector **y**, an eigenvalue $\mathbf{f}^{\mathrm{T}}\mathbf{y}$, corresponding to the average fitness of **y**, can be computed. As UF has only positive values, there is only one eigenvector in Λ, corresponding to the eigenvalue with the largest absolute value [13]. Then, all trajectories in Λ converge to this fixed point, i.e. the system is asymptotically stable [8, 22].

The remaining eigenvectors are not properly fixed points, as, for example, they can lie outside the simplex, and are called metastable states in the dynamical system approach [13]. However, they play an important role in the evolutionary process as they can change the trajectory in the simplex and can trap finite populations for several generations [21]. The metastable states that are encountered during a trajectory depends on the initial population, the dynamics of the GA, and random events related to the stochastic operators. In DOPs, as the generational operator changes during the evolutionary process, we do not use the term fixed points for the points where $\mathcal{G}(\mathbf{y}) = \mathbf{y}$, using instead the term metastable states. For the point **y** with the current largest average fitness (i.e., $\mathbf{f}^{\mathrm{T}}\mathbf{y}$), which is inside the simplex, we refer it as the current main metastable state.

When the crossover is added to the GA, the generational operator is given by:

$$\mathcal{G} = \mathcal{C} \circ \mathcal{U} \circ \mathcal{F}, \tag{10.10}$$

where each element k of the crossover operator is computed by:

$$\mathcal{C}_k(\mathbf{p}) = \mathbf{p}^{\mathrm{T}}\mathcal{M}_k\mathbf{p}, \tag{10.11}$$

where $\mathcal{M}_k$ is a matrix formed by the probabilities of generating individual $k \in \chi$ from crossing two individuals of the population. By equations 10.9-10.11, the generational operator can be now computed by:

$$\mathcal{G}_k(\mathbf{p}) = \frac{\mathbf{p}^{\mathrm{T}}FU^{\mathrm{T}}\mathcal{M}_kUF\,\mathbf{p}}{(\mathbf{f}^{\mathrm{T}}\mathbf{p})^2}. \tag{10.12}$$

As Eq. (10.9), Eq. (10.12) can be used to investigate the metastable states of the GA. In this case, the metastable states can be obtained by using a linearization procedure in states near the fixed points of the generational operator described by Eq. (10.12) [22].

10.3 Dynamic Optimization Problems

In this section, we define how the discrete dynamical system of the GA can be adapted to DOPs. For this purpose, it is necessary to define some aspects of the changing optimization problem in the context of the dynamical systems approach. In this way, we can define some classes of DOPs, analyzing how the dynamical system of the GA is modified during the evolutionary process. Here we focus on DOPs where the fitness landscape is modified by the changes intrinsic to the optimization problem. Before defining a DOP in the context of GAs (Definition 10.1), it is necessary to define changes in the dynamical system approach.

Definition 10.2 (Change in the GA). Consider a GA (Definition 10.1) where the generational operator $\mathcal{G}(.,t)$ at generation t is hyperbolic for all $t \geq 1$ and has n_f points $\mathbf{y}_i(t) = \mathcal{G}(\mathbf{y}_i(t),t)$ for $i = 1,\ldots,n_f$, i.e., $\mathbf{y}_i(t)$ is the i-th metastable state of $\mathcal{G}(.,t)$. A change in the GA occurs at generation t when

$$\mathbf{y}_i(t) \neq \mathbf{y}_i(t-1) = \mathcal{G}(\mathbf{y}_i(t-1),t) \tag{10.13}$$

for at least one $\mathbf{y}_i(t)$ for $i = 1,\ldots,n_f$, i.e., at least one metastable state of $\mathcal{G}(.,t)$ is not preserved.

Here we consider that changes occur only between the application of two consecutive generational operators. One can observe that not all modifications in the generational operator $\mathcal{G}$ (see Eq. (10.12) for the standard GA with crossover) can be described as a change according to Definition 10.2. Other important observation is that a change does not necessarily imply a modification in the current population or in its current trajectory. For example, if the change does not modify the current trajectory of the population to the metastable points, no effect will be observed in the evolutionary process. The same occurs if the population has converged to a given metastable state and this one is not modified by the change. Interestingly, the usual way to detect a change by reevaluating the fitness of only a few best or sentinel solutions [4] may be inefficient in cases where those solutions are not affected by the change.

Based on Definition 10.2, DOPs in the context of GAs can be defined according to the dynamical system approach for the standard GA.

Definition 10.3 (DOP). A DOP in the context of GAs (Definition 10.1) is an optimization problem where at least one change (Definition 10.2) occurs during the evolutionary process.

As the generation operator is modified after a change, Eq. (10.5) is not valid anymore for every generation t in a DOP. We now define change cycle.

Definition 10.4 (Change cycle). A change cycle is a series of generational operations between two consecutive changes. The first change cycle begins in the first
generation of the evolutionary process and ends one generation before the first
change, while the last change cycle begins in the generation after the last change
and ends in the last generation of the evolutionary process.

The change cycle duration d_e is the number of consecutive generations in change
cycle e. If change cycle e begins at generation t, then

$$\mathcal{G}(.,t) = \mathcal{G}(,t+1) = \mathcal{G}(.,t+2) = \ldots = \mathcal{G}(.,t+d_e-1), \tag{10.14}$$

where $d_e \in \mathbb{N}^+$. In abuse of notation, we define now $\mathcal{G}(.,e)$ as the generational operator in change cycle e. In this way, for the infinite population case, the population
in generation t is now given by:

$$\mathbf{p}(t) = \mathcal{G}^{(t-\sum_{i=1}^{e-1} d_i)}(.,e) \circ \mathcal{G}^{d_e-1}(.,e-1) \ldots \mathcal{G}^{d_2}(.,2) \circ \mathcal{G}^{d_1}\big(\mathbf{p}(0),1\big), \tag{10.15}$$

where $e > 0$. It can be observed that a DOP can be viewed as a sequence of stationary processes, where the initial population in the i-th change cycle is the last
population generated in the change cycle $i-1$. The minimum value of d_i is one
generation, which is the case where the generational operator is modified just one
generation after the prior change, while the maximum value of d_i is equal to the
index of the current generation, which is the case where the problem is stationary
(until the current generation) and Eq. (10.15) reproduces Eq. (10.5).

If the standard GA is employed, a change modifies at least one term of the generational operator defined by Eq. (10.12). In general, changes in the mutation matrix
U and in the crossover matrices $\mathcal{M}_k$ are related to changes in the algorithm, e.g., if
the mutation rate is increased, like in the case of an adaptive mutation operator, the
mutation matrix U is modified. According to Definition 10.3, both changes in the
GA induced by the programmer or intrinsic to the optimization problem can cause
DOPs. In fact, changes in the operators (e.g., the use of adaptive operators) can reduce the effects of changes intrinsic to the optimization problem. This property can
be used to develop and/or understand adaptive approaches used in EAs for DOPs.

Some changes intrinsic to the optimization problem can modify the mutation and
crossover operators too, e.g., in some algorithms, changes in the constraints of the
problem can imply a modification in the mutation and crossover matrices as some
solutions are not anymore allowed. In this work, we are not interested in problems
with such changes. In this way, we define DOPs with fitness landscape changes.

Definition 10.5 (DOP with fitness landscape changes). A DOP with fitness landscape changes is a DOP (Definition 10.3) where the fitness landscape is modified by
a change (Definition 10.2) at least one time during the optimization process, i.e.,

$$\mathbf{f}(e) \neq \mathbf{f}(e-1), \tag{10.16}$$

in at least one change cycle $e > 1$, where $\mathbf{f}(e)$ is the fitness vector in change cycle e.

It can be observed that in a DOP with fitness landscape changes (Definition 10.5), not all elements of $\mathbf{f}(e)$ are necessarily modified during the evolutionary process, e.g., the change can affect only a small fraction of the fitness landscape. According to Definition 10.2, only changes in $\mathbf{f}(e)$ that modify at least one metastable state of the GA before the change are considered. Definition 10.5 is very general, and not all problems with fitness landscape changes has attracted attention to the evolutionary computation community.

In general, researchers investigate problems where the fitness landscape changes according to a specific rule, stochastic or deterministic. For example, in continuous dynamic optimization, a traditional benchmark is the moving peaks benchmark [2, 10], where the dynamism of the problem is defined by changing a set of fitness landscape parameters (aggregated here in a vector named system control vector $\phi(e)$), e.g., the location and width of each peak. The system control vector $\phi(e)$ is computed in each change cycle e using its past values according to a given rule. As an example, the vector $\phi(e)$ can be obtained adding a random deviation, taken from a normal distribution, to the vector $\phi(e-1)$. In this way, we can classify the DOP according to the rule that modifies the fitness landscape. In a first moment, we can define single time-dependent DOPs, which compose a interesting subset of DOPs with fitness landscape changes.

Definition 10.6 (Single time-dependent DOP). A single time-dependent DOP is a DOP with fitness landscape changes (Definition 10.5) where the fitness landscape in change cycle e depends on the fitness landscape in change cycle $e - g$, i.e.,

$$\mathbf{f}(e) = h\big(\mathbf{f}(e-g), \phi(e)\big), \tag{10.17}$$

where $g \in \mathbb{N}^+$, $e - g \geq 1$, $h(.)$ is a real function, and $\phi(e)$ is the system control vector. The changing in the fitness landscape is defined by tuning $\phi(e)$.

It can be observed that the previous definition of single time-dependent DOP (Definition 10.6) is different from the definition of single time-dependent systems in [18]. By changing the function $h(., e)$ in Definition 10.6, we can define other DOPs, like periodic DOPs.

Definition 10.7 (Periodic DOP). Assume a single time-dependent DOP (Definition 10.6) where the fitness landscape in change cycle e is equal to the fitness landscape in change cycle $e - g$, i.e.,

$$\mathbf{f}(e) = \mathbf{f}(e-g), \tag{10.18}$$

where $g \in \mathbb{N}^+$ and $e - g \geq 1$. Such a DOP is called a periodic DOP.

In periodic DOPs (Definition 10.7), the changes are deterministic and predictable. As a consequence, memory-based approaches [9] can be, in general, successfully employed. Another special case of single time-dependent DOP (Definition 10.6) is when $g = 1$.

Definition 10.8 (Last environment dependent DOP). Assume a single time-dependent DOP (Definition 10.6) where the fitness landscape in change cycle e depends only on the fitness landscape in change cycle $e - 1$, i.e.,

$$\mathbf{f}(e) = h\big(\mathbf{f}(e-1), \phi(e)\big), \tag{10.19}$$

where $e > 1$, $h(.)$ is a real function, and $\phi(e)$ is the system control vector. Such first order DOP is called a last environment dependent DOP.

From changing the function h in Definition 10.8, we can define many last environment dependent DOP subsets. Here, we will present some special cases.

Definition 10.9 (Linear DOP). A linear DOP is a last environment dependent DOP (Definition 10.8) where the fitness landscape in change cycle $e - 1$ is modified according to a linear transformation, i.e.,

$$\mathbf{f}(e) = \mathbf{A}(\phi(e))\mathbf{f}(e-1) + \mathbf{b}(\phi(e)),$$

where $\mathbf{A}(.) \in \mathbb{R}^{n \times n}$ and $\mathbf{b}(.) \in \mathbb{R}^n$.

If the linear system is homogeneous and the matrix $A(\phi(e))$ is orthogonal, another last environment dependent DOP subset can be defined.

Definition 10.10 (Orthogonal DOP). An orthogonal DOP is a homogeneous linear DOP (Definition 10.9) where the fitness landscape in change cycle $e - 1$ is modified according to an orthogonal matrix, i.e.,

$$\mathbf{f}(e) = \mathbf{P}(\phi(e))\mathbf{f}(e-1), \tag{10.20}$$

where $\mathbf{P}(\phi(e)) \in \mathbb{R}^{n \times n}$ is an orthogonal matrix, i.e.,

$$\mathbf{P}(\phi(e))^{\mathsf{T}}\mathbf{P}(\phi(e)) = \mathbf{P}(\phi(e))\mathbf{P}(\phi(e))^{\mathsf{T}} = \mathbf{I}, \tag{10.21}$$

where $\mathbf{I}$ denotes the identity matrix.

An orthogonal matrix has some important properties [12], e.g.:

i) if $\mathbf{A}$ and $\mathbf{B}$ are orthogonal, then $\mathbf{AB}$ is orthogonal too.

ii) if λ is an eigenvalue of the orthogonal matrix $\mathbf{A}$, then $|\lambda| = 1$.

iii) if $\mathbf{A}$ is an orthogonal matrix, then $\|\mathbf{Ax}\|_2 = \|\mathbf{x}\|_2$ for every vector $\mathbf{x}$, where $\|\mathbf{x}\|_2$ denotes the Euclidean norm of the vector $\mathbf{x}$.

A geometrical interpretation of item iii) is that the linear transformation $\mathbf{A}$ preserves the angles and magnitudes of vector $\mathbf{x}$, i.e., the linear transformation behaves like a rotation in the space. A special case of an orthogonal matrix is a permutation matrix [22].

Definition 10.11 (DOP with permutation). A DOP with permutation is an orthogonal DOP (Definition 10.10) where the fitness landscape in change cycle $e - 1$ is modified according to a permutation matrix, i.e.,

$$\mathbf{f}(e) = \sigma(\phi(e))\mathbf{f}(e-1), \tag{10.22}$$

where $\sigma(\phi(e))$ is a permutation matrix defined by the control system parameter vector $\phi(.)$ at change cycle e. The permutation matrix maps the elements of vector $\mathbf{f}(e-1)$ to the elements of vector $\mathbf{f}(e)$.

In a DOP with permutation (Definition 10.11), the fitness values are preserved in the search space, i.e., they are only resorted. In [18], DOPs with regular changes, which are a special subset of DOPs with permutation (Definition 10.11), are defined.

Definition 10.12 (DOP with regular changes). A DOP with regular changes [18] is a DOP with permutation (Definition 10.11) where the transitional rule is deterministic and belongs to a permutation group where $\sigma(\phi(e+t)) = (\sigma(\phi(e)))^t$ for $t \in \mathbb{N}^+$.

Periodic (Definition 10.7) changes and the stationary case are special cases of regular changes. In [18], the standard GA without crossover applied to DOPs with regular changes are analyzed by using the dynamic system approach. In this special case, the fixed points can be computed and, in [18], the asymptotic states are analyzed in the sense of time-dependent quasispecies [7].

As a DOP can be viewed as a sequence of stationary environments (see Eq. (10.15)), the analysis of how the metastable states for each stationary environment are changed can provide insights in understanding GA's behavior on DOPs. Here, the state of the DOP in a change cycle corresponding to the fixed point in the respective stationary environment is called main metastable state. It is important to observe that the metastable states (including the main metastable state) are generally not fixed points of the DOP, as the problem changes and the population of the standard GA generally does not converge to a fixed point. However, the metastable states control the trajectory of the population during each change cycle and their analysis helps the understanding of the dynamics of the GA in DOPs. In the next section, some examples of DOPs are analyzed using the dynamical systems approach.

10.4 Examples

Here, we analyse DOPs produced by the XOR DOP Generator (Section 10.4.1), the Dynamic Environment Generator Based on Problem Difficulty (Section 10.4.2), and three variations of the Dynamic 0-1 Knapsack Problem (Section 10.4.3). Some simulations, where Eq. (10.15) is employed to generate the population vector $\mathbf{p}(t)$, i.e., the exact model with infinite population is employed in order to generate the expected next population during the evolutionary process, are presented. For simplicity, we will consider the standard GA with mutation and proportional selection as it allows the direct computation of the metastable points of the dynamical system (GA). In all simulations, the initial population ($\mathbf{p}_0$) is uniformly distributed, the mutation rate employed is equal to 0.01, and $l = 8$.

10.4.1 The XOR DOP Generator

The XOR DOP Generator generates DOPs from any binary encoded problem [23, 27]. Given a stationary problem with fitness function $f(\mathbf{x})$, where the solution $\mathbf{x} \in \{0,1\}^l$, the fitness function $f(\mathbf{x}, e)$ of an environment, periodically changed every τ generations, is computed by:

$$f(\mathbf{x}, e) = f(\mathbf{x} \oplus \mathbf{m}(e)), \tag{10.23}$$

where $e = \lceil t/\tau \rceil$ is the change cycle index, t is the generation index, and $\mathbf{m}(e) \in \{0,1\}^l$ is a binary mask for change cycle e, which is incrementally generated by:

$$\mathbf{m}(e) = \mathbf{m}(e-1) \oplus \mathbf{r}(e), \tag{10.24}$$

where $\mathbf{r}(e) \in \{0,1\}^l$ is a binary template randomly created for change cycle e containing $\lfloor \rho \times l \rfloor$ ones, and $\{\rho \in \mathbb{R} \mid 0.0 \leq \rho \leq 1.0\}$ controls the degree of change for the DOP. For $e = 1$, $\mathbf{m}(1)$ is equal to the zero vector.

In environments created by the XOR DOP Generator, each individual of the current population is moved to a new location in the fitness landscape before being evaluated [19]: instead of evaluating the fitness of the individual at $\mathbf{x}$, the fitness is evaluated at $\mathbf{x} \oplus \mathbf{m}(e)$. In [20], the XOR DOP Generator was analyzed based on the dynamical systems approach. The XOR DOP Generator produces a special kind of DOP with permutation (Definition 10.11) where the permutations in the fitness space are ruled by

$$\mathbf{f}(e) = \sigma_{\mathbf{r}(e)} \mathbf{f}(e-1), \tag{10.25}$$

where $\sigma_{\mathbf{r}(e)}$ is a permutation matrix at change cycle e, which maps the element at position $\mathbf{i}$ of the vector $\mathbf{f}(e-1)$ to the element at position $\mathbf{i} \oplus \mathbf{r}(e)$ of the vector $\mathbf{f}(e)$, where $\oplus$ is the bitwise exclusive-or (XOR), or addition modulo 2, operator. The vector $\mathbf{i} \in \{0,1\}^l$ indicates the position of the element in the fitness vector. The vector $\mathbf{r}(e) \in \{0,1\}^l$ controls the permutation of the elements of the fitness vector (see Eq. (10.24)), i.e., $\mathbf{r}(e)$ is the system control vector ($\phi(e)$) in Definition 10.11, controlling the changing in the fitness landscape. Permutation matrices $\sigma_{\mathbf{r}(e)}$ have the following properties:

i) $\sigma_{\mathbf{r}(e)}$ are symmetric matrices.
ii) $\sigma_{\mathbf{r}(i)} \sigma_{\mathbf{r}(j)} = \sigma_{\mathbf{r}(i) \oplus \mathbf{r}(j)}$.
iii) $\sigma_{\mathbf{r}(e)}$ commute and are self-inverse.

In this way, the i-th eigenvector $\mathbf{y}_i(e)$ of $U(e)F(e)$ in change cycle e, which defines a metastable state, can be obtained by the permutation of the respective eigenvector for the environment in change cycle $e-1$. Besides, the eigenvalues of $U(e)F(e)$ for two environments defined by change cycles $e-1$ and e are the same, which implies that the average fitness at the main metastable state remains the same.

The authors in [20] also proposed that the dynamical system of the GA with mutation and fitness proportional selection in a DOP with permutation (Definition 10.11) ruled by Eq. (10.25), as in the environments created by the XOR DOP Generator, is similar to the dynamical system of the same GA in a stationary environment where the population is changed according to the same permutation matrices used in the DOP with permutation in every d_e generations, where d_e is the duration of change cycle e for the DOP with permutation. As a consequence, the XOR DOP Generator can be simplified: instead of computing the fitness of each individual of the population at the new position $\mathbf{x}_t \oplus \mathbf{m}_e$ in every generation, each individual of the initial population in change cycle e is moved to $\mathbf{x}_t = \mathbf{x}_t \oplus \mathbf{r}_e$, i.e., the population

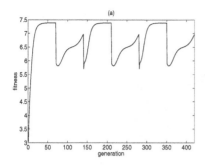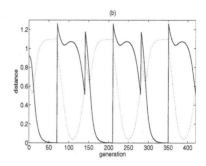

Fig. 10.1 Trajectory of the population during six change cycles with $\tau = 70$ and $\tau = 0.875$ in the DOP created by the XOR DOP Generator: mean fitness of the population (a); distance to the current first (solid) and second (dotted) main metastable states (b)

is moved only one time, and the fitness is computed as $f(\mathbf{x}_t)$, like in the stationary environment, which reduces the complexity of the procedure.

We present here simulations for the evolution of the population in a DOP created by the XOR DOP Generator from a simple stationary fitness function, chosen here due its simplicity. The stationary fitness function used here has two optima, and is defined by:

$$f(\mathbf{x}) = \begin{cases} l, & \text{if } u(\mathbf{x}) = l \\ (l-1) - u(\mathbf{x}), & \text{otherwise,} \end{cases} \tag{10.26}$$

where $u(\mathbf{x})$ is the unitation function of a binary vector $\mathbf{x}$ of length l, which is given by the number of ones in this vector. In the simulations presented here, seven values of τ (from $\tau = 10$ to $\tau = 70$) and seven values of ρ (from $\rho = 0.125$ to $\rho = 0.875$ with a step size 0.125) are considered. In this way, 49 simulations were executed, one for each pair of τ and ρ. Each evolutionary process is simulated for 30 change cycles of the infinite population model.

Figure 10.1 shows a simulation with $\rho = 0.875$ and $\tau = 70$, where the mean fitness of the population during the evolution and the Euclidean distance between the population in the current generation and the two eigenvectors with the largest eigenvalues are presented. The first eigenvector corresponds to the current main metastable state (where the number of individuals of the population at the global optimum is larger than the number of individuals at any other place), while the second eigenvector is the metastable state with the second largest eigenvalue (where the number of individuals of the population at the local optimum is larger than the individuals at any other place). It can be observed that, in some change cycles, the population goes to the neighborhood of the second metastable state and, after some generations, goes to the main metastable state.

Figure 10.2 presents the results for all simulations. From Fig. 10.2, some observations can be made. When τ is close to 70 generations, i.e., in slow changing environments, the population always reaches the main metastable state after changes with small degree of change ρ. When τ is large, there is enough time for the

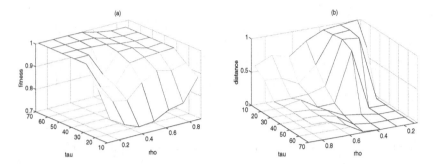

Fig. 10.2 Normalized fitness (a) and distance to the current main metastable state (b) in the DOP created by by the XOR DOP Generator for different τ and ρ. The values are relative to the average (over 30 change cycles) obtained by the population vector in the generation before the change.

population to go from the neighborhood of the main metastable state (where most of the population is at the global optimum) in change cycle $e-1$ to the neighborhood of the main metastable state in change cycle e. As a consequence, the fitness in the end of each change cycle is higher when τ is large and ρ is small. In the XOR DOP Generator, the parameter ρ controls the degree of change. As ρ controls the percentage of changed bits from template $\mathbf{m}_{e-1}$ to template $\mathbf{m}_e$, the hamming distance between $\mathbf{m}_{e-1}$ and $\mathbf{m}_e$ is $h(\mathbf{m}_e, \mathbf{m}_{e-1}) = \lfloor \rho \times l \rfloor$. In this way, larger ρ implies larger Hamming distance between the optima in two consecutive change cycles and in longer trajectories of the population in the simplex, and, thus, more time to reach the neighborhood of the main metastable point.

However, it can be observed that a higher degree of modification in the fitness landscape (larger ρ) does not necessarily imply a worse performance of the GA in the DOP for medium and small τ. One can observe that for medium and small τ, the simulations with $\rho = 0.375$ presented worse performance than those for larger ρ. This behavior can be found in experiments with the XOR DOP Generator for different algorithms (for example, see [26]). The performance of the GA is related to trajectories of the population in the simplex, and the trajectories are related to the fitness vector and the transformation operators. In a medium velocity or fast changing environment, generally, when the population reaches the neighborhood of the main metastable point in change cycle $e-2$, the population after the change is closer to the second metastable state in the next change cycle when ρ is large. In this case, the population does not have enough time to be closer to the new main metastable state neighborhood in change cycle $e-1$ than to the old main metastable neighborhood. However, when the problem changes again, the population is close to the neighborhood of the main metastable state in change cycle e for ρ close to 1. The mean Hamming distance of template $\mathbf{m}_e$ between two change cycles, which is given by $\bar{h}(\mathbf{m}_e, \mathbf{m}_{e-2}) = 2l(\rho - \rho^2)$, explains the behavior of the GA in this case. It can be observed that the mean fitness generally alternates between two different values for larger ρ and medium or small τ. One can observe that the values of distance in Fig. 10.2 are higher for ρ close to one than for ρ close to zero, as in part of the change cycles, the population remains in the neighborhood of the second metastable state for larger ρ.

Two observations can be made for the previous analysis. First, a higher degree of modification (ρ) in the templates $\mathbf{r}_e$ does not necessarily imply a worse performance of the GA. This result has been observed in several experiments with the XOR DOP Generator. The performance of the GA is related to the trajectories of the population in the simplex, which makes more complex the analysis of the performance of the algorithms.

Second, the metrics used to compare the algorithms in DOPs cannot be adequate for some problems. For example, in the problem investigated here, an algorithm that keeps the population close to the second metastable neighborhood for a high degree of change in a fast changing environment can have higher mean fitness than an algorithm that allows the population escaping from the local optima, but does not have enough time to reach the main metastable state neighborhood.

10.4.2 The Dynamic Environment Generator Based on Problem Difficulty

In another approach, Yang [24] proposed a dynamic environment generator based on unitation and trap functions. A trap function is defined as follows:

$$f(\mathbf{x}) = \begin{cases} \frac{a}{z}(z - u(\mathbf{x})), & \text{if } u(\mathbf{x}) \leq z \\ \frac{c}{l-z}(u(\mathbf{x}) - z), & \text{otherwise} \end{cases} \tag{10.27}$$

where $u(\mathbf{x})$ is the unitation function of the binary vector $\mathbf{x}$, a is the local and possibly deceptive optimum, c is the global optimum, and z is the slope-change location which separates the attraction basins of the two optima. A trap function is deceptive on average if the ratio of the fitness of the local optimum to that of the global optimum is constrained by the following relation: [5].

$$r \geq \frac{2 - 1/(l - z)}{2 - 1/z} \tag{10.28}$$

In the generator proposed in [24], the deception difficulty can be modified by changing, in every τ generations, the parameters a, c, and z in Eq. (10.27). The difficulty of the problem can also be controlled by changing the scaling of the function, what is not done here. We consider that the deception difficulty is modified in two consecutive change cycles by setting a between two values, a_{min} and a_{max}, where $0 < a_{min} < c$ and $a_{max} > c$.

In this way, the global optimum changes in each change cycle between c (located at $\mathbf{x} = \mathbf{1}$) and a_{max} (located at $\mathbf{x} = \mathbf{0}$). The dynamic fitness function can be defined as follows:

$$f(\mathbf{x}, e) = \begin{cases} \frac{a(e)}{z}(z - u(\mathbf{x})), & \text{if } u(\mathbf{x}) \leq z \\ \frac{c}{l-z}(u(\mathbf{x}) - z), & \text{otherwise} \end{cases} \tag{10.29}$$

where $a(e) = a_{min}$ if $a(e-1) = a_{max}$ and $a(e) = a_{max}$ if $a(e-1) = a_{min}$. One can observe that the fitness for the subspace of solutions where $u(\mathbf{x}) > z$ does not change during the whole evolutionary process, i.e., for $e > 1$:

$$f(\mathbf{x}, e) = f(\mathbf{x}, e-1) \text{ if } u(\mathbf{x}) > z \tag{10.30}$$

The fitness of the remaining solutions is changed for $e > 1$ according to:

$$f(\mathbf{x}, e) = \begin{cases} \frac{a_{min}}{a_{max}} f(\mathbf{x}, e-1), & \text{if } u(\mathbf{x}) \leq z \text{ and } a(e-1) = a_{max} \\ \frac{a_{max}}{a_{min}} f(\mathbf{x}, e-1), & \text{if } u(\mathbf{x}) \leq z \text{ and } a(e-1) = a_{min} \end{cases} \tag{10.31}$$

In this way, we can write:

$$\mathbf{f}(e) = \mathbf{B}(e)\mathbf{f}(e-1) \tag{10.32}$$

where $\mathbf{B}(e) = \text{diag}(\mathbf{b}(e))$ is a diagonal matrix. The i-th element of $\mathbf{b}(e)$, for $e > 1$, is given by:

$$b_i(e) = \frac{1}{b_i(e-1)} \tag{10.33}$$

where:

$$b_i(1) = \begin{cases} \frac{a_{min}}{a_{max}}, & \text{if } u(\mathbf{x}_i) \leq z \\ 1, & \text{if } u(\mathbf{x}_i) > z \end{cases} \tag{10.34}$$

From Eq. (10.32), one can observe that the Dynamic Environment Generator Based on Problem Difficulty generates deterministic homogeneous linear DOPs (Definition 10.9) with diagonal matrices, where the parameter $a(e)$ controls the changing in the problem (i.e., it is the control system parameter).

Figure 10.3 shows six change cycles of a simulation of a DOP created by the generator proposed in [24] for $\tau = 110$ and $a_{min} = 0.78$ or $a_{min} = 0.94$. In the simulations presented in this section, $l = 8$, $c = 1.0$, $a_{max} = 1.5$ and $z = 5$ (see Eq. (10.29)). While $a_{min} = 0.78$ creates environments that are non-deceptive on average, $a_{min} = 0.94$ creates environments that are deceptive on average (Eq. (10.28)). Figure 10.3 shows the mean fitness of the population during the simulation and the Euclidean distance between the population in the current generation and two metastable points (eigenvectors of $UF(e)$). The two metastable points correspond to locations in the population space where most individuals are at the optima positions $\mathbf{0}$ and $\mathbf{1}$. From previous equations, we can observe that $f(\mathbf{1}) = c$ and $f(\mathbf{0})$ is switched in consecutive change cycles between a_{min} and a_{max}.

It can be observed that for the environments that are non-deceptive on average ($a_{min} = 0.78$), the population goes to the neighborhood of the metastable states where most of the population are at the local optima and, after some generations, goes to the main metastable state. However, in the change cycles where the environments are deceptive on average ($a_{min} = 0.94$), the population vector remains in the metastable state where most of the population are at the local optima, except for the first change cycle ($e = 1$). Such behavior can be observed in Fig. 10.4, where the normalized fitness and the distance to the current main metastable state for different values of τ and a_{min} are plotted. From Eq. (10.28), the minimum value where the

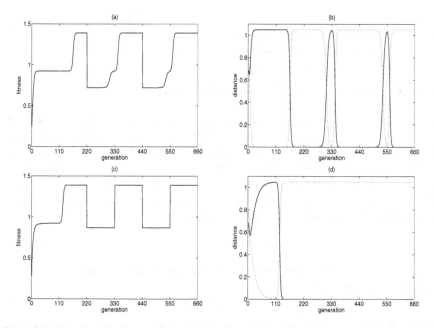

Fig. 10.3 Trajectory of the population during six change cycles with $\tau = 110$ in the DOP created by the generator proposed in [24]: mean fitness of the population for $a_{min} = 0.78$ (a) and $a_{min} = 0.94$ (c); distance to the current metastable states where most of the population are located at the optima positions **0** (dotted) and **1** (solid) for $a_{min} = 0.78$ (b) and $a_{min} = 0.94$ (d)

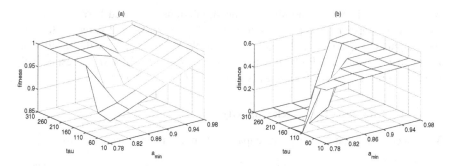

Fig. 10.4 Normalized fitness (a) and distance to the current main metastable state (b) in the DOP created by the generator proposed in [24] for different τ and a_{min}. The values are relative to the average (over 30 change cycles) obtained by the population vector in the generation before the change.

trap function is deceptive on average is $r = 0.9259$. We can observe that the population reaches the main metastable state for values of $a_{min} < r$ and large τ. However, a shorter distance to the main metastable state does not necessarily imply larger fitness for the population. One can observe that the fitness is higher for $a_{min} = 0.98$

when compared to the values obtained for $a_{min} = 0.94$. This fact is explained because the population vector remains closer to the metastable state where most of the population are in the local optima for $a_{min} = 0.98$.

10.4.3 The Dynamic 0-1 Knapsack Problem

In the 0-1 knapsack problem, the subset of items with the highest sum of profits, and in which total sum of weights is less than the knapsack capacity, should be found. In the dynamic version, the 0-1 knapsack problem is a DOP where the weights and profits of the items, the total number of items, and the knapsack capacity can change (see [15] for a dynamic benchmark problem based upon the 0-1 knapsack problem). Here, we will consider that the total number of items is static along the evolutionary optimization process and that the fitness is given by:

$$f(\mathbf{x},e) = c(\mathbf{x},e)v(\mathbf{x},e) \tag{10.35}$$

where $\mathbf{x} \in \{0,1\}^l$. The sum of the profits of the items in the knapsack in change cycle e is computed by:

$$v(\mathbf{x},e) = \mathbf{p}^{\mathrm{T}}(e)\mathbf{x}, \tag{10.36}$$

$\mathbf{p}(e) \in \mathbb{R}^{+l}$ is the vector with the profits of all items in change cycle e, and the constraint $c(\mathbf{x},e)$ is given by:

$$c(\mathbf{x},e) = \begin{cases} E, & \text{if } \mathbf{w}^{\mathrm{T}}(e)\mathbf{x} > C(e) \\ 1, & \text{otherwise} \end{cases} \tag{10.37}$$

where $E \in \mathbb{R}^+$ is a small number (in the simulations presented here, $E = 10^{-5}$), $C(e)$ is the knapsack capacity in change cycle e, and $\mathbf{w}(e) \in \mathbb{R}^{+l}$ is the vector with the weights of all items in change cycle e. In the following, we will consider the Dynamic 0-1 Knapsack Problem with changes in $C(e)$, $\mathbf{p}(e)$, or $\mathbf{w}(e)$, respectively (i.e., the control system vector $\phi(e)$ is formed by $C(e)$, $\mathbf{p}(e)$, or $\mathbf{w}(e)$).

10.4.3.1 Changing Knapsack Capacity

We can consider that the knapsack capacity change for $e > 1$ according to:

$$C(e) = C(e-1) + r(e), \tag{10.38}$$

where $\left(r(e) \in \mathbb{R} : r(e) + C(e-1) > 0\right)$ gives the change in the knapsack capacity in change cycle e, e.g., $r(e)$ can be obtained by a random sample taken from a normal distribution. Here, the weights and profits of the items remain stationary during the evolutionary optimization process. Thus, we can write:

$$c(\mathbf{x},e) = a\left(C(e-1),r(e),\mathbf{x}\right)c(\mathbf{x},e-1), \tag{10.39}$$

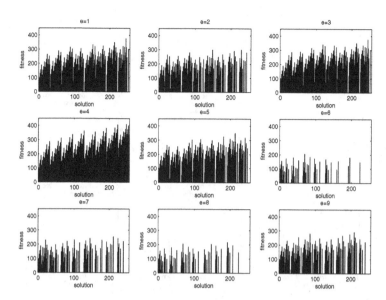

Fig. 10.5 Fitness for all possible solutions during nine change cycles in a simulation with $\tau = 100$ and $\rho = 0.2$ of the 0-1 Knapsack Problem with changing knapsack capacity

where:

$$a\big(C(e-1), r(e), \mathbf{x}\big) = \begin{cases} E, & \text{if } \mathbf{w}^{\mathrm{T}}\mathbf{x} \leq C(e-1) \text{ and } \mathbf{w}^{\mathrm{T}}\mathbf{x} > C(e-1) + r(e) \\ 1/E, & \text{if } \mathbf{w}^{\mathrm{T}}\mathbf{x} > C(e-1) \text{ and } \mathbf{w}^{\mathrm{T}}\mathbf{x} \leq C(e-1) + r(e) \\ 1, & \text{otherwise} \end{cases}$$

(10.40)

In this way, the fitness function in change cycle e is given by:

$$f(\mathbf{x}, e) = a\big(C(e-1), r(e), \mathbf{x}\big) c(\mathbf{x}, e-1) v(\mathbf{x}) = a\big(C(e-1), r(e), \mathbf{x}\big) f(\mathbf{x}, e-1)$$

(10.41)

and the fitness vector:

$$\mathbf{f}(e) = \mathbf{B}(e)\mathbf{f}(e-1)$$

(10.42)

where $\mathbf{B}(e) = \mathrm{diag}(\mathbf{b}(e))$ is a diagonal matrix. The i-th element of $\mathbf{b}(e)$ is given by:

$$b_i(e) = a\big(C(e-1), r(e), \mathbf{x}_i\big).$$

(10.43)

In this way, the 0-1 Knapsack Problem with changing knapsack capacity (defined by Eq. (10.38)) is a last environment dependent DOP (Definition 10.8) with fitness landscape changes ruled by Eq. (10.42).

Figures 10.5-10.7 show nine change cycles (all with duration given by τ) of a simulation of the 0-1 Knapsack Problem with changing knapsack capacity. In the simulations, the total number of items (l) is 8 and the weights and profits of each item are randomly generated from uniform distribution in the initial generation. The weights are real numbers generated in the range $[50, 150]$, while the profits

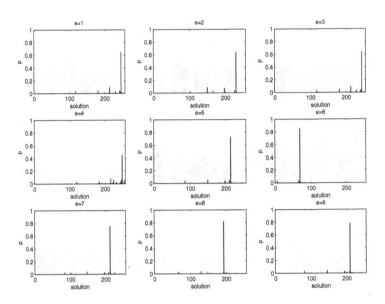

Fig. 10.6 Population vector in the last generation of each change cycle in a simulation with $\tau = 100$ and $\rho = 0.2$ of the 0-1 Knapsack Problem with changing knapsack capacity

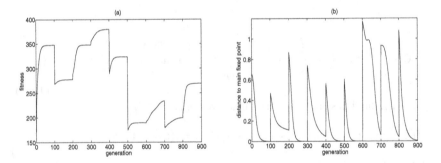

Fig. 10.7 Trajectory of the population during nine change cycles with $\tau = 100$ and $\rho = 0.2$ of the 0-1 Knapsack Problem with changing knapsack capacity: mean fitness of the population (a); distance to the current main metastable state (b)

are generated in the range $]0, 100]$. The initial knapsack capacity $(C(1))$ is equal to $0.7w_{total}$, where w_{total} is the sum of the weights of all items. The knapsack capacity changes in every change cycle according to a random deviation generated from a normal distribution with standard deviation equal to ρw_{total}. In this way, ρ controls the degree of change in the knapsack capacity, while τ controls the frequency of change.

Figure 10.5 shows the fitness of all solutions (**f**) in each change cycle. One can observe that the fitness of some solutions (including some optima) are drastically changed according to the value of C_e. Figure 10.6 shows the population vector (**p**)

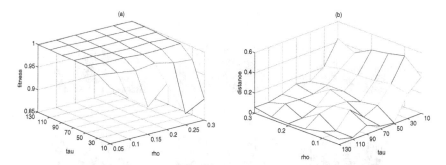

Fig. 10.8 Normalized Fitness (a) and distance to the current main metastable state (b) in the 0-1 Knapsack Problem with changing knapsack capacity for different τ and ρ. The values are relative to the average (over 30 change cycles) obtained by the population vector in the generation before the change.

in the end of each change cycle. It is possible to observe that, in this simulation, the global optima were found in every change cycle e. In the same simulation, Fig. 10.7 shows the mean fitness of the population and the Euclidean distance between the population in the current generation and the main metastable point (eigenvector of $UF(e)$ with the largest eigenvalue) in each change cycle.

Figure 10.8 shows the normalized fitness and the distance to the current main metastable state for different values of τ and ρ. We can observe that, despite some random fluctuation, the population reaches the main metastable state for values of $\tau > 10$.

10.4.3.2 Changing the Weights of the Items

In this case, only the weights of the items change during the evolutionary optimization process. We can consider that the weights of the items are modified for $e > 1$ according to:

$$\mathbf{w}(e) = \mathbf{w}(e-1) + \mathbf{r}(e), \tag{10.44}$$

where $\left(\mathbf{r}(e) \in \mathbb{R}^l : \mathbf{w}(e-1) + \mathbf{r}(e) > \mathbf{0}\right)$ gives the changes in the weights of the items in change cycle e. The deviation $\mathbf{r}(e)$ can be obtained by random samples $r_i(e)$ for $i = 1, \ldots, l$. If only n_c items are changed, then $l - n_c$ elements of $\mathbf{r}(e)$ are equal to zero, while the other n_c elements are randomly generated. Thus, we can write:

$$c(\mathbf{x}, e) = a\big(\mathbf{w}(e-1), \mathbf{r}(e), \mathbf{x}\big) c(\mathbf{x}, e-1), \tag{10.45}$$

where:

$$a\big(\mathbf{w}(e-1), \mathbf{r}(e), \mathbf{x}\big) = \begin{cases} E, & \text{if } \mathbf{w}^{\mathrm{T}}(e-1)\mathbf{x} \le C \text{ and } \big(\mathbf{w}(e-1)+\mathbf{r}(e)\big)^{\mathrm{T}}\mathbf{x} > C \\ 1/E, & \text{if } \mathbf{w}^{\mathrm{T}}(e-1)\mathbf{x} > C \text{ and } \big(\mathbf{w}(e-1)+\mathbf{r}(e)\big)^{\mathrm{T}}\mathbf{x} \le C \\ 1, & \text{otherwise} \end{cases}$$

$$\tag{10.46}$$

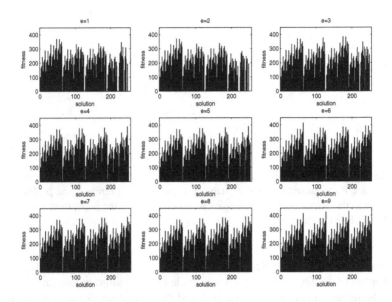

Fig. 10.9 Fitness for all possible solutions during nine change cycles in a simulation with $\tau = 100$ and $n_c = 3$ of the 0-1 Knapsack Problem with changing weights of the items

In this way, the fitness function in change cycle e is given by:

$$f(\mathbf{x}, e) = a\big(\mathbf{w}(e-1), \mathbf{r}(e), \mathbf{x}\big)c(\mathbf{x}, e-1)v(\mathbf{x}) = a\big(\mathbf{w}(e-1), \mathbf{r}(e), \mathbf{x}\big)f(\mathbf{x}, e-1) \tag{10.47}$$

and the fitness vector:

$$\mathbf{f}(e) = \mathbf{B}(e)\mathbf{f}(e-1) \tag{10.48}$$

where $\mathbf{B}(e) = \mathrm{diag}(\mathbf{b}(e))$ is a diagonal matrix. The i-th element of $\mathbf{b}(e)$ is given by:

$$b_i(e) = a\big(\mathbf{w}(e-1), \mathbf{r}(e), \mathbf{x}_i\big). \tag{10.49}$$

In this way, this 0-1 Knapsack Problem with changing weights of the items is a last environment dependent DOP (Definition 10.8) with fitness landscape changes ruled by Eq. (10.48).

Figures 10.9-10.11 show nine change cycles of a simulation of the 0-1 Knapsack Problem with changing weights of the items. The same parameters presented in last section were used. In each change cycle, the weights of n_c random items change by adding random samples taken from normal distribution with zero mean and standard deviation equal to 30 (Eq. (10.44)).

Figure 10.9 shows the fitness of all solutions ($\mathbf{f}$) in each change cycle. Figure 10.10 shows the population vector ($\mathbf{p}$) in the end of each change cycle. In the same simulation, Fig. 10.11 shows the mean fitness of the population and the Euclidean

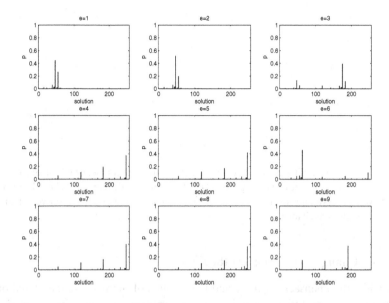

Fig. 10.10 Population vector in the last generation of each change cycle in a simulation with $\tau = 100$ and $n_c = 3$ of the 0-1 Knapsack Problem with changing weights of the items

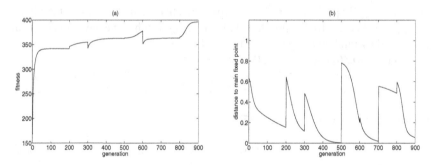

Fig. 10.11 Trajectory of the population during nine change cycles with $\tau = 100$ and $n_c = 3$ of the 0-1 Knapsack Problem with changing weights of the items: mean fitness of the population (a); distance to the current main metastable state (b)

distance between the population in the current generation and the main metastable point in each change cycle.

Figure 10.12 shows the normalized fitness and the distance to the current main metastable state for different values of τ and n_c. We can observe that, as for the problem with changing knapsack capacity, the population reaches the main metastable state for values of $\tau > 10$.

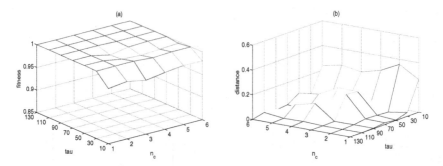

Fig. 10.12 Normalized Fitness (a) and distance to the current main metastable state (b) in the 0-1 Knapsack Problem with changing weights of the items for different τ and n_c. The values are relative to theaverage (over 30 change cycles) obtained by the population vector in the generation before the change.

10.4.3.3   Changing the Profits of the Items

Here, while the knapsack capacity and the weights of the items remain fixed during the evolutionary optimization process, the profits of the items change for $e > 1$ according to:

$$\mathbf{p}(e) = \mathbf{p}(e-1) + \mathbf{r}(e), \tag{10.50}$$

where $\left(\mathbf{r}(e) \in \mathbb{R}^l : \mathbf{p}(e-1) + \mathbf{r}(e) > \mathbf{0}\right)$ gives the changes in the profits of the items in change cycle e. As for the problem with changing weights, the deviation $\mathbf{r}(e)$ can be obtained by random samples $r_i(e)$ for $i = 1, \ldots, l$.

From Eq. (10.36) and Eq. (10.50), we can write:

$$v(\mathbf{x}, e) = \left(\mathbf{p}(e-1) + \mathbf{r}(e)\right)^{\mathrm{T}}\mathbf{x} = v(\mathbf{x}, e-1) + \mathbf{r}^{\mathrm{T}}(e)\mathbf{x}. \tag{10.51}$$

Substituting Eq. (10.51) in Eq. (10.35):

$$f(\mathbf{x}, e) = f(\mathbf{x}, e-1) + c(\mathbf{x})\mathbf{r}^{\mathrm{T}}(e)\mathbf{x}. \tag{10.52}$$

The fitness vector can be written as:

$$\mathbf{f}(e) = \mathbf{f}(e-1) + \mathbf{b}(\mathbf{p}(e-1), \mathbf{r}(e)), \tag{10.53}$$

where the i-th element of $\mathbf{b}(\mathbf{p}(e-1), \mathbf{r}(e))$ is equal to $c(\mathbf{x}_i)\mathbf{r}^{\mathrm{T}}(e)\mathbf{x}_i$.

One can observe that the 0-1 Knapsack Problem with changing profits of the items is a last environment dependent DOP (Definition 10.8) with fitness landscape changes ruled by Eq. (10.53). When only $n_c < l$ items are changed, the percentage of the fitness space (fitness vector) that are modified is equal to $(1 - 2^{-n_c})100\%$.

Figures 10.13-10.15 show nine change cycles of a simulation of the 0-1 Knapsack Problem with changing profits of the items. The same parameters presented in last section were used. In each change cycle, the profits of n_c random items changes by

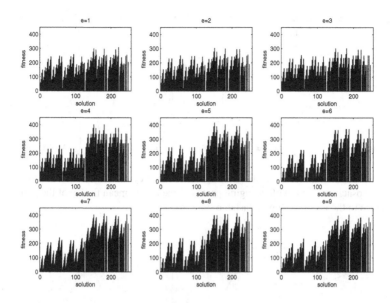

Fig. 10.13 Fitness for all possible solutions during nine change cycles in a simulation with $\tau = 100$ and $n_c = 3$ of the 0-1 Knapsack Problem with changing profits of the items

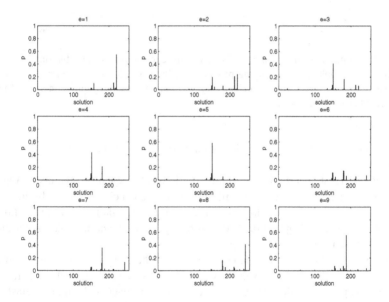

Fig. 10.14 Population vector in the last generation of each change cycle in a simulation with $\tau = 100$ and $n_c = 3$ of the 0-1 Knapsack Problem with changing profits of the items

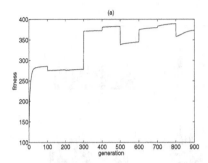
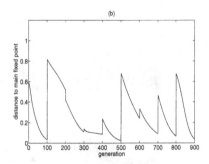

Fig. 10.15 Trajectory of the population during nine change cycles with $\tau = 100$ and $n_c = 3$ of the 0-1 Knapsack Problem with changing profits of the items: mean fitness of the population (a); distance to the current main metastable state (b)

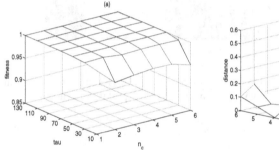
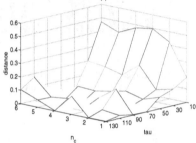

Fig. 10.16 Normalized Fitness (a) and distance to the current main metastable state (b) in the 0-1 Knapsack Problem with changing profits of the items for different τ and n_c. The values are relative to the average (over 30 change cycles) obtained by the population vector in the generation before the change.

adding random samples (Eq. (10.50)) with zero mean and standard deviation equal to 30.

Figure 10.13 shows the fitness of all solutions (**f**) in each change cycle. Figure 10.14 shows the population vector (**p**) in the end of each change cycle. In the same simulation, Fig. 10.15 shows the mean fitness of the population and the Euclidean distance between the population in the current generation and the main metastable point in each change cycle.

Figure 10.16 shows the normalized fitness and the distance to the current main metastable state for different values of τ and n_c. We can observe that, as for the previous knapsack problem instances, the population reaches the main metastable state for values of $\tau > 10$.

10.5   Conclusion and Future Work

In this chapter, dynamic changes and DOPs are defined based on the dynamical system approach (or exact model) of GAs [22]. Some classes of DOPs with fitness changes are also defined. Such definitions, and others that can be defined based on the same approach, can be useful to classify real-world DOPs and, as a consequence, to allow a systematic analysis of such problems based on the properties of each class. The performance of EAs for DOPs is, generally, related in problems of the same class. For example, if an algorithm has a good performance in problems of a class of DOPs (e.g., periodic DOPs), it can be tested in a new DOP identified as belonging to the same class. New DOP generators can be developed to explore the properties of such DOPs, e.g., generators for linear DOPs. However, it is important that the proposed classes of DOPs reproduce properties of real-world DOPs, which should be identified in a systematic approach by the study of real-world DOPs.

It can be observed that algorithms exploring the properties described on the analysis of the DOPs can be proposed. However, it is not clear if such algorithms are useful in real-world DOPs. To answer this question, the class of the real world DOP should be identified and similarities with well-known DOPs, if they exist, should be described. In this way, a very relevant future investigation is to analyze real-world DOPs, to classify them according to their properties, and to develop DOPs generators based on the identified class of DOPs. New forms of detecting changes and measuring the performance of different algorithms can still be investigated based on the analysis presented here. Another relevant future work is to analyze algorithms proposed for DOPs, e.g., GA with hypermutation and GA with random immigrants, according to approach presented here.

The theoretical analysis of EAs in DOPs is a relatively recent topic in Evolutionary Computation. Works in this area can significantly help researchers to develop and analyse new algorithms and real-world problems. Like for EAs in stationary problems, the theoretical works are, in general, based on different approaches, that are not necessary independent. A unified view combining different theoretical approaches, like those presented in this book, is a natural step to understand the performance of EAs in DOPs in near future.

Acknowledgements. This work was supported by CNPq and FAPESP of Brazil under Grant 2010/09273-1, and by the Engineering and Physical Sciences Research Council of U.K. under Grant EP/E060722/1 and Grant EP/E060722/2.

References

[1] Arnold, D., Beyer, H.-G.: Random dynamics optimum tracking with evolution strategies. In: Guervós, J.J.M., Adamidis, P.A., Beyer, H.-G., Fernández-Villacañas, J.-L., Schwefel, H.-P. (eds.) PPSN 2002. LNCS, vol. 2439, pp. 3–12. Springer, Heidelberg (2002)

[2]   Branke, J.: Evolutionary Optimization in Dynamic Environments. Kluwer Academic Publishers, Dordrecht (2001)

[3]   Cob, H.G., Grefenstette, J.J.: Genetic algorithms for tracking changing environments. In: Forrest, S. (ed.) Proc. 5th Int. Conf. on Genetic Algorithms, pp. 523–530 (1993)

[4]   Cruz, C., González, J., Pelta, D.: Optimization in dynamic environments: a survey on problems, methods and measures. Soft Comput. 15(7) (2011) 1432–7643

[5]   Deb, K., Goldberg, D.E.: Analyzing deception in trap functions. In: Foundation of Genetic Algorithms 2, pp. 93–108 (1993)

[6]   Droste, S.: Analysis of the (1+ 1) EA for a dynamically changing onemax-variant. In: Proc. 2002 IEEE Congr. Evol. Comput., pp. 55–60 (2002)

[7]   Eigen, M., McCaskill, J., Schuster, P.: Molecular quasi-species. J. of Physical Chemistry 92(24), 6881–6891 (1988)

[8]   Hayes, C., Gedeon, T.: Hyperbolicity of the fixed point set for the simple genetic algorithm. Theoretical Computer Science 411(25), 2368–2383 (2010)

[9]   Jin, Y., Branke, J.: Evolutionary optimization in uncertain environments – a survey. IEEE Trans. Evol. Comput. 9(3), 303–317 (2005)

[10] Morrison, R.W.: Designing evolutionary algorithms for dynamic environments. Springer-Verlag New York Inc. (2004)

[11] Nguyen, T.T., Yang, S., Branke, J.: Evolutionary dynamic optimization: A survey of the state of the art. Swarm and Evol. Comput. 6, 1–24 (2012)

[12] Noble, B., Daniel, J.W.: Applied Linear Algebra. Prentice-Hall (1977)

[13] Reeves, C.R., Rowe, J.E.: Genetic Algorithms - Principles and Perspectives: A Guide to GA Theory. Kluwer Academic Publishers (2003)

[14] Rohlfshagen, P., Lehre, P.K., Yao, X.: Dynamic evolutionary optimization: an analysis of frequency and magnitude of change. In: Proc. 2009 Genetic and Evol. Comput. Conf., pp. 1713–1720 (2009)

[15] Rohlfshagen, P., Yao, X.: The dynamic knapsack problem revisited: A new benchmark problem for dynamic combinatorial optimisation. In: Giacobini, M., et al. (eds.) EvoWorkshops 2009. LNCS, vol. 5484, pp. 745–754. Springer, Heidelberg (2009)

[16] Rowe, J.E.: Finding attractors for periodic fitness functions. In: Proc. 1999 Genetic and Evol. Comput. Conf., pp. 557–563 (1999)

[17] Rowe, J.E.: Cyclic attractors and quasispecies adaptability. In: Kellel, L., Naudts, B., Rogers, A. (eds.) Theoretical Aspects of Evolutionary Computing, pp. 251–259. Springer (2001)

[18] Ronnewinkel, C., Wilke, C.O., Martinetz, T.: Genetic algorithms in time-dependent environments. In: Kellel, L., Naudts, B., Rogers, A. (eds.) Theoretical Aspects of Evolutionary Computing, pp. 263–288. Springer (2001)

[19] Tinós, R., Yang, S.: Continuous dynamic problem generators for evolutionary algorithms. In: Proc. 2007 IEEE Congr. Evol. Comput., pp. 236–243 (2007)

[20] Tinós, R., Yang, S.: An analysis of the xor dynamic problem generator based on the dynamical system. In: Schaefer, R., Cotta, C., Kołodziej, J., Rudolph, G. (eds.) PPSN XI, Part I. LNCS, vol. 6238, pp. 274–283. Springer, Heidelberg (2010)

[21] Van Nimwegen, E., Crutchfield, J.P., Mitchell, M.: Finite populations induce metastability in evolutionary search. Physics Letters A 229(3), 144–150 (1997)

[22] Vose, M.D.: The Simple Genetic Algorithm: Foundations and Theory. The MIT Press (1999)

[23] Yang, S.: Non-stationary problem optimization using the primal-dual genetic algorithm. In: Proc. 2003 IEEE Congr. Evol. Comput., vol. 3, pp. 2246–2253 (2003)

[24] Yang, S.: Constructing dynamic test environments for genetic algorithms based on problem difficulty. In: Proc. 2004 IEEE Congr. Evol. Comput., vol. 2, pp. 1262–1269 (2004)

[25] Yang, S., Jiang, Y., Nguyen, T.T.: Metaheuristics for dynamic combinatorial optimization problems. IMA J. of Management Mathematics (2012),
doi:10.1093/imaman/DPS021

[26] Yang, S., Tinós, R.: A hybrid immigrants scheme for genetic algorithms in dynamic environments. Int. J. of Autom. and Comput. 4(3), 243–254 (2007)

[27] Yang, S., Yao, X.: Experimental study on population-based incremental learning algorithms for dynamic optimization problems. Soft Comput. 9(11), 815–834 (2005)

Chapter 11
Dynamic Fitness Landscape Analysis

Hendrik Richter

Abstract. Solving optimization problems with time varying objective functions by methods of evolutionary computation can be grounded on the theoretical framework of dynamic fitness landscapes. In this chapter, we define such dynamic fitness landscapes and discuss their properties. To this end, analyzing tools for measuring topological and dynamical landscape properties are studied. Based on these landscape measures we obtain an approach for drawing conclusion regarding characteristic features of a given optimization problem. This may allow to address the question of how difficult the problem is for an evolutionary search, and what type of algorithm is most likely to solve it successfully. The methodology is illustrated using a well–known example, the moving peaks.

11.1 Introduction

Evolutionary algorithms obtain their considerable problem–solving powers by connecting rather simple elements. They integrate parallelization due to the intrinsic population-based approach with an intensive dependence on random and the corrective guidance of evaluating and utilizing fitness. These simple elements finally come into effect by repeating them for a large number of instances. For the simplicity of its ingredients, an evolutionary algorithm could, at least in principle, be executed by pencil-and-paper. However, any of its practical applications are unthinkable without using a digital computer for calculation and thereby carrying out a numerical experiment. In fact, every numerical calculation using a computer can be seen as constituting an experiment. In conclusion, studies in evolutionary computation are frequently, and if so massively, driven by numerical experiments. For

Hendrik Richter
Department of Measurement Technology and Control Engineering,
Faculty of Electrical Engineering and Information Technology, HTWK Leipzig
University of Applied Sciences, D–04251 Leipzig, Germany
e-mail: richter@eit.htwk-leipzig.de

S. Yang and X. Yao (Eds.): *Evolutionary Computation for DOPs*, SCI 490, pp. 269–297.
DOI: 10.1007/978-3-642-38416-5_11   © Springer-Verlag Berlin Heidelberg 2013

instance, if a specific problem needs to be solved and an evolutionary approach is used to do so, or if a specific variant of an evolutionary algorithm is improved, and the improvement is tested and verified using different problem instances. As evolutionary computation is so clearly experiment-centered, a complementary theoretical approach is both important and useful to guide the experiments and to offer possible explanations of their outcome.

A theoretical understanding in evolutionary computation can come from to two starting points: the optimization problems to be solved or the class of evolutionary algorithms considered. The latter approach frequently leads to a theoretical modeling of the algorithm, for instance to mathematical descriptions using the framework of Markov chains, infinite population models, or ideas based on thermodynamics, statistical mechanics, or population dynamics, see for instance other chapters of this book. These mathematical models are subsequently employed to study questions such as the convergence and run-time behavior, or more generally to help the designer of evolutionary algorithms in employing appropriate parameters and operators and in constructing useful modifications. Ideally this lead to mathematical proofs for the studied topic.

A second starting point for an theoretical outlook is the problem to be solved by the evolutionary algorithm and the related question of what behavior and performance is likely to occur in solving. It explicitly includes to offer clues for the working principles and working modes of evolutionary algorithms. A key element in such a problem-driven theoretical approach is the concept of fitness landscapes. Fitness landscapes were introduced in theoretical biology to explain the effects that variations of the genetic make-up of individuals or species have on survival or perish [62]. They were later adopted by computer science to have a conceptual framework for posing questions such as how difficult is a specific optimization problem for an evolutionary algorithm, how is the behavior of the algorithm related to properties of the optimization problem, or ideally what performance of the algorithm can be expected [11, 14, 22, 23, 27, 51, 59]. The individuals of the evolutionary algorithm can be regarded as inhabiting the fitness landscape, and moving in it in order to find high–fitness regions, thereby solving the posed optimization problem. So, the explanatory power of a fitness landscape approach is mainly the result of giving a framework for studying how the population of the evolutionary algorithm interacts with the fitness landscape and hence with the optimization problem. Based on studying this interaction we may conclude what likely evolutionary paths for the individuals are and hence how probable in a given frame of time it is that high-fitness regions that may present solutions of the optimization problem are found.

The large majority of works on fitness landscapes has been done for the static case, that is for landscapes that do not change their topology during the run time of the evolutionary search. As solving dynamic optimization problems became more and more important [6, 8, 17, 28, 64], it appeared desirable to advance the treatment toward dynamic fitness landscapes. So far, this topic is rather underrepresented in the literature on evolutionary computation; some primary works are [10, 15, 38, 41]. This chapter also intends to summarizes these findings and serve as a starting point for further research. Therefore, in the next section we define dynamic fitness landscapes and discuss their properties. Dynamic fitness landscapes have topological and dynamical features that interdependent. So, we are particularly interested how the topological features (that can be understood as static in itself) interact with the dynamics. As a result of this interaction, dynamic fitness landscapes express spatio-temporal dynamics and should be regarded as spatial extended dynamical systems. Following this discussion, we study topological and dynamical landscape measures usable to characterize fitness landscapes and so the corresponding optimization problem. These measures are illustrated by numerical experiments using a well-known example, the moving peaks.

11.2 Dynamic Fitness Landscapes: Definitions and Properties

11.2.1 Introductory Example: The Moving Peaks

Before we come to defining dynamic fitness landscapes and to studying their properties, we start with recalling a well-known example, the so-called moving peaks benchmark [5, 29]. It uses as fitness function a static n-dimensional field of peaks (or cones) on a zero plane. The peaks have randomly chosen heights and slopes and are distributed across the landscape. So, we get

$$f(x) = \max\left\{ 0, \max_{1 \leq i \leq N} [h_i - s_i \|x - c_i\|] \right\}, \tag{11.1}$$

where $x \in S$ is an element in search space $S \subset \mathbb{R}^n$, N is the number of peaks (or cones) in the landscape, c_i are the coordinates of the i-th cone, and h_i, s_i specify its height and slope, see Fig. 11.1 for typical landscapes in $\mathbb{R}^2$.

This fitness function is tunable in several ways, which allows to specify the degree of hardness that the associated optimization problem poses. Apart from the dimensionality and the size (extension) of the search space, general factors contributing to the problem hardness of Eq. (11.1) are all of its defining elements: the number of cones N, their distribution in the search space (and hence the coordinates) and the slopes and heights (or more precisely the slope/height ratio). Intuitively obvious is that for a given state space dimension, an evolutionary algorithm has more difficulties in finding the maximum value if the number of cones N increases. At the same time it is also clear that this relationship cannot hold for an arbitrary number of cones. If the search space extension is limited, then a further increasing of an already large number most likely leads to cones starting to nest in each other. So, smaller cones are hidden by larger ones and do not continue to effect the search process. For instance, all landscapes in Fig. 11.1 are for randomly chosen cones, where c_i is normally distributed, and h_i and s_i are uniformly distributed, and there are always $N = 7$ peaks on a search space with limits $x_{min} = -5$, $x_{max} = 5$ (maximal extension $E_{max} = \|(x_{max}) - x_{min}\| = 14.14$). To have all seven peaks clearly appearing is a rare event if the c_i, h_i, s_i are numerically generated as realizations of a random process, even for such a small number of cones and dimension. In Fig. 11.1a we can distinguish 5 peaks, in Fig. 11.1b only 2 peaks. The reason for this

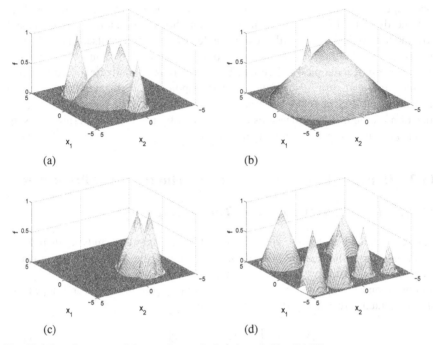

Fig. 11.1 Four instances of the moving peaks benchmark (Eq. (11.1))

behavior is that larger cones with small slopes cover smaller ones with larger slopes. If a random selection of h_i, s_i leads to certain $h_i - s_i$-combinations this effect becomes more or less prominent. This is also the reason why the $h_i - s_i$-ratio is more important for the landscape's appearance that the absolute values. Also note that for a small number of realizations of a random variable, the statistical properties of the sample might (possibly largely) differ from the statistical properties of the underlying distribution. This effect clearly depends on the random number generator used, but is always present to a larger or smaller degree. To summarize, this means that for a formally same landscape (same dimensionality, same maximal extension, same number of peak), a very different problem hardness for a potential evolutionary algorithm used to solve the problem is obtained. This somehow questions the comparability of seemingly very similar fitness landscapes. We will formalize this discussion in the Sec. 11.3 where measures for fitness landscapes are discussed.

Apart from the number of cones in the landscape, another aspect of problem hardness is how the cones are distributed. They can rather group, such as in Fig. 11.1c, or rather being scattered, as in Fig. 11.1d. In particular if the optima are scattered and separated by low-fitness or even by same-fitness areas, an evolutionary algorithm usually has problems to operate successfully. This is because a low-fitness or same-fitness area is hard to pass for the individuals of an evolutionary algorithm as they prefer to follow increasing fitness values and hence there are usually difficulties to draw them to such regions. The most promising way to do so is to have a strong random component or more generally a prolonged exploration phase, for instance be a high mutation-rate and mutation-strength. However, this strong random drive is disadvantageous for finding the actual optima once the most promising region has been found. In other words, optimization problems with such a fitness landscape require a special setting of the balance between exploration and exploitation.

Up to now the problem hardness of an optimization problem is entirely due to the static appearance of the fitness landscape. To transform the static fitness function into a dynamic one, there is the need to change the static features of Eq. (11.1) with the run-time of the evolutionary algorithm. This means that N cones are to move in terms of coordinates, heights and slopes by using a discrete time variable k, for instance $k \in \mathbb{Z}$. By defining dynamic sequences for coordinates $c(k)$, heights $h(k)$ and slopes $s(k)$, a dynamic fitness landscape

$$f(x,k) = \max \left\{ 0, \max_{1 \leq i \leq N} [h_i(k) - s_i(k)\|x - c_i(k)\|] \right\} \qquad (11.2)$$

can be obtained. A 2D visualization can also be imagined from Fig. 11.1, where now the cones move across the landscape, changing their heights and slopes at the same time. From the picture possible problems in finding the optima get visible. Finding the maximum means finding the highest peak; so for the dynamic case this is tracking the highest peak.

11.2.2 Definition of Dynamic Fitness Landscapes

As with the introductory examples of the moving peaks, dynamic fitness landscapes are frequently constructed from their static counterpart. So, a static fitness landscape Λ_S can be defined by

$$\Lambda_S = (S, n, f), \qquad (11.3)$$

where S is the search space with elements $x \in S$, $n(x)$ is a neighborhood function which orders for every $x \in S$ a set of direct and possibly also more distant neighbors, and $f(x) : S \to \mathbb{R}$ is the fitness function giving every search space element a fitness value, e.g., see [19, 48]. The search space is either the product of a genotype–to–fitness mapping or constructed from encoding and distancing the set of all possible solutions of an optimization problem. Either way it is basically the representation that the evolutionary algorithm uses (for instance binary, integer, real, tree etc.) and the design of the genetic operators that defines the search space and also its neighborhood structure. This is in line with the general understanding that the concept of fitness landscapes is particularly useful for studying how the evolutionary search algorithm interrelates with the fitness function. Moreover, as shown in [18], the neighborhood structure of a fitness landscape may vary with variation of the genetic operators. Hence, an analysis of the fitness landscape can be helpful for designing genetic operators as it gives insight into which design is more likely than others to belong to the landscape easiest searchable in average [51]. If the representation is fixed, for instance as real numbers, then the search space frequently has a metric and the neighborhood structure is inherent. The exact design and the parameters of the genetic operators, for instance the mutation strength, defines which points can be reached in average from a given starting point in the landscape within one generation.

The geometrical interpretation that is central to the intuitive understanding of fitness landscapes is particularly visible if $S \subseteq \mathbb{R}^2$, see Fig. 11.1. It means that fitness can be viewed as elevation of the search space. In other words, fitness is a property belonging to the search space as its orthogonal projection. Therefore, search space points with high fitness appear as peaks, while low fitness regions are valleys, and points with the same fitness plateaus. Solving the corresponding optimization problem hence means to find the highest peak

$$f_S = \max_{x \in S} f(x), \qquad (11.4)$$

and its location $x_S = arg\ f_S$. The individuals of an evolutionary algorithm used to solve the given optimization problem can be thought of as to populate the fitness landscape. The design of the genetic operators (basically selection, recombination, and mutation) is meant to organize that they (at least in average) perform a climbing of the hills in the fitness landscape and ideally find the highest one, even in the presence of other (but smaller) hills. With the movement the individuals contribute to the dynamics of the population and hence to the evolutionary dynamics. Note that in analyzing these processes there can be made a distinction between the dynamics generated by genetic variation (mutation and recombination) and the dynamics generated by corrective guidance (selection), which is of considerable interest for fine–tuning the genetic operators.

For defining a dynamic fitness landscape there is the need to set how the element in Eq. (11.3) change over time. So, all of its three defining ingredients – search space S, fitness function $f(x)$, and neighborhood structure $n(x)$ – can, at least in principle, be dynamically changing. Hence, we additionally need for description a time set and mappings that tell how S, $f(x)$ and/or $n(x)$ evolve with time [37, 38, 41]. Dynamic optimization problems considered in the literature so far address all these possibilities of change to some extend. Whereas a real alteration of the fundamental components of a search space such as dimensionality or representation (binary, integer, discrete, real, etc.) is really rare, a change in the feasibility of individuals is another and less substantial kind of a dynamic search space and is discussed within the problem setting of dynamic constraints [34, 42, 43]. The works on dynamic routing can partially be interpreted as a changing neighborhood structures [4, 54], while most of the work so far has been devoted to time-dependent fitness functions [26, 30, 36, 39, 40, 45, 53, 63]. For these reasons a dynamically changing search space and neighborhood structure is omitted in the following definition, while the fitness function is time–dependent.

We can next define a dynamic fitness landscape as

$$\Lambda_D = (S, n, K, F, \phi_f). \qquad (11.5)$$

Equivalent to the static landscape (Eq. (11.3)) the search space S represents all possible solutions of the optimization problem and the neighborhood function $n(x)$ gives a set of neighbors to every search space point. The time set $K \subseteq \mathbb{Z}$ provides a scale for measuring and ordering dynamic changes; F is the set of all fitness functions in time $k \in K$ and so every $f \in F$ with $f : S \times K \to \mathbb{R}$ also depends on time and gives fitness values to every search space point for any $k \in K$. The transition map $\phi_f : F \times S \times K \to F$ defines how the fitness function changes over time. It must satisfy the temporal identity and composition conditions, that is $\phi_f(f,x,0) = f(x,0)$ and $\phi_f(f,x,k_1 + k_2) = \phi_f(\phi_f(f,x,k_1),x,k_2), \forall f \in F, \forall x \in S, \forall k_1,k_2 \in K$ and the spatial boundary conditions $\phi_f(f,x_{bound},k) = f(x_{bound},k), \forall f \in F, \forall k \in K$ and x_{bound} being the boundary set of search space S. With these definitions we assume that the changes in the fitness landscape happen (or come into effect) at discrete points in time and are the result of comparing the landscape for points in time k to the following points $k + 1$. This is in line with fitness landscapes being a tool for analyzing the behavior of evolutionary algorithms. The population dynamics of evolutionary algorithms develops along discrete generations τ. A generation of an evolutionary algorithm can be defined as the time interval between subsequent fitness function evaluations of the whole population. In other words, a generation indicates the time between serial and self–contained steps in the solution finding process. As fitness evaluation in an evolutionary algorithm usually takes place just once in a generation, a difference in fitness can only be noticed by the algorithm as discrete points in time. Hence, if we model the changes by a fitness landscape, the most natural and straightforward time regime is discrete time.

The intuitive geometrical interpretation of a static fitness landscape as pointed out above still holds to some extend for the dynamic case. The main difference is that the hills and valleys move within the search space and/or change their topological form. This includes that hills grow and shrink, valleys deepen or flatten, or the landscape completely or partially turns inside out. The corresponding dynamic optimization problem now reads

$$f_S(k) = \max_{x \in S} f(x,k), \qquad \forall k \geq 0, \qquad (11.6)$$

which yields the temporarily highest fitness $f_S(k)$ and its solution trajectory

$$x_S(k) = arg\, f_S(k), \qquad \forall k \geq 0. \qquad (11.7)$$

As before the individuals of the evolutionary algorithm are meant to climb the hills, and moreover to follow if they are moving and find hills that dynamically appear. Even from such a simplifying picture it becomes obvious that the standard genetic operators (selection, recombination, mutation) might not be sufficient to perform the task. Indeed, there exists a multitude of modification to deal with the changes induced by a dynamic fitness landscape, which are discussed in other chapters of this book. With Eq. (11.6) and Eq. (11.7) we have a description of the dynamic optimization problem and its solutions. However, for evaluating the performance and the behavior of an evolutionary algorithm used to deliver such solutions, other quantities can be more interesting, meaningful and significant. These quantities usually generalize the notion (Eq. (11.6)) and (Eq. (11.7)) over the run time and/or runs of the evolutionary algorithm, may include data from the evolving population's fitness and make them statistically evaluable, see [1, 58] and references cited there for an overview.

11.2.3 Dynamics and Fitness Landscapes

Dynamic fitness landscapes differ from static ones in that their topological features changes with time. The changes relevant for discussions in evolutionary computing must be on a time scale comparable to the generations (or iterations) of the search algorithm. If the dynamics is very fast, for instance several changes of the landscape within one generation, then the evolutionary algorithm would have to solve a different optimization problem every generation. As a fundamental working principle of evolutionary approaches is parallelized population-based, fitness-and-random driven search with improvements over a certain number of generations, an one-generation optimum finding seems unlikely and in every sense contradicts those basic ideas. The only context where such a scenario seems to be fitting is if the changes are very light (that is have s small dynamic severity, see Sec. 11.3.2) and hence the evolutionary search is only lightly interrupted by the dynamics of the fitness landscape. On the other hand, if the changes are slow (depending on problem hardness and dimension a general number is slower that every 50–100 generations), then the problem is no longer to be considered a dynamic one, but a series of static problems, where one solution instance has no influence on the next.

For describing static fitness landscapes algebraic equations can be used, see Eq. (11.3). For dynamic fitness landscapes we need to add a mathematical prescription for evolving the fitness values forward in time. In principle, there are two ways for doing so. A first is to take the algebraic description of a static fitness function, select some terms from these equations, and define how these terms depend on an additional (and usually external defined) time regime. Geometrically speaking, this means that we define dynamics laws for how selected topological features in the landscape evolve with time. This only implicitly results in explaining the landscape's dynamics in the whole. On the other hand, this also implies that we in fact have dynamics only for certain elements in the landscape, for instance in the moving peaks benchmark (Eq. (11.2)) the peaks can change by the coordinates $c_i(k)$, heights $h_i(k)$ and slopes $s_i(k)$ of the cones. By changing these elements dynamically, we implicitly also describe how the neighborhood of the peaks behave with time. We will call this kind of dynamics generation external.

In a second approach for describing the dynamic changes we may formulate a general law for the fitness landscape's time evolution that applies for all fitness values in the search landscape. In it the fitness of every point $f(x, k+1)$ may depend on the fitness one time step before, $f(x, k)$ and the (element–wise) fitness values of all of its neighbors, $f(n(x), k)$. So, we can write

$$f(x, k+1) = \psi(f(x, k), f(n(x), k)). \tag{11.8}$$

Table 11.1 Hierarchy of fitness landscapes; S static, D discrete, C continuous

Class	Space	Time	Possible model	Dynamics
1	D	S	Discrete fitness function	no
2	C	S	Continuous fitness function	no
3	D	D	Discrete fitness function with discrete dynamics	internal/external
4	C	D	Continuous fitness function with discrete dynamics	external
5	D	C	Lattice of coupled ordinary differential equation (ODE)	internal
6	C	C	Partial differential equation (PDE)	internal

With a description (Eq. (11.8)), we have formulated a dynamic fitness landscape as spatially extended dynamical system. To have such description means that the topology and the dynamics of the fitness landscape are generated simultaneously and by the same equation. In other words, the dynamics is here internal to the fitness landscapes. An example for such an evolution law are fitness landscapes constructed from Ordinary Differential Equations (ODE), Partial Differential Equations (PDE) and Coupled Map Lattices (CML). A special property of this type of dynamic fitness landscape is that not only time is discrete but also the search space has a countable number of elements. Such a search space characteristic corresponds strictly speaking to combinatorial optimization problems, see [38, 41] for further discussion.

The distinction between external and internal dynamics has, apart from characterizing the source of changes in the fitness landscape, implications for the specification of the dynamics of neighboring points in the search space. For internal dynamics, the landscape's time evolution law (Eq. (11.8)) states how the fitness of all points in the search space and all their neighboring points change with time. In other words, the changes are explained for every search space point. With an external dynamics, we only define changes of selected and characteristic features in the landscape. Here, the changes are explicitly explained for only a discrete subset of search space points. Their neighborhood may change too, but according the to the same time regime as the points themselves.

To put these facts into a wider context there has been a attempt to draw a connection to spatially extended systems and to establish a hierarchy of fitness landscapes [41], which is based on a hierarchy of spatio-temporal dynamics [7, 20], see Table 11.1 . The hierarchy comes form the different combinations of discretizing space and time in the fitness landscape. For the landscape being static, the search space can be either continuous or discrete (binary), which results in the problem classes 1 and 2. If the discrete search space consists additionally of a finite number of elements, the corresponding optimization problem is a combinatorial one. The class 3 are discrete fitness functions with discrete dynamics. This dynamics can be external, as for instance in the XOR DOP generator [63], or dynamic combinatorial optimization problems such a dynamic knapsack, dynamic royal road or dynamic bit-matching [49]. An example for internal dynamics are the before mentioned fitness landscapes constructed from Coupled Map Lattices (CML). Continuous fitness functions with discrete dynamics form class 4. Examples with external dynamics include the moving peaks benchmark, but also other similar problems such a dynamic sphere, dynamic Ackley, dynamic Rosenbrook etc. To define an internal dynamics for class 4 problems would mean to have a description (Eq. (11.8)) for a non–countable number of points in the fitness landscape, which is impractical for any numerical calculation. Therefore, class 4 problems with internal dynamics do not play a role in the literature. The class 5 and 6 fitness landscapes have continuous time and discrete or continuous search spaces. Possible models for such dynamic fitness landscapes are lattices of ODEs or PDEs. Such models might be suitable for formulating general fitness landscapes and mainly found in modeling and studying fundamental properties of evolution [24, 31–33, 61], but rarely in studies in evolutionary computation (see e.g. [3] for such a work). The reason for this might be that, as mentioned before, the discrete population dynamics of an evolutionary algorithm is best linked to fitness landscapes with discrete time characteristics. In addition, both types of mathematical description do, at least not in general, have an analytic solution. So, any numerical calculation involves a discretization of time and/or space, for instance in numerical integrating ODEs and PDEs. So, both classes can, at least from a numerical point of view, be reduced to class 3 problems.

If a dynamic fitness landscape relies on external dynamics questions of how to generate dynamic sequences arise. A first step is to select terms in the algebraic fitness landscape description that are to change with time. For the moving peak (Eq. (11.2)), these are $c_i(k)$, $h_i(k)$ and $s_i(k)$. The dynamic changes are induced by moving sequences $x(k)$, that is

$$c_i(k) = x_{ci}(k), \quad h_i(k) = x_{hi}(k), \quad s_i(k) = x_{si}(k). \tag{11.9}$$

In principle, the dynamic changes can be of three types:

- regular dynamics
- chaotic dynamics
- random dynamics.

According to these types, the moving sequences $x(k)$ can be generated. Regular changes are usually obtained by analytic coordinate transformations, for instance cyclic dynamics where each $x(k)$ repeats itself after a certain period of time and shows recurrence or translatory dynamics where the quantities ascribe a pre–defined track or tour. The period of the recurrence, the cycle width, and the step–length are to be adjusted and normalized so that moving sequences become comparable. Chaotic changes can be generated by a discrete-time dynamical system $x(k+1) = g(x(k))$. Such systems are known to show chaotic behavior for certain parameter values and initial states $x(0)$, for instance the generalized Hénon map, see [35, 36] for details of the generation process. For using these moving sequences in numerical experiments, there might be the need for preprocessing as depending on the dynamical systems used the amplitudes $x(k+1)$ might be not unitary. If so, a re–normalization should take place. Random changes we get if we select that each $c_i(k)$, $h_i(k)$, $s_i(k)$ for each k is an independent realization of, for example, a normally or uniformly distributed random variable. Again, the statistical properties of the random variable should guarantee (maybe after renormalization) comparability. A general feature of the three types of dynamics is that regular dynamics is completely predictable, chaotic dynamics is short-term predictable, and random dynamics unpredictable.

11.3 Analysis Tools for Dynamic Fitness Landscapes

In the previous section, we have defined dynamic fitness landscapes. The main content of analyzing fitness landscapes is to have a notion of how the population of an evolutionary algorithm interrelates with properties of the optimization problem under study. In doing so, we may get information about how difficult a certain problem is for the search algorithm, or at least we may determine some characteristic features of the problems usable for recommending which type of algorithm is most likely to succeed in solving it. We can view a static fitness landscape as heights orthogonal to the search space. These heights form geometrical structures that can be thought of (in visualizable dimension) as hills, valleys, plateaus or ridges. If the fitness landscape gives indication for problem hardness, then this information should be found in properties of the geometrical structures. If the geometrical structures change with time, as in dynamical fitness landscapes, then properties of these changes should be related, too. For this reasons, we can distinguish two types of fitness landscape properties, which we will call *topological properties* and *dynamical properties*. For both types of properties, there exist analyzing tools. These analyzing tools account for so-called landscape measures.

Landscapes measures serve as quantitative evaluation as a visual inspection of a fitness landscape is only possible in a 2D search space (for an attempt to visualize higher-dimensional fitness landscapes, see [60]) and can be misleading even then. For static fitness landscapes, measures have been intensively studied and a large variety of different measures has been proposed and studied [11, 14, 19, 47, 48, 56]. All these measures of landscape analysis intend to catch properties that are related to static problem hardness. Such properties are basically the number and distribution of local optima, and the accessibility (or basin of attraction) of these optima. Considering the working principles of evolutionary algorithms, it appears easily understandable that the number of optima scales with the hardness for the algorithm to find the global one amongst them. The individuals of the evolutionary algorithm are initially set in the fitness landscape and are meant to move in it, ideally finding the global optima on their way. They move by using their fitness values and random as driving forces. In average, the movement means an increase in fitness. If there are several optima then there are around them locally increasing fitness values which may attract individuals. Therefore, the larger the number of optima, the more likely it is that individuals get (at least temporally) occupied on them. In short, the larger the number of optima, the more difficult it should be in average. Distribution of the optima plays a role insofar as optima widely scattered on an otherwise neutral (or almost neutral) landscape may be more challenging than optima grouped. Also, accessibility (that is the availability and easy localizability of an evolutionary path form the start population to the global optima) is important as neutrality [50] or long–path problems show [16]. To account for these fitness landscape features and their interdependence, there are landscape measures. The majority of these measure are analytical in that they calculate a quantity, usually a number, for the fitness landscape under study. There are also descriptive measures, as for instance GA–deceptiveness [9]. As pointed out before problem hardness cannot be attributed to a single quantity, for instance modality, but depends on how other factors such as distribution or accessibility contribute, modify, intensify or weaken the effect of the sheer number of optima. In the same way, no single landscape measure can be sufficient to express problem hardness, let alone for all types of fitness landscapes. As has been shown [2], a single measure can sometimes lead to an incorrect prediction of problem hardness, in particular if the global optimum is not known (as is usually the case with real–world problems). For a class of static fitness landscapes, the so–called combinatorial SAT problem, and a class of evolutionary algorithms, the $(1 + 1)$ evolution strategy, it has been even shown by proof that there can be no single measure that catches all aspects of problem difficulty and allows prediction of performance [11]. For these reasons it is sensible to have a variety of measures. An application of these topological measures to dynamic fitness landscapes has been suggested in [37, 38, 42], where the main focus was to analyze CML-based fitness landscapes, which are of class 3 with internal dynamics. In the following these measures are recalled and modifications to consider class 4 problems with external dynamics are given. So, these measures generalize the static ones by considering their time dependence.

11.3.1 Analysis of Topological Properties

11.3.1.1 Modality

A first and most straightforward measure is modality with counts the number of optima. Obviously this measure can only be easily calculated for constructed fitness landscapes for which an equation–like mathematical description is available. Therefore, this measure also serves as a quantity of comparison to other measures. On the other hand, it has been clearly shown that modality is an influential factor on problem hardness for a considerable number of fitness landscapes. A dynamic fitness landscape has a local optima at search space point x and time k if all points in its neighborhood do not exceed its fitness value

$$f(x,k) \geq f(n(x),k). \tag{11.10}$$

For the moving peaks benchmark (Eq. (11.2)) modality can be calculated by enumeration. At time k, the i-th peak with coordinates $c_i(k)$, slope $s_i(k)$ and height $h_i(k)$ is a maxima if

$$\max_{\substack{1 \leq j \leq N \\ j \neq i}} [h_j(k) - s_j(k)\|c_i(k) - c_j(k)\|] < h_i(k). \tag{11.11}$$

Note that not all peaks must be optima which is due to the effect of nesting, described in Sec. 11.2.1. We denote $\#_{LM}(k)$ the number of local maxima at time k, which is the number of cones for which condition (Eq. (11.11)) holds. In a dynamic fitness landscape, obviously, this number may vary with time and can be regarded as a time series. Therefore, we consider this quantity statistically and will analyze its time average $\#_{LM}$.

11.3.1.2   Ruggedness

As given above modality is a good indicator for the probability that an evolutionary search is easy or not. On the other hand, modality is impractical to calculate for a given fitness landscape. The essence of modality is the appearance of hills and valley, which means regions with fitness values strictly higher or lower than the surrounding ones. In other words, modality can be viewed as the degree of alteration between high and low fitness values in a given segment of the landscape, or as the question of how predictable it is if the fitness value of a neighboring point is higher or lower than the own. Consider for instance a single sphere that curves down uniformly into all search space directions, or alternatively a large collection of spiky peaks laying closely to each other. The former appears smooth and it is highly predictable from a given sequence of points and fitness values if the next point in the sequence will have a higher or lower value than the current one. The latter is rugged and a like–wise prediction much harder. This characteristics even applies for a superficially rugged landscape where repeatedly but predictable high fitness values follow low fitness values, but there is a high correlation if we are to move on this landscape. So, to have a quantitative measure, correlation between fitness values on a tour on the fitness landscape that can be used. Starting point for calculating a measure for ruggedness is a random walk on the fitness landscape. This random walk should have the length T and the step size t_S. So, we can obtain a random walk on the fitness landscape as

$$x(j+1) = x(j) + t_s \cdot \text{rand} \tag{11.12}$$

with rand being independent realizations of a random variable, usually normally or uniformly distributed. The random walk starts from a random position and must be corrected if it oversteps the bounds x_{bound} of the search space. In this case, we reset the walk to a randomly selected point within the search space. For the random walk (Eq. (11.12)) we can record the fitness value as time series:

$$f(j,k) = f(x(j),k)), j = 1,2,\ldots T. \tag{11.13}$$

From these fitness values we can calculate the spatial correlation which is widely used in determining ruggedness of static landscapes [14, 47, 59]. The spatial correlation $r(t_L, k)$ can be obtained from the autocorrelation function of the time series with time lag t_L, also called random walk correlation function:

$$r(t_L, k) = \frac{\sum\limits_{j=1}^{T-t_L} \left(f(j,k) - \bar{f}(k) \right) \left(f(j+t_L, k) - \bar{f}(k) \right)}{\sum\limits_{j=1}^{T} \left(f(j,k) - \bar{f}(k) \right)^2}, \tag{11.14}$$

where $\bar{f}(k) = \frac{1}{T} \sum_{j=1}^{T} f(j,k)$ and $T \gg t_L > 0$. The spatial random walk correlation function measures the correlation between different segments of the fitness landscape for a fixed k. As $r(t_L,k)$ changes over time, we may consider its time average $r(t_L)$. Next, the correlation of the lag t_L

$$\lambda_R(t_L) = -\frac{1}{\ln(|r(t_L)|)} \tag{11.15}$$

can be calculated. It has been shown that ruggedness is best expressed by the correlation length with lag $t_L = 1$, [47]:

$$\lambda_R = -\frac{1}{\ln(|r(1)|)}. \tag{11.16}$$

The lower the value of λ_R, the lower the correlation and therefore, the more rugged is the landscape.

11.3.1.3   Information Content

In contrast to the correlation based measure ruggedness, the information content accounts an entropic quantity [25, 52, 55, 56]. Starting point for this method to evaluate landscapes is again, as for the correlation structure considered above, a time series (Eq. (11.13)), $f(j,k)$, which is generated by a random walk on the dynamic landscape for a fixed time k. Taking this time series, we express the differences in fitness between two consecutive walking steps by a code using the symbols $s_j \in \mathbb{S}$, $j = 1,2,\ldots,T-1$, taken from the set $\mathbb{S} = \{-1,0,1\}$. These symbols are obtained by

$$s_j(e,k) = \begin{cases} -1, & \text{if } f(j+1,k) - f(j,k) < -e \\ 0, & \text{if } |f(j+1,k) - f(j,k)| \leq e \\ 1, & \text{if } f(j+1,k) - f(j,k) > e \end{cases} \tag{11.17}$$

for a fixed $e \in [0,L]$, where L is the maximum difference between two fitness values. The obtained symbols are concatenated to a string

$$S(e,k) = s_1 s_2 \ldots s_{T-1}. \tag{11.18}$$

The parameter e defines the sensitivity by which the string $S(e,k)$ accounts for differences in the fitness values. For $e = 0$, the string $S(e,k)$ contains the symbol $s_j = 0$ only if the random walk has reached an area with neutrality. It hence discriminates very sensitively between increasing and decreasing fitness values. On the other hand, for $e = L$, the string only contains the symbol $s_j = 0$, which makes evaluating the structure of the landscape futile. So, with a fixed value of e with $0 < e < L$, we define a level of detail for the information content of the landscape. The string (Eq. (11.18)) expresses this information depending on e and codes it by subblocks over the set $\mathbb{S}$. In other words, varying the sensitivity e allows to zoom in on or to zoom out of the information structure of the landscape.

For defining entropic measures of the landscape, we analyze the distribution of subblocks of length two, $s_j s_{j+1}$, $j = 1, 2, \ldots T - 2$, within the string (11.18). These subblocks express local patterns in the landscape. We denote the probability of the occurrence of the pattern $\delta_1 \delta_2$ with $\delta_1, \delta_2 \in \mathbb{S}$ and $\delta_1 \neq \delta_2$ by $p_{\delta_1 \delta_2}$. For numerical calculation, we approximate this probability by the relative frequency of the patterns within the string $S(e, k)$. As the set $\mathbb{S}$ consists of three elements, we find 6 different kinds of subblock $s_j s_{j+1} = \delta_1 \delta_2$ with $\delta_1 \neq \delta_2$ within the string. From their probabilities at a fixed time k and a given sensitivity level e we calculate the entropic measure [56]

$$h_{IC}(e, k) = - \sum_{\substack{\delta_1, \delta_2 \in \mathbb{S} \\ \delta_1 \neq \delta_2}} p_{\delta_1 \delta_2}(e, k) \log_6 \left(p_{\delta_1 \delta_2}(e, k) \right), \qquad (11.19)$$

which is called information content of the fitness landscape. Note that by taking the logarithm in Eq. (11.19) with the base 6, the information content is scaled to the interval $[0, 1]$. As for the other landscape measures, for evaluating dynamic fitness landscapes, we consider the time average $h'_{IC}(e)$.

11.3.1.4 Fitness Distance Correlation

With the landscape measure fitness distance correlation we determine how closely fitness values and distance to the nearest optimum are related. Again, we start with a random walk resulting in a time series (Eq. (11.13)), $f(j, k)$. Also, for the random walk (Eq. (11.12)), $x(j)$ we record the minimum distance $d_j(k) = \|x(j) - x_S(k)\|$ to the global optimum using the solution $x_S(k)$ of the dynamic optimization problem, see eq. (11.7). We then can calculate the quantity

$$\rho(k) = \frac{1}{\sigma_f \sigma_d T} \sum_{j=1}^{T} \left(f(j, k) - \bar{f}(k) \right) \left(d_j(k) - \bar{d}(k) \right) \qquad (11.20)$$

where $\bar{f}(k) = \frac{1}{T} \sum_{j=1}^{T} f(j, k)$, $\bar{d}(k) = \frac{1}{T} \sum_{j=1}^{T} d(k)$ and σ_f, σ_d are the standard deviations of $f(j, k)$ and $d(k)$, respectively. The problem should be easy if a decrease of distance to the optimum relates strongly to an increase of fitness. This may suggest that an easily localizable evolutionary path to the optimum exists via increasing fitness values, which poses no obstacles for an evolutionary search. Such a strong inverse relation we obtain for $\rho(k) = -1$, thus indicating low problem hardness. A $\rho(k) = 1$ would show the exact opposite. In the numerical experiments we again consider the time average ρ.

11.3.2 Analysis of Dynamical Properties

Next to topological properties that have effect on the difficulties an evolutionary search has in a fitness landscape, features of the dynamics play an important role.

While topological landscape measures are an established topic in evolutionary computing, a like–wise treatment of the dynamic effects plays still a considerably minor role; some primary works can be found in [10, 15, 38, 41]. It is, again intuitively, clear that two main features of dynamics should have a strong influence on the performance of the evolutionary search: change frequency and dynamic severity. Change frequency indicates the speed with which the fitness landscapes change relatively to the evolutionary algorithm. Dynamic severity signifies how fundamental these changes are in terms of their magnitude. Change frequency should be important as a high speed (the fitness landscape changes its topology frequently) gives the evolutionary algorithm just a few generations to find the optimum. A high dynamic severity results in optima moving considerable distances in the search space, and hence recovering it might be complicated or takes time.

11.3.2.1   Change Frequency

An evolutionary search routine takes information out of the fitness landscape and therefore actively interacts with it by evaluating fitness at selected search space points, that are the places the individuals of its population occupy at a given generation. As this is the only form of information extraction, it is also a (repeated but otherwise) isolated instance of interplay with the dynamics of the fitness landscape. In most evolutionary search algorithms, fitness function evaluation takes place once in a generation (iteration); in all algorithms it happens at disjunct points in time. In this sense, the evolutionary algorithm samples the fitness landscape at *discrete points in space and time*.

For these facts, it is reasonable to measure the speed of the fitness landscape relatively to the evolutionary search algorithm. This can be done by counting the number of fitness function evaluations or the generations from one landscape change to the next. For the number of individuals being constant, both countings approximately scale directly linear. In the following we prefer to count landscape speed by the number of generation between changes; this number we call change frequency $\gamma \in \mathbb{N}$. By linking the speed of the fitness landscape to the generations of the evolutionary algorithm, we also establish a connection in real time. The dynamics of the landscape is a results of the dynamic optimization problem undergoing changes. For a real–world problem these changes are in real time. Via the linear link the change frequency can be counted as generations. On the other hand, generations of an evolutionary algorithm reflect the CPU time for doing the fitness evaluation and performing the genetic operators. For a given hardware and implementation, we hence get an approximation of the computer's real running time, too. An evolutionary algorithm should have a certain number of generations (depending on problem hardness and population size) for finding the optimum to perform well. If we define this number of generations for a real-world dynamic optimization problem, we in fact define our requirements on hardware and implementation for the evolutionary algorithm. Via the linear link, the realized running time (and hence the number of generation calculated) can be checked against the change pattern of the problem.

The population dynamics of the evolutionary algorithm is counted by the discrete generations τ. So, with the change frequency γ we have the link

$$\tau = \gamma k \tag{11.21}$$

to the landscape time k. Change frequency γ is usually considered as constant over the run time of the algorithm, but in general γ might also by varying for an ongoing search, or even be a random number to be considered as the realization of an integer random process.

11.3.2.2 Dynamic Severity

The second influential factor on dynamic problem hardness is dynamic severity. It measure the (relative) strength of the landscape change by comparing the landscape before and after a change. If a change happens and the optimum moves, the individuals momentarily lose it. This may go along with a drop in the individuals' fitness. For the evolutionary search it hence makes a difference if the optimum moves a long or a short way from its current position. In the former it might be complicated and time-consuming to recover and track it, in the latter the individuals might be still close enough to catch it again quickly. However, this general observation is not without exceptional cases, as it is easy to construct counterexamples, for instance XOR dynamic pseudo-Boolean functions, and show for the $(1+1)$ evolution strategy a contrary behavior [44]. Dynamic severity intends to measure this dynamic property, which can be done using several notations [6, 36, 57]. They all account for the (relative and time average) distance the optimum moves from one point in landscape time k to the next $k+1$. Using the notation of solution $x_S(k)$ of the dynamic optimization problem (Eq. (11.7)) we can put dynamic severity $\eta(k)$ as

$$\eta(k+1) = \|x_S(k+1) - x_S(k)\|. \tag{11.22}$$

As this quantity varies over time k, we calculate the time average severity

$$\eta = \lim_{K \to \infty} \frac{1}{K} \sum_{k=0}^{K-1} \eta(k). \tag{11.23}$$

Change frequency and dynamic severity are features that are highly influential on the success an evolutionary search has in solving the corresponding dynamic optimization problem. However, in some sense both quantities are much more adjustable parameters than measures if the fitness landscape has an external dynamics. Change frequency is not a direct property of the fitness landscape but a comparison of its time variable to the generations of the evolutionary algorithm. Dynamic severity depends entirely on the moving sequence (Eq. (11.9)) and is hence much more a property of the external dynamics. So, the question is of interest how the (external) dynamics interacts with topological features of the landscape, particularly such that are influential on problem hardness. In principle, this question can be addressed by measures for spatio–temporal dynamics such as complexity. In [41], for instance,

Lyapunov exponents and bred vector dimensions have been studied for CML-based dynamic fitness landscapes. However, for dynamic fitness landscapes with external dynamics this approach is not accessible. The reason is that both quantities measure the effect that small changes in the landscape's fitness have on the time evolution of fitness values in the neighborhood of the perturbed point. For fitness landscapes with external dynamics local perturbability is not fully given. As discussed in Sec. 11.2.3, external dynamics means to have dynamics for selected elements in the landscape only. In consequence, the general law (Eq. (11.8)) does not apply unrestricted. The time evolution of a fitness value does not explicitly depend on the fitness values of its neighbors. Therefore, perturbations afflicted on its neighbors do not have effect on how the fitness value moves forward in time. Therefore, all spatio-temporal complexity measures based on perturbation ideas are not applicable for fitness landscapes with external dynamics.

For analysis accessible, however, is the external dynamics itself. The moving sequence (Eq. (11.9)) can be evaluated using tools form nonlinear and/or statistical time series analysis, see e.g. [21] for an overview. For instance, for all three types of considered dynamics (regular, chaotic, random) a Lyapunov exponent λ of the time series can be calculated (for cyclic dynamics, we obtain $\lambda = 0$, for chaotic dynamics, $\lambda > 0$ and for random dynamics $\lambda = \infty$). Also, the time series can be analyzed statistically. The obtained quantities can be compared to measures directly extracted from the fitness landscape. In this, we have a way to address the question of how dynamical properties of the external drive are reflected by the fitness landscape.

11.4 Numerical Experiments

In this section, we analyze a dynamic fitness landscape using the measures presented and discussed in Sec. 11.3. As fitness landscape we consider the moving peaks (Eq. (11.2)). This landscape is rather simple and gives the experimenter a large degree of control as all topological landscape features can be adjusted. The control extends even to the dynamics which is defined by an external moving sequence. Nonetheless, we belief that considering this dynamic fitness landscape here is instructive as it allows to illustrate the usage of these landscape measures for a well–known and rather easy understandable environment. It should be made clear that the following experimental study is not meant to offer a comprehensive treatment. This is not the focus of the chapter and also, such a study would require to link the obtained data to a specific evolutionary search algorithm, which is not our intention here.

Further specification for the moving peaks are dimension $n = 2$ and we only let the peaks' coordinates $c_i(k)$ have dynamics. If the number of cones is not varied in the experiments, we use $N = 20$. The slopes s_i and heights h_i remain constant over the runs. Unless otherwise stated, the s_i and h_i are realizations of a random variable uniformly distributed on the interval $[0, 1]$.

As pointed out before, the dynamics of the landscape is defined externally as moving sequence. In the experiments, we study 6 different forms of dynamics. Two

types of regular dynamics, cycle and closed–loop linear track, chaotic dynamics
with the generating equations being the Hénon map [12]

$$x(k+1) = \begin{pmatrix} 1.4 - x_1(k)^2 + 0.3x_2(k) \\ x_1(k) \end{pmatrix} \qquad (11.24)$$

with Lyapunov exponent $\lambda = 0.38$ and the Holmes map [13]

$$x(k+1) = \begin{pmatrix} x_2(k) \\ -0.2x_1(k) + 2.77x_2(k) - x_2(k)^3 \end{pmatrix}, \qquad (11.25)$$

with Lyapunov exponent $\lambda = 0.59$, and random dynamics which is generated by a
stochastic process with normal distribution $\mathcal{N}(0, 0.75)$ and uniform distribution on
the interval $[-2, 2]$. Figure 11.2 shows the 6 different dynamics as plots of the points
in the search space the coordinates $c_i(k)$ can have. Hence, every such point can be
seen as a possible location for optima at a given point in time k. The point sets in Fig.
11.2 can hence be understood as to depict timely realizations of the spatial optima
distribution in the dynamic fitness landscape. In Fig. 11.3, the dynamic severity
η, Eq. (11.23), is given over the number of cones N. It is almost constant for the
number of cones increasing, but not the same for the different dynamics. This allows
on the one hand to test if other (topological) measures are capable of distinguishing
between different severity. Also, the point sets can be subsequently normalized to
obtain equal severity for testing how the dynamics itself scales to the measures.

For the random walks that are used to calculate ruggedness, information content and fitness distance correlation the listed parameters were used following a
recommendation in a similar study for static fitness landscapes [25]: walk length
$T = 10000$ and step size $t_s = E_{max}/100$, where E_{max} is the maximal extension in the
search space.

In a first experiment we look at the temporal distribution of modality for the
moving cones, depending on the maximal number of cones N, see Fig. 11.4. In this
figures, histograms for the relative frequency h of optima over the normalized number of optima $\#_{LM}/N$ for different N and the different dynamics are shown, using
10 bins in the histogram. The histograms give the distributions for 2500 different
dynamic realizations of the landscape for each considered dynamics. It can be seen
that the distributions of the number of optima is not Gaussian. In particular, Holmes
chaos (which has a larger Lyapunov exponent as Hénon chaos and hence a higher
chaoticity) gives a distribution that largely differs from normal. Further, it can be
seen that increasing the number of cones leads to getting a smaller and smaller proportion of high ratios $\#_{LM}/N$. This means that the larger the number of cones N
is, the less likely is it that this number is realized as optima in the landscape, the
nesting effect mentioned in Sec. 11.2.1. Note that with 2500 dynamic realizations
of the fitness landscape we considered a rather larger number that fairly represents
the long term dynamic behavior. For a smaller number the relative frequencies might
be distributed largely different.

Next, in Fig. 11.5 we give the average number of optima $\#_{LM}$ for different combinations of slopes s_i and heights h_i. Therefore, the s_i and h_i are generated as

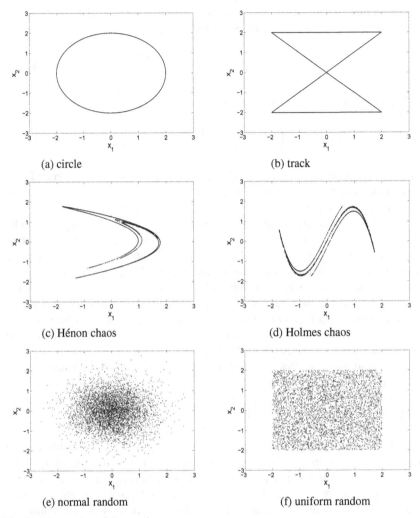

Fig. 11.2 Dynamics of the moving sequence

realizations of a random variable uniformly distributed on the interval $[0, h_{max}]$ and $[0, s_{max}]$, respectively. The results in Fig. 11.5a are for Hénon chaos and $N = 15$, and again 2500 dynamic realizations are averaged. Only for selected combinations of s_i and h_i the average number of optima $\#_{LM}$ reaches the number of cones N. For other combinations only a fraction is obtained. This again can be attributed to the nesting effect for the moving cones. This becomes even more visible if we consider the ratio $\frac{h_{max}}{s_{max}}$, see Fig. 11.5b, where $\#_{LM}$ is depicted over $\frac{h_{max}}{s_{max}}$ and different numbers of cones N. Only for $\frac{h_{max}}{s_{max}}$ close to zero, the quantity $\#_{LM}$ reaches the maximum value N; for $\frac{h_{max}}{s_{max}}$ getting larger $\#_{LM}$ soon takes very small values, meaning that only a fraction of

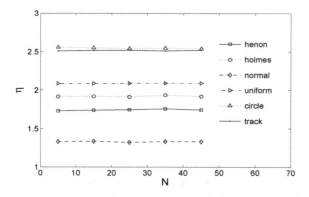

Fig. 11.3 Severity η for the 6 different dynamics of moving sequences

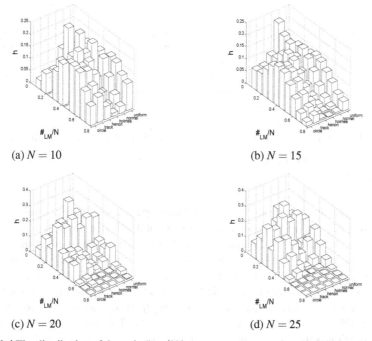

(a) $N = 10$ (b) $N = 15$

(c) $N = 20$ (d) $N = 25$

Fig. 11.4 The distribution of the ratio $\#_{LM}/N$ between average number of optima and number of cones for the 6 different dynamics considered and different N

the intended optima are actually obtained. Further it can be seen that increasing h_{max} generally leads to a decrease in the average number of optima $\#_{LM}$, while increasing s_{max} mildly increases $\#_{LM}$ too. The results for the other dynamics look very similar and are therefore not depicted here.

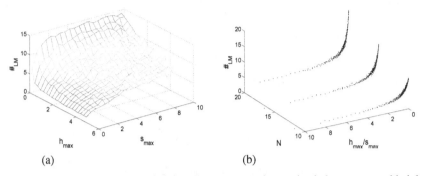

(a)   (b)

Fig. 11.5 The average number of optima $\#_{LM}$: (a) over the maximal slopes s_{max} and heights h_{max}, (b) over the ratio $\frac{h_{max}}{s_{max}}$ and different numbers of cones N

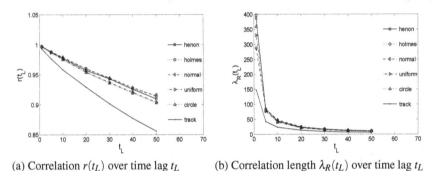

(a) Correlation $r(t_L)$ over time lag t_L (b) Correlation length $\lambda_R(t_L)$ over time lag t_L

Fig. 11.6 The spatial correlations (Eq. (11.14)) and (Eq. (11.15)) over the time lag t_L in correlograms

In Sec. 11.3.1 measures for topological landscape properties are discussed. In the following, we present results for ruggedness λ_R and information content h_{ic}, where we analyze the influence of parameters used in calculating these quantities. These are the time lag t_L for ruggedness in Eq. (11.14) and Eq. (11.15) and the sensitivity e for information content (Eq. (11.19)). In Fig. 11.6 the correlograms for different dynamics and $N = 20$ are given, again for 2500 different dynamic realizations. The correlation $r(t_L)$ gradually falls with the time lag t_L; this effect is even more visible of the correlation of length t_L, $\lambda_R(t_L)$, see Fig. 11.6b. This is the expected result as the correlation between the fitness of two points in the landscape weakens for the random walks on it getting a larger spatial difference. Generally it can be noticed that the spatial correlation in the moving peak landscape is high and remains high, even for larger time lags t_L. Also, there is little difference in the graphs for the different types of dynamics. The information content $h_{IC}(e)$ over sensitivity e is shown in Fig. 11.7. For increasing e we see a temporal increasing or constant $h_{IC}(e)$ that tapers off for increasing e further. Based in these experiment we fix the value $t_L = 1$ and $e = 0.01$ for the subsequent numerical experiments.

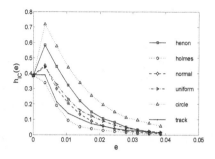

Fig. 11.7 Information content h_{IC} over the sensitivity e

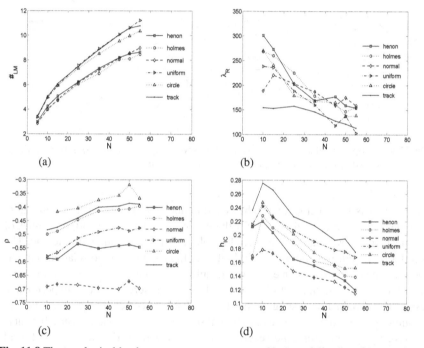

Fig. 11.8 The topological landscape measures over cones N. a) modality $\#_{LM}$, b) ruggedness λ_R, c) fitness distance correlation ρ, d) information content h_{IC}

In Fig. 11.8 the analyzed topological landscape measure modality $\#_{LM}$, ruggedness λ_R, fitness distance correlation ρ and information content h_{IC} are given for varying number of cones N in the landscape and the different dynamics. The results are for 1500 different dynamic realization, which were averaged. We can see that the average number of optima increases with N, but we do not get a linear relationship and the increase wears off for increasing N. The fitness distance correlation is negative and slightly increasing, while ruggedness and information content are generally

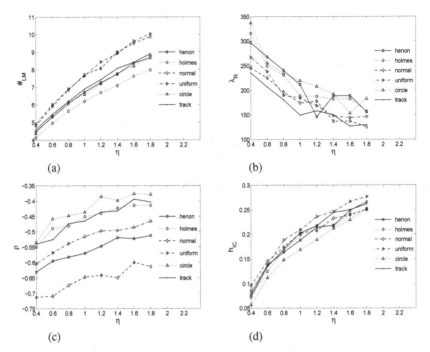

Fig. 11.9 The topological landscape measures over severity η. a) modality $\#_{LM}$, b) ruggedness λ_R, c) fitness distance correlation ρ, d) information content h_{IC}

falling. This is in agreement with the general understanding that a rise in modality also means a more difficult search for the optima. On the other hand, the values of the measures show that the considered problem is indeed a rather simple one. The correlation length remains relatively high, the fitness distance correlation remains negative, meaning there is an easily localizable evolutionary path to the optima, and the information content is rather low. The data shown here, and other experimental results not depicted here suggest that fitness distance correlation ρ scales best to dynamic severity. In a second experimental set we calculated the measure for varying severity η. Therefore, the severity was normalized to the value $\eta = 2$ for all considered dynamics and then collectively varied, see Fig. 11.9. Again, modality and information content rise with increasing severity, ruggedness falls, while fitness distance correlation is negative and very slightly rising. So, it can be concluded for the considered moving peaks that there is a clear and evaluable relation between the considered measures and topological problem hardness also for dynamic fitness landscapes.

Interestingly, there is no clear cut conclusion regarding the considered dynamics to be gain from the topological measures. So, for dynamic fitness landscapes with internal dynamics, dynamics measures are needed as discussed and applied for CML-based landscapes in [38, 41]. For external dynamics these measures are no applicable as discussed above. Based on the topological measures given, the

behavior of an evolutionary algorithm used to solve such dynamic optimization problems can be analyzed by additionally considering quantities to measure the external
dynamics.

11.5   Conclusion

In this chapter we presented an introduction in dynamic fitness landscape analysis. It was shown that solving a dynamic optimization problem using an evolutionary search algorithm can be grounded on the theoretical framework of dynamic fitness landscapes. These landscapes were defined and their characteristic features discussed. Moreover, tools for analyzing dynamic fitness landscapes to account for both topological and dynamical properties were introduced. We considered modality, ruggedness, fitness distance correlation and information content as topological measures and dynamic severity as dynamical measure. The usage and calculation of these measures were illustrated using a well-known example, the moving peaks.

Unlike studying static fitness landscapes, which has meanwhile reached a certain maturity, the treatment of their dynamic counterparts is still a new topic in evolutionary computation. Our hope is that the methodology presented in this chapter can serve as a starting point for further research. Although some foundations stones for defining and measuring dynamic fitness landscapes are laid, many important issues remain unresolved. A main issue is how the landscape measures are linked to real performance and behavior data of an evolutionary search algorithm employed to solve the corresponding dynamic optimization problem. A considerable difficulty with establishing a reasonably strong, meaningful and convincing relation is that there are a large variety of evolutionary search algorithms, and an even larger variety of the algorithms' parameters. As the performance solving a given dynamic optimization problem sensitively depends on this choice of algorithm and parameters, there would be the need for a large data base over an exhaustive variation of both. This is still not available for dynamic optimization and might be hard to come by. This is additionally complicated by the fact that simple and theoretically analyzable evolutionary algorithms such as the $(1 + 1)$ evolution strategy mostly deliver poor results in numerical problem solving. However, that this is a possible way to go is shown in studies of static fitness landscapes where the question was answered for certain algorithm-problem combinations [27, 51]. A first step towards this aim might be to have a general set of rules that allow to distinguish different generic types of dynamic problems and link these types to expectable behavior of certain types of evolutionary algorithms regardless of the actual performance. A second interesting issue is how measures of the external dynamics are reflected by landscape measures and the behavior of evolutionary algorithms. Some qualitative works have been done on this topic [10, 36, 40, 46, 57], but so far they are clearly centered on specific implementation variants of evolutionary algorithms. Again, it might be illuminating to use a simple evolutionary algorithm and analyze the behavior unconcerned about the performance. Another open topic is how the fitness landscape approach are to adapt to other types of populations-based search algorithm. As has been shown other types of algorithms such as PSO, differential evolution or memetic algorithms are sometimes more successful in solving dynamic optimization problems. In principle, the arguments made for the interplay between the individuals of an evolutionary algorithm with the fitness landscape should apply for them in a largely similar way. A final point surely is applying the approach to real-world problems. So far, analyzing dynamic fitness landscapes is using constructed landscape models, but the analyzing tools are usable for all kind of dynamic optimization problems.

References

[1] Alba, E., Sarasola, B.: Measuring fitness degradation in dynamic optimization problems. In: Di Chio, C., et al. (eds.) EvoApplicatons 2010, Part I. LNCS, vol. 6024, pp. 572–581. Springer, Heidelberg (2010)

[2] Altenberg, L.: Fitness distance correlation analysis: An instructive counterexample. In: Bäck, T. (ed.) Proc. 7th Int. Conf. on Genetic Algorithms, pp. 57–64. Morgan Kaufmann, San Francisco (1997)

[3] Asselmeyer, T., Ebeling, W., Rosé, H.: Analytical and numerical investigations of evolutionary algorithms in continuous spaces. In: Voigt, H.-M., Ebeling, W., Rechenberg, I., Schwefel, H.-P. (eds.) PPSN 1996. LNCS, vol. 1141, pp. 111–121. Springer, Heidelberg (1996)

[4] Bosman, P.A.N., La Poutré, H.: Computationally intelligent online dynamic vehicle routing by explicit load prediction in an evolutionary algorithm. In: Runarsson, T.P., Beyer, H.-G., Burke, E., Merelo-Guervós, J.J., Whitley, L.D., Yao, X. (eds.) PPSN 2006. LNCS, vol. 4193, pp. 312–321. Springer, Heidelberg (2006)

[5] Branke, J.: Memory enhanced evolutionary algorithms for changing optimization problems. In: Angeline, P.J., Michalewicz, Z., Schoenauer, M., Yao, X., Zalzala, A. (eds.) Proc. 1999 IEEE Congr. Evol. Comput., pp. 1875–1882. IEEE Press, Piscataway (1999)

[6] Branke, J.: Evolutionary Optimization in Dynamic Environments. Kluwer Academic Publishers, Dordrecht (2001)

[7] Crutchfield, J.P., Kaneko, K.: Phenomenology of spatiotemporal chaos. In: Hao, B. (ed.) Directions in Chaos, vol. 1, pp. 272–353. World Scientific, Singapore (1987)

[8] Cruz, C., Gonzlez, J.R., Pelta, D.A.: Optimization in dynamic environments: a survey on problems, methods and measures. Soft Comput. 15, 1427–1448 (2011)

[9] Deb, K., Goldberg, D.E.: Sufficient conditions for deceptive and easy binary functions. Ann. Math. Artif. Intell. 10, 385–408 (1994)

[10] Eriksson, R., Olsson, B.: On the performance of evolutionary algorithms with life-time adaptation in dynamic fitness landscapes. In: Greenwood, G.W. (ed.) Proc. 2004 IEEE Congr. Evol. Comput., pp. 1293–1300. IEEE Press, Piscataway (2004)

[11] He, J., Reeves, C., Witt, C., Yao, X.: A note on problem difficulty measures in black-box optimization: classification, realizations and predictability. Evolut. Comput. 15, 435–443 (2007)

[12] Hénon, M.: A two-dimensional mapping with a strange attractor. Commun. Math. Phys. 50, 69–77 (1976)

[13] Holmes, P.J.: A nonlinear oscillator with a strange attractor. Philos. Trans. R. Soc. London A 292, 419–448 (1979)

[14] Hordijk, W.: A measure of landscapes. Evolut. Comput. 4, 335–360 (1996)

[15] Hordijk, W., Kauffman, S.A.: Correlation analysis of coupled fitness landscapes. Complexity 10, 42–49 (2005)

[16] Horn, J., Goldberg, D.E., Deb, K.: Long path problems. In: Davidor, Y., Schwefel, H.-P., Männer, R. (eds.) PPSN 1994. LNCS, vol. 866, pp. 149–158. Springer, Heidelberg (1994)

[17] Jin, Y., Branke, J.: Evolutionary optimization in uncertain environments – A survey. IEEE Trans. Evol. Comput. 9, 303–317 (2005)

[18] Jones, T.: Evolutionary algorithms, fitness landscape and search. PhD thesis, The University of New Mexico, Albuquerque (1995)

[19] Kallel, L., Naudts, B., Reeves, C.R.: Properties of fitness functions and search landscapes. In: Kallel, L., Naudts, B., Rogers, A. (eds.) Theoretical Aspects of Evolutionary Computing, pp. 177–208. Springer, Heidelberg (2001)

[20] Kaneko, K., Tsuda, I.: Complex Systems: Chaos and Beyond. Springer, Heidelberg (2001)

[21] Kantz, H., Schreiber, T.: Nonlinear Time Series Analysis. Cambridge University Press, Cambridge (1999)

[22] Kauffman, S.A., Levin, S.: Towards a general theory of adaptive walks on rugged landscapes. J. Theor. Biology 128, 11–45 (1987)

[23] Kauffman, S.A.: The Origin of Order: Self-Organization and Selection in Evolution. Oxford University Press, New York (1993)

[24] Li, Y., Wilke, C.O.: Digital evolution in time–dependent fitness landscapes. Artificial Life 10, 123–134 (2004)

[25] Malan, K., Engelbrecht, A.P.: Quantifying ruggedness of continuous landscapes using entropy. In: Tyrrell, A. (ed.) Proc. 2009 IEEE Congr. Evol. Comput., pp. 1440–1447. IEEE Press, Piscataway (2009)

[26] Mendes, R., Mohais, A.: DynDE: Differential Evolution for dynamic optimization problems. In: Corne, D. (ed.) Proc. 2005 IEEE Congr. Evol. Comput., pp. 2808–2815. IEEE Press, Piscataway (2005)

[27] Merz, P.: Advanced fitness landscape analysis and the performance of memetic algorithms. Evolut. Comput. 12, 303–325 (2004)

[28] Morrison, R.W.: Designing Evolutionary Algorithms for Dynamic Environments. Springer, Heidelberg (2004)

[29] Morrison, R.W., De Jong, K.A.: A test problem generator for non–stationary environments. In: Angeline, P.J., Michalewicz, Z., Schoenauer, M., Yao, X., Zalzala, A. (eds.) Proc. 1999 IEEE Congr. Evol. Comput., Piscataway, NJ, pp. 2047–2053 (1999)

[30] Morrison, R.W., De Jong, K.A.: Triggered hypermutation revisited. In: Zalzala, A., et al. (eds.) Proc. 2000 IEEE Congr. Evol. Comput., pp. 1025–1032. IEEE Press, Piscataway (2000)

[31] Nilsson, M., Snoad, N.: Error thresholds for quasi-species on dynamic fitness landscapes. Phys. Rev. Lett. 84, 191–194 (2000)

[32] Nilsson, M., Snoad, N.: Quasi-species evolution on dynamic fitness landscapes. In: Crutchfield, J.P., Schuster, P. (eds.) Evolutionary Dynamics: Exploring the Interplay of Selection, Accident, Neutrality and Function. Santa Fe Institute Studies in the Sciences of Complexity Series, pp. 275–290. Oxford University Press, New York (2002)

[33] Nilsson, M., Snoad, N.: Optimal mutation rates in dynamic environments. Bull. Math. Biol. 64, 1033–1043 (2002)

[34] Nguyen, T.T., Yao, X.: Benchmarking and solving dynamic constrained problems. In: Tyrrell, A. (ed.) Proc. 2009 IEEE Congr. Evol. Comput., pp. 690–697. IEEE Press, Piscataway (2009)

[35] Richter, H.: Behavior of evolutionary algorithms in chaotically changing fitness landscapes. In: Yao, X., et al. (eds.) PPSN 2004. LNCS, vol. 3242, pp. 111–120. Springer, Heidelberg (2004)

[36] Richter, H.: A study of dynamic severity in chaotic fitness landscapes. In: Corne, D. (ed.) Proc. 2005 IEEE Congr. Evol. Comput., pp. 2824–2831 (2005)

[37] Richter, H.: Evolutionary optimization in spatio–temporal fitness landscapes. In: Runarsson, T.P., Beyer, H.-G., Burke, E.K., Merelo-Guervós, J.J., Whitley, L.D., Yao, X. (eds.) PPSN 2006. LNCS, vol. 4193, pp. 1–10. Springer, Heidelberg (2006)

[38] Richter, H.: Coupled map lattices as spatio-temporal fitness functions: Landscape measures and evolutionary optimization. Physica D237, 167–186 (2008)

[39] Richter, H., Yang, S.: Memory based on abstraction for dynamic fitness functions. In: Giacobini, M., et al. (eds.) EvoWorkshops 2008. LNCS, vol. 4974, pp. 596–605. Springer, Heidelberg (2008)

[40] Richter, H., Yang, S.: Learning behavior in abstract memory schemes for dynamic optimization problems. Soft Comput. 13, 1163–1173 (2009)

[41] Richter, H.: Evolutionary optimization and dynamic fitness landscapes: From reaction–diffusion systems to chaotic CML. In: Zelinka, I., Celikovsky, S., Richter, H., Chen, G. (eds.) Evolutionary Algorithms and Chaotic Systems. SCI, vol. 267, pp. 409–446. Springer, Heidelberg (2010)

[42] Richter, H.: Memory design for constrained dynamic optimization problems. In: Di Chio, C., et al. (eds.) EvoApplicatons 2010, Part I. LNCS, vol. 6024, pp. 552–561. Springer, Heidelberg (2010)

[43] Richter, H., Dietel, F.: Solving dynamic constrained optimization problems with asynchronous change pattern. In: Di Chio, C., et al. (eds.) EvoApplications 2011, Part I. LNCS, vol. 6624, pp. 334–343. Springer, Heidelberg (2011)

[44] Rohlfshagen, P., Lehre, P.K., Yao, Y.: Dynamic evolutionary optimisation: An analysis of frequency and magnitude of change. In: Rothlauf, F. (ed.) Proc. 2009 Genetic and Evol. Comput. Conf., pp. 1713–1720. ACM, Seattle (2009)

[45] Simões, A., Costa, E.: Variable-size memory evolutionary algorithm to deal with dynamic environments. In: Giacobini, M. (ed.) EvoWorkshops 2007. LNCS, vol. 4448, pp. 617–626. Springer, Heidelberg (2007)

[46] Simões, A., Costa, E.: The influence of population and memory sizes on the evolutionary algorithm's performance for dynamic environments. In: Giacobini, M., et al. (eds.) EvoWorkshops 2009. LNCS, vol. 5484, pp. 705–714. Springer, Heidelberg (2009)

[47] Stadler, P.F.: Landscapes and their correlation functions. J. Math. Chem. 20, 1–45 (1996)

[48] Stadler, P.F., Stephens, C.R.: Landscapes and effective fitness. Comm. Theor. Biol. 8, 389–431 (2003)

[49] Stanhope, S.A., Daida, J.M.: (1+1) Genetic algorithm fitness dynamics in a changing environment. In: Angeline, P.J., Michalewicz, Z., Schoenauer, M., Yao, X., Zalzala, A. (eds.) Proc. 1999 IEEE Congr. Evol. Comput., pp. 1851–1858 (1999)

[50] Smith, T., Husbands, P., Layzell, P., O'Shea, M.: Fitness landscapes and evolvability. Evolut. Comput. 10(1), 1–34 (2002)

[51] Tavares, J., Pereira, F.B., Costa, E.: Multidimensional knapsack problem: a fitness landscape analysis. IEEE Trans. Syst., Man, and Cyber. Part B: Cybern. 38, 604–616 (2008)

[52] Teo, J., Abbass, H.A.: An information–theoretic landscape analysis of neuro-controlled embodied organisms. Neural Comput. Appl. 13, 80–89 (2004)

[53] Tinós, R., Yang, S.: A self–organizing random immigrants genetic algorithm for dynamic optimization problems. Genetic Programming and Evolvable Machines 8, 255–286 (2007)

[54] van Hemert, J.I., La Poutré, J.A.: Dynamic routing problems with fruitful regions: Models and evolutionary computation. In: Yao, X., et al. (eds.) PPSN 2004. LNCS, vol. 3242, pp. 692–701. Springer, Heidelberg (2004)

[55] Vassilev, V.K.: Information analysis of fitness landscapes. In: Husbands, P., Harvey, I. (eds.) Proc. 4th European Conf. on Artificial Life, pp. 116–124. MIT Press, Cambridge (1997)

[56] Vassilev, V.K., Fogarty, T.C., Miller, J.F.: Information characteristics and the structure of landscapes. Evol. Comput. 8(1), 31–60 (2000)

[57] Weicker, K.: An analysis of dynamic severity and population size. In: Deb, K., Rudolph, G., Lutton, E., Merelo, J.J., Schoenauer, M., Schwefel, H.-P., Yao, X. (eds.) PPSN 2000. LNCS, vol. 1917, pp. 159–168. Springer, Heidelberg (2000)

[58] Weicker, K.: Performance measures for dynamic environments. In: Guervós, J.J.M., Adamidis, P.A., Beyer, H.-G., Fernández-Villacañas, J.-L., Schwefel, H.-P. (eds.) PPSN 2002. LNCS, vol. 2439, pp. 64–73. Springer, Heidelberg (2002)

[59] Weinberger, E.D.: Correlated and uncorrelated fitness landscapes and how to tell the difference. Biol. Cybern. 63, 325–336 (1990)

[60] Wiles, J., Tonkes, B.: Hyperspace geography: Visualizing fitness landscapes beyond 4D. Artificial Life 12, 211–216 (2006)

[61] Wilke, C.O., Ronnewinkel, C., Martinetz, T.: Dynamic fitness landscapes in molecular evolution. Phys. Rep. 349, 395–446 (2001)

[62] Wright, S.: The roles of mutation, inbreeding, crossbreeding and selection in evolution. In: Jones, D.F. (ed.) Proc. 6th Int. Congr. on Genetics, pp. 356–366 (1932)

[63] Yang, S., Yao, X.: Experimental study on population–based incremental learning algorithms for dynamic optimization problems. Soft Comput. 9(11), 815–834 (2005)

[64] Yang, S., Ong, Y.S., Jin, Y. (eds.): Evolutionary Computation in Dynamic and Uncertain Environments. Springer, Heidelberg (2007)

Chapter 12
Dynamics in the Multi-objective Subset Sum: Analysing the Behavior of Population Based Algorithms

Iulia Maria Comsa, Crina Grosan, and Shengxiang Yang

Abstract. Real-world problems often present two characteristics that are challenging to examine theoretically: 1) they are dynamic (they change over time), and 2) they have multiple objectives. However, current research in the field of dynamic multi-objective optimization (DMO) is relatively sparse. In this chapter, we review recent work in this field and present our analysis of the subset sum problem in the DMO variant. Our approach uses a genetic algorithm with an external archive and a combination of Pareto dominance and aggregated fitness function. We show that the algorithm performs better on a smaller number of objectives, on type III dynamicity problems, and sometimes, counter-intuitively, on a larger data set.

12.1 Introduction

Many real-life optimization problems present two characteristics that are challenging for theoretical study: firstly, they are dynamic (i.e., they change over time), and secondly, they require the optimization of multiple objectives at the same time.

Iulia Maria Comsa
Department of Computer Science, Babes-Bolyai University, Kogalniceanu 1,
Cluj-Napoca 400084, Romania
e-mail: iulia.m.comsa@gmail.com

Crina Grosan
Department of Computer Science, Babes-Bolyai University, Kogalniceanu 1,
Cluj-Napoca 400084, Romania, and Department of Information Systems and Computing,
Brunel University, Uxbridge, Middlesex UB8 3PH, U.K.
e-mail: crina.grosan@brunel.ac.uk

Shengxiang Yang
Centre for Computational Intelligence (CCI), School of Computer Science and
Informatics, De Montfort University, The Gateway, Leicester LE1 9BH, U.K.
e-mail: syang@dmu.ac.uk

S. Yang and X. Yao (Eds.): *Evolutionary Computation for DOPs*, SCI 490, pp. 299–313.
DOI: 10.1007/978-3-642-38416-5_12   © Springer-Verlag Berlin Heidelberg 2013

This chapter aims to study theoretically a dynamic and multi-objective version of the subset sum problem. The theoretical study is original, as no other study on the dynamic multi-objective subset sum problem has been made (there are currently very few studies on any dynamic multi-objective problem).

The subset sum problem arises in practice whenever a set of objects have to be stored efficiently in a container with a finite capacity. One example of the subset sum problem applications is transporting in a limited weight truck several items with certain weights. The problem becomes dynamic when, for example, the storage capacities or the objects to be stored change over time. For multiple trucks, it becomes a multi-objective problem.

In dynamic problems, either the Pareto set, the Pareto front, or both, can change with time, giving rise to three types of dynamicity. We study all three types for the subset sum problem with 2, 3 and 4 objectives, on two sets of data, with a genetic algorithm (GA). The GA uses an external archive to preserve the best (non-dominated) solutions and assure elitism, and an aggregate fitness function.

The rest of the chapter is organized as follows. Section 12.2 briefly reviews the domain of dynamic optimization. Section 12.3 discuses the multi-objective aspect of it. Section 12.4 describes the multi-objective subset sum version considered in this work. Section 12.5 is dedicated to numerical experiments and discussions followed by conclusions in Section 12.6.

12.2 Dynamic Optimization

Real life optimization problems often involve a degree of uncertainty. There are four types of uncertainty that a problem can exhibit [16]: noise (the fitness function is noisy, such as the measurement errors of sensors), environmental changes after finding an optimal solution (which requires the found solution to be robust against such changes), fitness approximation errors (since accurate fitness functions may be too expensive, approximate functions, also called meta-models, are used instead, which inevitably suffer from approximation errors), time-dependence of the optimum (the fitness value is deterministic, but changes in time, so the algorithm should be able to track its changes). In this chapter, we are interested in the fourth type of uncertainty for problems solved using evolutionary algorithms (EAs). An optimization problem that exhibits this type of uncertainty is called a *dynamic problem*.

If a static optimization problem can be defined as:

$$\text{Find } x \in X^m \text{ that minimizes } y = f(x),$$

a dynamic optimization problem can be defined as:

$$\text{Find the sequence } x_k \in X^m \text{ that minimizes } y = f(x_k), k = 1, 2, ...$$

The simplest way of dealing with dynamic behavior is restarting the algorithm whenever a change occurs. However, since the algorithm has already sampled the search space while searching for the old solution, restarting the algorithm leads to the loss of potentially valuable information. Moreover, in some cases, attempting to detect at every time step whether a change has occurred in the problem may not be practical. Thus, we are interested in reusing the information already gained by the algorithm. This information may speed up the process of finding the new optimal solution. We take this approach assuming that the new optimum is somewhat close to the previous optimum; in the case of a more dramatic change, restarting the algorithm may be a better choice.

In general, there are four strategies for coping with dynamic problems when using EAs [16]: increasing the diversity, maintaining the diversity, using multiple populations, and keeping archives. The first strategy is to increase significantly the population diversity, for example by a significant increase in the mutation rate (hypermutation) after a change occurs. This would lead to the current population spreading away from the previous optimum, while still keeping some information (genes) close to the previous optimum. An example is the macromutation used in [3], which involves generating several new random individuals and increasing significantly the mutation rates (recrudescence) in order to reorganize the individual fitness landscape. A disadvantage of this strategy is the difficulty of achieving the right amount of diversification. Too much diversification leads to a high chance that potentially useful information will be lost; while too little diversification may lead to the premature convergence of the solution.

Another useful strategy is maintaining good diversity in the population during the search process. This could help identify promising search space regions more easily in the event of a change. An example of an algorithm emphasizing diversity is the thermodynamical GA (TDGA) [18]. This algorithm employs a temperature parameter that regulates diversification. However, it has been noted that there are cases when enforcing diversity hinders the search [2].

The third strategy for dealing with dynamic problems is the use of multiple populations to track promising areas of the search space. Such an approach is presented in [4], where a main population is searching for promising peaks, while multiple smaller populations are keeping track of each peak. This approach, as well as maintaining diversity, can be especially useful when it is not possible to detect whether a change in the problem has occurred.

The fourth strategy is to use an archive (memory) to store the best solutions from the current population over time and reuse these stored solutions when a change occurs [5]. The memory based strategy can greatly improve the performance of EAs when the dynamic problem is subject to cyclic dynamic environments, i.e., old environments may re-appear in the future.

Performance assessment is somewhat difficult in dynamic environments. Weiker [22] described three qualities of a dynamic optimization algorithm. Firstly, the algorithm should be accurate: the best solution in the population should be as close as possible to the optimal solution. Secondly, the population should be stable – its accuracy should not fluctuate severely in the event of a change. Thirdly, the algorithm should be able to react quickly to changes in order to recover its accuracy when a change occurs. An example of a simple performance measurement is the offline performance index. This is calculated by averaging the best solutions found at each time step, where a new time step marks a dynamic change in the problem.

12.3 Multi-objective Aspect

The domain of dynamic multi-objective optimization is still in its infancy and there are few publications in the field so far. Proposed algorithms include evolutionary (usually hybrid) algorithms [10, 14, 15, 23], immunity-inspired algorithms [21, 24] or other innovative methods [1].

A DMO problem can be defined as:

$$\text{Minimize } y = (y_1, y_2, ..., y_n) = f(x, t) \in Y^n$$
$$\text{s.t. } x = (x_1, x_2, ..., x_m) \in X^m,$$
$$\text{possibly under certain constraints } g(x) > 0, h(x) = 0.$$

The Pareto-optimal set (POS) and the Pareto-optimal front (POF) of a DMO problem can change at every time step. Farina *et al.* [10] described four types of DMO problems depending on the POF and POS dynamics, as presented in Table 12.1. The problems of interest in this chapter are of Type I, Type II and Type III.

Table 12.1 Types of DMO problems

	Static POS	Dynamic POS
Static POF	Type IV	Type I
Dynamic POF	Type III	Type II

In order to solve DMO problems with EAs, ones has to pay attention to issues from both multi-objective optimization and dynamic optimization problems. Several approaches are common for both types of problems, such as diversity preservation and usage of multiple populations. Nevertheless, designing a good EA for DMO problems requires more effort than in a simpler case. Although, ideally, an EA for a DMO problem should work well on any search space, it is often easier to design algorithms for specific search spaces.

An extensive suite of DMO test problems was proposed in [10]. The proposed test problems give rise to different challenges regarding the time-dependent search space, such as nonconvexity, disconnectedness, or deceptiveness. The suite includes a few mathematically derived problems (including FDA1), DMO versions of the knapsack problem and the traveling salesman problem, and a DMO controller problem. Other test problems can be created based on the given construction procedure.

Farina *et al.* [10] proposed a direction-based search method for tackling DMO problems. The algorithm is a dynamic adaptation of the hybrid algorithm for multi-objective problems found in [9]. The direction-based search method employs EAs to achieve a global search on the decision space. Their research suggests the need for more efficient algorithms, perhaps adapted from the state of the art multi-objective EAs, such as the nondominated sorting genetic algorithm II (NSGA-II) [7], strength Pareto EA 2 (SPEA2) [25], and Pareto envelope-based selection algorithm (PESA) [6].

Shang *et al.* [21] proposes an innovative algorithm inspired by the human immune system – the clonal selection algorithm (CSADMO). The algorithm structurally resembles an EA, but defines some unique operators. The clonal selection is implemented on a population consisting of antibodies to allow information exchange among individuals. Mutation is nonuniform, as it depends on the current generation: at the beginning of the algorithm, it promotes vast search space exploration, but as the algorithm advances, its effect is limited to solution fine-tuning. Finally, a distance method is used to assure the diversity of the population. This algorithm outperforms the direction-based algorithm [10]. Another immunity-inspired algorithm is described in [24].

An interesting algorithm inspired by artificial life (ALife) was proposed in [1], which provides an approximation of the Pareto-optimal front. Under the tag line "the dumbest smart thing you can do is stay alive", the idea of the algorithm is to consider a population of social chromosomes, and to allow them to interact. When two individuals meet, they will either reproduce to create offspring that are added to the population, or the stronger individual will kill the other. Individuals can also reproduce asexually when they do not meet another individual, which results in a mutated copy of the parent. The algorithm tracks successfully changes of the Pareto set.

Hatzakis and Wallace [15] presented an EA for DMO problems merged with a forecasting module that aims to predict the future location of the optimum. The forecasting module makes its prediction based on information regarding the previous evolution of the optimum. In response to the forecast, several individuals are placed in the estimated region.

Zeng *et al.* [23] proposed a simple genetic algorithm, called DOMOEA. This algorithm works on continuous search spaces and uses the results obtained at a time step as a starting population for the next time step. Between two consecutive time steps, the algorithm employs an "orthogonal design method" to improve the population fitness.

A real-life application of DMO was presented in [8]. The NSGA-II algorithm is adapted for dynamic environment and applied to the hydrothermal power scheduling problem. This algorithm does not automatically detect changes, but rather reevaluates 10% of the population at every generation in order to detect changes.

Goh and Tan [14] proposed a new evolutionary paradigm involving multiple populations. Their algorithm combines co-evolution with competition in order to allow a flexible, adaptive decomposition of the problem. Several subpopulations compete for representing a particular subcomponent of the problem; in the end, the winners cooperate for producing globally optimal solutions.

The majority of these approaches are derived from static multi-objective problems, and are tested against benchmark functions. They are often based on sophisticated or hybridized methods. The domain of DMO remains open for further research.

12.4   The Multi-objective Subset Sum Problem

The subset sum problem is a classic NP-complete problem [13]. Its widely studied decision variant can be stated as follows:

> Given a set of items (numbers) $W = \{w_1, w_2, ..., w_n\}$, find a subset of W whose sum of elements equals a given sum S.

No exact solution in polynomial time exists for the problem. However, there have been attempts to find exact solutions in pseudo-polynomial time (that run in practice as if they were polynomial) by Flaxman and Przydatek [11] and Galil and Margalit [12], or even in linear time for certain constraints by Pisinger [19].

We are interested in the combinatorial version of the problem:

> Given a set of items (numbers) $W = \{w_1, w_2, ..., w_n\}$, find a subset of W whose sum of elements is as close as possible to, without exceeding, a given sum S.

Although real-life applications exist for the subset sum problem, it is also useful to employ it as a toy problem for testing and analyzing the behavior of existing algorithms. Khuri *et al.* [17] were the first to study this problem, in its combinatorial version, using GAs. They performed an analysis of a standard GA on instances of size (number of elements in W) 100 and 1000, with very good results. An interesting finding of their study is that the problem size does not influence the results of the algorithm.

Rohlfshagen and Yao [20] presented an extensive analysis of the subset sum problem in its dynamic form. Their analysis showed that, in non-extreme cases, the severity of change affecting the parameters (the actual magnitude of change) is proportional to the distance between successive global optima (the observed rate of change).

12.5   Analysis of the Dynamic Multi-objective Subset Sum Problem

We are interested in the following dynamic and multi-objective version of the subset sum problem:

> Given a set of n positive numbers $W = \{w_1, w_2, ..., w_n\}$ and a set of m positive integers $S = \{S_1, S_2, ..., S_m\}$, find m disjoint subsets of W, such as the sum of elements in subset i $(i = 1, 2, ..., m)$ is as close as possible to, without exceeding, S_i.

We will study the behavior of a GA whose aim is to approximate the Pareto-optimal front, i.e. to find as many Pareto-optimal solutions as possible. We will analyze

Table 12.2 The algorithm parameters

Representation	$X = \{x_1 x_2 ... x_n\}$, $x_i \in \{0, 1, ..., m\}$ with the meaning : w_i belongs to subset j if $x_i = j, i = 1, ..., n, j = 1, ..., m$
Fitness function	$f(X) = (f_1(X), ... f_m(X));$ $f_i(X) = k_i * (S_i - \sum_{j=1}^{n} w_j * p_{ji}) + (1 - k_i) * (\sum_{j=1}^{n} w_j * p_{ji}),$ where $p_{ji} = \begin{cases} 1, if x_j = i \\ 0, \text{otherwise} \end{cases}$, $k_i = \begin{cases} 1, \text{ if } \sum_{j=1}^{m} (w_j * p_{ji}) \leq S_i \\ 0, \text{ otherwise (the chromosome is invalid)} \end{cases}$
Selection	Binary tournament selection
Crossover rate	0.8
Mutation rate	$0.5 / n$
Population size	Small data set: 50 Large data set: 100
Iterations until change	Small data set: 20 Large data set: 50

situations of Type I, Type II, and Type III (as defined in Table 12.1). For the subset sum problem, the dynamics are defined as follows:

Type I. The set W changes over time, while the set S of objective sums does not change.

Type II. The set W does not change, while the S of objective sums change with time, such that the Pareto-optimal set changes at every time step.

Type III. The set W does not change, while the set S of objective sums change with time, but the Pareto-optimal set does not change.

12.5.1 Algorithm Description

We use a standard GA to search for Pareto-optimal solutions. The algorithm parameters are summarized in Table 12.2. A chromosome X is encoded as a set of n integers $X = \{x_1 x_2 \cdots x_n\}$, where $x_i \in \{1, 2, \cdots, m\}$ $(i = 1, 2, \cdots, n)$ and $x_i = j$ means w_i belongs to subset j. A chromosome that encodes a solution where at least one of the objective sums is exceeded is considered invalid. The strategy for dealing with invalid chromosomes is a penalty incorporated in the fitness function. This approach is often used in conjunction with the subset sum problem [17, 20].

When comparing two chromosomes, we first use Pareto dominance: a solution dominates another when its fitness is better (lower) with respect to one objective and equal or better with respect to the other objectives. If neither of the solutions is dominant, the chromosomes are non-dominant and therefore considered equally good.

Table 12.3 Data at time step $t = 0$ for two objectives

2 objective sums	Small data set		Large data set	
	W	S	W	S
Type I problem	1, 2, 3, 20, 21, 80	10, 46	1, 2, 3, 4, 5, 30, 31, 32, 135, 150, 200	15, 108
Type II problem	1, 2, 3, 4, 5, 30	4, 25	1, 2, 3, 4, 5, 6, 7, 8, 9, 10, 100	6, 65
Type III problem	1, 2, 3, 20, 21, 80	6, 70	1, 2, 3, 4, 5, 30, 31, 32, 135, 150, 200	15, 118

This comparison did not yield very good results on the second large data set (dynamic Pareto set), as it found around 70% of the Pareto optimal solutions. Therefore, we further compare the non-dominant solutions using the sum of fitness values with respect to every objective, considering as better the chromosome which minimizes this sum. The results improved (which was also influenced by the crossover and mutation rates), with more than 90% of the Pareto optimal solutions found.

Chromosome comparison is performed during the analysis according to a modified form of the Pareto dominance relation. We experimentally observed that the algorithm performs better if the comparison between chromosomes favors: a) the valid chromosome, if the other chromosome is invalid, and b), in the case of non-dominance, the chromosome with smaller absolute value of the sum of fitness function components. Therefore, the objective function values are aggregated into a single value for the purpose of selection.

We use an external memory in the form of a separate archive, updated at every generation with new found solutions. The archive contains the best set of non-dominated solutions found so far, without influencing the rest of the population. Such a memory was needed for multi-objective optimization: since the population was completely replaced at every iteration and mutations were involved, there was little chance that all the solutions would be present in the population at the last iteration before the change. Thus, the GA can be classified as elitist.

The archive is completely erased at every time change. The population is not modified, but it is re-evaluated. Therefore, the past knowledge to be reused with respect to the problem dynamicity consists of the current population.

The result of the algorithm at every time step is considered to be the set of solutions found in the archive at the last iteration.

12.5.2 Numerical Results and Discussions

We constructed our own data sets for each type of problem, using 2, 3 and 4 objective sums. The data for the 2, 3, and 4 objective problems are described in Table 12.3, Table 12.4, and Table 12.5, respectively. The dynamics are given in Table 12.6. An analysis regarding the size of the search space in each case is shown in Table 12.7.

Table 12.4 Data at time step $t = 0$ for three objectives

3 objective sums	Small data set		Large data set	
	W	S	W	S
Type I problem	15, 16, 17, 32, 35, 8	25, 55, 65	18, 35, 24, 30, 74, 56, 58, 11, 2, 3, 12	25, 55, 175
Type II problem	2, 4, 14, 15, 21, 25	6, 20, 15	14, 15, 21, 55, 32, 87, 90, 107, 112, 2, 3	215, 98, 130
Type III problem	20 45 60 100 300 1	66, 236, 301	20, 50, 65, 70, 400, 345, 250, 900, 950, 1, 1200	101, 511, 936

Table 12.5 Data at time step $t = 0$ for four objectives

4 objective sums	Small data set		Large data set	
	W	S	W	S
Type I problem	52, 13, 11, 61, 5, 6	75, 27, 58, 22	11, 24, 38, 174, 112, 113, 256, 32, 34, 76, 78	145, 201, 98, 77
Type II problem	18, 52, 45, 3, 7, 8	17, 22, 46, 80	21, 2, 47, 58, 65, 97, 101, 23, 11, 78, 81	45, 87, 210, 153
Type III problem	1, 2, 3, 20, 21, 80	6, 70, 86, 16	20, 21, 22, 23, 60, 61, 600, 1000, 1001, 1002, 2000	45, 55, 127, 1657

We performed 10 runs for every test data. Since the data was relatively small, we were able to compute the real Pareto-optimal set using exhaustive search. For each time step of every run, we averaged the percentage of Pareto-optimal solutions found by the algorithm. The results are summarized in Table 12.8.

Since the average may not fully reflect particularities of the GA behavior as the time step changes, we present in Figs. 12.1, 12.2, and 12.3 the evolution of the algorithm in terms of Pareto-optimal solutions found at every time step, for one data set for each problem type. We show the best and the worst number of solutions discovered during the 10 runs. It can be seen that the differences between runs are often considerable, ranging from a close approximation at one run to almost no solutions at another run.

The GA performed visibly better in the tests for two objective sums and for the problems of Type III, while the performance for Type II problems and more than two objectives was low. This behavior does not seem to depend on the size of the search space, which confirms the remark found in [17] for the subset sum problem in its single objective, static version – that the performance does not seem to depend on the problem size. We also found that doubling the number of generations or the

Table 12.6 Data dynamics

Problem type	Number of objective sums		
	2	3	4
Type I	$S_i(t+1) = S_i(t), i = 1,2$ *Small data set:* $W_i(t+1) = W_i(t)+1,$ $i=1,2,3$ $W_i(t+1) = W_i(t) - 1, i= 4,5$ $W_i(t+1) = W_i(t), i = 6$ *Large data set:* $W_i(t+1) = W_i(t)+1,$ $i=1,....,5$ $W_i(t+1) = W_i(t) - 1, i= 6,7$ $W_i(t+1) = W_i(t), i = 8,...,11$	$S_i(t+1) = S_i(t), i = 1,2,3$ *Small data set:* $W_i(t+1) = W_i(t)+ 1, i = 1,4$ $W_i(t+1) = W_i(t) - 1, i = 2,3$ $W_i(t+1) = W_i(t), i = 5,6$ *Large data set:* $W_i(t+1) = W_i(t)+ 1, i = 2,3$ $W_i(t+1) = W_i(t) - 1, i = 1,4$ $W_i(t+1) = W_i(t), i = 5,...,11$	$S_i(t+1) = S_i(t), i = 1,2,3,4$ $W_i(t+1) = W_i(t)+ 1, i = 2,3$ $W_i(t+1) = W_i(t) - 1, i = 1,4$ $W_i(t+1) = W_i(t), i = 5,...,n$
Type II	$W_i(t+1) = W_i(t), i = 1,...,n$ $S_1(t+1) = S_1(t) + 1$ $S_2(t+1) = S_2(t) - 1$	$W_i(t+1) = W_i(t), i = 1,...,n$ $S_i(t+1) = S_i(t)+ 1, i = 1,3$ $S_2(t+1) = S_2(t) - 1$	$W_i(t+1) = W_i(t), i = 1,...,n$ $S_i(t+1) = S_i(t)+ 1, i = 1,3$ $S_i(t+1) = S_i(t) - 1, i = 2,4$
Type III	$W_i(t+1) = W_i(t), i = 1,...,n$ $S_1(t+1) = S_1(t) + 1$ $S_2(t+1) = S_2(t) - 1$	$W_i(t+1) = W_i(t), i = 1,...,n$ $S_i(t+1) = S_i(t)+ 1, i = 1,3$ $S_2(t+1) = S_2(t) - 1$	$W_i(t+1) = W_i(t), i = 1,...,n$ $S_i(t+1) = S_i(t)+ 1, i = 1,3$ $S_i(t+1) = S_i(t) - 1, i = 2,4$

Table 12.7 The search space size

Number of objective sums	Small data set	Large data set
2	729	177 147
3	4 096	4 194 304
4	15 625	48 828 125

population size did not influence the results for the instances where the performance was low.

To further explain the results, we calculated two statistics for every test data: the average Hamming distance between the Pareto-optimal solutions at the same time step, and the average Hamming distance between every Pareto-optimal solution at time step $t+1$ and every Pareto-optimal solution at time step t. In the case of three and four objective sums, a larger difference between solutions at consecutive time steps (the second statistic) is consistently correlated with a lower performance of the GA. The average difference between solutions at the same time step (the first

Table 12.8 Results and statistics for each problem, where Column A is the percentage of real Pareto-optimal solutions found by the GA, Column B is the average Hamming distance between the Pareto-optimal solutions (averaged for every time step), and Column C is the average Hamming distance between Pareto-optimal solutions at consecutive time steps

| Problem type | 2 objective sums | | | | | |
| | Small | | | Large | | |
	A	B	C	A	B	C
Type I	79.53	1.63	1.4	90.39	2.08	1.91
Type II	94.02	2.36	2.25	96.03	3.04	2.97
Type III	86.02	1.71	1.5	99.57	2.58	2.5

| Problem type | 3 objective sums | | | | | |
| | Small | | | Large | | |
	A	B	C	A	B	C
Type I	34.87	3.69	3.61	2.66	4.40	5.82
Type II	43,02	1.82	2.46	0.22	5.78	6.96
Type III	96.12	2.08	1.94	27.36	5.05	4.88

| Problem type | 4 objective sums | | | | | |
| | Small | | | Large | | |
	A	B	C	A	B	C
Type I	33.19	3.51	3.57	1.13	4.56	5.26
Type II	43.33	3.17	3.19	0,43	6.21	6.67
Type III	84.09	2.28	2.25	46.42	3.27	3.22

statistic) is smaller when the GA performs better in the case of type I and type II problems, for three and four objective sums. It is interesting to note that in the case of two objective sums, when the algorithm performs considerably better, the results are sometimes better when the differences between solutions are greater, both at the same time step and at consecutive time steps.

12.6   Conclusions

This chapter has described the most important aspects of dynamic multi-objective optimization and has analyzed a combinatorial optimization problem in the multi-objective and dynamic form.

Dynamic multi-objective optimization is a relatively recent field of study. Algorithms for solving this class of problems are usually adapted from versions employed for static multi-objective problems. General techniques that are used include maintaining diversity, elitism, multiple subpopulations, and mutation increase after changes.

The analysis was done on the dynamic subset sum problem with two, three and four objectives, using a standard genetic algorithm. An external memory in form of an archive was used to ensure elitism and the fitness function was aggregated. We found that the performance of the algorithm was significantly better in the case of

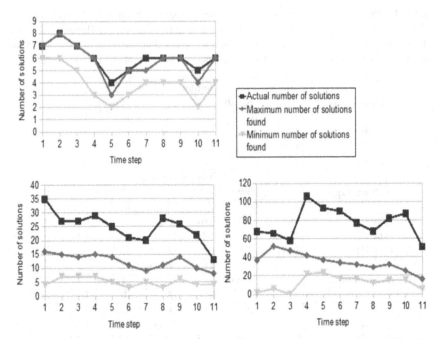

Fig. 12.1 Algorithm behavior in the case of the Type I problem, small data set, for 2 objectives (top), 3 objectives (bottom-left) and 4 objectives (bottom-right)

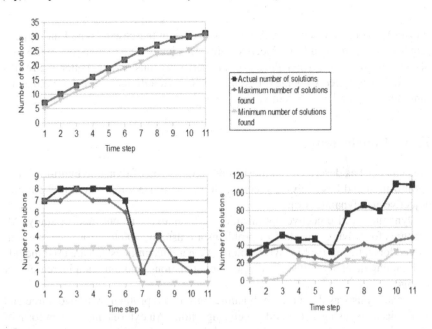

Fig. 12.2 Algorithm behavior in the case of the Type II problem, small data set, for 2 objectives (top), 3 objectives (bottom-left) and 4 objectives (bottom-right)

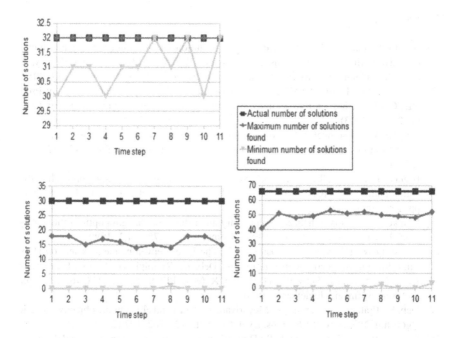

Fig. 12.3 Algorithm behavior in the case of the Type III problem, large data set, for 2 objectives (top), 3 objectives (bottom-left) and 4 objectives (bottom-right)

two objectives and problems of type III. We also found that, counter-intuitively, the performance of the algorithm was better for a larger data set than for a smaller data set. This may be caused by the larger population and larger number of iterations for the large data set, and may confirm the remark found in [17] about the independence of the problem size and algorithm generations in terms of performance. We also found that the performance of the algorithm fluctuates considerably during different runs. This, coupled with the low performance obtained on three and four objectives for problems of type I and II, calls for more advanced algorithms.

Further work that we envisage is studying dynamic multi-objective problems with improved algorithms, including evolutionary models with multiple populations. Another desirable improvement is exploiting further the information about the search space gathered before a change, perhaps by inserting certain individuals in a separate archive using decision making models.

Acknowledgements. This work was supported by the Engineering and Physical Sciences Research Council (EPSRC) of U.K. under Grant numbers EP/E060722/1, EP/E060722/2, and EP/K001310/1.

References

[1] Amato, P., Farina, M.: An ALife-inspired evolutionary algorithm for dynamic multi-objective optimization problems. In: Hoffmann, F., Köppen, M., Klawonn, F., Roy, R. (eds.) Soft Computing: Methodologies and Applications, Part III, pp. 113–125 (2005)

[2] Andrews, M., Tuson, A.: Diversity does not necessarily imply adaptability. In: Proc. GECCO Workshop on Evolutionary Algorithms for Dynamic Optimization Problems, pp. 24–28 (2003)

[3] Aragón, V.S., Esquivel, S.C.: An evolutionary algorithm to track changes of optimum value locations in dynamic environments. J. of Comput. Sci. and Tech. 4(3), 127–134 (2004)

[4] Branke, J., Kaußler, T., Schmidt, C., Schmeck, H.: A multi-population approach to dynamic optimization problems. In: Proc. 4th Int. Conf. Adaptive Comput. Des. Manuf., pp. 299–308 (2000)

[5] Branke, J.: Memory enhanced evolutionary algorithms for changing optimization problems. In: Proc. 1999 IEEE Congr. Evol. Comput., pp. 1875–1882 (1999)

[6] Corne, D.W., Knowles, J.D., Oates, M.J.: The Pareto envelope-based selection algorithm for multi-objective optimization. In: Deb, K., Rudolph, G., Lutton, E., Merelo, J.J., Schoenauer, M., Schwefel, H.-P., Yao, X. (eds.) PPSN 2000. LNCS, vol. 1917, pp. 839–848. Springer, Heidelberg (2000)

[7] Deb, K., Pratap, A., Agarwal, S., Meyarivan, T.: A fast and elitist multi-objective genetic algorithm: NSGA-II. IEEE Trans. Evol. Comput. 6(2), 182–197 (2002)

[8] Deb, K., Rao N., U.B., Karthik, S.: Dynamic multi-objective optimization and decision-making using modified NSGA-II: A case study on hydro-thermal power scheduling. In: Obayashi, S., Deb, K., Poloni, C., Hiroyasu, T., Murata, T. (eds.) EMO 2007. LNCS, vol. 4403, pp. 803–817. Springer, Heidelberg (2007)

[9] Farina, M.: A minimal cost hybrid strategy for Pareto optimal front approximation. Evol. Optim. 3(1), 41–52 (2001)

[10] Farina, M., Deb, K., Amato, P.: Dynamic multi-objective optimization problems: test cases, approximations, and applications. IEEE Trans. Evol. Comput. 8(5), 425–442 (2004)

[11] Flaxman, A.D., Przydatek, B.: Solving medium-density subset sum problems in expected polynomial time. In: Diekert, V., Durand, B. (eds.) STACS 2005. LNCS, vol. 3404, pp. 305–314. Springer, Heidelberg (2005)

[12] Galil, Z., Margalit, O.: An almost linear-time algorithm for the dense subset-sum problem. SIAM J. Computing 20(6), 1157–1189 (1991)

[13] Garey, M.R., Johnson, D.S.: Computers and Intractability: A Guide to the Theory of NP-completeness. WH Freeman & Co., New York (1979)

[14] Goh, C.K., Tan, K.C.: A competitive-cooperative coevolutionary paradigm for dynamic multi-objective optimization. IEEE Trans. Evol. Comput. 13(1), 103–127 (2009)

[15] Hatzakis, I., Wallace, D.: Dynamic multi-objective optimization with evolutionary algorithms: A forward-looking approach. In: Proc. 8th Annual Conf. on Genetic and Evol. Comput., pp. 1201–1208 (2006)

[16] Jin, Y., Branke, J.: Evolutionary optimization in uncertain environments - a survey. IEEE Trans. Evol. Comput. 9(3), 303–317 (2005)

[17] Khuri, S., Bäck, T., Heitkötter, J.: An evolutionary approach to combinatorial optimization problems. In: Proc. 22nd Annual ACM Computer Science Conf., pp. 66–73 (1994)

[18] Mori, N., Kita, H., Nishikawa, Y.: Adaptation to a changing environment by means of the thermodynamical genetic algorithm. In: Ebeling, W., Rechenberg, I., Voigt, H.-M., Schwefel, H.-P. (eds.) PPSN 1996. LNCS, vol. 1141, pp. 513–522. Springer, Heidelberg (1996)

[19] Pisinger, D.: Linear time algorithms for knapsack problems with bounded weights. J. of Algorithms 33, 1–14 (1999)

[20] Rohlfshagen, P., Yao, X.: Dynamic combinatorial optimisation problems: An analysis of the subset sum problem. Soft Comput. 15(9), 1723–1734 (2011)

[21] Shang, R., Jiao, L., Gong, M., Lu, B.: Clonal selection algorithm for dynamic multiobjective optimization. In: Hao, Y., Liu, J., Wang, Y.-P., Cheung, Y.-M., Yin, H., Jiao, L., Ma, J., Jiao, Y.-C. (eds.) CIS 2005. LNCS (LNAI), vol. 3801, pp. 846–851. Springer, Heidelberg (2005)

[22] Weicker, K.: Performance measures for dynamic environments. In: Guervós, J.J.M., Adamidis, P.A., Beyer, H.-G., Fernández-Villacañas, J.-L., Schwefel, H.-P. (eds.) PPSN 2002. LNCS, vol. 2439, pp. 64–73. Springer, Heidelberg (2002)

[23] Zeng, S., Chen, G., Zheng, L., Shi, H., de Garis, H., Ding, L., Kang, L.: A dynamic multi-objective evolutionary algorithm based on an orthogonal design. In: Proc. 2006 IEEE Congr. on Evol. Comput., pp. 573–580 (2006)

[24] Zhang, Z.: Multi-objective optimization immune algorithm in dynamic environments and its application to greenhouse control. Applied Soft Comput 8, 959–971 (2008)

[25] Zitzler, E., Laumanns, M., Thiele, L.: SPEA2: Improving the strength Pareto evolutionary algorithm for multi-objective optimization. In: Proc. Evol. Methods Des., Optimisation Control., pp. 95–100 (2002)

Part IV
Applications

Chapter 13
Ant Colony Optimization Algorithms with Immigrants Schemes for the Dynamic Travelling Salesman Problem

Michalis Mavrovouniotis and Shengxiang Yang

Abstract. Ant colony optimization (ACO) algorithms have proved to be powerful methods to address dynamic optimization problems (DOPs). However, once the population converges to a solution and a dynamic change occurs, it is difficult for the population to adapt to the new environment since high levels of pheromone will be generated to a single trail and force the ants to follow it even after a dynamic change. A good solution is to maintain the diversity via transferring knowledge from previous environments to the pheromone trails using immigrants. In this chapter, we investigate ACO algorithms with different immigrants schemes for two types of dynamic travelling salesman problems (DTSPs) with traffic factor, i.e., under random and cyclic dynamic changes. The experimental results based on different DTSP test cases show that the investigated algorithms outperform other peer ACO algorithms and that different immigrants schemes are beneficial on different environmental cases

13.1 Introduction

Ant colony optimization (ACO) algorithms are inspired from the behaviour of real ant colonies when they search for food from their nest to food sources. A colony of ants communicates via the pheromone trails in order to complete their food-searching task as efficiently as possible. ACO algorithms have proved that they are good meta-heuristics to many difficult optimization problems [11, 12, 15, 36].

The first optimization problem addressed by ACO algorithms was the travelling salesman problem (TSP), where a population of ants is placed on each city randomly and walk to the edges of the cities until each ant generates a feasible tour, in which all customers are satisfied [13]. Each ant writes pheromone to the trail of its tour for the other ants to read it while they construct their tours.

Michalis Mavrovouniotis · Shengxiang Yang
Centre for Computational Intelligence (CCI), School of Computer Science and Informatics,
De Montfort University, The Gateway, Leicester LE1 9BH, U.K.
e-mail: {mmavrovouniotis,syang}@dmu.ac.uk

S. Yang and X. Yao (Eds.): *Evolutionary Computation for DOPs*, SCI 490, pp. 317–341.
DOI: 10.1007/978-3-642-38416-5_13   © Springer-Verlag Berlin Heidelberg 2013

Researchers have mainly focused on ACO for stationary optimization problems (SOPs), where the environment remains fixed during the execution of the algorithm [2, 3, 28]. However, in many real-world applications we have to deal with dynamic optimization problems (DOPs), where the problem, including the objective function, the variables, the problem instance, the constraints, and so on, may change over time [26]. Usually, such uncertainties cause the optimum to move. For example, a dynamic version of the TSP can be generated where the cost of the edges between two cities may increased, representing potential traffic jams. The objective of the dynamic TSP (DTSP) is not only to converge and output a near optimum (or the optimum) solution quickly, as in the static TSP, but to also track and output the moving optimum.

Considering the DTSP, traditional ACO algorithms may face a serious challenge due to the fact that the pheromone trails of the previous environment will not be compatible with the new environment when a dynamic change occurs. A simple way to address this problem is to re-initialize the pheromone trails with an equal amount and consider every dynamic change as the arrival of a new problem that needs to be solved from scratch. This strategy acts as a restart of the algorithm which is computationally expensive and usually not efficient. Moreover, in order to perform this action, the dynamic change needs to be detected which usually is not possible on DOPs [31].

However, it is believed that ACO algorithms can adapt to DOPs since they are inspired from nature which is a continuous adaptation process [5, 26]. Since ACO algorithms have been designed for SOPs lose their adaptation capabilities quickly because of stagnation behaviour, where all ants follow the same path from the early stages of the execution. Recently, several approaches have been proposed to avoid stagnation behaviour and address DTSPs, which includes: (1) local and global restart strategies [21]; (2) pheromone manipulation schemes to maintain diversity [16]; (3) increase diversity via immigrants schemes [29, 31]; (4) memory-based approaches [19, 22]; (5) and memetic algorithms [30].

Among these approaches, immigrants schemes have been found beneficial when integrated with ACO algorithms for different DTSPs. Every iteration, immigrant ants are generated and replace a small portion of the worst ants in the current population. This action is performed before pheromone is updated, in order to bias the ants of the next iteration with the diversity and knowledge transferred from the immigrant ants. Immigrants schemes mainly differ on the way immigrant ants are generated.

In this chapter, all different ACO algorithms based on immigrants schemes are examined intensively, and compared with other peer ACO algorithms for different DTSPs cases. The contents of this chapter are categorized as follows. Section 13.2 describes the DTSPs used in the experiments. Section 13.3 describes traditional ACO algorithms for the DTSP, whereas Section 13.4 gives details of the investigated ACO algorithms based on immigrants schemes. Section 13.5 presents the experimental results and analysis. Finally, Section 13.6 concludes this contribution and points out future work.

13.2 Dynamic Travelling Salesman Problem with Traffic Factor

The TSP is the most fundamental, popular and well-studied NP-hard combinatorial optimization problem. It can be described as follows: Given a collection of cities, we need to find the shortest path that starts from one city and visits each of the other cities once and only once before returning to the starting city. Usually, the problem is represented by a fully connected weighted graph $G = (V, E)$, where V is a set of n vertices and E is a set of edges. The collection of cities is represented by the set V and the connections between them by the set E. Each connection is associated with a cost D_{ij}, which represents the distance (or travel time) between cities i and j.

Many algorithms, either exact algorithms or approximation algorithms, including ACO have been proposed to solve the static TSP [13, 27, 33]. Although exact algorithms guarantee to provide the global optimum solution, in the case of NP-hard problems, they need, in the worst case, exponential time to find it. On the other hand, approximation algorithms can provide a solution efficiently but cannot guarantee the global optimum [24, 35].

The TSP becomes more challenging and realistic if it is subject to a dynamic environment. For example, a salesman wants to distribute items sold in different cities starting from his home city and returning after he visited all the cities to his home city again. The task is to optimize his time and plan his tour as efficiently as possible. Therefore, by considering the distances between cities it can generate the route and start the tour. However, it is difficult to consider traffic delays that may affect the route. Traffic delays may change the time planned beforehand, and the salesman will need a new alternative route fast to avoid long traffic delays and optimize his time again.

There are several variations of DTSPs considered in the literature, such as changing the topology of cities by replacing cities [19, 21, 29, 30], and changing the distances between cities by adding traffic factors to the links between cities [16, 31, 32]. In DTSPs where cities are replaced, each city has a probability m to be replaced regularly in time, usually measured in a certain number of iterations of running an algorithm. On the other hand, in DTSPs with traffic factors, each link has a probability m to add or deduce traffic regularly in time.

13.2.1 DTSP with Random Traffic

In this chapter, we generate DTSPs with traffic factors as follows. We assume that the cost of the link between cities i and j is $D_{ij} = D_{ij} \times F_{ij}$, where D_{ij} is the normal travelled distance and F_{ij} is the traffic factor between cities i and j. Every f iterations of running an algorithm, a random number in $[F_L, F_U]$ is generated probabilistically to represent the traffic factor between cities, where F_L and F_U are the lower and upper bounds of the traffic factor, respectively. Each link has a probability m to add traffic every f iteration, where the traffic factor F_{ij} of the remaining links is set to 1, which indicates no traffic.

For example, a dynamic case with high traffic is constructed by setting traffic factor values closer to F_U with a higher probability to be generated, while for a

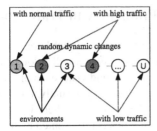
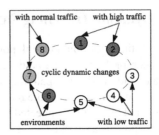

Fig. 13.1 Illustration of a random dynamic environment with unlimited states and a cyclic dynamic environment with 8 states. Each node represents a different environment where white, light grey, and dark grey, represent low, medium, and high traffic jams, respectively

dynamic case with low traffic, a higher probability is given to traffic factor values closer to F_L. This type of environments are denoted *random DTSPs* in this chapter because previously visited environments are not guaranteed to reappear.

13.2.2 DTSP with Cyclic Traffic

Another variation of the DTSP with traffic factors is the DTSP where the dynamic changes occur with a cyclic pattern. In other words, previous environments will appear again in the future. Such environments are more realistic since they represent a 24-hour traffic jam situation in a day.

A cyclic environment can be constructed by generating different dynamic cases with traffic factors as the base states, representing DTSP environments where each link has a probability m to add low, normal, or high traffic as in random DTSPs. Then, the environment cycles among these base states, every f iteration, in a fixed logical ring as in Fig. 13.1. Depending on the period of the day, dynamic cases with different traffic factors can be generated. For example, during the rush hour periods, a higher probability is given to the traffic factors closer to F_U, whereas during evening hour periods, a lower probability is given to F_U and a higher probability to F_L. This type of environments are denoted as *cyclic DTSPs* in this chapter because previously visited environments will reappear several times.

13.3 Ant Colony Optimization for the DTSP

ACO consists of a population of μ ants that construct solutions, i.e., tours in the TSP, and update their trails with pheromone according to the solution quality [9]. Considering the TSP, the ants "walk" on the links between the cities, where they "read" pheromone from the links or "write" additional pheromone to the links.

Initially, all trails are assigned with an equal amount of pheromone, i.e., τ_{init}, and each ant is placed on a randomly selected city. With a probability $1 - q_0$, where $0 \le q_0 \le 1$ is a parameter of the decision rule, an ant k chooses the next city j from city i, probabilistically, as follows:

$$p_{ij}^k = \frac{[\tau_{ij}]^\alpha [\eta_{ij}]^\beta}{\sum_{l \in N_i^k} [\tau_{il}]^\alpha [\eta_{il}]^\beta}, \text{if } j \in N_i^k, \tag{13.1}$$

where τ_{ij} and $\eta_{ij} = 1/D_{ij}$ is the existing pheromone trail and heuristic information available a priori, respectively, where D_{ij} is the cost between cities i and j (including the traffic factor). N_i^k denotes the neighbourhood of cities of ant k that have not yet been visited when its current city is i. α and β are the two parameters that determine the relative influence of the pheromone trail and heuristic information, respectively. With the probability q_0, ant k chooses the next city, i.e, z, with the maximum probability which satisfies the following formula:

$$z = \underset{j \in N_i^k}{\text{argmax}} \; [\tau_{ij}]^\alpha [\eta_{ij}]^\beta . \tag{13.2}$$

This process continues until each ant has visited all cities once. Thereafter, the ants update their pheromone trails. The different variations of ACO algorithms mainly differ in the way pheromone trails are updated [4, 10, 13, 14, 37].

13.3.1 Standard ACO

The state-of-the-art ACO on the static TSP is the MAX-MIN ant system (MMAS) [38]. In MMAS, the ants construct solutions using Eq. (13.1) and only the best ants are allowed to retrace their solution and deposit pheromone as follows:

$$\tau_{ij} \leftarrow \tau_{ij} + \Delta \tau_{ij}^{best}, \forall \, (i,j) \in T^{best}, \tag{13.3}$$

where T^{best} is the tour of the best ant and $\Delta \tau_{ij}^{best} = 1/C^{best}$, where C^{best} is the cost of the tour T^{best}. However, the ant allowed to deposit pheromone may be either the best-so-far ant, in which case $\Delta \tau_{ij}^{best} = 1/C^{bs}$, where C^{bs} is the tour cost of the best-so-far ant, or the iteration-best ant, in which case $\Delta \tau_{ij}^{best} = 1/C^{ib}$, where C^{ib} is the tour cost of the best ant of the iteration. Both update rules are used in an alternative way under a pre-defined criteria (for more details see [38]).

In addition, a constant amount of pheromone is deducted from all trails due to the pheromone evaporation, which is defined as:

$$\tau_{ij} \leftarrow (1 - \rho) \, \tau_{ij}, \forall \, (i,j), \tag{13.4}$$

where $0 < \rho \leq 1$ is the rate of evaporation.

Moreover, the pheromone trail values in MMAS are kept to the interval $[\tau_{min}, \tau_{max}]$ and they are re-initialized to τ_{max} every time the algorithm shows a stagnation behaviour, where all ants follow the same path or when no improved tour has been found for several iterations. The MMAS, is denoted as S-ACO, and it is used in the experimental study later in this chapter.

13.3.2 Population-Based ACO (P-ACO)

The P-ACO algorithm was first applied on the static TSP [20]. Later on, it has been applied to the DTSP where a small portion of cities is replaced by other new ones [19, 22]. In P-ACO, ants construct solutions using Eq. (13.1). However, it differs from the S-ACO, since it maintains a population-list (memory) of ants (solutions), denoted k_{long} of limited size K_l, and stores the iteration-best ant in every iteration.

The pheromone update depends on k_{long}, where every time the iteration-best ant enters k_{long}, a positive constant update is added to its corresponding pheromone trails, which is defined as follows:

$$\tau_{ij} \leftarrow \tau_{ij} + \Delta \tau_{ij}^{ib}, \forall\, (i,j) \in T^{ib}, \tag{13.5}$$

where $\Delta \tau_{ij}^{ib} = (\tau_{max} - \tau_{init})/K_l$ and T^{ib} is the tour of the iteration-best ant. More-over, τ_{max} and τ_{init} denote the maximum and initial pheromone amount, respectively. When k_{long} is full, the iteration-best ant needs to replace an ant k stored in k_{long}, and, thus, a negative constant update to its corresponding pheromone trails is done, which is defined as follows:

$$\tau_{ij} \leftarrow \tau_{ij} - \Delta \tau_{ij}^{k}, \forall\, (i,j) \in T^{k}, \tag{13.6}$$

where $\Delta \tau_{ij}^{k}$ is defined as in Eq. (13.5) and T^{k} is the tour of the ant to be replaced. Pheromone evaporation is not used in the P-ACO algorithm.

Several strategies regarding which ant should the iteration-best ant replace in k_{long} have been proposed, such as *Age*, *Prob*, and *Quality* [19]. In the default strat-egy, i.e., *Age*, the iteration-best ant replaces the ant which has entered k_{long} first. In the *Prob* strategy, the iteration-best ant can replace any ant probabilistically, and in the *Quality* strategy, the worst ant is replaced. Experiments show that the *Age* strat-egy is more consistent and performs better than the others, since other strategies have more chances to maintain identical ants into k_{long}, which leads the algorithm to the stagnation behaviour [19]. This is due to the fact that high levels of pheromone will be generated into a single trail and dominate the search space. Therefore, the P-ACO algorithm with the *Age* strategy is used in the experimental study later in this chapter.

13.3.3 React to Dynamic Changes

In Bonabeau *et al.* [5], it was discussed that traditional ACO algorithms may have good performance for DTSPs, since they are very robust algorithms. The mechanism which enables ACO algorithms to adapt to DOPs is the pheromone evaporation. Lowering the pheromone values enables the algorithm to forget bad decisions made in previous iterations. Moreover, when a dynamic change occurs, it will eliminate the pheromone trails of the previous environment that are not useful in the new environment, where the ants may be biased and not adapt well.

The S-ACO algorithm can be applied directly to the proposed DTSPs with traffic factors, either random or cyclic, without any modifications, apart from the heuristic information where the traffic factor needs to be considered. Further special measures when a dynamic change occurs are not required.

Similar to S-ACO, P-ACO can be applied directly to the proposed DTSPs. The ants stored in the k_{long} are re-evaluated in every iteration to be consistent with the changing environments.

13.4   Investigated ACO Algorithms with Immigrants Schemes

ACO algorithms are constructive heuristics, where every iteration ants move from one city to the next city probabilistically, until they generate feasible solutions as described in Section 13.3. At the end of each iteration the constructed solutions are cleared to generate new ones, but every ant deposits pheromone and leave a trail to the corresponding solutions, e.g., the links between the cities of a TSP tour as in Eq. 13.3. In contrast, genetic algorithms (GAs) are based on a pre-defined set of feasible solutions (population of individuals), e.g., a set of tours for TSP. The population is directly transferred from one iteration to the next using the fittest individuals [25, 33]. Search operators, i.e., crossover and mutation, are used to generate the new population of solutions, which is usually better than the previous one.

The P-ACO framework is based on ants that construct solutions, where the best ant of each iteration is stored in an actual population as in a GA and they are transferred directly to the next iteration. The solutions in the population are then used to update the pheromone information for the new ants of the new iteration. The population-list is updated every iteration as described in Section 13.3.

13.4.1   General Framework of ACO with Immigrants Schemes

The framework of ACO algorithms with immigrants schemes is inspired from the GA characteristics of the P-ACO framework and the good performance of immigrants schemes in GAs for binary-encoded DOPs [39, 40, 43, 45]. Considering that P-ACO maintains a population of solutions, immigrant ants can be generated and replace ants in the current population. The aim of the proposed framework is to maintain the diversity within the population and transfer knowledge from previous environments to the pheromone trails of the new environment.

The main idea is to generate the pheromone information for every iteration considering information from the pheromone trails of the previous environment and extra information from the immigrant ants generated. Therefore, instead of using a long-term memory k_{long} as in P-ACO, a short-term memory is used, denoted k_{short}, where all ants stored from iteration $t - 1$ are replaced by the first K_s best ants of the current iteration t, where K_s is the size of k_{short}, instead of only replacing the oldest one as in P-ACO. Moreover, immigrant ants are generated and replace the worst ants in k_{short} with the replacement rate r, usually small. Therefore, when ants are removed, a negative update is made to their pheromone trails as in Eq. (13.6), and when new ants are added, a positive update is made to their pheromone trails as in Eq. (13.5). This process is repeated as represented in Fig. 13.2.

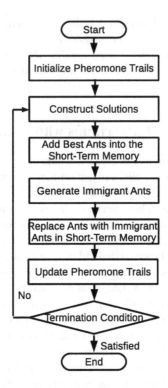

Fig. 13.2 General framework of ACO algorithms with immigrants schemes

The benefits of using k_{short} are closely related to the survival of ants in a dynamic environment, where no ant can survive in more than one iteration. For example, in iteration t, if ants are stored from iteration $t-2$ and an environmental change occurred in iteration $t-1$, then the solutions may not be feasible for the current environment in iteration t, and hence need to be repaired. Usually, a repair procedure is computationally expensive, and requires prior knowledge of the problem. Furthermore, this action can be taken only if the environmental changes can be detected, which is usually not applicable in real-world applications. As discussed previously, the S-ACO algorithm with a re-initialization of pheromone trails may not be a sufficient choice on DOPs where the frequency of change is not available beforehand.

The investigated algorithms follow the framework described above, but they differ on the way immigrant ants are generated. The algorithms have been applied on different DTSPs as follows. Random immigrants ACO (RIACO), elitism-based immigrants ACO (EIACO) and hybrid immigrants ACO (HIACO) were applied on a DTSP were cities are added/removed [29]. Memory-based immigrants ACO (MIACO) were applied on a DTSP with cyclic traffic factor as described in Section 13.2 [31]. Environmental-information immigrants ACO (EIIACO) were applied on the DTSP with random traffic factor as described in Section 13.2 [32]. In this chapter, we re-investigate and compare all the algorithms on the same DTSPs, i.e., on both DTSP with random and cyclic traffic factor.

13.4.2   ACO with Random Immigrants

The traditional random immigrants have been found beneficial for ACO for the DTSP, since they maintain a certain level of diversity during the execution [29]. The principle is to introduce new randomly generated immigrant ants to the population. Therefore, before the pheromone trails are updated, a set S_{ri} of $r \times K_s$ immigrants are randomly generated to replace the worst ants in k_{short}, where r is the replacement rate and K_s is the size of k_{short}.

The RIACO algorithm was proposed to address DTSPs with significantly changing environments. This is because it was claimed that the adaptation of ACO algorithms makes sense only when the environmental changes of a problem are small to medium [6, 26]. This is due to the fact that a new environment has more chance to be similar with the old one. After a change occurs, transferring knowledge from the old environment to the pheromone trails may move the ants into promising areas in the new environment.

Considering the above argument, when the changing environments are not similar or when their is not enough time to gain knowledge from the previous environment, i.e., fast changing environment, the knowledge transferred may misguide the ants from tracking the optimum. Therefore, in such environmental cases is better to generate random diversity, instead of guided diversity by transferring knowledge. However, there is a high risk of randomization, if too much diversity is generated from the immigrants.

13.4.3   ACO with Elitism-Based Immigrants

Differently from RIACO, which generates diversity randomly, EIACO generates guided diversity by transferring knowledge from previous environments and it was proposed to address DTSPs with slowly and slightly changing environments [29]. For each iteration t, within EIACO, the elite from the previous environment, i.e., the best ant from $k_{short}(t-1)$, is used as the base to generate a set S_{ei} of $r \times K_s$ elitism-based immigrants, where r is the replacement rate and K_s is the size of the k_{short} memory.

An elitism-based immigrant is generated using the inversion operator based on the inver-over operator as follows [23]. First, one city, i.e., c, is selected randomly from the best ant of $k_{short}(t-1)$; then with probability p (usually 0.02) the second city c' is selected from $k_{short}(t-1)$; otherwise another ant from μ is randomly selected and assign as the second city c' the next city to the city c. The segment from the next city of c to city c' is reversed and c is set to c'. This process continues until the selected second city c' appears next or previous to the first city c. From the resulting tour an elitism-based immigrant is generated, which inherits some segments from the elite of the previous environment and some random segments from other ants.

The EIACO algorithm is beneficial in cases where the changing environments are similar, e.g., slightly changing environments, and when the population has sufficient time to converge into a good solution and gain knowledge in the previous environment, e.g., slowly changing environments. Transferring the knowledge gained from the previous environment, to the pheromone trails of the new environment will make sense and guide the population of ants to promising areas.

However, if too much information is transferred, the run basically starts near a local optimum, and get stuck there. Therefore, in some cases with slightly changing environments, EIACO may not perform well. On the contrast, RIACO may generate high level of diversity in slightly changing environments, and degrade the performance of ACO.

13.4.4 ACO with Hybrid Immigrants

The HIACO algorithm uses an immigrants scheme that combines both random and elitism-based immigrants [29]. For each iteration t within HIACO, a set $S_{hi} = S_{ri} + S_{ei}$ hybrid immigrants are generated, where S_{ri} and S_{ei} are two sets of $(r \times K_s)/2$ random and elitism-based immigrants, respectively, r is the replacement rate and K_s is the size of the k_{short} memory. HIACO attempts to combine the merits of both RIACO and EIACO, where one is good on slowly and slightly changing environments and the other on fast and significantly changing environments.

Considering the fact that RIACO face the risk of randomization because of too much diversity, and the fact the EIACO face the risk of too much transferred knowledge, the HIACO may promote the performance of both algorithms. The two types of immigrants may cooperate to address all cases of dynamic environments. For example, in cases the random immigrant will generate high levels of diversity, the elitism-based immigrants will decrease the levels of diversity. On the other hand, if too much knowledge is transferred from elitism-based immigrants and the population gets trapped in local optimum, the random immigrants will help the population to escape from it.

13.4.5 ACO with Memory-Based Immigrants

Differently from EIACO, where the best ant from the previous environment is used as the base to generate immigrants, MIACO uses the best ant from several environments as the base to generate immigrants [31]. The only difference between MIACO and EIACO lies in that MIACO uses both k_{short} and k_{long}, where the first type of memory is updated and used as in RIACO and EIACO. The second type of memory is initialized with random ants and updated by replacing any of the randomly initialized ants if they still exists in the memory, with the best-so-far ant; otherwise, the closest ant in the memory is replaced with the best-so-far ant if it is better. Note that the update strategy of k_{long} in MIACO is different from P-ACO regarding which ant to replace, since in MIACO the most similar memory updating strategy is used [6], whereas in P-ACO, the new ant replaces the oldest one. In MIACO, a metric of how close ant i is to ant j is used and defined as follows:

$$M_{ij} = 1 - \frac{CE_{ij}}{n}, \tag{13.7}$$

where CE_{ij} is defined as the number of common edges between ant i and ant j, and n is the number of cities. A value M_{ij} closer to 0 means that the ants are closer since they are more similar [1].

Apart from which ant is replaced in k_{long}, the update strategy of MIACO is different from the one used in P-ACO with respect to when an ant is replaced. In P-ACO, the update occurs every iteration, whereas in MIACO the update occurs whenever a dynamic change is detected in order to store useful solutions from different environments.

For each iteration within MIACO, the ants in k_{long} are re-evaluated in order to be valid with the new environment and to detect an environmental change. An environmental change is detected if there is a change in the total cost of ants currently stored in k_{long}. Then, the best ant from k_{long} is selected and used as the base to generate a set S_{mi} of $r \times K_s$ memory-based immigrants, where r is the replacement rate and K_s is the size of the k_{short} memory. A memory-based immigrant is generated using the inver-over operator as in EIACO, but instead of selecting the best ant from $k_{short}(t-1)$, the best ant from $k_{long}(t)$ is selected.

MIACO inherits the advantages of the memory scheme to guide the population directly to an old environment already visited and maintains diversity with immigrants in order to avoid the stagnation behaviour of ACO algorithms. It is very important to store different solutions in k_{long} which represent good solutions for the different environments that may be useful in the future. The key idea behind MIACO is to provide guided diversity into the pheromone trails in order to avoid the disruption of the optimization process [41].

MIACO may be beneficial on the same environmental cases with EIACO since it is a generalized version of EIACO. However, it may be also advantageous in cases where the previous environments will reappear in the future, e.g., cyclic DTSPs.

13.4.6 ACO with Environmental-Information Immigrants

The information obtained from EIACO and MIACO to transfer knowledge is based on individual information, i.e., the best ant from k_{short} and k_{long}, respectively. The EIIACO algorithm generates immigrants using environmental information, i.e., a population of best ants, to transfer knowledge from the previous environment to the new one, in order to address slowly and slightly changing environments [32]. The knowledge transferred from EIIACO contains much more information than the EIACO and MIACO algorithms. EIIACO follows the same framework with other ACO algorithms based on immigrants schemes.

Environmental information-based immigrants are generated using all the ants stored in k_{short} of the previous environment and replace the worst ants in the current k_{short}. Within EIIACO, a probabilistic distribution based on the frequency of cities is extracted, representing information of the previous environment, and is used as the base to generate immigrant ants. The frequency vector of each city i, i.e, $\mathbf{D}_{c_i}$, is constructed by taking the ants of k_{short} as a dataset and locating city c_i from them. The successor and predecessor cities, i.e., c_{i-1} and c_{i+1}, respectively, of city c_i are obtained and update $\mathbf{D}_{c_i}$ accordingly. Note that both cities are recorded since the TSP solution is cyclic. For example, one is added to the corresponding position $i-1$ and $i+1$ in $\mathbf{D}_{c_i}$. The process is repeated for all cities and a table $S = (\mathbf{D}_{c_1},\dots,\mathbf{D}_{c_n})$ is generated (where n is the number of cities) .

An environmental information-based immigrant ant, i.e., $A_{eii} = (c_1,\dots,c_n)$, is generated as follows. First, randomly select the start city c_1; then, the probabilistic distribution of $\mathbf{D}_{c_{i-1}} = (d_1,\dots,d_n)$ is used to select the next city c_i probabilistically as follows:

$$p_i = \frac{d_i}{\sum_{j \in \mathbf{D}_{c_{i-1}}} d_j}, \text{if } i \in \mathbf{D}_{c_{i-1}}, \tag{13.8}$$

where d_i is the frequency number where city c_i appears before or after city c_{i-1}. Note that all cities currently selected and stored in A_{eii} have a probability of 0.0 to be selected since they are already visited. In the case where the sum of $p_i = 0.0$, which means that all cities in $\mathbf{D}_{c_i}$ are visited, a random city j that has not been visited yet is selected. This probabilistic selection is repeated until all cities are used in order to generate a valid immigrant ant based on the environmental information.

13.5 Experiments

13.5.1 Experimental Setup

The investigated algorithms were tested on the DTSP instances that are constructed from three static benchmark TSP instances taken from TSPLIB[1], i.e., pr76, pr152, pr264, indicating small, medium, and large scale problem instances in this chapter, respectively, in order to investigate the effect of the corresponding immigrants schemes on ACO algorithms for the DTSP.

Our implementation follows the guidelines of the ACOTSP[2] application. Using the methods described in Section 13.2, we have generated two kinds of DTSPs, with random and cyclic traffic factors, respectively, with $F_L = 0$ and $F_U = 5$. For cyclic DTSP, four cyclic states are used. For both types of DTSPs, the value of f was set to 5 and 50, indicating fast and slow environmental changes, respectively. The value of m was set to $0.1, 0.25, 0.5$, and 0.75, indicating the degree of environmental changes from small, to medium, and large, respectively. As a result, eight dynamic test DTSPs, i.e., two values of $f \times$ four values of m, were generated from each

[1] See http://comopt.ifi.uni-heidelberg.de/software/TSPLIB95/
[2] See http://www.aco-metaheuristic.org/aco-code

static TSP instance. Therefore, in order to systematically analyse the adaptation and searching capabilities of each algorithm on the DTSP, 24 dynamic test cases are used, i.e., three problem instances × eight cases each, for each type of DTSP, i.e., with random and cyclic traffic factors.

For each algorithm on a DTSP, 30 independent runs were executed on the same environmental changes. The algorithms were executed for 1000 iterations and one observation was taken on each iteration. The overall performance of an algorithm on a DTSP instance is defined as follows:

$$\bar{P}^{best} = \frac{1}{G} \sum_{i=1}^{G} \left(\frac{1}{N} \sum_{j=1}^{N} P_{ij}^{best} \right), \tag{13.9}$$

where G is the total number of iterations, N is the number of runs and P_{ij}^{best} is the tour cost of the best-so-far ant, after a change, of iteration i of run j, respectively.

13.5.2   Parameter Settings

The parameters of the investigated algorithms are chosen from our preliminary experiments and some of them are taken from the literature [29, 31]. For all algorithms $\alpha = 1$, $\beta = 5$, $q_0 = 0.0$, $r = 0.3$, $K_s = 10$, and for MIACO $K_l = 4$.

The population of ants μ for each algorithm varies in order to have the same number of evaluations every iteration, i.e., 25. The population of RIACO and EI-IACO was set to $\mu = 25$, for EIACO and HIACO was set to $\mu = 24$ and for MIACO was set to $\mu = 21$. This is because MIACO has a k_{long} memory of size K_l where the solutions need to be re-evaluated on every change to detect dynamic changes, and, thus, $\mu = \mu - K_l$, whereas EIACO and HIACO re-evaluates the best ant from the previous iteration, which counts as a single evaluation, and, thus, $\mu = \mu - 1$. The ants in the k_{short} memory, including the generated immigrants, do not count as evaluations since they are removed every iteration. Moreover, the pheromone they deposit is not based on the quality of the solution as in the S-ACO. Instead, it is a constant value as in P-ACO.

13.5.3   Experimental Results and Analysis of the Investigated Algorithms

The experimental results regarding the offline performance of the investigated algorithms in both DTSPs with random and cyclic traffic factors are presented in Tables 13.1 and 13.2, respectively. The corresponding two-tailed t-test results with 58 degrees of freedom at a 0.05 level of significance are presented in Table 13.3. In the comparisons, "+" or "−" indicates that the first algorithm is significantly better or the second algorithm is significantly better, respectively, and "∼" indicates no significance between the algorithms. Moreover, to better understand the dynamic behaviour of algorithms, the offline performance against the first 500 iterations is plotted in Fig. 13.3 for random DTSPs for $f = 50$ and $m = 0.10$ and $m = 0.75$,

Table 13.1 Experimental results of algorithms regarding the offline performance for random DTSPs

Alg. & Inst.	pr76							
	$f = 5$				$f = 50$			
$m \Rightarrow$	0.10	0.25	0.50	0.75	0.10	0.25	0.50	0.75
RIACO	53405.77	61545.11	78939.34	109341.65	50484.60	56943.36	71997.99	95742.96
EIACO	53481.98	61995.73	79939.40	111215.18	50740.69	57488.74	72295.83	96720.11
HIACO	53299.32	61708.22	79278.76	109990.85	50282.14	56778.05	71587.04	95796.19
MIACO	53669.59	62367.39	80503.22	111982.58	50786.29	57582.64	72551.06	96996.51
EIIACO	53856.99	62156.46	79909.29	110854.31	50745.09	57363.72	72405.07	96330.06

Alg. & Inst.	pr152							
	$f = 5$				$f = 50$			
$m \Rightarrow$	0.10	0.25	0.50	0.75	0.10	0.25	0.50	0.75
RIACO	43731.12	49856.78	61702.52	84089.01	40606.23	45251.32	55209.46	73016.88
EIACO	43582.79	49869.53	62185.09	84449.34	41013.54	45154.71	54581.22	71855.33
HIACO	43579.22	49922.58	61935.99	84262.97	40561.09	44913.56	54486.51	71830.50
MIACO	43777.26	50226.28	62633.86	85252.96	41047.21	45387.66	54854.02	72307.76
EIIACO	43975.13	50167.45	62216.30	84511.91	41087.96	45643.79	55308.69	72920.66

Alg. & Inst.	pr264							
	$f = 5$				$f = 50$			
$m \Rightarrow$	0.10	0.25	0.50	0.75	0.10	0.25	0.50	0.75
RIACO	27803.32	32317.14	41440.08	56978.59	25707.36	29019.14	36155.71	48289.91
EIACO	27631.00	32284.07	41686.12	57496.55	25636.91	28906.24	35834.23	47431.27
HIACO	27801.55	32452.96	41686.40	57321.89	25547.00	28849.66	35877.25	47629.74
MIACO	27720.96	32463.54	41914.54	57863.13	25697.97	29022.73	36038.26	47866.07
EIIACO	27757.27	32338.73	41637.13	57420.54	25788.50	29080.56	36237.60	48362.34

and the offline performance against the first 100 iterations is plotted in Fig. 13.4 for cyclic DTSPs for $f = 5$ and $m = 0.10$ and $m = 0.75$. From the experimental results, several observations can be made by comparing the behaviour of the algorithms.

First, RIACO outperforms EIACO, MIACO, EIIACO in almost all dynamic cases with $f = 5$ and $m = 0.75$ in both random and cyclic DTSPs; see the comparisons of RIACO $\Leftrightarrow$ EIACO, RIACO $\Leftrightarrow$ MIACO and RIACO $\Leftrightarrow$ EIIACO in Table 13.3. This is because both EIACO, MIACO and EIIACO use knowledge, either individual- or environmental-based information, from previous environments to generate immigrant ants, and thus, when not enough time to converge to a good solution is available, it is difficult to transfer useful knowledge, except if the magnitude of change is small, i.e., $m = 0.10$. RIACO generates diversity randomly that is more useful on dynamic cases with $m = 0.50$ and $m = 0.75$, where the changing environments are not similar.

Second, EIACO outperforms RIACO in almost all dynamic cases with $f = 50$ and $m = 0.10$, and $m = 0.25$ in both random and cyclic DTSPs. This is because transferring knowledge makes more sense when the environments are similar. However, if too much knowledge is transferred from the previous environments may lead the population to start from a local optimum solution and get stuck to it, as in the

Table 13.2 Experimental results of algorithms regarding the offline performance for cylic DTSPs

Alg. & Inst.	pr76							
	$f = 5$				$f = 50$			
$m \Rightarrow$	0.10	0.25	0.50	0.75	0.10	0.25	0.50	0.75
RIACO	54082.16	58731.59	69896.53	103643.91	51862.08	54594.27	63953.72	92123.94
EIACO	53560.05	58794.42	70804.71	105641.64	52171.97	54801.34	64591.92	92765.91
HIACO	53601.65	58631.55	70239.99	104357.97	51696.62	54252.89	63789.63	91961.38
MIACO	53698.48	58781.13	70505.11	104851.77	52203.14	54819.25	64559.57	92717.01
EIIACO	53901.79	59191.25	70597.36	105385.63	52162.16	55046.11	64417.57	92590.94

Alg. & Inst.	pr152							
	$f = 5$				$f = 50$			
$m \Rightarrow$	0.10	0.25	0.50	0.75	0.10	0.25	0.50	0.75
RIACO	43123.17	50036.28	59777.25	68348.05	40685.18	45311.61	53565.86	60826.87
EIACO	42440.88	50000.39	59599.67	68250.80	40715.22	45520.32	52889.69	60479.81
HIACO	42714.18	50100.98	59754.37	68432.10	40571.04	45090.22	52844.89	60292.85
MIACO	42397.92	50130.10	59610.09	68134.46	40759.33	45545.47	53002.74	60343.23
EIIACO	42774.22	50365.58	59941.62	68406.41	40925.42	45778.35	53769.09	60479.81

Alg. & Inst.	pr264							
	$f = 5$				$f = 50$			
$m \Rightarrow$	0.10	0.25	0.50	0.75	0.10	0.25	0.50	0.75
RIACO	27118.23	31176.89	38557.06	52909.78	25413.82	28437.01	34043.66	46071.37
EIACO	26463.93	31166.71	38589.28	53356.77	25351.78	28485.02	33875.45	45133.16
HIACO	26850.52	31287.26	38750.83	53190.87	25314.59	28293.39	33877.79	45433.54
MIACO	26510.63	31180.66	38447.30	52951.17	25357.61	28442.63	33914.93	45122.77
EIIACO	26750.38	31192.29	38595.38	53278.29	25556.66	28620.66	34255.57	45930.70

case of pr152 when $f = 50$ and $m = 0.10$ where the RIACO is significantly better than EIACO; see the comparison of RIACO $\Leftrightarrow$ EIACO in Table 13.3. Moreover, RIACO outperforms EIACO in almost all cases of the smallest problem instance, i.e., pr76. This behaviour may have several reasons: (1) the elitism mechanism used in EIACO may not be effective since the environmental changes in a smaller search space have a higher probability to affect the solution that is used to generate guided immigrants for the new environment; (2) random immigrants have a higher probability to hit the optimum in a smaller search space and the risk of randomization is limited, whereas on larger search space it is dangerous; and (3) too much knowledge transferred from previous environments.

Third, HIACO outperforms both RIACO and EIACO in almost all dynamic cases with $f = 50$ in both random and cyclic DTSPs; see comparisons of RIACO $\Leftrightarrow$ HIACO and EIACO $\Leftrightarrow$ HIACO in Table 13.3. This behaviour shows that HIACO inherited the merit of EIACO which is beneficial on slowly changing environments. It can be also observed that the HIACO is significantly better than RIACO even on the smallest problem instance in which EIACO is outperformed. This behaviour shows that HIACO inherited the merit of RIACO. However, HIACO is significantly better than EIACO because it may possibly achieve a good balance between the

Table 13.3 Statistical test results regarding the offline performance of the algorithms for random and cyclic DTSPs

Alg. & Inst.	pr76				pr152				pr264			
Random DTSPs												
$f=5, m \Rightarrow$	0.1	0.25	0.5	0.75	0.1	0.25	0.5	0.75	0.1	0.25	0.5	0.75
RIACO⇔EIACO	~	+	+	+	−	~	+	+	−	−	+	+
RIACO⇔HIACO	−	+	+	+	−	+	+	+	~	+	+	+
RIACO⇔MIACO	+	+	+	+	~	+	+	+	−	+	+	+
RIACO⇔EIIACO	+	+	+	+	+	+	+	+	−	~	+	+
EIACO⇔HIACO	−	−	−	−	~	~	−	−	+	+	~	−
EIACO⇔MIACO	+	+	+	+	+	+	+	+	+	+	+	+
EIACO⇔EIIACO	+	+	~	−	+	+	~	~	+	+	−	−
HIACO⇔MIACO	+	+	+	+	+	+	+	+	−	~	+	+
HIACO⇔EIIACO	+	+	+	+	+	+	+	+	−	−	−	+
MIACO⇔EIIACO	+	−	−	−	+	~	−	−	+	−	−	−
$f=50, m \Rightarrow$	0.1	0.25	0.5	0.75	0.1	0.25	0.5	0.75	0.1	0.25	0.5	0.75
RIACO⇔EIACO	+	+	+	+	+	−	−	−	−	−	−	−
RIACO⇔HIACO	−	−	−	~	~	−	−	−	−	−	−	−
RIACO⇔MIACO	+	+	+	+	+	+	−	−	~	~	−	−
RIACO⇔EIIACO	+	+	+	+	+	+	~	~	+	+	+	~
EIACO⇔HIACO	−	−	−	−	−	−	~	~	−	−	~	+
EIACO⇔MIACO	~	~	+	+	~	+	+	+	+	+	+	+
EIACO⇔EIIACO	~	−	~	−	~	+	+	+	+	+	+	+
HIACO⇔MIACO	+	+	+	+	+	+	+	+	+	+	+	+
HIACO⇔EIIACO	+	+	+	+	+	+	+	+	+	+	+	+
MIACO⇔EIIACO	~	−	~	−	~	+	+	+	+	+	+	+
Cyclic DTSPs												
$f=5, m \Rightarrow$	0.1	0.25	0.5	0.75	0.1	0.25	0.5	0.75	0.1	0.25	0.5	0.75
RIACO⇔EIACO	−	~	+	+	−	~	−	−	−	~	+	+
RIACO⇔HIACO	−	−	+	+	−	+	~	+	−	+	+	+
RIACO⇔MIACO	−	~	+	+	−	+	−	−	−	~	−	~
RIACO⇔EIIACO	−	+	+	+	−	+	+	~	−	~	+	+
EIACO⇔HIACO	~	−	−	−	+	+	+	+	+	+	+	−
EIACO⇔MIACO	+	~	−	−	~	+	~	−	~	~	−	−
EIACO⇔EIIACO	+	+	−	−	+	+	+	+	+	+	~	−
HIACO⇔MIACO	+	+	+	+	−	~	−	−	−	−	−	−
HIACO⇔EIIACO	+	+	+	+	+	+	+	~	−	−	−	+
MIACO⇔EIIACO	+	+	+	+	+	+	+	+	+	~	+	+
$f=50, m \Rightarrow$	0.1	0.25	0.5	0.75	0.1	0.25	0.5	0.75	0.1	0.25	0.5	0.75
RIACO⇔EIACO	+	+	+	+	~	+	−	−	−	+	−	−
RIACO⇔HIACO	−	−	−	−	−	−	−	−	−	−	−	−
RIACO⇔MIACO	+	+	+	+	+	+	−	−	−	~	−	−
RIACO⇔EIIACO	+	+	+	+	+	+	+	+	+	+	+	−
EIACO⇔HIACO	−	−	−	−	−	−	~	−	~	−	~	+
EIACO⇔MIACO	~	~	~	~	~	~	~	~	~	~	~	~
EIACO⇔EIIACO	~	+	−	−	+	+	+	+	+	+	+	+
HIACO⇔MIACO	+	+	+	+	+	+	+	~	~	+	+	−
HIACO⇔EIIACO	+	+	+	+	+	+	+	+	+	+	+	+
MIACO⇔EIIACO	~	+	~	~	+	+	+	+	+	+	+	+

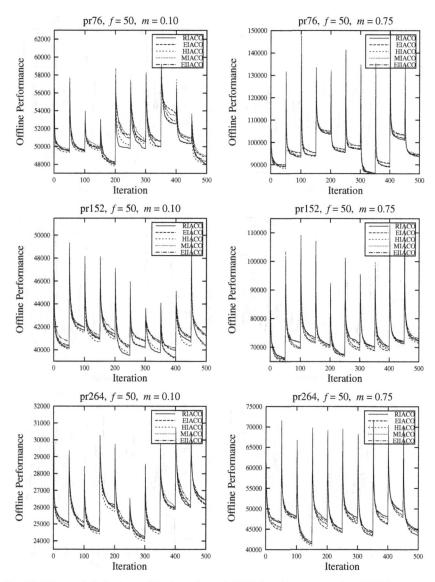

Fig. 13.3 Dynamic behaviour of the investigated ACO algorithms on random DTSPs

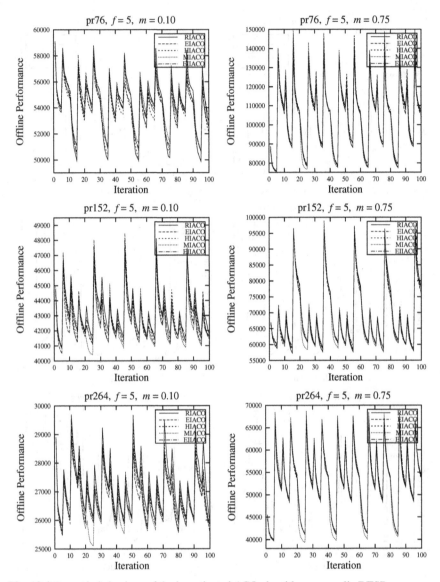

Fig. 13.4 Dynamic behaviour of the investigated ACO algorithms on cyclic DTSPs

diversity level and the knowledge transferred. Moreover, in cases with $f = 5$, HI-ACO outperforms EIACO, whereas it has a similar behaviour with RIACO, but slightly degraded. On the other hand, HIACO outperforms EIIACO in all dynamic cases except on the largest problem instance, i.e., pr264, with $m = 0.10, 0.25$ and 0.5; see the comparisons of HIACO $\Leftrightarrow$ EIIACO in Table 13.3. A similar observation of this behaviour has been found in [32], where the performance of

EIIACO was promoted as the size of the problem instance increased. This because of the diversity level generated from the environmental-information based immigrants, which gather more information than the individual-information based immigrants, i.e., in EIACO and MIACO.

Fourth, MIACO has a similar behaviour with EIACO when compared with other algorithms, in random DTSPs. However, it is outperformed by EIACO and HIACO in almost all dynamic cases; see the results regarding MIACO $\Leftrightarrow$ EIACO and HIACO $\Leftrightarrow$ MIACO in Table 13.3 and the results in Table 13.1. However, in cyclic DTSPs, MIACO outperforms EIACO and HIACO in cases where $f = 5$ and $m = 0.50$ and 0.75, whereas it is underperformed in cases where $f = 5$ and $m = 0.10$ and 0.25; see the results regarding MIACO $\Leftrightarrow$ EIACO and HIACO $\Leftrightarrow$ MIACO in Table 13.3 and the results in Table 13.2. Furthermore, MIACO and EIACO are insignificant different in cyclic DTSPs where $f = 50$, while HIACO significantly outperforms them. This is because EIACO is beneficial when the changing environment is similar, either in cyclic or random DTSPs, and MIACO is beneficial when the changing environment re-appears, i.e., cyclic DTSPs. The reason why EIACO is effective in slightly changing environments is explained above. The reason why MIACO is effective in cyclic DTSPs is that it can move the population directly to a previously visited environment. MIACO stores the best solutions for all cyclic base states and reuses them by generating memory-based immigrants. Moreover, on the smallest problem instance, i.e., pr76, RIACO outperforms MIACO in almost all dynamic cases, either random or cyclic DTSPs. This is because of the same reasons discussed on EIACO above, since both algorithms use the elitist mechanism to generate immigrants.

Finally, EIIACO outperforms RIACO in almost all dynamic cases with $f = 5$ and $m = 0.10$, while it is underperformed in dynamic cases with $f = 50$; see the comparisons of RIACO $\Leftrightarrow$ EIIACO in Table 13.3 for both random and cyclic DTSPs. This is because EIIACO transfers knowledge from previous environment and it is beneficial when the changing environments are similar. On the other hand, if too much knowledge is transferred the performance is degraded as in EIACO. Moreover, EIIACO outperforms EIACO and MIACO in $f = 5$ and $m = 0.50$ and 0.75, while it is underperformed in dynamic cases with $f = 50$. However, on the smallest problem instance, i.e., pr76, EIIACO overcomes the issue of EIACO and MIACO, which they have a degraded performance, since it is significantly better. However, in cyclic DTSPs, MIACO outperforms EIIACO in almost all dynamic cases, as expected; see the comparisons of EIACO $\Leftrightarrow$ EIIACO and MIACO $\Leftrightarrow$ EIIACO in Table 13.3 for both random and cyclic DTSPs.

13.5.4   Experimental Results and Analysis of the Investigated Algorithms with Other Peer ACO

In this section, we compare the offline performance of the investigated algorithms above, with several other existing peer ACO proposed in the literature for different DTSPs. These peer ACO algorithms are S-ACO and P-ACO described in Section

Table 13.4 Experimental results regarding offline performance on the DTSP with $m = rand[0, 1]$ and $f = rand[1, 100]$ in lin318

Algorithms	S-ACO	P-ACO	M-ACO	RIACO	EIACO	HIACO	MIACO	EIIACO
Offline Performance								
Mean	35008.25	34960.87	34845.93	34200.89	33794.34	33875.42	33904.18	34215.37
(Std Dev)	96.07	186.63	141.75	140.23	155.60	148.64	163.62	169.32
t-test Results								
S-ACO		~	+	+	+	+	+	+
P-ACO	~		+	+	+	+	+	+
M-ACO	−	−		+	+	+	+	+
RIACO	−	−	−		+	+	+	~
EIACO	−	−	−	−		−	−	−
HIACO	−	−	−	−	+		~	−
MIACO	−	−	−	−	+	~		−
EIIACO	−	−	−	~	+	+	+	

13.3, and Memetic ACO (M-ACO) [30], which is a hybridization of P-ACO and an adaptive inversion local search.

In the previous experiments, we have investigated the offline performance and dynamic behaviour of ACO algorithms with immigrants schemes under two kinds of dynamic environments, i.e., random and cyclic DTSPs, but with fixed values of f and m. However, in real-world problems both f and m may vary during the execution of the algorithm. In order to investigate the behaviour of the investigated algorithms and compare them with existing algorithms in such kinds of environments, further experiments were carried out in lin318. The values f and m were generated randomly with a uniform distribution in $[1, 100]$ and $[0, 1]$, respectively. Since the time interval of such kind of environment varies, many existing approaches used in ACO algorithms for DTSPs, e.g. global and local restart strategies and diversity maintenance schemes [16, 21], cannot be applied, since they do not have any mechanism to detect dynamic changes.

The experimental settings and performance measure were the same as in previous experiments. The experimental results regarding the offline performance are presented in Table 13.4 with the corresponding two-tailed t-test results with 58 degrees of freedom at a 0.05 level of significance, "+" or "−" indicates that the algorithm in the column is significantly better or the algorithm in the row is significantly better, respectively, where $\sim$ indicates no significance between the algorithms. Moreover, the values of varying f and m are plotted in Fig. 13.5 and the corresponding dynamic behaviour of the algorithms is plotted in Fig. 13.6. From the experimental results, several observations can be drawn.

First, the results in Table 13.4 almost match the analysis of our previous experiments, where EIACO and MIACO are significantly better than RIACO, whereas EIIACO is significantly worst than RIACO. HIACO improves the performance of RIACO, whereas it is worst than EIACO. EIACO is the champion algorithm from all the ACO algorithms with immigrants schemes. This is because the generated environment is a DTSP with random traffic factors, where EIACO perform well.

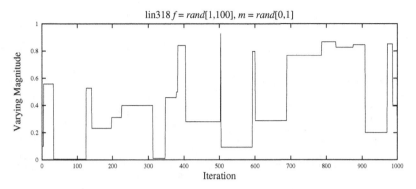

Fig. 13.5 Varying values for $m = rand[0,1]$ and $f = rand[1,100]$ used for the DTSP

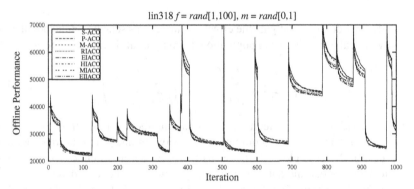

Fig. 13.6 Dynamic behaviour of the investigated ACO algorithms in comparison with other peer ACO algorithms on the DTSP with $m = rand[0,1]$ and $f = rand[1,100]$

MIACO is not performing well since it is not guaranteed that previously visited environments will reappear again.

Second, all investigated algorithms outperform other peer ACO algorithms. This is because the S-ACO uses only pheromone evaporation to eliminate pheromone trails from the previous environment that are not useful to the new one, and thus, needs sufficient time to adapt to the changing environments. On the other hand, P-ACO eliminates pheromone trails directly if an ant is removed from k_{long}. However, if identical ants are stored in the k_{long}, then the algorithm will reach stagnation behaviour, and thus, needs sufficient time to escape from it. M-ACO is significantly better than both S-ACO and P-ACO since the local search that is integrated with ACO promotes exploitation to improve the solution quality, and the risk of stagnation is eliminated using a diversity maintenance scheme based on traditional immigrants. Whenever, k_{long} reaches stagnation behaviour a random immigrant replaces an ant until the algorithm generates sufficient diversity.

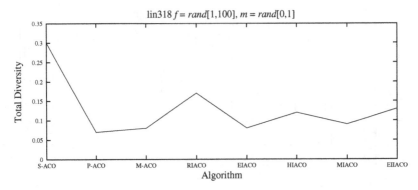

Fig. 13.7 Total diversity of the investigated ACO algorithms in comparison with other peer ACO algorithms on the DTSP with $m = rand[0, 1]$ and $f = rand[1, 100]$

Finally, in order to investigate the effect of immigrants scheme on the population diversity, we calculate the mean population diversity of all iterations as follows:

$$Div = \frac{1}{G} \sum_{i=1}^{G} \left(\frac{1}{N} \sum_{j=1}^{N} \left(\frac{1}{\mu(\mu - 1)} \sum_{p=1}^{\mu} \sum_{q \neq p}^{\mu} M_{ij} \right) \right), \quad (13.10)$$

where G is the number of iterations, N is the number of runs, μ is the size of the population and M_{ij} is the common edges between ant p and ant q as defined in Eq. (13.7). A value closer to 0 means that the ants are identical and a value closer to 1 means that the ants are completely different. The total diversity results for the different dynamic cases are presented in Fig. 13.7. It can be observed that S-ACO has a higher diversity than all algorithms. The P-ACO algorithm has the lowest diversity level which shows the effect when identical ants are stored in the population-list. RIACO maintains the highest diversity among the remaining algorithms with different immigrants schemes, since diversity is generated randomly, whereas the remaining algorithms generate guided diversity via transferring knowledge. However, EIIACO maintains higher diversity than both EIACO and MIACO which shows that the environmental-information immigrants generate higher diversity than individual-information immigrants. HIACO has a lower diversity than RIACO but higher than EIACO which shows that it has inherited the merits of the two immigrants schemes. Considering the results of the total diversity with the those of the offline performance shows that ACO algorithms that maintain high diversity levels do not always achieve better performance than other ACO algorithms for the DTSP; see Table 13.4 and Fig. 13.6.

13.6 Conclusions and Future Work

Several immigrants schemes have been successfully applied to ACO algorithms to address different DTSPs [29, 31, 32]. Immigrant ants are generated to transfer knowledge to the pheromone trails and maintain diversity. In this chapter, we re-investigate those algorithms, on the way they generate immigrant ants, and apply them to DTSP with traffic factors. We generate two types of dynamic environments: (1) the traffic factor changes randomly; and (2) the traffic factor changes in a cyclic pattern where the environments will reappear.

From the experimental results of comparing the investigated algorithms on different cases of DTSPs and with other peer ACO algorithms, the following concluding remarks can be drawn. First, immigrants schemes enhance the performance of ACO for DTSPs. Second, RIACO is advantageous in fast and significantly changing environments. Third, EIACO is advantageous in slowly and slightly changing environments. Fourth, HIACO promotes the performance of EIACO in slowly changing environments, while it slightly degrades the performance of RIACO in fast changing environments. Fourth, MIACO is advantageous in cyclic changing environments, where previous environments will re-appear. Fifth, EIIACO promotes the performance of EIACO in environments with significant changes. Sixth, transferring too much knowledge from previous environments may degrade the performance. Finally, a high level of diversity do not always enhance the performance of ACO in DTSPs.

In general, almost all ACO algorithms based on immigrants schemes outperform other peer ACO algorithms. Furthermore, different immigrants schemes are beneficial for different dynamic environmental cases.

For future work, it will be interesting to hybridize more types of immigrants schemes, to achieve a good balance between the knowledge transferred and the diversity maintenance. Another future work is to apply the proposed algorithms to other relevant problems, e.g., the dynamic vehicle routing problem, and in more challenging environments, where, apart from dynamic changes, a small amount of noise may be generated every iteration.

Acknowledgements. This work was supported by the Engineering and Physical Sciences Research Council (EPSRC) of U.K. under Grant numbers EP/E060722/1, EP/E060722/2, and EP/K001310/1.

References

[1] Angus, D.: Niching for population-based ant colony optimization. In: Proc. of the 2nd IEEE Inter. Conf. on e-Science and Grid Comp., pp. 15–22 (2006)

[2] Bell, J.E., McMullen, P.R.: Ant colony optimization techniques for the vehicle routing problem. Advanced Engineering Informatics 18, 41-48 (2004)

[3] Bullnheimer, B., Haïti, R., Strauss, C.: An improved ant system algorithm for the vehicle routing problem. Ann. Oper. Res. **89**(1), 319-328 (1999)

[4] Bullnheimer, B., Hartl, R.F., Strauss, C.: A new rank based version of the ant system - a computational study. Central Eur. J. for Oper. Res. in Economics **7**(1), 25–38 (1999)

[5] Bonabeau, E., Dorigo, M., Theraulaz, G.: Swarm Intelligence: From Natural to Artificial Systems. Oxford University Press, New York (1999)

[6] Branke, J.: Memory enhanced evolutionary algorithms for changing optimization problems. In: Proc. 1999 IEEE Congr. Evol. Comput. vol. 3, pp. 1875–1882 (1999)

[7] Cruz, C., González, J.R., Pelta, D.A.: Optimization in dynamic environments: a survey on problems, methods and measures. Soft Comput., **15**(7), 1427–1448 (2011)

[8] Cobb, H.G., Grefenstette, J.J.: Genetic algorithms for tracking changing environments. In: Proc. 5th Int. Conf. on Genetic Algorithm, pp. 523–530 (1993)

[9] Colorni, A., Dorigo, M., Maniezzo, V.: Distributed optimization by ant colonies. In: Proc. 1st Europ. Conf. on Artif. Life, pp. 134–142 (1992)

[10] Cordón, O., de Viana, I.F., Herrera, F., Moreno, L.: A new ACO model integrating evolutionary computation concepts: The best worst Ant System. In: Proc. 2nd Int. Workshop on Ant Algorithms, pp. 22-29 (2000)

[11] Di Caro, G., Dorigo, M.: Ant Net: Distributed Stigmergetic Control for Communications Networks. J. of Artif. Intell. Res. **9**(1), 317–365 (1998)

[12] Di Caro, G., Ducatelle, F., Gambardella, L.M.: AntHocNet: An ant-based hybrid routing algorithm for mobile ad hoc networks. In: Proc. 8th Int. Conf. on Parallel Problem Solving from Nature, LNCS 3242, pp. 461–470 (2004)

[13] Dorigo, M., Maniezzo, V., Colorni, A.: Ant system: optimization by a colony of cooperating agents. IEEE Trans. Syst., Man and Cybern., Part B: Cybern. **26**(1), 29–41 (1996)

[14] Dorigo, M., Gambardella, L.M.: Ant colony system: a cooperative learning approach to the traveling salesman problem. IEEE Trans. Evol. Comput. **1**(1), 53–66 (1997)

[15] Dorigo, M., Stützle, T.: Ant Colony Optimization. The MIT Press, London (2004)

[16] Eyckelhof, C.J., Snoek, M.: Ant Systems for a Dynamic TSP. In: Proc. 3rd Int. Workshop on Ant Algorithms, pp. 88–99 (2002)

[17] Gambardella, L.M., Taillard, E.D., Dorigo, M.: Ant colonies for the quadratic assignment problem. J. of the Oper Res Society **50**, 167–176 (1999)

[18] Grefenestette, J.J.: Genetic algorithms for changing environments. In: Proc. 2nd Int. Conf. on Parallel Problem Solving from Nature, pp. 137–144 (1992)

[19] Guntsch, M., Middendorf, M.: Applying population based ACO to dynamic optimization problems. In: Proc. 3rd Int. Workshop on Ant Algorithms, LNCS 2463, pp. 111–122 (2002)

[20] Guntsch, M., Middendorf, M.: A population based approach for ACO. In: EvoWorkshops 2002: Appl. of Evol. Comput., pp. 72–81 (2002)

[21] Guntsch, M., Middendorf, M.: Pheromone modification strategies for ant algorithms applied to dynamic TSP. In: EvoWorkshops 2001: Appl. of Evol. Comput., pp. 213-222 (2001)

[22] Guntsch, M., Middendorf, M., Schmeck, H.: An ant colony optimization approach to dynamic TSP. In: Proc. 2001 Genetic and Evol. Comput. Conf., pp. 860–867 (2001)

[23] Guo, T., Michalewicz, Z.: Inver-over operator for the TSP. In: Proc. 5th Int. Conf. on Parallel Problem Solving from Nature, LNCS 1498, pp. 803–812 (1998)

[24] He, J., Yao, X.: From an individual to a population: An analysis of the first hitting time of population-based evolutionary algorithms. IEEE Trans. Evol. Comput. **6**(5), 495–511 (2002)

[25] Holland, J.: Adaption in Natural and Artificial Systems, University of Michigan Press (1975)

[26] Jin, Y., Branke, J.: Evolutionary optimization in uncertain environments - a survey. IEEE Trans. Evol. Comput. **9**(3), 303–317 (2005)

[27] Lin, S., Kerninghan, B.W.: An effective heuristic algorithm for the traveling salesman problem. Oper. Res. **21**(2), 498–516 (1973)

[28] Maniezzo, V., Colorni, A.: The ant system applied to the quadratic assignment problem. IEEE Trans. Knowledge and Data Engineering **9**(5), 769–778 (1999)

[29] Mavrovouniotis, M., Yang, S.: Ant colony optimization with immigrants schemes for dynamic environments. In: Proc. 11th Int. Conf. on Parallel Problem Solving from Nature, LNCS 6239, pp. 371–380 (2010)

[30] Mavrovouniotis, M., Yang, S.: A memetic ant colony optimization algorithm for the dynamic travelling salesman problem. Soft Comput. **15**(7), 1405–1425 (2011)

[31] Mavrovouniotis, M., Yang, S.: Memory-based immigrants for ant colony optimization in changing environments. In: EvoWorkshops 2011: Appl. of Evol. Comput., LNCS 6624, pp. 324–333 (2011)

[32] Mavrovouniotis, M., Yang, S.: An immigrants scheme based on environmental information for ant colony optimization for the dynamic travelling salesman problem. In: Proc. 10th Int. Conf. Evolution Artificial, LNCS 7401, pp. 1-12 (2011)

[33] Michalewicz, Z.: Genetic Algorithms + Data Structures = Evolution Programs, Springer-Verlag, Berlin, third edition (1999)

[34] Montemanni, R., Gambardella, L., Rizzoli, A., Donati, A.: An new algorithm for a dynamic vehicle routing problem based on ant colony system. In: Proc. 2nd Int. Workshop on Freight Transportation and Logistics, pp. 27–30 (2003)

[35] Neumann, F., Witt, C.: Runtime analysis of a simple ant colony optimization algorithm. Algorithmica **54**(2), 243–255 (2009)

[36] Rizzoli, A. E., Montemanni, R., Lucibello, E., Gambardella, L. M.: Ant colony optimization for real-world vehicle routing problems - from theory to applications. Swarm Intell. **1**(2), 135–151 (2007)

[37] Stützle, T., Hoos, H.: The MAX-MIN ant system and local search for the traveling salesman problem. In: Proc. 1997 IEEE Int. Conf. Evol. Comput., pp. 309–314 (1997)

[38] Stützle, T., Hoos, H.: MAX-MIN Ant System. Future Generation Computer Systems **8**(16), 889–914 (2000)

[39] Yang, S.: Memory-based immigrants for genetic algorithms in dynamic environments. In: Proc. 2005 Genetic and Evol. Conf., vol. 2, pp. 1115–1122 (2005)

[40] Yang, S.: Genetic algorithms with elitism based immigrants for changing optimization problems. In: EvoWorkshops 2007: Appl. of Evol. Comput., LNCS 4448, pp. 627–636 (2007)

[41] Yang, S.: Genetic algorithms with memory and elitism based immigrants in dynamic environments. Evol. Comput. **16**(3), 385–416 (2008)

[42] Yang, S., Cheng, H., Wang, F.: Genetic algorithms with immigrants and memory schemes for dynamic shortest path routing problems in mobile ad hoc networks. IEEE Trans. Syst., Man, and Cybern. Part C: Appl. and Rev. **40**(1), 52-63 (2010)

[43] Yang, S., Tinos, R.: A hybrid immigrants scheme for genetic algorithms in dynamic environments. Int. J. of Autom. and Comput. **4**(3), 243–254 (2007)

[44] Yang, S., Yao, X.: Population-based incremental learning with associative memory for dynamic environments. IEEE Trans. Evol. Comput. **12**(5), 542–561 (2008)

[45] Yu, X., Tang, K., Yao, X.: An immigrants scheme based on environmental information for genetic algorithms in changing environments. In: Proc. 2008 IEEE Congr. Evol. Comput., pp. 1141–1147 (2008)

[46] Yu, X., Tang, K., Yao, X.: Immigrant schemes for evolutionary algorithms in dynamic environments: Adapting the replacement rate. Sci. China Series F: Inf. Sci. **53**(1), 1–11 (2010)

[47] Yu, X., Tang, K., Chen, T., Yao, X.: Empirical analysis of evolutionary algorithms with immigrants schemes for dynamic optimization. Memetic Comput. **1**(1), 3–24 (2009)

Chapter 14
Genetic Algorithms for Dynamic Routing Problems in Mobile Ad Hoc Networks

Hui Cheng and Shengxiang Yang

Abstract. Routing plays an important role in various types of networks. There are two main ways to route the packets, i.e., unicast and multicast. In most cases, the unicast routing problem is to find the shortest path between two nodes in the network and the multicast routing problem is to find an optimal tree spanning the source and all the destinations. In recent years, both the shortest path routing and the multicast routing have been well addressed using intelligent optimization techniques. With the advancement in wireless communications, more and more mobile wireless networks appear, e.g., mobile ad hoc networks (MANETs). One of the most important characteristics in MANETs is the topology dynamics, that is, the network topology changes over time due to energy conservation or node mobility. Therefore, both routing problems turn out to be dynamic optimization problems in MANETs. In this chapter, we investigate a series of dynamic genetic algorithms to solve both the dynamic shortest path routing problem and the dynamic multicast routing problem in MANETs. The experimental results show that these specifically designed dynamic genetic algorithms can quickly adapt to environmental changes (i.e., the network topology changes) and produce high quality solutions after each change.

14.1 Introduction

Routing plays an important role in various types of networks. There are two main ways to route the packets. One is unicast and the other is multicast. Unicast refers

Hui Cheng
Department of Computer Science and Technology, University of Bedfordshire, Park Square, Luton LU1 3JU, U.K.
e-mail: hui.cheng@beds.ac.uk

Shengxiang Yang
Centre for Computational Intelligence (CCI), School of Computer Science and Informatics, De Montfort University, The Gateway, Leicester LE1 9BH, U.K.
e-mail: syang@dmu.ac.uk

S. Yang and X. Yao (Eds.): *Evolutionary Computation for DOPs*, SCI 490, pp. 343–375.
DOI: 10.1007/978-3-642-38416-5_14   © Springer-Verlag Berlin Heidelberg 2013

to one-to-one communication between a source and a destination. Multicast refers to one-to-many communication where the same source sends the same packets to a set of destinations. In most cases, the unicast routing problem is to find the shortest path between two nodes in the network and the multicast routing problem is to find an optimal tree spanning the source and all the destinations. In recent years, both the shortest path routing and the multicast routing have been well addressed using intelligent optimization techniques, e.g., artificial neural networks, genetic algorithms (GAs), and particle swarm optimization (PSO), etc.

With the advancement in wireless communications, more and more mobile wireless networks appear, e.g., mobile ad hoc networks (MANETs) and wireless sensor networks . A MANET [26] is a self-organizing and self-configuring multi-hop wireless network, which is comprised of a set of mobile hosts (MHs) that can move around freely and cooperate in relaying packets on behalf of one another. A MANET supports robust and efficient operations by incorporating the routing functionality into MHs. In multi-hop networks, routing is one of the most important issues that has a significant impact on the network's performance. In a MANET, each mobile node is a router and forwards packets on behalf of other nodes. Multi-hop forwarding paths are established for nodes beyond the direct wireless communication range. Routing protocols for MANETs must discover such paths and maintain connectivity when links in these paths break due to effects such as the node movement, battery drainage, and wireless interference.

In this chapter, we investigate two main routing problems in MANETs, i.e., the shortest path (SP) routing problem and the multicast routing problem. The SP routing problem aims to find the shortest path from a source to a destination in a given network while minimizing the total cost associated with the path. The SP routing problem is a classical combinatorial optimization problem arising in many design and planning contexts [3]. There are several deterministic search algorithms for the SP problem: the Dijkstra's algorithm, the breadth-first search algorithm, and the Bellman-Ford algorithm, etc. All these algorithms have a polynomial time complexity. They are effective in fixed infrastructure wireless or wired networks. But, they exhibit an unacceptable high computational complexity for real-time communications involving rapidly changing network topologies [4]. Therefore, for the dynamic shortest path routing problem (DSPRP) in a changing network environment, we need to employ appropriate new approaches. The DSPRP has become a topic of interest in recent years.

Multicast [38] is an important network service, which is the delivery of information from a source to multiple destinations simultaneously using the most efficient strategy to deliver the messages over each link of the network only once, creating copies only when the links to the destinations split. It provides underlying network support for collaborative group communications, such as the video conference, distant education, and content distribution. Quality-of-service (QoS) requirements [38] proposed by different network applications are often versatile. Among them, the end-to-end delay [18, 32] is a pretty important QoS metric since the real-time delivery of multimedia data is required. An efficient QoS multicast algorithm should construct a multicast routing tree, by which the data can be transmitted from the source to all the destinations with a guaranteed QoS. The multicast tree cost, which is used to evaluate the utilization of network resources, is also an important metric especially in wireless mobile networks where limited radio resources are available.

The QoS multicast routing problem also involves a classical combinatorial optimization problem [29]. In a MANET, the network topology keeps changing due to its inherent characteristics. Therefore, for the dynamic multicast routing problem (DMRP), an effective multicast algorithm should track the topological changes and adapt the best multicast tree to the changes accordingly. There are mainly two types of algorithms for the multicast routing problem: the deterministic algorithms and the search heuristics. Given the multicast routing request, only one multicast tree is constructed for a given topology by a deterministic algorithm, e.g., the shortest path tree (SPT) [27] algorithm. However, by the search heuristics, such as GAs [4] and simulated annealing (SA) algorithms [38], lots of multicast trees are searched and the best one is selected as the final result.

Both the DSPRP and the DMRP are real-world dynamic optimization problems (DOPs) in MANETs. In recent years, studying evolutionary algorithms (EAs) for DOPs has attracted a growing interest due to its importance in real-world applications of EAs [49]. The simplest way of addressing DOPs is to restart EAs from scratch whenever an environment change is detected. Though the restart scheme really works for some cases [48], for many DOPs it is more efficient to develop other approaches that make use of knowledge gathered from old environments. Over the years, several approaches have been developed for EAs to address dynamic environments [45], such as maintaining diversity during the run via random immigrants [13, 37], increasing diversity after a change [9], using memory schemes to reuse stored useful information [5, 35, 41], applying multi-population and speciation schemes to search in different regions of the search space [6, 30], and adapting (the parameters of) operators to quickly respond to a new environment [25, 47].

In this chapter, we adapt and investigate several genetic algorithms that are developed to deal with general DOPs to solve the DSPRP [44] and the DMRP [8] in MANETs. First, we design the components of the standard GA specifically for each problem. Then, we integrate several immigrants and memory schemes and their combination into the GA to enhance its searching capacity for the shortest paths in dynamic environments. We also integrate several improved immigrants schemes into the GA to enhance its searching capacity for the optimal multicast trees in dynamic environments. Once the topology is changed, new immigrants or the useful information stored in the memory can help guide the search of good solutions in the new environment. For comparison purposes, for each problem we also implement two traditional GA schemes, i.e., Standard GA and Restart GA, as the peer algorithms. Via simulation experiments, we evaluate these GAs under different parameter settings to find the best combinations. More importantly, we evaluate them under various settings of dynamic environments to see their performance and find the best match between algorithms and environmental characteristics. Generally speaking, the investigated well-designed dynamic GAs work well for both the DSPRP and the DMRP in MANETs.

The rest of this chapter is organized as follows. We discuss related work in Section 14.2. The MANET network model, the DSPRP model and the DMRP model are described in Section 14.3. Section 14.4 presents the design of specialized GAs for the static SP routing problem and the static multicast routing problem, respectively. The investigated dynamic GAs that are the integration of several immigrants and/or memory schemes are described in Section 14.5. The extensive experimental study and relevant analysis are presented in Section 14.6. Finally, Section 14.7 concludes this chapter.

14.2 Related Work

Both the shortest path routing problem and the multicast routing problem have been extensively investigated in the literature.

14.2.1 Shortest Path Routing

Since deterministic algorithms with a polynomial time complexity are not suitable for the real-time computation of shortest paths, quite a few research works have been conducted to solve the SP problem using artificial intelligence techniques, e.g., artificial neural networks [3], GAs [4], and PSO [23].

In [3], a near-optimal routing algorithm employing a modified Hopfield neural network (HNN) was proposed. It uses every piece of information that is available at the peripheral neurons, in addition to the highly correlated information that is available at the local neuron. Therefore, it can achieve a faster convergence and a better route optimality than other HNN based algorithms. In [4], a genetic algorithmic approach was presented to the SP routing problem. Computer simulations showed that the GA based SP algorithm exhibits a much better quality of solution (i.e., the route optimality) and a much higher rate of convergence than other algorithms. A population-sizing equation that facilitates a solution with the desired quality was also developed. In [23], a PSO-based search algorithm was proposed. A priority-based indirect path-encoding scheme is used to widen the scope of the search space and a heuristic operator is used to reduce the probability of invalid loop creations during the path construction procedure. It was claimed that the PSO-based SP algorithm is superior to those using GAs including the one in [4].

However, all these algorithms address the static SP routing problem only. When the network topology changes, they will regard it as a new network and restart the algorithms over the new topology. It is well known that the topology changes rapidly in MANETs due to the characteristics of wireless networks. Therefore, for the DSPRP in MANETs, these algorithms are not good choices since they require frequent restart and cannot meet the real-time requirement. In this regard, immigrants and memory enhanced GAs have their inherent advantages. These GAs use the immigrants or the useful information stored in the memory to help the population quickly adapt to the new environment after a change occurs. Hence, these algorithms can keep running over the continuously changing topologies and avoid the expensive and inefficient restart.

14.2.2   Multicast Routing

Multicast routing trees produced by deterministic algorithms can be classified into two types, i.e., Steiner minimum tree (SMT) [16] and shortest path tree [27]. An SMT is also the minimum-cost multicast tree. An SPT is constructed by applying the shortest path algorithm to find the shortest (e.g., minimum cost or delay) path from the source to each destination and then merging them. The problem of finding an SMT has been proved to be NP-complete [17] and lots of approximation algorithms [2, 14, 33] have been developed. An SPT provides a good solution for finding delay-constrained multicast tree because it determines the minimum delay path from the source to each destination. Inspired by SMT and SPT, many heuristic algorithms [18, 19, 32] have been proposed to construct a QoS-aware multicast tree by making a tradeoff between them. QoS multicast routing is still a challenging problem due to its intractability and comprehensive application backgrounds.

Intelligent search heuristics is a type of promising techniques to solve combinatorial optimization problems [31] including the SMT problem. GAs are a type of stochastic meta-heuristic optimization methods that model the biological principles of Darwinian theory of evolution and Mendelian principles of inheritance [15]. GAs have been extensively used in solving the QoS multicast problems in various networks such as the wired multimedia networks [38] and optical networks [12]. In [38], we also developed a unified framework for producing QoS multicast trees using intelligent search heuristics and proposed three QoS multicast algorithms based on GAs, simulated annealing, and tabu search, separately.

In [38], the binary encoding is adopted where each bit of the binary string corresponds to a different node in the network. For each binary string, a graph G' is derived from the network topology G by including all the nodes appearing in the string and the links connecting these nodes. Then, the minimum spanning tree T of G' acts as the candidate multicast tree represented by the binary string. This encoding method is a bit complicated and each binary string cannot directly represent a candidate solution. A multicast tree is a union of the routing paths from the source to each receiver. Hence, it is a natural choice to adopt the path-oriented encoding method [4, 12] instead of the binary encoding.

In MANETs, a number of multicast routing protocols, using a variety of basic routing algorithms and techniques, have been proposed [10]. However, they mainly focus on the discovery of the optimal multicast forwarding structure (i.e., tree or mesh) spanning mobile nodes and do not consider the everlasting changes in the network topologies. Topology dynamics is the inherent characteristics in wireless mobile networks. For example, at time T_1, the network topology is G_1. At time T_2, the network topology may change to G_2. Although G_1 and G_2 are different, they are highly relevant since each change alters part of the topology only. Therefore, the solutions obtained on G_1 could benefit the search of good solutions on G_2.

An effective multicast algorithm should track the topological changes and adapt the multicast trees to the changes accordingly. We are not aware of any other work that considers the dynamic multicast routing in the environment where the network topology keeps changing in a continuous way, although there are quite a few works that are related to some relevant aspects. For example, some researchers have investigated the multicast problem where the dynamic group membership exists [1, 50]. In a dynamic group, nodes are allowed to join or leave it.

14.3 Network and Problem Models

In this section, we first present our network model and then formulate both the DSPRP and the DMRP.

14.3.1 Mobile Ad Hoc Network Model

We consider a mobile ad hoc network operating within a fixed geographical region. We model it by an undirected and connected topology graph $G_0(V_0, E_0)$, where V_0 represents the set of wireless nodes (i.e., routers) and E_0 represents the set of communication links connecting two neighboring routers falling into the radio transmission range. A communication link (i, j) can not be used for packet transmission unless both node i and node j have a radio interface each with a common channel. However, the channel assignment is beyond the scope of this chapter. In addition, message transmission on a wireless communication link will incur remarkable delay and cost.

Here, we summarize some notations that we use throughout this chapter.

- $G_0(V_0, E_0)$, the initial MANET topology graph.
- $G_i(V_i, E_i)$, the MANET topology graph after the i-th change.
- s, the source node.
- r, the destination node.
- $P_i(s, r)$, a path from s to r on the graph G_i.
- $C(P_i)$, the total cost of the path P_i.
- $R = \{r_0, r_1, \ldots r_m\}$, the set of receivers of the multicast routing request.
- $T_i(V_{T_i}, E_{T_i})$, a multicast tree with nodes V_{T_i} and links E_{T_i}.
- $P_{T_i}(s, r_j)$, a path from s to r_j on the tree T_i.
- C_{T_i}, the cost of the tree T_i.
- d_l, the transmission delay on the communication link l.
- c_l, the cost on the communication link l.
- $\Delta(P_i)$, the total transmission delay on the path P_i.

14.3.2 Dynamic Shortest Path Routing Problem Model

The DSPRP can be informally described as follows. Initially, given a network of wireless routers, a delay upper bound, a source node and a destination node, we wish to find a delay-bounded least cost loop-free path on the topology graph. Since the end-to-end delay [32] is a pretty important QoS metric to guarantee the real-time data delivery, we also require the routing path to satisfy the delay constraint. Then, periodically or stochastically, due to energy conservation or some other issues, some nodes are scheduled to sleep or some sleeping nodes are scheduled to wake up. Therefore, the network topology changes from time to time. The objective of the dynamic SP routing problem is to quickly find the new optimal delay-constrained least cost acyclic path after each topology change.

More formally, consider a mobile ad hoc network $G(V, E)$ and a unicast communication request from the source node s to the destination node r with the delay upper bound Δ. The *dynamic delay-constrained shortest path problem* is to find a series of paths $\{P_i | i \in \{0, 1, ...\}\}$ over a series of graphs $\{G_i | i \in \{0, 1, ...\}\}$, which satisfy the delay constraint as shown in Eq. (14.1) and have the least path cost as shown in Eq. (14.2).

$$\Delta(P_i) = \sum_{l \in P_i(s,r)} d_l \leq \Delta \tag{14.1}$$

$$C(P_i) = \min_{P \in G_i} \left\{ \sum_{l \in P(s,r)} c_l \right\} \tag{14.2}$$

14.3.3 Dynamic Multicast Routing Problem Model

The DMRP can be informally described as follows. Initially, given a MANET consisting of wireless routers, a delay upper bound, and a multicast communication request from a source node to a set of receiver nodes, we wish to find a delay-bounded least cost loop-free multicast tree on the topology graph. Then, after each topology change, the objective of our problem is to quickly find the new delay-constrained least cost acyclic tree.

It is an extremely difficult job to completely model the network dynamics in a single way. Here, we propose two models to describe it and they are named as the general dynamics model and the worst dynamics model, respectively. In the general model, periodically or stochastically, due to energy conservation or other reasons, some nodes are scheduled to sleep or some sleeping nodes are scheduled to wake up. Therefore, the network topology changes from time to time. Since in most cases, the selected nodes may not belong to the present best multicast tree, the topological changes have relatively moderate effect on the routing problem. In the worst model, each change is generated manually by removing a few links on the present best multicast tree. Thus, the topological changes will destroy the present best solution and thereby cause the worst effect on the problem. Although these two models cannot cover the full cases of network dynamics, they correspond to the general scenario and the worst scenario, respectively. Based on these two representative models, the multicast routing problem can be investigated in a relatively thorough way.

More formally, we consider a MANET $G(V, E)$ and a multicast communication request from the source node s to the set of receivers R with the delay upper bound

Δ. The *dynamic delay-constrained multicast routing problem* is to find a series of trees $\{T_i | i \in \{0, 1, ...\}\}$ over a series of graphs $\{G_i | i \in \{0, 1, ...\}\}$, which satisfy the delay constraint as shown in Eq. (14.3) and have the least tree cost as shown in Eq. (14.4).

$$\max_{r_j \in R} \left\{ \sum_{l \in P_T(s, r_j)} d_l \right\} \leq \Delta . \tag{14.3}$$

$$C(T_i) = \min_{T \in G_i} \left\{ \sum_{l \in T(V_T, E_T)} c_l \right\} . \tag{14.4}$$

14.4 Specialized GAs for the Routing Problems

This section describes the design of the GA for both the shortest path routing problem and the multicast routing problem. The design of the GA involves several key components: genetic representation, population initialization, fitness function, selection scheme, crossover and mutation.

14.4.1 Specialized GA for the Shortest Path Routing Problem

A routing path consists of a sequence of adjacent nodes in the network. Hence, it is a natural choice to adopt the path-oriented encoding method. For the routing problem, the path-oriented encoding and the path-based crossover and mutation are also very popular [4, 12].

14.4.1.1 Genetic Representation

A routing path is encoded by a string of positive integers that represent the IDs of nodes through which the path passes. Each locus of the string represents an order of a node (indicated by the gene of the locus). The gene of the first locus is for the source node and the gene of the last locus is for the destination node. The length of a routing path should not exceed the maximum length $|V_0|$, where V_0 is the set of nodes in the MANET. Chromosomes are encoded under the delay constraint. In case it is violated, the encoding process is usually repeated so as to satisfy the delay constraint.

14.4.1.2 Population Initialization

In the GA, each chromosome corresponds to a potential solution. The initial population Q is composed of a certain number of, say, q, chromosomes. To promote the genetic diversity, in our algorithm, for each chromosome in the initial population, the corresponding routing path is randomly generated. We start to search a random path from s to r by randomly selecting a node v_1 from $N(s)$, the neighborhood of s. Then, we randomly select a node v_2 from $N(v_1)$. This process is repeated until r is reached. Since the path should be loop-free, those nodes that are already included in the current path are excluded from being selected as the next node to be added into the path, thereby avoiding reentry of the same node into a path. In this way, we get a random path $P(s, r) = \{s, v_1, v_2, ..., r\}$. Repeating this process for q times, the initial population $Q = \{Ch_0, Ch_1, ..., Ch_{q-1}\}$ can be obtained.

14.4.1.3   Fitness Function

Given a solution, we should accurately evaluate its quality (i.e., the fitness value), which is determined by the fitness function. In our algorithm, we aim to find the least cost path between the source and the destination. Our primary criterion of solution quality is the path cost. Therefore, among a set of candidate solutions (i.e., unicast paths), we choose the one with the least path cost. The fitness value of chromosome Ch_i (representing the path P), denoted as $f(Ch_i)$, is given by:

$$f(Ch_i) = [\sum_{l \in P(s,r)} c_l]^{-1} \tag{14.5}$$

The proposed fitness function is to be maximized and only involves the total path cost. As mentioned above, the delay constraint is checked for each chromosome during the evolutionary process.

14.4.1.4   Selection Scheme

Selection plays an important role in improving the average quality of the population by passing the high quality chromosomes to the next generation. The selection of chromosome is based on the fitness value. We adopt the scheme of pair-wise tournament selection without replacement [20] as it is simple and effective. The tournament size is 2.

14.4.1.5   Crossover and Mutation

GA relies on two basic genetic operators - crossover and mutation. Crossover processes the current solutions so as to find better ones. Mutation helps a GA keep away from local optima [4]. The performance of GA very much depends on them. The type and implementation of operators depend on encoding as well as the problem in hand.

In our algorithm, since chromosomes are expressed by the path structure, we adopt the single point crossover to exchange partial chromosomes (sub-paths) at positionally independent crossing sites between two chromosomes [4]. With the crossover probability, each time we select two chromosomes Ch_i and Ch_j for crossover. Ch_i and Ch_j should possess at least one common node. Among all the common nodes, one node, denoted as v, is randomly selected. In Ch_i, there is a path consisting of two parts: $(s \xrightarrow{Ch_i} v)$ and $(v \xrightarrow{Ch_i} r)$. In Ch_j, there is a path consisting of two parts: $(s \xrightarrow{Ch_j} v)$ and $(v \xrightarrow{Ch_j} r)$. The crossover operation exchanges the subpaths $(v \xrightarrow{Ch_i} r)$ and $(v \xrightarrow{Ch_j} r)$.

The population will undergo the mutation operation after the crossover operation. With the mutation probability, each time we select one chromosome Ch_i on which one gene is randomly selected as the mutation point (i.e., mutation node), denoted as v. The mutation will replace the subpath $(v \xrightarrow{Ch_i} r)$ by a new random subpath.

Both crossover and mutation may produce new chromosomes which represent infeasible solutions. Therefore, we check if the path represented by a new chromosome is acyclic. If not, a repair function [28] will be applied to eliminate the loops. The delay checking is incorporated into the crossover and mutation operations to guarantee that all new chromosomes produced satisfy the delay constraint.

14.4.2 Specialized GA for the Multicast Routing Problem

A multicast tree is a union of the routing paths from the source to each receiver. Hence, it is also a natural choice to adopt the path-oriented encoding method [4, 12].

14.4.2.1 Genetic Representation

For a multicast tree T spanning the source s and the set of receivers R, there are $|R|$ routing paths all originating from s. Therefore, a tree is encoded by an integer array in which each row encodes a routing path along the tree. For example, for a tree T that spans s and R, the j-th row in the corresponding array A lists up node IDs on the routing path from s to r_j along T. Therefore, A is an array of $|R|$ rows. All the solutions are encoded under the delay constraint. In case the delay constraint is violated, the encoding process is usually repeated so that it is satisfied.

14.4.2.2 Population Initialization

Similarly, the initial population Q is assumed to have q chromosomes. We use the same method in the specialized GA for the SP problem to search a random path $P_T(s, r_j)$ from s to $r_j \in R$. Since no loop is allowed on the multicast tree, the nodes that are already included in the current tree are excluded, thereby avoiding reentry of the same node. In this way, the initial population $Q = \{Ch_0, Ch_1, ..., Ch_{q-1}\}$ is obtained. The pseudo-code is shown in Algorithm 1.

14.4.2.3 Fitness Function

We aim to find the least cost multicast tree from the source to a set of receivers. The criterion used to evaluate the solution quality is the tree cost. Therefore, among a set of candidate solutions (i.e., multicast trees), we choose the one with the minimal tree cost. The fitness value of chromosome Ch_i (representing the tree T), denoted as $f(Ch_i)$, is given by:

Algorithm 1 Population Initialization

1: $i =: 0$;
2: **while** $i < q$ **do**
3:    // Generate chromosome Ch_i
4:    $j =: 0$;
5:    $V_T := E_T := \emptyset$;
6:    **while** $j < |R|$ **do**
7:       Search a random path $P_T(s, r_j)$ which can guarantee $T \cup P_T$ be an acyclic graph;
8:       Add all the nodes and links in P_T into V_T and E_T, respectively;
9:       j++;
10:   **end while**
11:   i++;
12: **end while**

$$f(Ch_i) = [\sum_{l \in T(V_T, E_T)} c_l]^{-1} \tag{14.6}$$

The proposed fitness function is to be maximized and only involves the total tree cost. As aforementioned, the delay constraint is checked for each chromosome during the evolutionary process.

14.4.2.4   Selection Scheme

We also adopt the pair-wise tournament selection without replacement [20].

14.4.2.5   Crossover and Mutation

In our algorithm, since a chromosome is expressed by a tree data structure, we adopt a single point crossover to exchange partial chromosomes (sub-trees) at positionally independent crossing sites between two chromosomes [4]. With a crossover probability, each time we select two chromosomes Ch_i and Ch_j for crossover. To at least one receiver, Ch_i and Ch_j should possess at least one common node from which one, denoted as v, is randomly selected. In Ch_i, there is a path consisting of two parts: ($s \xrightarrow{Ch_i} v$) and ($v \xrightarrow{Ch_i} r_k$). In Ch_j, there is a path consisting of two parts: ($s \xrightarrow{Ch_j} v$) and ($v \xrightarrow{Ch_j} r_k$). The crossover operation exchanges the paths ($v \xrightarrow{Ch_i} r_k$) and ($v \xrightarrow{Ch_j} r_k$).

The population will undergo the mutation operation after the crossover operation. With a mutation probability, each time we select one chromosome Ch_i on which one receiver r_k is randomly selected. On the path ($s \xrightarrow{Ch_i} r_k$) one gene is selected as the mutation point (i.e., mutation node) denoted as v. The mutation will replace the path ($v \xrightarrow{Ch_i} r_k$) by a new random path.

Both crossover and mutation may produce new chromosomes which represent infeasible solutions. Therefore, we check if the multicast trees represented by the new chromosomes are acyclic. If not, the repair function used in [28] will be applied to eliminate the loops. The delay checking is incorporated into both the crossover and mutation operations to guarantee that all the new chromosomes produced satisfy the delay constraint.

14.5 Investigated GAs for the Dynamic Routing Problems

In this chapter, we investigate both traditional GAs and several dynamic GAs for the two dynamic routing optimization problems.

14.5.1 Traditional GAs

For the two dynamic routing problems, we can still address them using the specialized GAs described above with two variants, denoted *Standard GA (SGA)* and *Restart GA*. In the SGA, when an environmental change leads to infeasible solutions, SGA handles them by taking the measure of penalty. That is, infeasible solutions are set to a very low fitness. In this way, the population in SGA can keep evolving even in a continuously changing environment. In the Restart GA, once a change is detected, the population will be re-initialized based on the new network topology.

14.5.2 GAs with Immigrants Schemes

In stationary environments, convergence at a proper pace is really what we expect for GAs to locate the optimum solutions for many optimization problems. However, for DOPs, convergence usually becomes a big problem for GAs because changing environments usually require GAs to keep a certain population diversity level to maintain their adaptability. To address this problem, the random immigrants approach is a quite natural and simple way [13, 34, 46, 51, 52]. It was proposed by Grefenstette with the inspiration from the flux of immigrants that wander in and out of a population between two generations in nature. It maintains the diversity level of the population through replacing some individuals of the current population with random individuals, called *random immigrants*, every generation. As to which individuals in the population should be replaced, usually there are two strategies: replacing random individuals or replacing the worst ones [37]. In order to avoid that random immigrants disrupt the ongoing search progress too much, especially during the period when the environment does not change, the ratio of the number of random immigrants to the population size is usually set to a small value, e.g., 0.2.

However, in a slowly changing environment, the introduced random immigrants may divert the searching force of the GA during each environment before a change occurs and hence may degrade the performance. On the other hand, if the environment only changes slightly in terms of severity of changes, random immigrants may not have any actual effect even when a change occurs because individuals in the previous environment may still be quite fit in the new environment. Based on the above consideration, an immigrants approach, called elitism-based immigrants, was proposed for GAs to address DOPs [42].

Within the elitism-based immigrants GA (EIGA), for each generation t, after the normal genetic operations (i.e., selection and recombination), the elite $E(t-1)$ from previous generation is used as the base to create immigrants. From $E(t-1)$, a set of $r_{ei} * n$ individuals are iteratively generated by mutating $E(t-1)$ with a probability p_m^i, where n is the population size and r_{ei} is the ratio of the number of elitism-based immigrants to the population size. The generated individuals then act as immigrants and replace the worst individuals in the current population. The elitism-based immigrants scheme combines the idea of elitism with traditional random immigrants scheme. It uses the elite from previous population to guide the immigrants toward the current environment, which is expected to improve the performance of GAs in dynamic environments.

In order to address significant changes that a DOP may suffer, the elitism-based immigrants can be hybridized with traditional random immigrants scheme. The new scheme is called hybrid immigrants GA (HIGA). Within HIGA, in addition to the $r_{ei} * n$ immigrants created from the elite of previous generation, $r_{ri} * n$ immigrants are also randomly created, where r_{ri} is the ratio of the number of random immigrants to the population size. These two sets of immigrants will then replace the worst individuals in the current population.

14.5.3   Improved GAs with Immigrants Schemes

In the DMRP, when a change is caused by removing a few links from the present optimal multicast tree, the present population undergoes dramatic changes and some individuals become infeasible. We name this type of environmental changes as the worst model of network dynamics. The general immigrants based GAs do not consider this case and thereby cannot perform well under the worst dynamics model. We propose improved RIGA, EIGA, and HIGA, denoted as *iRIGA*, *iEIGA*, and *iHIGA*, respectively, to address these difficulties.

When there is no environmental changes detected, the above three improved immigrants based GAs just follow the procedures of their corresponding original GAs, respectively. When a change occurs, in iRIGA, all the infeasible individuals are replaced by random immigrants. In iEIGA, all the infeasible individuals are repaired to become feasible and then the elitism is re-selected. In iHIGA, when the environmental change occurs, for each infeasible solution, we either replace it by a random immigrant or repair it with an equal probability of 0.5.

In iEIGA, since the infeasible solutions are previous elitisms, it is required to keep as many feasible components in them as possible. Therefore, the repair should result in the least change to the tree structure. The proposed repair method works as follows. For each removed link, we search a random path starting from its downstream node. Once an existing tree node is encountered, the search ends. This random path is added to the tree to solve the unconnected problem caused by that removed link. After all the removed links are dealt with, the tree becomes feasible again. Intuitively, this simple method can repair an infeasible tree with the least cost added.

14.5.4   GAs with Memory Schemes

While the immigrants schemes use random immigrants or elitism-based immigrants to maintain the population diversity to adapt to the changing environments, the memory scheme aims to enhance the performance of GAs for DOPs in a different way. It works by storing useful information from the current environment, either implicitly through redundant representations [11, 21, 36] or explicitly by storing good (usually best) solutions of the current population in an extra memory [5, 22, 24]. The stored information can be reused later in new environments. For example, for the explicit memory scheme, when the environment changes, old solutions in the memory that fit the new environment well will be reactivated and hence may adapt GAs to the new environment more directly than random immigrants would do. Especially, when the environment changes cyclically, memory can work very well. This is because in cyclic dynamic environments, with time going, the environment will return to some old environment precisely and the solution in the memory, which has been optimized with respect to the old environment, will instantaneously move the GA to the reappeared optimum of that environment.

The GA with the memory scheme studied in this chapter is called *memory-enhanced GA* (MEGA) [43]. MEGA (and other memory based GAs used in this chapter) uses a memory of size m. The memory in MEGA is re-evaluated every generation to detect environmental changes. The environment is detected as changed if the fitness of at least one individual in the memory has been detected to have changed its fitness. If an environmental change is detected, the memory is merged with the current population and the best $n - m$ individuals are selected as an interim population to undergo genetic operations for a new population while the memory remains unchanged.

14.5.5   GAs with Memory and Immigrants Schemes

The random immigrants approach aims to improve GA's performance in dynamic environments through maintaining the population diversity level with random immigrants and the memory approach aims to move the GA directly to an old environment that is similar to the new one through reusing old good solutions. It is straightforward that the random immigrants and memory approaches can be combined into GAs to deal with DOPs [35]. Therefore, the GA with memory and random immigrants was developed, denoted as MRIGA [43]. MRIGA differs from MEGA only in that in MRIGA before entering the next generation, $r_i * n$ random immigrants are swapped into the population to replace those worst individuals in the population.

However, a more efficient approach of hybridizing memory and random immigrants for GAs to deal with dynamic environments is the memory-based immigrants scheme, denoted as MIGA [41]. MIGA uses the same memory updating scheme as MEGA and MRIGA. However, the memory retrieval does not depend on the detection of environmental changes and is hybridized with the random immigrants scheme via the mutation mechanism. For each generation, the memory is reevaluated and the best memory point is retrieved as the base to create immigrants.

A set of $r_{ri} * n$ individuals are iteratively generated by performing mutation with a probability p_m^i on the best memory point. The generated individuals then act as immigrants and replace the worst $r_{ri} * n$ individuals in the population. In summary, the key idea behind MIGA is that the memory is used to guide the immigrants to make them more biased to the current environment (be it a new one or not) than random immigrants.

14.6 Experimental Study

In the simulation experiments, we implement both the traditional GAs (i.e., SGA and Restart GA) and the dynamic GAs (i.e., RIGA, EIGA, HIGA, iRIGA, iEIGA, iHIGA, MEGA, MRIGA, MIGA). In SGA, if the change makes one individual in the current population become infeasible (e.g., one or more links in the correspond-ing path are lost), we add a penalty value to that individual. By simulation experi-ments, we evaluate their performance in a continuously changing wireless network.

14.6.1 Dynamic Test Environment

The initial network topology is generated using the following method. We first spec-ify a square region with the area of $200 * 200$ that has the width $[0, 200]$ on the x axis and the height $[0, 200]$ on the y axis. Then we generate 100 nodes and the posi-tion (x, y) of each node is randomly specified within the square area. If the distance between two nodes falls into the radio transmission range D, a link will be added to connect them and both the cost and the delay of this link are randomly assigned within the corresponding ranges. Finally, we check if the generated topology is con-nected. If not, the above process is repeated until a connected topology is generated. In the experiments, D is given a reasonable value 50.

All the algorithms start from the initial network topology. Then, after a certain number (say, R) of generations (i.e., the change interval), a certain number (say, M) of nodes are scheduled to sleep or wake up depending on their current status. It means that the selected working nodes will be turned off to sleep and the se-lected sleeping nodes will be turned on to work. Therefore, the network topology is changed accordingly since some links are lost and some other links appear again. By this means, we create a series of network topologies corresponding to the con-tinuous network changes. Furthermore, these adjacent topologies are highly related since each time the changes affect only part of the nodes. It can be seen that R and M determine the change frequency and severity respectively. The larger the value of R, the slower the changes. The larger the value of M, the more severe the changes.

14.6.2 Experimental Study for the DSPRP

14.6.2.1 Parameter Setting

In the following experiments, we set R to 5, 10 and 15 respectively to see the impact of the change frequency on the performance of GAs. We also set M to 2, 3, and 4 respectively. Thus, by the number of nodes changed per time, we have three different series of topologies. When M is set to 2, 3 and 4, we generate the topology series #2, #3, and #4 respectively. Each of these three series has 21 different topologies. In addition, since memory schemes are claimed to work well in the cyclicly changing environment, we set M to 2 and generate a cyclic topology series, named as #1. In topology series #1, topology 1 is the same as topology 21 and the sub-series from topology 1 to 21 is repeated 5 times. Therefore, the cyclic topology series consist of 101 topologies in total. All the experiments are based on the four topology series.

In all the experiments, the population size is set to 50 and the mutation probability is set to 0.1. For RIGA and EIGA, the ratios of the number of immigrants to the population size, r_{ri} and r_{ei}, are set to 0.2. However, to guarantee the comparison fairness (i.e., the same number of immigrants are introduced every generation) in HIGA, r_{ri} and r_{ei} are set to 0.1. In EIGA, HIGA, and MIGA, the mutation probability p_m^i for generating new immigrants, is set to 0.8. Both the source and destination nodes are randomly selected and they are not allowed to be scheduled in any change. The delay upper bound Δ is set to be 2 times of the minimum end-to-end delay.

At each generation, for each algorithm, we select the best individual from the current population and output the cost of the shortest path represented by it. We first set up basic experiments to evaluate the impact of the change interval and the change severity, and the improvements over traditional GAs using RIGA, EIGA and HIGA. Then, since the memory related schemes (i.e., MEGA, MRIGA, and MIGA) are mainly designed for dynamic environments where changes occur in a cyclic way, we set up cyclic environments to evaluate their performance.

For each experiment of an algorithm on a dynamic problem, 10 independent runs are executed with the same set of random seeds. For each run, 21 environmental changes in acyclic dynamic environments and 101 environmental changes in cyclic dynamic environments are allowed. In acyclic dynamic environments, they are equivalent to 105, 210, and 315 generations for $R = 5$, 10, and 15, respectively. In cyclic dynamic environments, they are equivalent to 505, 1010, and 1515 generations for $R = 5$, 10, and 15, respectively. For each run, at each generation we record the best-of-generation fitness which is averaged over the 10 runs.

14.6.2.2 Basic Experimental Results and Analysis

First, we investigate the impact of the change interval on the algorithm performance. When the change interval is 5, the population evolves only 5 generations between two sequential changes. Intuitively, a larger interval will give the population more time to evolve and search better solutions than what a smaller interval does. We take HIGA as an example to compare the quality of solutions obtained at different intervals. However, one problem is that the total generations are different for different intervals, i.e., 105, 210 and 315 versus the interval 5, 10, and 15. Since the number of change points (i.e., the generation at which a new topology is applied) is the same for all the intervals, we take the data at each change point and its left two and right two generations. Thus, the three data sets can be aligned over the three intervals. Figs. 14.1(a) and (b) show the results over topology series #3 and #4 respectively.

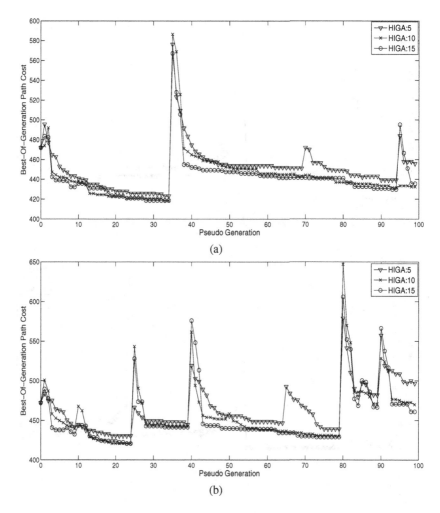

Fig. 14.1 Comparison results of the quality of solution for HIGA with different change intervals over: (a) topology series #3 and (b) topology series #4

Since the generation number does not correspond to the actual number when the interval is 10 or 15, we rename it as pseudo generation. From the two subfigures, it can be seen that the solution quality becomes better when the change interval becomes larger. Therefore, in a relatively slowly changing environment, the studied GAs can achieve a good performance.

Second, we investigate the impact of the change severity on the performance of algorithms. In our problem, the change severity is reflected by the number of nodes involved per change. Therefore, we choose topology series #2 and #4 as the two environments with different change severity. This time we pick up RIGA, EIGA, and HIGA together as the examples. To see the reaction of the algorithms to the

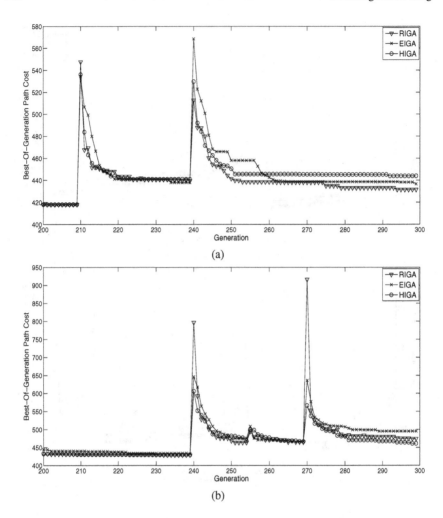

Fig. 14.2 Comparison results of the response speed to changes for RIGA, EIGA, and HIGA over: (a) topology series #2 and (b) topology series #4

changes clearly, we set the interval to 15. Figs. 14.2(a) and (b) show the results over topology series #2 and #4, respectively.

From Fig. 14.2(a), it can be seen that there are two drastic change points: one is at generation 210 and the other is at generation 240. We count the number of the generations that the population spends to find the best solution before the next change. For generation 210, it is 10, 9, and 11 for RIGA, EIGA, and HIGA, respectively. For generation 240, it is 13, 10, and 11 for them, respectively. The total average value is 10.67. From Fig. 14.2(b), we can see that there are three remarkable change points at generation 240, 255, and 270, respectively. We also count the above mentioned number. For generation 240, it is 11, 10, and 14 for RIGA, EIGA, and HIGA, respectively. For generation 255, it is 12, 13, and 14, respectively. For generation 270, it is 15, 12, and 13, respectively. The total average value is 12.67. In average, two more generations are spent for achieving good solutions in the topology series where more severe changes occur. Therefore, we can conclude that these dynamic GAs respond to the environmental changes in a reasonable speed and the more severe the changes, the longer the response time.

Third, we compare the dynamic GAs with the traditional GAs over the dynamic shortest path problem. Since the dynamic GAs are designed for the dynamic environments, they should show better performance than the traditional GAs over our problem. We compare RIGA, EIGA, and HIGA with SGA and Restart GA. We choose topology series #2 and #3 as the two dynamic environments. The interval is set to 10 here and also in the following experiments. Figs. 14.3(a) and (b) show the comparison results over topology series #2 and #3 respectively. From Figs. 14.3(a) and (b), it can be seen that the Restart GA exhibits the worst performance even when the changes have trivial impacts on the current population. The reason is that the Restart GA does not exploit any useful information in the old environment and that the frequent restart sacrifices its evolving capability. Although SGA is much better than the Restart GA, the best solutions that it can find in the new environment are not competitive to those found by any of the three GAs with immigrants schemes. The immigrants bring more diversity to the populations in RIGA, EIGA and HIGA and therefore enhance their search capabilities.

Fourth, we compare the immigrants based GAs with the memory related GAs (i.e., MEGA, MRIGA, and MIGA) in the acyclic environments. According to the above experiments, HIGA is a good representative of the three immigrants based GAs. Therefore, we evaluate the quality of solutions for HIGA, MEGA, MRIGA, and MIGA over topology series #2 and #3 respectively. The memory size is set to 20. Figs. 14.4(a) and (b) show the results. In Fig. 14.4(a), it can be seen that HIGA and MIGA show a competitive performance. MEGA performs the worst among all the memory based GAs. The reason is that in our problem, when a change occurs, the best individual in the memory may become infeasible. Therefore, the memory scheme may lose its power. However, in MRIGA and MIGA, the random immigrants are added or the individuals in the memory are just used to generate immigrants by mutation. Therefore, the infeasible solutions from the memory have a very high probability to be replaced by feasible solutions. In Fig. 14.4(b), it can be seen that although HIGA is not always the best, the memory degrades the algorithm performance when changes occur. Therefore, we conclude that the memory related schemes have no advantages in acyclic dynamic environments.

14.6.2.3   Experimental Results and Analysis in Cyclic Dynamic Environments

In this section, we focus on the cyclic dynamic environments and topology series #1 build such an environment for our experiments. We repeat 20 different toplogies 5 times and the memory schemes will show more power when the same environments

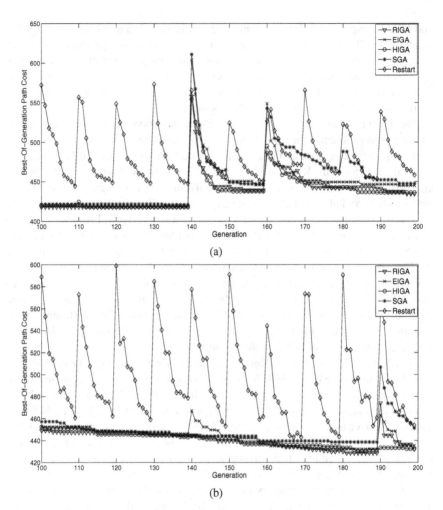

Fig. 14.3 Comparison results of the quality of solution for RIGA, EIGA, HIGA, SGA, and Restart GA over: (a) topology series #2 and (b) topology series #3

are visited more times. Therefore, we sample the data from the latter part of the evolutionary process in MEGA, MRIGA and MIGA. The memory size is set to 20.

First, we compare the memory related schemes with the traditional GAs in the cyclic dynamic environment. Since the immigrants based GAs beat both SGA and Restart GA in the acyclic dynamic environment, we also want to know if the traditional GAs are suitable for cyclic dynamic environments. We evaluate MEGA, MRIGA, MIGA, SGA, and Restart GA over topology series #1. Figs. 14.5(a) and (b) show the results. From Figs. 14.5(a) and (b), it can be seen that the results are similar as the ones in Fig. 14.3. The Restart GA always exhibits the worst performance. The frequent restart severely sacrifices its capability of searching the good

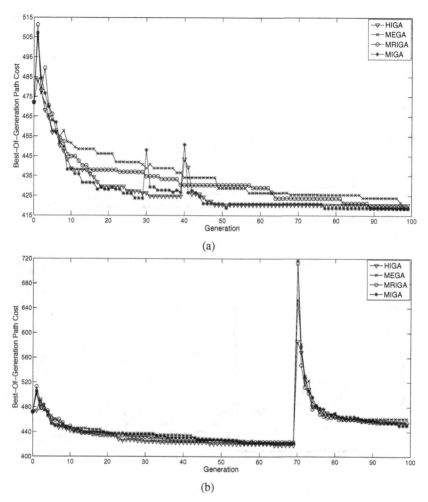

Fig. 14.4 Comparison results of the quality of solution for HIGA, MEGA, MRIGA, and MIGA over: (a) topology series #2 and (b) topology series #3

solutions. Although SGA is much better than the Restart GA, the best solutions that it can find in the new environment are not competitive to those found by any of the three memory related GAs. Therefore, the traditional GAs do not work well in a cyclic dynamic environment, either.

Second, we compare the three memory related GAs with the immigrants based GAs in the cyclic environments. We also pick up HIGA as the representative of the immigrants schemes. Figs. 14.6(a) and (b) show the results. We can see that in both subfigures the three memory related schemes perform better than HIGA. It is contrary to the results shown in Fig. 14.4. In Fig. 14.6(a), when a change occurs, MEGA just takes 1 generation to find the good solution while RIGA takes 8 generations but still finds a worse solution. In Fig. 14.6(b), MEGA, MRIGA and MIGA also takes 1

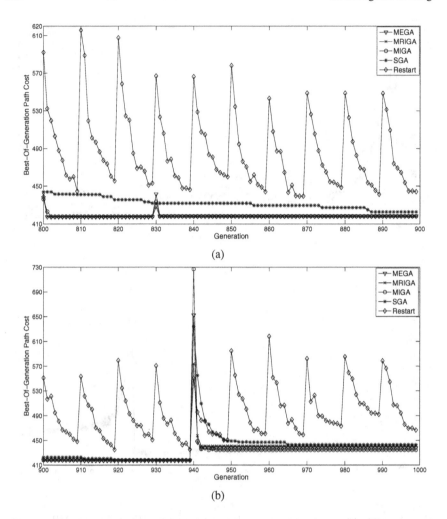

Fig. 14.5 Comparison results of the quality of solution for MEGA, MRIGA, MIGA, SGA, and Restart GA in the cyclic topology series from: (a) generation 800 to 899 and (b) generation 900 to 999

generation to find the good solution for the new environment. The reason is that the good solution stored in the memory can be retrieved immediately when the population enters an environment which has been visited before. Therefore, the memory related schemes are really suitable for cyclic dynamic environments.

14.6.3 Experimental Study for the DMRP

For the DMRP, we implement the two traditional GAs (i.e., SGA and Restart GA) and the six immigrants based GAs (i.e., RIGA, EIGA, HIGA, iRIGA, iEIGA, and

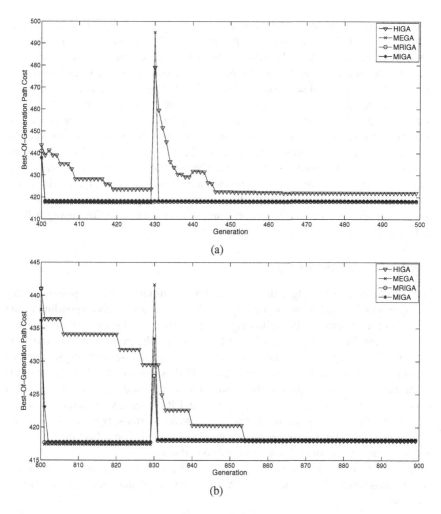

Fig. 14.6 Comparison results of the quality of solution for HIGA, MEGA, MRIGA, and MIGA in the cyclic topology series from: (a) generation 400 to 499 and (b) generation 800 to 899

iHIGA). There are two dynamics models, i.e., general and worst. In the general dynamics model, every R generations, M nodes are scheduled to sleep or wake up depending on their current status. The network topology is changed accordingly since some links are lost and some other links appear again. The nodes are randomly selected and thereby the affected links may belong to the present multicast tree or not. The source and all the destinations are not allowed to be scheduled in any change. All the experiments under the general model are based on topology series #2 and #4 because they represent different change severity. We set up experiments to evaluate the improvements over traditional GAs using RIGA, EIGA and HIGA.

In the worst dynamics model, every R generations, the present best multicast tree is first identified. Then, a certain number (say, U) of links on the tree are selected for removal. It means that the selected links will be forced to be removed from the network topology. Just before the next change occurs, the network topology is recovered to its original state and ready for the coming change. The population is severely affected by each topology change since the optimal solution and possibly some other good solutions become infeasible suddenly. To be fair, at most one link is allowed to be removed on the tree path from the source to each receiver. We let U range from 1 to 3 to see the effect of the change severity. Under the worst dynamics model, the topology series cannot be generated in advance because every change is correlated with the algorithm running. However, similarly, we also allow 20 changes. We set up the experiments to evaluate the impact of the change interval and the change severity, and the improvements over traditional GAs using iRIGA, iEIGA and iHIGA.

14.6.3.1 Parameter Setting

In both models, all the algorithms start from the initial network topology. Every change interval R, the network topology is changed in a way corresponding to the dynamics model used. In the following experiments, we set R to 5, 10 and 15 separately to see the impact of the change frequency on the performance of dynamic GAs. In all the experiments, the crossover probability was set to 0.95 and the mutation probability was set to 0.05. For RIGA, iRIGA, EIGA, and iEIGA, the ratios of the number of immigrants to the population size, r_{ri} and r_{ei}, were set to 0.2. In HIGA and iHIGA, r_{ri} and r_{ei} were set to 0.1. In EIGA, iEIGA, HIGA, and iHIGA, the mutation probability p_m^i for generating new immigrants, was set to 0.8. Both the source and destination nodes were randomly selected. The delay upper bound Δ was set to be 2 times of the minimum end-to-end delay.

In order to have fair comparisons among GAs, the population size and immigrants ratios were set such that each GA has 60 fitness evaluations per generation as follows:

$$(1 + r_i) * n = 60, \tag{14.7}$$

where n is the whole population size, which was set to 50. Hence, we have $n = 60$ for SGA and Restart GA, and $n = 50$ for RIGA, EIGA, HIGA, iRIGA, iEIGA, and iHIGA. At each generation, for each algorithm, we select the best individual from the current population and output the cost of the optimal tree represented by it. For each experiment of an algorithm on a dynamic problem, 10 independent runs were executed with the same set of random seeds. For each run, at each generation we record the best-of-generation fitness which is averaged over the 10 runs.

14.6.3.2 Under the General Dynamics Model

Since under the general dynamics model, the environment changes do not have significant effects on the GAs, the investigation on both the change interval and the change severity are put under the worst dynamics model. However, under this

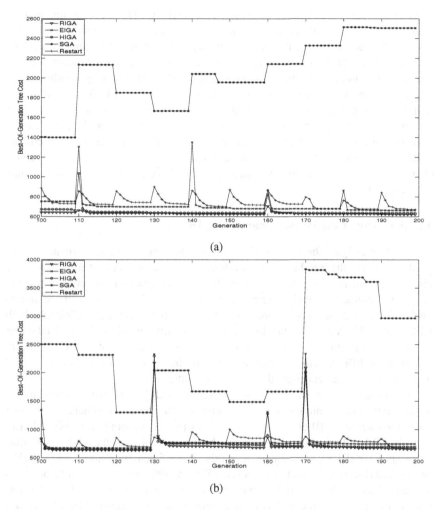

Fig. 14.7 Comparison results of the quality of solution for RIGA, EIGA, HIGA, SGA, and Restart GA over: (a) topology series #2 and (b) topology series #4

model, we are still interested in the comparison between the dynamic GAs with the traditional GAs over the dynamic multicast routing problem. Since the dynamic GAs are designed for the dynamic environments, they should show a better performance than the traditional GAs over our problem. We compared RIGA, EIGA, and HIGA with SGA and Restart GA in the experiments using topology series #2 and #4 as the two dynamic environments. The interval of changes was set to 10 here.

Figs. 14.7(a) and (b) show the comparison results over topology series #2 and #4, respectively. From Figs. 14.7(a) and (b), it can be seen that SGA always exhibits the worst performance. When the topology is changed and infeasible solutions occupy the population, SGA cannot recover the population by generating new feasible solutions through the standard evolutionary operations. Therefore, simple penalty cannot make the population adapt to the complicated environmental changes. On average, the Restart GA is also worse than any of the three immigrants based GAs. The reason is that the Restart GA does not exploit any useful information in the old environment and that the frequent restart sacrifices its evolving capability. Immigrants bring more diversity to the populations in RIGA, EIGA and HIGA and therefore enhance their search capabilities. Among the three dynamic GAs, EIGA achieves the worst performance in most of the time. The reason lies in that in EIGA, new immigrants are generated from the mutation of the elitism. In our problem, the mutation operation just changes a partial path on the tree. Thus, the new immigrants share most of the tree components and bring less diversity into the population than RIGA and HIGA.

14.6.3.3 Under the Worst Dynamics Model

First, we investigate the impact of the change interval on the performance of algorithms. Here, the number of links removed per change was set to 2. When the change interval is 5, the population evolves only 5 generations between two sequential changes. Intuitively, a larger interval will give the population more time to evolve and search better solutions than what a smaller interval does. We take both iRIGA and iHIGA as examples to compare the quality of solutions obtained under different change intervals. However, one problem is that the total generations are different for different change intervals. We use the same method in Section 14.6.2 to get the three data sets aligned over the three intervals.

Figs. 14.8(a) and (b) show the results regarding iRIGA and iHIGA, respectively. Since the generation number does not correspond to the actual generation number when the interval is 10 or 15, we rename it as pseudo generation. In Fig. 14.8(a), over the 20 topologies, the iRIGA with change interval 15 achieves 11 best solutions while the iRIGA with change intervals 10 and 5 achieve 6 and 3, respectively. In Fig. 14.8(b), over the 20 topologies, the iRIGA with change interval 15 achieves 16 best solutions while the iRIGA with change interval 10 achieves 4. It can be concluded that the solution quality becomes better when the change interval becomes larger. Therefore, in a relatively slowly changing environment, the improved immigrants based GAs can achieve a good performance.

Second, we investigate the impact of the change severity on the performance of algorithms. Under the worst dynamics model, the change severity is reflected by the number of links removed from the present optimal tree per change. Therefore, we generate two topology series by removing different number of links each change. One is to remove only one link each change and the other is to remove three links each change. These two topology series act as the two environments with different change severities. This time, we pick up iRIGA, iEIGA, and iHIGA together as the example algorithms and we set the change interval to 10. Figs. 14.9(a) and (b) show the results in the two different environments, respectively.

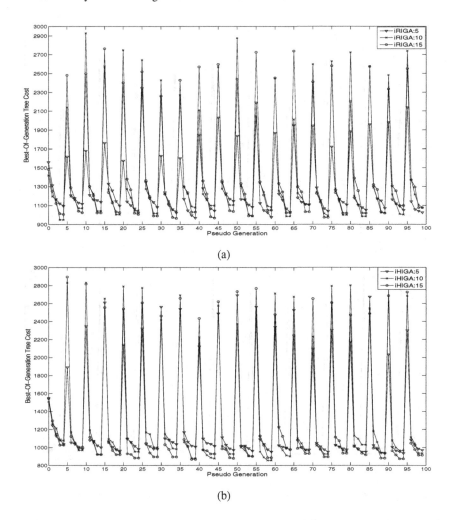

Fig. 14.8 Comparison results of the quality of solution under different change intervals for: (a) iRIGA and (b) iHIGA

From Fig. 14.9(a), it can be seen that iRIGA takes almost 9 generations to get its best solution after one change occurs. iEIGA takes about 5 generations to get its best solution. However, iHIGA can quickly adapt to the environmental changes and get the best solution among these three GAs in 90% of the time. Therefore, in the environment with a low change severity, iHIGA performs the best since it takes the advantages of both iRIGA and iEIGA. From Fig. 14.9(b), it can be seen that for both iRIGA and iHIGA, they need almost 9 generations to get their best solutions after one change occurs and iHIGA achieves a better solution quality than iRIGA. However, in this environment with a high change severity, iEIGA performs very well which takes 2 to 5 generations to get the overall best solution. The reason is

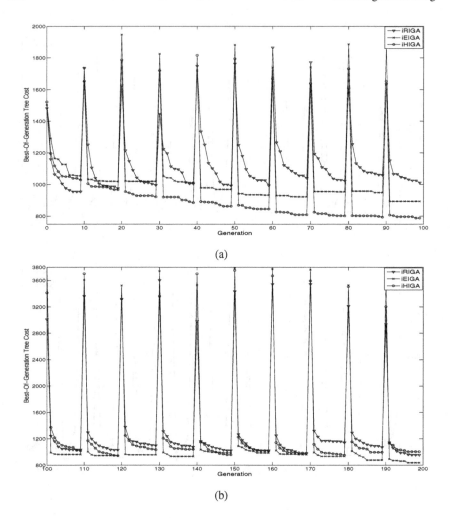

Fig. 14.9 Comparison results of the response speed to changes for iRIGA, iEIGA, and iHIGA over two different topology series where: (a) each change removes 1 link and (b) each change removes 3 links from the present optimal tree

that in the highly dynamic environment, after a change occurs, the proposed repair method can reserve the useful components of the elitism and repair the broken part with the least added cost. The new immigrants generated by repairing the elitisms can quickly adapt to the severe changes. However, as we have discussed under the general dynamics model, iEIGA brings a less diversity to the population compared to both iRIGA and iHIGA. Therefore, we can conclude that these dynamic GAs respond to the environmental changes in a reasonable speed and perform well.

Third, we compare the dynamic GAs with the traditional GAs under the worst dynamics model. Here, the number of links removed per change is set to 2 and

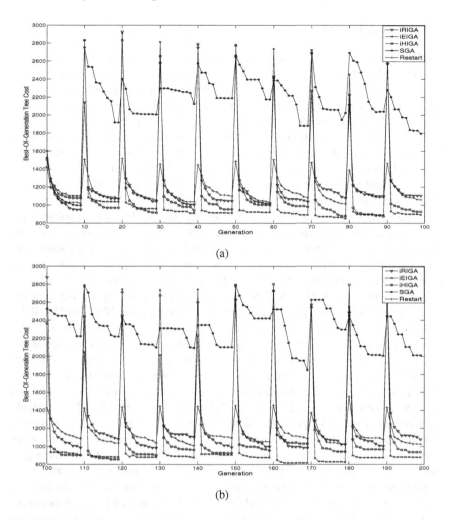

(a)

(b)

Fig. 14.10 Comparison results of the quality of solution for iRIGA, iEIGA, iHIGA, SGA, and Restart GA from: (a) generation 1 to 99 and (b) generation 100 to 199

the change interval is 10. Figs. 14.10(a) and (b) show the results. Similar as under the general model, SGA performs the worst since it does not explicitly handle the environmental changes. Most of the time, Restart GA performs worse than all the improved immigrants based GAs. However, occasionally, iRIGA is worse than it. Overall, iEIGA is better than iHIGA as it has shown in the environment with a high change severity. It can be concluded that the three improved immigrants based GAs greatly outperform the two standard GAs under the worst dynamics model.

14.7   Conclusion

Mobile ad hoc network is a self-organizing and self-configuring multi-hop wireless network, which has a wide usage nowadays. The shortest path routing problem aims to establish a multi-hop forwarding path from a source node to a destination node and is one important issue that significantly affects the network performance. MANETs have also seen various collaborative multimedia applications which require an efficient information delivery service from a designated source to multiple receivers. An QoS multicast tree is preferred to support this service. However, the optimal QoS multicast routing problem is proved to be NP-hard.

So far, quite some works have been done to address the shortest path routing problem and the multicast routing problem by genetic algorithms or other artificial intelligence techniques. These works consider the fixed network topology only. However, in a continuously changing network like MANETs, it is much more challenging to deal with the routing optimization problems than to solve the static ones in a fixed infrastructure. In this chapter, we formulate both the dynamic shortest path routing problem and the dynamic multicast routing problem in MANETs. By observing that dynamic GAs perform very well over many dynamic benchmark optimization problems, we apply them to both the DSPRP and the DMRP in MANETs.

Among approaches developed for GAs to deal with DOPs, immigrants schemes aim at maintaining the diversity of the population throughout the run via introducing random individuals into the current population, while memory schemes aim at storing useful information for possible reuse in a cyclic dynamic environment. Based on the characteristics of the multicast routing problem, we also propose three improved versions of immigrants based GAs, i.e., iRIGA, iEIGA, and iHIGA, to handle highly dynamic environments. Specialized GAs are designed for the shortest path routing problem and the multicast routing problem in MANETs. Several immigrants and/or memory schemes are adapted and integrated into the specialized GAs (which give several GA variants) to solve both problems. Extensive simulation experiments are conducted based on a large scale MANET constructed to evaluate various aspects of these GA variants for the DSPRP and the DMRP. Experimental results demonstrate that our algorithms can adapt to the environmental changes well and achieve better solutions after each change than the traditional GAs. Therefore, they are promising techniques for dealing with dynamic telecommunication optimization problems.

Acknowledgements. This work was supported by the Engineering and Physical Sciences Research Council (EPSRC) of U.K. under Grant numbers EP/E060722/1, EP/E060722/2, and EP/K001310/1, and partially by the State Key Laboratory of Synthetical Automation for Process Industries, Northeastern University, China.

References

[1] Adelstein, F., Richard, G., Schwiebert, L.: Distributed multicast tree generation with dynamic group membership. Comput. Commun. 26(10), 1105–1128 (2003)
[2] Aharoni, E., Cohen, R.: Restricted dynamic Steiner trees for scalable multicast in datagram networks. IEEE/ACM Trans. Netw. 6(3), 286–297 (1998)

[3] Ahn, C.W., Ramakrishna, R.S., Kang, C.G., Choi, I.C.: Shortest path routing algorithm using hopfield neural network. Electron. Lett. 37(19), 1176–1178 (2001)

[4] Ahn, C.W., Ramakrishna, R.S.: A genetic algorithm for shortest path routing problem and the sizing of populations. IEEE Trans. Evol. Comput. 6(6), 566–579 (2002)

[5] Branke, J.: Memory enhanced evolutionary algorithms for changing optimization problems. In: Proc. 1999 Congr. Evol. Comput., pp. 1875–1882 (1999)

[6] Branke, J., Kaußler, T., Schmidt, C., Schmeck, H.: A multi-population approach to dynamic optimization problems. In: Proc. 4th Int. Conf. Adaptive Comput. Des. Manuf., pp. 299–308 (2000)

[7] Cheng, H., Wang, X., Yang, S., Huang, M.: A multipopulation parallel genetic simulated annealing based QoS routing and wavelength assignment integration algorithm for multicast in optical networks. Appl. Soft Comput. 9(2), 677–684 (2009)

[8] Cheng, H., Yang, S.: Genetic algorithms with immigrants schemes for dynamic multicast problems in mobile ad hoc networks. Eng. Appl. Artif. Intel. 23(5), 806–819 (2010)

[9] Cobb, H.G., Grefenstette, J.J.: Genetic algorithms for tracking changing environments. In: Proc. 5th Int. Conf. Genetic Algorithms, pp. 523–530 (1993)

[10] Cordeiro, C., Gossain, H., Agrawal, D.: Multicast over wireless mobile ad hoc networks: present and future directions. IEEE Netw. 17(1), 52–59 (2003)

[11] Dasgupta, D., McGregor, D.: Nonstationary function optimization using the structured genetic algorithm. In: Proc. 2nd Int. Conf. Parallel Problem Solving from Nature, pp. 145–154 (1992)

[12] Din, D.: Anycast routing and wavelength assignment problem on WDM network. IEICE Trans. Commun. E88-B(10), 3941–3951 (2005)

[13] Grefenstette, J.J.: Genetic algorithms for changing environments. In: Proc. 2nd Int. Conf. Parallel Problem Solving from Nature, pp. 137–144 (1992)

[14] Helvig, C., Robins, G., Zelikovsky, A.: An improved approximation scheme for the group Steiner problem. Networks 37(1), 8–20 (2000)

[15] Holland, J.: Adaptation in Natural and Artificial Systems. University of Michigan Press, Ann Arbor (1975)

[16] Hwang, F., Richards, D.: Steiner tree problems. Networks 22(1), 55–89 (1992)

[17] Jia, X., Pissinou, N., Makki, K.: A real-time multicast routing algorithm for multimedia applications. Comput. Commun. 20(12), 1098–1106 (1997)

[18] Jia, X.: A distributed algorithm of delay-bounded multicast routing for multimedia applications in wide area networks. IEEE/ACM Trans. Netw. 6(6), 828–837 (1998)

[19] Khuller, S., Raghavachari, B., Young, N.: Balancing minimum spanning and shortest path trees. Algorithmica 14(4), 305–321 (1995)

[20] Lee, S., Soak, S., Kim, K., Park, H., Jeon, M.: Statistical properties analysis of real world tournament selection in genetic algorithms. Appl. Intel. 28(2), 195–205 (2008)

[21] Lewis, J., Hart, E., Ritchie, G.: A comparison of dominance mechanisms and simple mutation on non-stationary problems. In: Eiben, A.E., Bäck, T., Schoenauer, M., Schwefel, H.-P. (eds.) PPSN 1998. LNCS, vol. 1498, pp. 139–148. Springer, Heidelberg (1998)

[22] Louis, S., Xu, Z.: Genetic algorithms for open shop scheduling and re-scheduling. In: Proc. 11th ISCA Int. Conf. Comput. Their Appl., pp. 99–102 (1996)

[23] Mohemmed, A.W., Sahoo, N.C., Geok, T.K.: Solving shortest path problem using particle swarm optimization. Appl. Soft Comput. 8(4), 1643–1653 (2008)

[24] Mori, H., Nishikawa, Y.: Adaptation to changing environments by means of the memory based thermodynamical genetic algorithm. In: Proc. 7th Int. Conf. Genetic Algorithms, pp. 299–306 (1997)

[25] Morrison, R.W., De Jong, K.A.: Triggered hypermutation revisited. In: Proc. 2000 Congr. Evol. Comput., vol. 2, pp. 1025–1032 (2000)

[26] Siva Ram Murthy, C., Manoj, B.S.: Ad Hoc Wireless Networks: Architectures and Protocols. Prentice Hall PTR (2004)

[27] Narvaez, R., Siu, K.-Y., Tzeng, H.-Y.: New dynamic algorithms for shortest path tree computation. IEEE/ACM Trans. Netw. 8(6), 734–746 (2000)

[28] Oh, S., Ahn, C., Ramakrishna, R.: A genetic-inspired multicast routing optimization algorithm with bandwidth and end-to-end delay constraints. In: King, I., Wang, J., Chan, L.-W., Wang, D. (eds.) ICONIP 2006. LNCS, vol. 4234, pp. 807–816. Springer, Heidelberg (2006)

[29] Oliveira, C., Pardalos, P.: A survey of combinatorial optimization problems in multicast routing. Comput. & Oper. Res. 32(8), 1953–1981 (2005)

[30] Oppacher, F., Wineberg, M.: The shifting balance genetic algorithm: improving the GA in a dynamic environment. In: Proc. 1999 Genetic Evol. Comput. Conf., vol. 1, pp. 504–510 (1999)

[31] Papadimitriou, C., Steiglitz, K.: Combinatorial Optimization: Algorithms and Complexity. Dover Publications Inc., NY (1998)

[32] Parsa, M., Zhu, Q., Garcia-Luna-Aceves, J.: An iterative algorithm for delay-constrained minimum-cost multicasting. IEEE/ACM Trans. Netw. 6(4), 461–474 (1998)

[33] Robins, G., Zelikovsky, A.: Improved Steiner tree approximation in graphs. In: Proc. ACM/SIAM Symp. on Discrete Algorithms, pp. 770–779 (2000)

[34] Tinos, R., Yang, S.: A self-organizing random immigrants genetic algorithm for dynamic optimization problems. Genet. Program. Evolvable Mach. 8(3), 255–286 (2007)

[35] Trojanowski, K., Michalewicz, Z.: Evolutionary optimization in non-stationary environments. J. Comput. Sci. Tech. 1(2), 93–124 (2000)

[36] Uyar, A., Harmanci, A.: A new population based adaptive dominance change mechanism for diploid genetic algorithms in dynamic environments. Soft Comput. 9(11), 803–815 (2005)

[37] Vavak, F., Fogarty, T.C.: A comparative study of steady state and generational genetic algorithms for use in nonstationary environments. In: Fogarty, T.C. (ed.) AISB-WS 1996. LNCS, vol. 1143, pp. 297–304. Springer, Heidelberg (1996)

[38] Wang, X., Cao, J., Cheng, H., Huang, M.: QoS multicast routing for multimedia group communications using intelligent computational methods. Comput. Commun. 29(12), 2217–2229 (2006)

[39] Weicker, K.: Evolutionary Algorithms and Dynamic Optimization Problems. Der andere Verlag, Osnabrück (2003)

[40] Xu, Y., Salcedo-Sanz, S., Yao, X.: Metaheuristic approaches to traffic grooming in WDM optical networks. Int. J. of Comput. Intel. Appl. 5(2), 231–249 (2005)

[41] Yang, S.: Memory-based immigrants for genetic algorithms in dynamic environments. In: Proc. 2005 Genetic Evol. Comput. Conf., vol. 2, pp. 1115–1122 (2005)

[42] Yang, S.: Genetic algorithms with elitism-based immigrants for changing optimization problems. In: Giacobini, M. (ed.) EvoWorkshops 2007. LNCS, vol. 4448, pp. 627–636. Springer, Heidelberg (2007)

[43] Yang, S.: Genetic algorithms with memory- and elitism-based immigrants in dynamic environments. Evol. Comput. 16(3), 385–416 (2008)

[44] Yang, S., Cheng, H., Wang, F.: Genetic algorithms with immigrants and memory schemes for dynamic shortest path routing problems in mobile ad hoc networks. IEEE Trans. Syst., Man, Cybern. C, Appl. Rev. 40(1), 52–63 (2010)

[45] Yang, S., Ong, Y.-S., Jin, Y. (eds.): Evolutionary Computation in Dynamic and Uncertain Environments. Springer (2007)

[46] Yang, S., Tinos, R.: A hybrid immigrants scheme for genetic algorithms in dynamic environments. Int. J. Autom. Comput. 4(3), 243–254 (2007)

[47] Yang, S., Tinos, R.: Hyper-selection in dynamic environments. In: Proc. 2008 Congr. Evol. Comput., pp. 3185–3192 (2008)

[48] Yang, S., Yao, X.: Experimental study on population-based incremental learning algorithms for dynamic optimization problems. Soft Comput. 9(11), 815–834 (2005)

[49] Yang, S., Yao, X.: Population-based incremental learning with associative memory for dynamic environments. IEEE Trans. Evol. Comput. 12(5), 542–561 (2008)

[50] Yong, K., Poo, G., Cheng, T.: Proactive rearrangement in delay constrained dynamic membership multicast. Comput. Commun. 31(10), 2566–2580 (2008)

[51] Yu, X., Tang, K., Chen, T., Yao, X.: Empirical analysis of evolutionary algorithms with immigrants schemes for dynamic optimization. Memetic Comput. 1(1), 3–24 (2009)

[52] Yu, X., Tang, K., Yao, X.: An immigrants scheme based on environmental information for genetic algorithms in changing environments. In: Proc. 2008 Congr. Evol. Comput., pp. 1141–1147 (2008)

Chapter 15
Evolutionary Computation for Dynamic Capacitated Arc Routing Problem

Yi Mei, Ke Tang, and Xin Yao

Abstract. In this chapter, a new dynamic capacitated arc routing problem (CARP) is defined and investigated. Compared with the static CARP and other dynamic CARP investigated by the existing researches, the new dynamic CARP is more general and closer to reality, and thus is more worthwhile to be solved. Due to the stochastic factors included in the dynamic CARP, the objective is not to obtain the optimal solution in a specific environment, but to find a robust solution that shows good performance in all the possible environments. For the dynamic CARP, a robustness measure based on repair operator is defined. The corresponding repair operator is designed according to the real-world considerations. Then, the benchmark instances of the dynamic CARP are generated by extending from the static counterparts to facilitate evaluating potential approaches. After that, the preliminary analysis for the fitness landscape of the dynamic CARP is conducted by experimental studies.

Yi Mei
School of Computer Science and Information Technology, RMIT University,
Melbourne VIC 3001, Australia
e-mail: yi.mei@rmit.edu.au

Ke Tang
Nature Inspired Computation and Applications Laboratory (NICAL),
School of Computer Science, University of Science and Technology of China,
Hefei 230027, China
e-mail: ketang@ustc.edu.cn

Xin Yao
Nature Inspired Computation and Applications Laboratory (NICAL), The USTC-Birmingham Joint Research Institute in Intelligent Computation and Its Applications (UBRI), School of Computer Science, University of Science and Technology of China, Hefei 230027, China, and Centre of Excellence for Research in Computational Intelligence and Applications (CERCIA), School of Computer Science, University of Birmingham, Birmingham B15 2TT, U.K.
e-mail: x.yao@cs.bham.ac.uk

S. Yang and X. Yao (Eds.): *Evolutionary Computation for DOPs*, SCI 490, pp. 377–401.
DOI: 10.1007/978-3-642-38416-5_15   © Springer-Verlag Berlin Heidelberg 2013

15.1 Introduction

The capacitated arc routing problem (CARP) is a classic combinatorial optimization problem that has wide applications in the real world including the salt routing optimization [4, 23, 24, 28, 29, 42], urban waste collection [14, 16, 33, 34, 39] and snow removal [11, 41].

CARP can be described as follows: Given a connected graph, some edges (called the *tasks*) of the graph are required to be served by a vehicle fleet located at the *depot* vertex. The problem aims to determine a least-cost plan subject to the following constraints:

- Each vehicle must start and end at the depot vertex;
- Each task is served exactly once by one vehicle;
- The total demand of the tasks served by each route cannot exceed its capacity.

CARP has been proven to be NP-hard by Golden and Wong in [26]. That is, the computational complexity of finding the global optimum increases exponentially with the increase of problem size. On the other hand, the real-world problems often have quite large problem sizes, making it impractical to find the global optimum by the exact methods. In this situation, methods based on evolutionary computation are promising methods due to their capability of obtaining good sub-optimal solutions within a given time budget.

So far, intensive investigations have been conducted for solving CARP. Most of them are focused on the static CARP, in which all the problem parameters are exactly known in advance and do not change as time goes on. However, in the real world, the above assumption can hardly be guaranteed, and some or all of the problem parameters cannot be known in advance or change over time. For example, in the snow removal application, the amount of snow to be removed for each street cannot be known exactly until the truck finishes the removal, and the traffic jam or road maintenance influences the time needed to traverse a street.

When the problem contains stochastic or dynamic problem parameters, the corresponding CARP can be called the dynamic CARP . Although the dynamic CARP is closer to the reality than the static counterpart, it has been overlooked so far, and there are only a few related research done. Fleury *et al.* proposed a stochastic CARP model in [21]. In the model, the task demands are considered to be stochastic. In this case, a solution that is expected to be feasible may be actually infeasible since the actual total demand served by a route may be larger than expected and exceed the capacity. Therefore, the solution must be made feasible by some repair operator before the calculation of its total cost.

The methods proposed for solving the dynamic CARP are also quite few. Fleury *et al.* proposed an evolutionary algorithm in [21], which employs a simple repair operator and optimizes the total cost of the repaired solution. Christiansen *et al.* proposed a branch-and-price algorithm in [13]. Laporte *et al.* proposed an adaptive large neighborhood search heuristic in [34].

Compared with the dynamic CARP, the stochastic vehicle routing problem (SVRP) , which is the node routing counterpart of the dynamic CARP, has received much more research interest. Starting from the one-vehicle special case of VRP, i.e., the stochastic traveling salesman problem (TSP) , the presence of customers and the travel times between the customers are considered to be stochastic. The TSP with stochastic customers was proposed by Jaillet [31] along with a number of mathematical models, while there has been no mathematical model for the TSP with stochastic travel times. In the m-vehicle version of the TSP with stochastic travel times, there are m vehicles available instead of only one vehicle. In the problem, all the vehicles have to depart from and arrive at a common depot, and a deadline is imposed on each vehicle route. A penalty is induced for the completion delay. For stochastic VRPs, research works are focused on stochastic demand of customers [5, 18, 45] and on the stochastic presence of customers [5, 47], and both [6]. A comprehensive survey of the aforementioned problems is given in [22], and a dynamic replanning for VRP in case of unforseen situations such as traffic jams is considered in [48] and [49].

In summary, three stochastic factors have been considered in previous research work: (1) the presence of tasks (vertices in VRP and edges in CARP); (2) the demand of the tasks and (3) the deadheading costs (e.g., travel time) between the tasks. In fact, a fourth stochastic factor can be considered: the availability of the path between each pair of vertices. For example, when a street is on its maintenance, it becomes temporarily unavailable to be traversed and thus disappears from the graph. The above four stochastic factors can occur simultaneously in the problem. Unfortunately, a corresponding model has not been investigated. Most research works consider them separately or combine at most two of them together (e.g., the presence and demand of the tasks are combined in [6]). In this chapter, a more general dynamic CARP with all the above four stochastic factors embedded is defined. This dynamic CARP is different from the static CARP with respect to the inputs, outputs, objective and constraints. In the dynamic CARP, the inputs are random variables and the outputs include a solution and a repair operator that can make the solution feasible in any possible environment. The objective of the dynamic CARP switches from obtaining the optimal solution in a specific environment to finding a robust solution, i.e., a solution that shows relatively good performance in all possible environment. The goal of the constraints of the dynamic CARP is no longer to guarantee the feasibility of the solution, but to help improving the robustness of the obtained solution.

After the problem has been defined, preliminary investigations are conducted on it. First, in order to facilitate the potential approaches for the dynamic CARP, the benchmark instances are generated by extending from the static CARP benchmark instances. Then, a rough analysis and discussion about the fitness landscape of the dynamic CARP is conducted. It has been found that for the dynamic CARP, merely using the expected information does not necessarily lead to robust solutions, and the optimal solution for the corresponding static counterpart may be much less robust than other solutions for the dynamic CARP.

The rest of the chapter is organized as follows: First, the dynamic CARP is defined in Section 15.2. After that, the related work for solving the dynamic CARP and its static counterpart is comprehensively reviewed in Section 15.3. Then, the benchmark instances for dynamic CARP is generated in Section 15.4. The preliminary investigation of the fitness landscape of the dynamic CARP is conducted in Section 15.5. Finally, the conclusion and future work is given in Section 15.6.

15.2 Problem Definition

Before defining the dynamic CARP, the simpler and more basic static CARP is first introduced to help understanding. Then, the dynamic CARP is defined and compared with the static CARP.

15.2.1 Static Capacitated Arc Routing Problem

The static CARP is the most basic and simplest version of CARP. It is defined on a weighted connected graph $G(V, E, A)$, where V, E and A are the sets of vertices, undirected edges and directed edges, respectively. For the sake of convenience, the undirect edges will be called *edges* while the directed edges will be called *arcs*. All the edges $(v_i, v_j) \in E$ and arcs $\langle v_i, v_j \rangle \in A$ are associated with a nonnegative serving cost $sc(v_i, v_j)$, a positive deadheading cost $dc(v_i, v_j)$ and a nonnegative demand $d(v_i, v_j)$. The edges and arcs having positive demands are called the *tasks* and must be served. The serving cost of an edge or arc indicates the cost induced by serving it if it is a task, while the deadheading cost indicates the cost induced by traversing it without serving. The sets of edge tasks, arc tasks and total tasks are represented by $E_R = \{(v_i, v_j) \in E | d(v_i, v_j) > 0\}$, $A_R = \{\langle v_i, v_j \rangle \in A | d(v_i, v_j) > 0\}$ and $R = E_R \cup A_R$. A number of vehicles, each with capacity of Q, are located at the *depot* vertex $v_0 \in V$ to serve the tasks in R.

A CARP solution can be essentially represented by a route set $X = \{X_1, ..., X_m\}$ and a corresponding 0-1 vector set $Y = \{Y_1, ..., Y_m\}$. The kth route $X_k = (x_{k1}, ..., x_{kl_k})$, indicating the route traversed by the kth vehicle, is a sequence of vertices starting and ending at v_0, i.e., $x_{k1} = x_{kl_k} = v_0$. The corresponding 0-1 vector $Y_k = (y_{k1}, ..., y_{k(l_k-1)})$ is defined as follows: if $(x_{ki}, x_{k(i+1)})$ is a task and is served at the current position, then $y_{ki} = 1$; otherwise, $y_{ki} = 0$. Under such a solution representation scheme, the static CARP can be stated as follows:

$$\min\ tc(S) = \sum_{k=1}^{m} \sum_{i=1}^{l_k-1} \left(sc(x_{ki}, x_{k(i+1)}) \cdot y_{ki} + dc(x_{ki}, x_{k(i+1)}) \cdot (1 - y_{ki}) \right) \tag{15.1}$$

$$s.t.:\ x_{k1} = x_{kl_k} = v_0,\ k = 1, 2, ..., m \tag{15.2}$$

$$\sum_{k=1}^{m} \sum_{i=1}^{l_k-1} y_{ki} = |R| \tag{15.3}$$

$$(x_{ki}, x_{k(i+1)}) \in E_R \cup A_R,\ \forall y_{ki} = 1 \tag{15.4}$$

$$(x_{k_1 i_1}, x_{k_1 (i_1+1)}) \neq (x_{k_2 i_2}, x_{k_2 (i_2+1)}),\ \forall y_{k_1 i_1} = 1, y_{k_2 i_2} = 1, (k_1, i_1) \neq (k_2, i_2) \tag{15.5}$$

$$\left(x_{k_1 i_1}, x_{k_1(i_1+1)}\right) \neq \left(x_{k_2(i_2+1)}, x_{k_2 i_2}\right), \ \forall y_{k_1 i_1} = 1, y_{k_2 i_2} = 1, (k_1, i_1) \neq (k_2, i_2) \quad (15.6)$$

$$\sum_{i=1}^{l_k-1} d\left(x_{ki}, x_{k(i+1)}\right) \cdot y_{ki} \leqslant Q, \ k = 1, 2, \dots, m \quad (15.7)$$

$$x_{ki} \in V, \ dc\left(x_{ki}, x_{k(i+1)}\right) < \infty, \ y_{ki} = 0 \ or \ 1 \quad (15.8)$$

where $|R|$ is the number of tasks in R. In constraints (15.5) and (15.6), the inequality $(k_1, i_1) \neq (k_2, i_2)$ is satisfied if and only if at least one of the two inequalities $k_1 \neq k_2$ and $i_1 \neq i_2$ is satisfied. Objective (15.1) is to minimize the total cost $tc(S)$. Constraint (15.2) indicates all the routes start and end at the depot v_0. Constraints (15.3)–(15.6) guarantee that all the tasks are served exactly once. Constraint (15.7) is the capacity constraint, i.e., the total demands served by each route cannot exceed the capacity Q. Constraint (15.8) defines the domain of the variables.

15.2.2 Dynamic Capacitated Arc Routing Problem

The above static CARP can be characterized by the following four aspects: inputs, outputs, objective and constraints. To be specific, the inputs include the deadheading costs $dc(v_i, v_j)$, the serving costs $sc(v_i, v_j)$, the demand of the tasks $d(v_i, v_j)$ and the capacity Q. The output is a solution represented by a route vector X and a 0-1 vector Y. The objective is to minimize the total cost of the solution. The constraints include the basic constraints (each route starts and ends at the depot and each task is served exactly once) and the capacity constraint. Next, the dynamic CARP is described from the above four aspects as well, and compared with the static CARP with respect to each aspect.

15.2.2.1   Inputs of the Dynamic Capacitated Arc Routing Problem

In the static CARP, all the input parameters are assumed to be known in advantage and fixed over time. As mentioned in Section 15.1, in the real-world applications, this is usually not the case, and it is more appropriate to represent the input parameters as random variables rather than constants.

As mentioned in Section 15.1, there are four stochastic factors in the dynamic CARP: (1) presence of tasks; (2) demand of tasks; (3) presence of paths between vertices and (4) deadheading costs between vertices. These stochastic factors transform the deadheading costs $dc(v_i, v_j)$ and the demand of tasks $d(v_i, v_j)$ from constants to random variables. It should be noted that the serving costs $sc(v_i, v_j)$ are also affected by the stochastic factors. Here, for the sake of simplicity, one can assume that the serving costs are proportional to the deadheading costs, or even equal to them. Then, the two random variables $dc(v_i, v_j)$ and $sc(v_i, v_j)$ can be combined into a single random variable $dc(v_i, v_j)$.

When a task (v_i, v_j) is absent, one can consider that its demand $d(v_i, v_j)$ is zero. Similarly, $dc(v_i, v_j) = \infty$ implies that the edge (v_i, v_j) temporarily disappears.

Therefore, the distribution of the random variables can be seen as a combination of a constant (including infinity) for the case of absence and a random distribution for the case of presence.

In the real-world applications, an implementation process of a solution can be seen as a sample of the random variables. To be specific, during the implementation process, the presence and actual demand of each task is unknown until it has been served, and the presence and deadheading cost between each pair of vertices can only be known after the path has been traversed.

15.2.2.2   Outputs of the Dynamic Capacitated Arc Routing Problem

The stochastic nature of the input parameters in the dynamic CARP influence both the quality and feasibility of solutions. First, it is natural that the change of deadheading costs leads to the change of the total cost, which determines the solution quality. Second, the change of demand or the presence of the tasks that are expected to be absent will make the total demand served by a route larger than expected, and thus violates the capacity constraint. Third, the absence of a path existing in the solution makes the corresponding route disconnected so that the solution becomes illegal unless another path is found to connect the separated vertices. In summary, given a solution, the quality and feasibility changes in the dynamic environment. That is, a solution is feasible in one environment, but is infeasible in another environment. A high-quality solution in one environment may perform quite badly in another environment.

The change of feasibility can only be found during the implementation process. For example, a vehicle can only know whether its remaining capacity can afford the actual demand of the next task after serving it. If the change of feasibility has been detected, i.e., the solution becomes infeasible in the current environment, the solution must be modified (repaired) so that it becomes feasible again. For this purpose, a corresponding repair operator is needed to make the solution feasible whenever it becomes infeasible.

In summary, the outputs of the dynamic CARP should include a solution and a repair operator. The repair operator must be able to make the solution feasible in any possible environment.

15.2.2.3   Objective of the Dynamic Capacitated Arc Routing Problem

It is known that the quality and feasibility of solution are different in different environments. Therefore, it is meaningless to obtain the optimal solution in one specific environment since its performance may severely deteriorate when the environment changes. Instead, the objective of the dynamic CARP should be to obtain a robust solution, i.e., a solution that shows relatively good performance under all possible environments. To this end, a proper robustness measure for the dynamic CARP solutions is to be defined.

Taguchi proposed the concept of robustness optimization for the first time in [43], in which the quality of solution depends on a noise parameter ξ that is out of the control of the designer as well as the control parameter x. To evaluate a solution in this situation, Taguchi defined the following robustness measure:

$$MSD = \frac{1}{k} \sum_{i=1}^{k} (y(x, \xi_i) - \hat{y})^2 \tag{15.9}$$

where $y(x, \xi_i)$ is the actual performance under the control parameter x and the noise ξ_i), k indicates the number of all possible noises, and $\hat{y}$ is the target performance. MSD can be seen as the deviation of the actual performance y of the solution x, which is influenced by the noise ξ, from the target performance $\hat{y}$.

After that, the concept of robustness optimization in uncertain and dynamic environments has received the research interest in various scientific fields such as operations research and engineering design (e.g., [3] [19] [32]), and many other robustness measures have been proposed. Some of the representative measures are introduced below:

1. Worst-case Performance $R_w(x)$: Take minimizing the objective function $f(x)$ as an example, the worst-case performance is defined as:

$$R_w(x) = \sup_{x' \in \mathcal{N}(x)} f(x') \tag{15.10}$$

where $\mathcal{N}(x)$ stands for a predefined neighborhood of x. When the actual parameter fluctuates within the neighborhood $\mathcal{N}(x)$ of the solution x due to the stochastic nature, optimizing $R_w(x)$ means optimizing the performance of the solution in the worst case. Examples of the worst-case performance measure can be found in [20] [30] [35] [37].

2. Expected Performance $R_e(x)$: The expected performance measures the expectation of $f(x)$ with respect to the environmental parameter ξ. It can be stated as:

$$R_e(x) = E[f(x, \xi)|\xi] = \int f(x; \xi) d\xi \tag{15.11}$$

The expected performance has been adopted in [9] [10] [15] [50] to evaluate the robustness of solutions.

3. Threshold-based Robustness Measure $R_t(x)$: In many real-world cases, the objective is not to maximize the performance, but to meet the predefined quality threshold q. In such a situation, one can maximize the probability of reaching the quality threshold, i.e.,

$$R_t(x) = Pr[f(x) \leqslant q] \tag{15.12}$$

4. Reliability-based Robustness Measure: Unlike the above three measures, this measure is used to deal with the stochastic factors that appear in the constraints and change the practical feasibility of solution. Given the following optimization problem:

$$\min \ f(x) \tag{15.13}$$
$$s.t. \ g_i(x) \leqslant 0, \ i = 1,...,I \tag{15.14}$$
$$h_j(x) = 0, \ j = 1,...,J \tag{15.15}$$

in which there is a stochastic constraint $g_k(x;\xi) \leqslant 0$ affected by the environmental parameter ξ. For the same value of x, there may exist ξ_1 and ξ_2 so that $g_k(x;\xi_1) \leqslant 0$ and $g_k(x;\xi_2) > 0$. In other words, x is feasible in environment ξ_1, while becomes infeasible in environment ξ_2. In such a situation, the reliability-based robustness measure transforms the original constraint $g_k(x;\xi) \leqslant 0$ to the following constraint $Pr[g_k(x;\xi) \leqslant 0] \geqslant P_0$, which means the probability of satisfying the constraint is no less than the confidence probability P_0. This measure employs the same idea of the probabilistic constrained programming in stochastic programming [7], and has been widely used in the design optimization problems based on reliability [1, 2, 12, 27, 36, 40].

5. Repair-based Robustness Measure: This measure is used to deal with the stochastic factors appearing in constraints as well. As the name implies, the measure defines a repair operator Φ to make infeasible solutions feasible again. Given a solution x and a sample of the environmental parameter ξ, if x is feasible in the current environment, then it remains unchanged. Otherwise, it is modified by Φ to a feasible solution, i.e., $x \rightarrow \Phi(x,\xi)$. Here, the repair operator Φ has to be defined in such a way that for any solution x and environment sample ξ, $\Phi(x,\xi)$ must be a feasible solution. The measure was adopted in [21] in the context of the CARP with stochastic task demands. Under the assumption that the demand of each single task is much smaller than the capacity, the repair operator simply cut the infeasible routes before the last task.

The dynamic CARP has stochastic factors in both the objective function and the constraints. As mentioned above, the stochastic constraints are addressed by the repair operator, and thus the corresponding repair-based robustness measure is to be used. As for the stochastic objective function, measures 1–3 reflect different aspects and can be chosen according to practical consideration.

Then, the robustness measure for the dynamic CARP solutions can be defined as:

$$R(S) = R_*(\Phi(S,\xi)) \tag{15.16}$$

Here, R_* can be R_w, R_e or R_t, and $\Phi(S,\xi)$ is the feasible solution obtained by applying the repair operator Φ on the solution S according to the environment ξ.

15.2.2.4   Constraints of the Dynamic Capacitated Arc Routing Problem

Unlike the static CARP, the feasibility of solution cannot be guaranteed by imposing the constraints since it depends on the actual environment parameters. However, proper constraints can still help find more robust solutions. For example, one can impose a capacity constraint based on the expected demand of tasks so that the obtained solution is expected to satisfy the actual capacity constraint. In practice, different constraints can lead to different optimal solutions in terms of the defined robustness measure.

15.2.2.5 Summary

Table 15.1 summaries the differences between the dynamic CARP and the static CARP with respect to the above four aspects. In the table, "SCARP" and "DCARP" stand for the static and dynamic CARPs, respectively.

Table 15.1 Comparison between the dynamic and static CARPs

Aspect	SCARP	DCARP
Input	The inputs are constants	The inputs include constants and random variables
Output	A feasible solution in the given specific environment	A solution along with an operator to repair the solution whenever it becomes infeasible
Objective	Minimize the total cost	Optimize the robustness
Constraints	To guarantee the feasibility of solutions	To help obtain more robust solutions

Finally, by setting the robustness measures as R_e and imposing proper constraints, the dynamic CARP can be stated as follows:

$$\min R_e(\Phi(S,\xi)) \tag{15.17}$$

$$s.t. : x_{k1} = x_{kl_k} = v_0, \ k = 1,2,...,m \tag{15.18}$$

$$\sum_{k=1}^{m} \sum_{i=1}^{l_k-1} y_{ki} = |R| \tag{15.19}$$

$$(x_{ki}, x_{k(i+1)}) \in E_R \cup A_R, \ \forall y_{ki} = 1 \tag{15.20}$$

$$(x_{k_1 i_1}, x_{k_1(i_1+1)}) \neq (x_{k_2 i_2}, x_{k_2(i_2+1)}), \ \forall y_{k_1 i_1} = 1, y_{k_2 i_2} = 1, (k_1, i_1) \neq (k_2, i_2) \tag{15.21}$$

$$(x_{k_1 i_1}, x_{k_1(i_1+1)}) \neq (x_{k_2(i_2+1)}, x_{k_2 i_2}), \ \forall y_{k_1 i_1} = 1, y_{k_2 i_2} = 1, (k_1, i_1) \neq (k_2, i_2) \tag{15.22}$$

$$E[\sum_{i=1}^{l_k-1} d(x_{ki}, x_{k(i+1)}, \xi) \cdot y_{ki} | \xi] \leqslant Q, \ k = 1,2,...,m \tag{15.23}$$

$$x_{ki} \in V, \ dc(x_{ki}, x_{k(i+1)}, \xi) < \infty, \ y_{ki} = 0 \ or \ 1 \tag{15.24}$$

Constraints (15.18)–(15.22) and (15.24) are the basic constraints and domain of the variables, while constraint (15.23) implies that the expected total demand of each route does not exceed the capacity.

15.3 Evolutionary Computation for Dynamic Capacitated Arc Routing Problem

The challenges of solving the dynamic CARP come from both the complicatedness of CARP itself and the difficulties caused by the dynamic environment. Next, we will discuss how to address each of the two issues with evolutionary computation, respectively.

15.3.1 Addressing the Capacitated Arc Routing Problem Issues

In this sub-section, two competitive approaches proposed for CARP are introduced, i.e., the Repair-based Tabu Search (RTS) [38] and the Memetic Algorithm with Extended Neighborhood Search (MAENS) [44].

15.3.1.1 The Global Repair Operator and the Repair-Based Tabu Search

The capacity constraint is one of the most important constraints that lead to the complicatedness of CARP. Without the capacity constraint, the problem can be seen as a single routing problem. However, with the capacity constraint imposed, the problem becomes a combination of a routing sub-problem and a clustering sub-problem, both of which are difficult to solve. Therefore, it is important to tackle the capacity constraint properly.

For a solution $S = (X, Y)$, the total cost depends only on the route set X, while the 0-1 vector Y determines whether the solution satisfies the capacity constraint and if not, the extent of the violation. Since a task can be traversed multiple times but only served once in the vertex sequence, a single X can be associated with several different Y's. An example is given in Fig. 15.1. In Fig. 15.1, the vertex o represents the depot. All the 6 edges $\{(o,a), (o,b), (o,c), (a,b), (a,c), (b,c)\}$ are tasks. The capacity of each vehicle is 4. The number on each edge denotes its demand, e.g., $d(a,b) = 3$. The right part of Fig. 15.1 shows two different solutions S_1 and S_2, which share the same route set, but are different in the 0-1 vector. As a result, S_1 is infeasible with one unit capacity violation, i.e., $d(a,b) + d(b,o) = 3 + 2 = 5 > 4$, whereas S_2 is the global optimum.

During the search process, a low-cost route set and the corresponding feasible 0-1 vector are not easy to be obtained simultaneously, and an infeasible solution is usually discarded because of the bad 0-1 vector despite its promising route set. In order to address this issue, the Global Repair Operator (GRO) [38] is proposed.

Given an infeasible solution, GRO preserves its route set and re-assigns the 0-1 variables to minimize the constraint violation. In other words, GRO seeks the optimal assignment of 0-1 variables for a given route set. Such a repair process

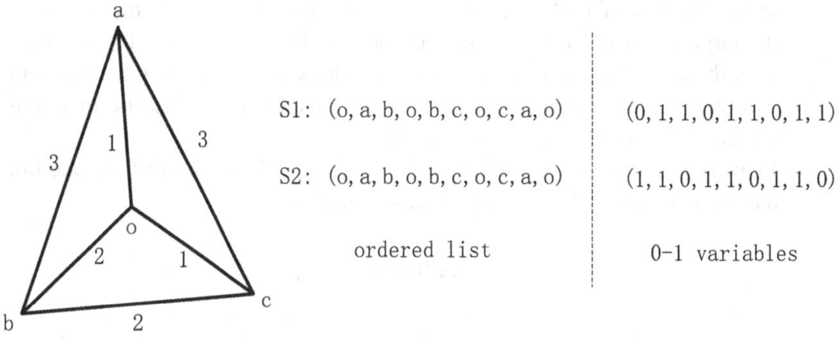

S1: (o, a, b, o, b, c, o, c, a, o)	(0, 1, 1, 0, 1, 1, 0, 1, 1)
S2: (o, a, b, o, b, c, o, c, a, o)	(1, 1, 0, 1, 1, 0, 1, 1, 0)
ordered list	0-1 variables

Fig. 15.1 An example of two different solutions sharing the same route set

takes into account all routes involved in the solution, and thus GRO can be viewed as a global operator.

Suppose we have an infeasible solution with m routes, re-assigning the 0-1 variables can be formulated as the following problem:

$$\min \sum_{i=1}^{m}\left(\max\left\{\sum_{j=1}^{N} s_j x_{ij} - Q, 0\right\}\right) \tag{15.25}$$

$$s.t.: \sum_{i \in \Omega(j)} x_{ij} = 1, \ \forall j = 1, 2, ..., N, \tag{15.26}$$

$$x_{ij} = 0 \text{ or } 1, \forall \ i = 1, 2, ..., m; \ j = 1, 2, ..., N. \tag{15.27}$$

where N is the total number of tasks and s_j denotes the serving cost for task j. x_{ij} is set to 1 if task j is served in the route i, and set to 0 otherwise. $\Omega(j)$ is defined as

$$\Omega(j) = \{ i \mid \text{task } j \text{ is traversed in route } i \text{ in S} \}$$

Given the vertex sequence, constraints (15.26) and (15.27) guarantee that each task is served only once among the routes it is traversed.

Let $\{a_1, a_2, ..., a_N\}$ and $\{b_1, b_2, ..., b_m\}$ be a set of items and bins, respectively. The above problem can be viewed as a bin-packing problem, where the size of the item j is s_j, and all the bins share an identical capacity Q. GRO employs an insertion heuristic followed by a short-term tabu search to solve the bin-packing problem. The general idea of the insertion heuristic is straightforward. We sequentially pick an item out of the whole set and insert it into a bin, until all the items have been inserted. Each item is inserted in a bin that minimizes the objective function (15.25). Such procedure can be described as follows:

Step 1: Initialize $x_{ij} = 0, \forall i, j$. Let $A = \{1, 2, ..., N\}$ and $cl(b_i) = 0, \forall i$. Here, $cl(b_i)$ is the current load of b_i. Then repeat step 2 to step 4 until $A = \emptyset$.

Step 2: For each $j \in A$, identify the set $\Omega'(j)$ satisfying $\Omega'(j) = \{i \in \Omega(j) | cl(b_i) + s_j \leqslant Q\}$. Select the item corresponding to the smallest $|\Omega'(j)|$ as the one to be inserted. If multiple items share the smallest $|\Omega'(j)|$, the one with the largest s_j will be selected. Then ties are broken by selecting the item with the smallest index (j). The selected index is j^*. Selecting the items in this way guarantees that the item with the least choice of insertion without violating the constraints is chosen first.

Step 3: Identify the b_i with the smallest $cl(b_i)$ from $\Omega'(j^*)$. If more than one bin has the smallest $cl(b_i)$, the one with the smallest

$$\sum_{j \in A} I_{\Omega(j)}(i) \cdot s_j$$

is selected, where $I_{\Omega(j)}(i)$ is an indicator function. $I_{\Omega(j)}(i) = 1$ if $i \in \Omega(j)$, and 0 otherwise. The above equation indicates that the bin available for the least untreated items is considered first. After that, ties are broken by selecting the bin with the smallest index (i). The selected index is i^*.

Step 4: Insert the selected item a_{j^*} in the chosen bin b_{i^*}. Set $x_{i^* j^*} = 1$, remove j^* from A and update $cl(b_{i^*})$ with s_{j^*}.

After obtaining the initial solution with the insertion heuristic, a standard tabu search is employed to further improve it. The tabu search is described in Algorithm 1. S_0 is the solution obtained by the insertion heuristic, and $f(S)$ is the objective function (15.25). The neighborhood $N(S)$ of solution S indicates the set of solutions that can be obtained by moving an item to another admissible bin. The tabu list is designed as follows: when an item is moved from one bin to another, it is not allowed to be moved back to its original bin in a certain number of subsequent iterations, unless the movement leads to a better solution than the current best solution. Here, the tabu tenure is set to $F/2$, where F is the number of items with more than one admissible bin ($|\Omega(j)| > 1$). The tabu search process terminates after N iterations or $N/2$ consecutive iterations without improvement.

Based on a solution of the bin-packing problem, a new solution of CARP can be directly obtained by updating the 0-1 variables according to x_{ij}'s. However, the tabu search process might generate multiple assignments of 0-1 variables that all correspond to feasible solutions of the CARP, which are saved in the archive A (lines 9–11 of Algorithm 1). A further refinement procedure is needed to select the best solution in A.

In the insertion heuristic and tabu search process, the total cost is not considered since the route set is assumed to be unchanged. However, after the change of 0-1 variables, the adjacent services may be connected with shorter paths. Hence, as the final step of the GRO, all the archived assignments of 0-1 variables are transformed to the corresponding solutions of CARP, and then the route sets of these solutions are refined by updating the vertices between each pair of adjacent services with the shortest path. Finally, the solution with the smallest cost is chosen as the output of the GRO.

To summarize, the major steps of the GRO are listed as follows:

1. Formulate the repair operation as a bin-packing problem and get a solution via the insertion heuristic;

Algorithm 1 $A = \text{TS}(f, S_0)$

1: Set the current solution $S = S_0$, the current best solution $S_b = S_0$. Set $A = \emptyset$;
2: **while** the stopping criteria are not satisfied **do**
3:    Set $f(S') = \infty$;
4:    **for all** $S'' \in N(S)$ **do**
5:       **if** S'' is not tabu and $f(S'') < f(S')$ **then**
6:          $S' = S'', f(S') = f(S'')$;
7:       **end if**
8:    **end for**
9:    **if** $f(S') = 0$ **then**
10:       $A = A \cup S'$;
11:    **end if**
12:    **if** $f(S') < f(S_b)$ **then**
13:       $S_b = S', f(S_b) = f(S')$;
14:    **end if**
15:    Update the tabu list and set $S = S'$;
16:    **if** $A = \emptyset$ **then**
17:       $A = \{S_b\}$;
18:    **end if**
19: **end while**
20: **return** A;

---

2. Utilize a tabu search process to further improve the solution obtained in the first step, and get an archive of candidate assignments of 0-1 variables.
3. Obtain new solutions of CARP based on the archived assignments of 0-1 variables and update these solutions with the further refinement procedure. The solution with the smallest cost is chosen as the output.

The GRO can be easily embedded in any search-based approach to enhance its search capability. The RTS is thus proposed by simply embedding the GRO into an existing competitive tabu search algorithm [8]. Specifically, for each infeasible solution with promising total cost (which is smaller than the total cost of the best feasible solution found so far), the GRO is applied to reduce its violation to the capacity constraint.

15.3.1.2   Memetic Algorithm with Extended Neighborhood Search

Another difficulty of CARP is that the existing search operators are with small step sizes, and thus have difficulty to explore in the large solution space. In this situation, a search operator with large step size is more desirable. However, it is not a trivial task to design such a search operator. An intuitive idea is to apply the traditional search operators for multiple times. Nevertheless, the neighborhood size increases exponentially with the number of times to apply the search operators. As a result, it is prohibitive to enumerate all the possible solutions in the neighborhood. One simple solution to this problem is to randomly sample a part of the huge neighborhood. However, it is often the case that some regions in the solution space are more promising than the others. Hence, random sampling is a bit blind and might waste a lot of computational resource. To summarize, although a large step-size local search can be beneficial, it cannot be implemented by simply extending the traditional move operators, and a more refined approach is required. For this purpose, a Merge-Split (MS) operator [44] is developed.

The MS operator aims to improve a given solution by modifying multiple routes of it. As indicated by its name (Merge-Split), this operator has two components, i.e., the Merge and Split components. Given a solution, the Merge component randomly selects p ($p > 1$) routes of it, combines them together to form an unordered list of tasks, which contains all the tasks of the selected routes. The Split component directly operates on the unordered list generated by the Merge component, which is composed of the path scanning (PS) heuristic [25] and Ulusoy's split procedure [46]. Given a set of tasks and the whole graph, PS is used to quickly generate a set of feasible routes that serve all the given tasks with a relatively low total cost. It starts by initializing an empty path. At each iteration, it seeks the tasks that do not violate the capacity constraint. If no task satisfies the constraint, it connects the end of the current path to the depot with the shortest path between them to form a route, and then initializes a new empty path. If a unique task satisfies the constraint, PS connects that task to the end of the current path (again, with the shortest path between them). If multiple tasks satisfy the constraint, the one closest to the end of the current path is chosen. If multiple tasks not only satisfy the capacity constraint, but also are the closest to the end of the current path, five rules are further adopted to determine which to choose: (1) maximize the distance from the head of task to the depot; (2) minimize the distance from the head of task to the depot; (3) maximize the term $d(t)/sc(t)$, where $d(t)$ and $sc(t)$ are demand and serving cost of task t, respectively; (4) minimize the term $d(t)/sc(t)$; (5) use rule (1) if the vehicle is less than half-full, otherwise use rule (2). If multiple tasks still remain, ties are broken arbitrarily. PS terminates when all the tasks in the unordered list have been selected. Note that PS does not use the five rules alternatively. Instead, it scans the unordered list of tasks for five times. In each scan, only one rule is used. Hence, PS will generate five route sets in total. Then, Ulusoy's split procedure is applied to all the five route sets to further improve them. Here, the route sets can be seen as an ordered list of tasks, and Ulusoy's split procedure can obtain the optimal feasible route sets for the ordered list of tasks.

To summarize, the MS operator first merges multiple routes to obtain an unordered list of tasks, then employs PS to sort the unordered list. After that, Ulusoy's splitting procedure is used to split the ordered lists into new routes in the optimal way. Finally, we may obtain five new solutions of CARP by embedding the new routes back into the original solution, and the best one is chosen as the output of the MS operator. Figure 15.2 demonstrates the whole process of the MS operator.

The advantages of the MS operator are twofold. First, it can generate new solutions that are significantly different from the current one as it conducts on routes instead of tasks. In general, the larger the p (i.e., the number of routes involved in MS), the more distant the new solution is from the current solution. Second, the new solutions obtained by the MS operator tend to have low total cost due to the

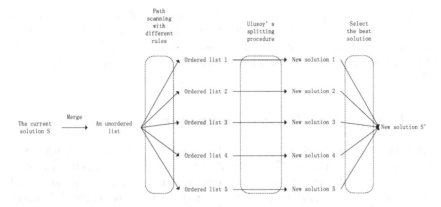

Fig. 15.2 The process of the MS operator

Algorithm 2 The brief description of MAENS

1: **Initialization:** Generate an initial population;
2: **while** Stopping criteria are not satisfied **do**
3:     Select the parent solutions and generate offsprings by the crossover operator;
4:     **for** each offspring **do**
5:         Perform extended neighborhood search around it with probability P_{ls};
6:         Select solutions from the original ones and the offsprings to form the population in the next generation.
7:     **end for**
8: **end while**

---

adoption of PS and Ulusoy's splitting procedure, both of which are known to be capable of generating relatively good solutions. On the other hand, the major drawback is its high computational complexity. Fortunately, such a drawback may be more or less alleviated by a careful coordination of the MS operator and other search operators. For this purpose, the memetic algorithm framework is adopted and the MS and traditional move operators are integrated to form the local search with extended neighborhood. The resultant algorithm is thus called the Memetic Algorithm with Extended Neighborhood Search (MAENS). A brief description of MAENS is given in Algorithm 2.

During the extended neighborhood search process, the MS and traditional search operators are employed in the following way: Given an offspring individual generated by the crossover operator, the traditional move operators (i.e., the single insertion, double insertion and swap) are applied to the individual until the local optimum is reached. After that, the MS operator is applied to this local optimal solution to form the second stage of the local search, and the local optimum with respect to the extended neighborhood is obtained. Finally, the traditional-neighborhood-based local search is again applied to further refine the local optimum obtained in the second stage and exploit the new local region.

392 Y. Mei, K. Tang, and X. Yao

It has been demonstrated in [38] and [44] that the RTS and MAENS are both competitive approaches for CARP. RTS can obtain solutions as well as the best-known ones in a shorter time than other state-of-the-art methods, while MAENS is able to find better solutions than the best-known ones at the cost of more computational efforts.

15.3.2 Tackling the Dynamic Environment

The dynamic environment changes the actual feasibility of solutions. As mentioned before, a repair operator must be designed to make the solution feasible whenever it becomes infeasible in the current environment. An ideal repair operator should not only be able to repair any infeasible solution, but also be practical. In many real-world applications, the departments in charge wish to keep the modification as small as possible during the repair process. Based on such practical consideration, the repair operator Φ should satisfy the following conditions:

- If all the constraints are satisfied, then Φ should not make any change on the solution;
- If the solution violates the capacity constraint, then Φ should cut the infeasible routes into several feasible routes;
- If the path between adjacent vertices in the solution becomes absent, then Φ connects them with another path in the graph.

All the above operations minimizes the modification of the solution.

The sampling values of the random variables can only be known during the implementation process. To be specific, the vehicle does not know whether the path from vertex x_{ki} to its successor $x_{k(i+1)}$ is connected until it reaches x_{ki}. The deadheading cost of $(x_{ki}, x_{k(i+1)})$ can only be known after the vehicle has arrived at $x_{k(i+1)}$. Besides, if $(x_{ki}, x_{k(i+1)})$ is a task, its demand cannot be known until the vehicle finishes its service and reaches $x_{k(i+1)}$. In this situation, the repair operation of Φ can be defined as follows: Given a solution $S = (\{X_1, ..., X_m\}, \{Y_1, ..., Y_m\})$, each vehicle starts from $x_{k1} = v_0$ and traverses according to the vertex sequence $X_k = (x_{k1}, ..., x_{kl_k})$. Once the vehicle arrives at a new vertex (including the starting vertex v_0), Φ checks the value of the first time occurring $y_{kj} = 1$ in the sub-vector $Y_k = (y_{ki}, ..., y_{k(l_k-1)})$ to locate the position of the next task $(x_{kj}, x_{k(j+1)})$. If there is no task left, then it returns the depot through the predefined path. Otherwise, Φ examines whether the current residue is enough to serve the expected demand $E[d(x_{kj}, x_{k(j+1)}), \xi)|\xi]$. If so, then the vehicle traverses to x_{kj}. Otherwise, the vehicle returns to the depot and update the capacity, then goes to x_{kj} again. Note that the practical demand of $(x_{kj}, x_{k(j+1)})$ may be larger than expected and exhaust the capacity on the way of its service. In this case, the vehicle neglects the remaining demand and returns to the depot to update the capacity, and then goes back to x_{kj} and finish the service of $(x_{kj}, x_{k(j+1)})$. All the above repair operations are done through the shortest paths between the origin and the target. In this way, the satisfactory of the capacity constraint can be guaranteed. On the other hand, if it is found that the path from the current vertex x_{ki} to the next vertex $x_{k(i+1)}$ disappears, Φ simply update the deadheading cost matrix by setting $dc(x_{ki}, x_{k(i+1)}) = \infty$, and calculate the shortest path under the updated cost matrix by Dijstra's alorithm [17]. Then, it replaces the interrupted path with the new shortest path.

Based on the above descriptions, Φ can be divided into a repair operator Φ_d to deal with stochastic task demands and a repair operator Φ_c to deal with stochastic presence of paths. Given an infeasible solution, it is first repaired by Φ_d, and then repaired by Φ_c. Algorithms 3 and 4 give the pseudo codes of the repair operators Φ_d and Φ_c, respectively.

In Algorithms 3 and 4, the function $X.push(a)$ indicates inserting the element a (can be a single element or a sequence) into the end of the sequence X, and $|X|$ stands for the length of X. 0_l represents the sequence composed with l 0's. $ESP(v_i, v_j)$ is the shortest path from v_i to v_j under the expected deadheading cost matrix, while $SP(v_i, v_j, \xi)$ is that under the practical deadheading cost matrix of the environmental parameter ξ.

Under the definition of Φ_d and Φ_c, an infeasible solution S is repaired at the following two steps: 1) $S' = \Phi_d(S)$; 2) $S^\xi = \Phi_c(S')$.

15.4   Benchmark for Dynamic Capacitated Arc Routing Problem

In order to evaluate the performance of potential approaches, benchmark instances of the dynamic CARP are needed. However, there has been no benchmark instance proposed so far. In [21], the well-known *gdb* test set with static CARP instances was extended to a stochastic CARP test set by replacing the deterministic demand $d(t_i)$ of each task t_i with a Gaussian distributed random variable $D(t_i) \sim N(d(t_i), \sigma_i^2)$, where the variance $\sigma_i = k \times d(t_i)$ is proportional to $d(t_i)$. Here, we generate the benchmark instances for the dynamic CARP similarly. In contrast with [21], we choose Gamma distribution instead of normal distribution. The reason is that all the random variables in the dynamic CARP is nonnegative, and Gamma distribution is one of the most commonly used distribution with nonnegative support set for simulating real environments. Besides, $d(v_i, v_j)$ equals zero with probability $1 - p_{ij}$, and $dc(v_i, v_j)$ equals infinity with probability $1 - p_{ij}$. Hence, in the dynamic CARP, the random variables should satisfy the following distributions:

$$D(v_i, v_j) \begin{cases} \sim G(k_{ij}^d, \theta_{ij}^d), & r < p_{ij}; \\ = 0, & \text{otherwise.} \end{cases} \tag{15.28}$$

$$DC(v_i, v_j) \begin{cases} \sim G(k_{ij}^c, \theta_{ij}^c), & r < q_{ij}; \\ = \infty, & \text{otherwise.} \end{cases} \tag{15.29}$$

where $G(k, \theta)$ is the Gamma distribution with the shape parameter k and the scale parameter θ. The probability density function of $G(k, \theta)$ is $pdf(x; k, \theta) = x^{k-1} \frac{e^{-x/\theta}}{\theta^k \Gamma(k)}$ for $x > 0$ and $k, \theta > 0$, where $\Gamma(k) = \int_0^\infty t^{k-1} e^t dt$. It is known that the mean of the Gamma distribution $G(k, \theta)$ is $\mu = k\theta$ and $G(k, \theta)$ converges to the

---

Algorithm 3 $S' = \Phi_d(S, \xi)$

1: **for** $k = 1 \rightarrow m$ **do**
2: Set $X'_k = (v_0)$, $Y'_k = ()$, $\Delta Q = Q$;
3: **for** $i = 1 \rightarrow |X_k| - 1$ **do**
4:   **if** $y_{ki} = 0$ **then**
5:    $X'_k.push(x_{k(i+1)})$, $Y'_k.push(0)$;
6:   **else**
7:    **if** $d(x_{ki}, x_{k(i+1)}, \xi) = 0$ **then**
8:     $X'_k.push(x_{k(i+1)})$;
9:     $Y'_k.push(0)$;
10:    **else if** $\Delta Q < d(x_{ki}, x_{k(i+1)}, \xi)$ **then**
11:     $X'_k.push(ESP(x_{k(i+1)}, v_0))$;
12:     $X'_k.push(v_0)$;
13:     $Y'_k.push(1)$;
14:     $Y'_k.push(0_{|ESP(x_{k(i+1)}, v_0)|})$;
15:     $X'_k.push(ESP(v_0, x_{ki}))$;
16:     $Y'_k.push(0_{|ESP(v_0, x_{ki})|})$;
17:     $X'_k.push(x_{ki})$, $Y_k.push(0)$;
18:     $\Delta Q \leftarrow \Delta Q + Q - d(x_{ki}, x_{k(i+1)}, \xi)$;
19:    **else**
20:     $X'_k.push(x_{k(i+1)})$, $Y'_k.push(1)$, $\Delta Q \leftarrow \Delta Q - d(x_{ki}, x_{k(i+1)}, \xi)$;
21:    **end if**
22:    **for** $j = i + 1 \rightarrow |X_k| - 1$ **do**
23:     **if** $Y_{kj} = 1$ **then**
24:      **break**;
25:     **end if**
26:    **end for**
27:    **if** $Y_{kj} = 1 \& \Delta Q < E[d(x_{kj}, x_{k(j+1)}, \xi)|\xi]$ **then**
28:     $X'_k.push(ESP(x_{k(i+1)}, v_0))$;
29:     $X'_k.push(v_0)$;
30:     $Y'_k.push(0_{|ESP(x_{k(i+1)}, v_0)|+1})$;
31:     $X'_k.push(ESP(v_0, x_{kj}))$;
32:     $X'_k.push(x_{kj})$;
33:     $Y'_k.push(0_{|ESP(v_0, x_{kj})|+1})$;
34:     $i \leftarrow j$, $\Delta Q \leftarrow Q$;
35:    **end if**
36:   **end if**
37: **end for**
38: **end for**
39: **return** $S' = (\{X'_1, ..., X'_m\}, \{Y'_1, ..., Y'_m\})$;

Gaussian distribution when the shape parameter k becomes infinite. According to the idea of the dynamic CARP instance generation in [21], the random variables in the dynamic CARP should have the following properties:

Property 1: The Gamma distribution is close to Gaussian distribution;
Property 2: The expected value of the random variables equals their static values;

Algorithm 4 $S^\xi = \Phi_c(S', \xi)$

1: **for** $k = 1 \to m$ **do**
2: Set $X_k^\xi = (v_0), Y_k^\xi = ()$;
3: **for** $i = 1 \to |X_k'| - 1$ **do**
4:   **if** $dc(x_{ki}', x_{k(i+1)}', \xi) < \infty$ **then**
5:    $X_k^\xi.push(x_{k(i+1)}'), Y_k^\xi.push(y_{ki}')$;
6:   **else**
7:    $X_k^\xi.push(SP(x_{ki}', x_{k(i+1)}', \xi)), X_k^\xi.push(x_{k(i+1)}')$;
8:    $Y_k^\xi.push(0_{|SP(x_{ki}', x_{k(i+1)}', \xi)|+1})$;
9:   **end if**
10: **end for**
11: **end for**
12: **return** $S^\xi = (\{X_1^\xi, ..., X_m^\xi\}, \{Y_1^\xi, ..., Y_m^\xi\})$;

Property 1 can be realized by setting a sufficiently large k. For Property 2, on the other hand, the realizations of the stochastic demands and deadheading costs are different. For the task demand $D(v_i, v_j)$, Property 2 can be realized by directly setting

$$E[D(v_i, v_j)] = p_{ij}k_{ij}^d\theta_{ij}^d + (1 - p_{ij}) \times 0 = d(v_i, v_j) \qquad (15.30)$$

However, the above equation is not available for the deadheading cost $DC(v_i, v_j)$ since it is likely to become infinity. Therefore, we neglect such case and only set

$$E[DC(v_i, v_j)] = k_{ij}^c\theta_{ij}^c = dc(v_i, v_j) \qquad (15.31)$$

In practice, the shape parameters k_{ij}^d and k_{ij}^c are set to 20. Figure 15.3 gives the probability density functions of the Gamma distribution with $k = 20$ and $\theta = 1.0, 1.5, 2.0$. The settings of θ_{ij}^d and θ_{ij}^c can be derived from Eq. (15.30) and (15.31) as follows:

$$\theta_{ij}^d = \frac{d(v_i, v_j)}{p_{ij}k_{ij}^d} \qquad (15.32)$$

$$\theta_{ij}^c = \frac{dc(v_i, v_j)}{k_{ij}^c} \qquad (15.33)$$

Finally, p_i and q_{ij} can be intuitively set to 0.9 and 0.95, respectively.

The C++ source code of the instance generator for extending the static CARP instances to the dynamic CARP instances can be downloaded from the website at http://goanna.cs.rmit.edu.au/~e04499, along with the dynamic CARP instances, namely the *Dgdb*, *Dval* and *Degl* sets, respectively. For each static instance, 30 sampling instances were generated one by one by the instance generator with the starting random seed of 0. If necessary, users can also generate more samplings with other random seeds.

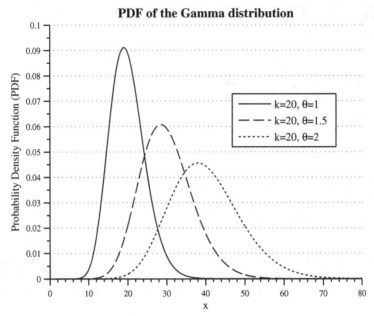

Fig. 15.3 Probability density functions of Gamma distribution with $k = 20$ and $\theta = 1.0, 1.5, 2.0$

15.5 Preliminary Investigation of the Fitness Landscape

The stochastic characteristic of the random variables makes the fitness landscape of the dynamic CARP much more complicated than the static CARP. To investigate the fitness landscape of the dynamic CARP and the impact of the stochastic factors on the performance of the search algorithms, the two static CARP approaches introduced in Section 15.3.1, i.e., RTS and MAENS, were applied to the *Dgdb* set, which is the simplest and smallest test set among the three dynamic CARP benchmark sets generated in Section 15.4. In this way, it is easier to observe how the performance of the algorithms are influenced by the dynamic environment, but not the complicatedness of the problem itself. As demonstrated in [38] and [44], the two selected algorithms are able to reach the global optima for all the static version of the *Dgdb* instances, i.e., the *gdb* instances.

In the experiments, RTS and MAENS were implemented once on all the *gdb* instances, and the sequences of best feasible solutions updated during the search process were recorded. Recalling that the expected values of the random variables of the *Dgdb* instances are equal to the corresponding static values of the *gdb* instances. Thus, by solving the *gdb* instances, the algorithms can be seen as solving the *Dgdb* instances by utilizing the expectation of the random variables. For each solution recorded in the best feasible solution sequence obtained by RTS and MAENS, denoted as $(S_{11}, ..., S_{1l_1})$ and $(S_{21}, ..., S_{2l_2})$, the robustness $R(S_{ij})$ is calculated in terms of the average total cost of the 30 corresponding *Dgdb* instance samples generated in Section 15.4, i.e.,

$$R(S_{ij}) = \frac{1}{30} \sum_{k=1}^{30} tc(S_{ij,\xi_k}) \tag{15.34}$$

where $S_{ij,\xi_k} = \Phi(S_{ij}, \xi_k)$ is the feasible solution obtained by applying Φ to S_{ij} under ξ_k, which is the environmental parameter of the kth sample.

Table 15.2 presents the experimental results. In the table, the columns headed "tcBest" and "RBest" stand for the solutions with the lowest $tc(S)$ and the lowest $R(S)$ among all the recorded solutions. The column headed "RBK" presents the best-known solution with respect to $R(S)$. The columns headed "$tc(S)$" and "$R(S)$" are the objective functions defined for the static and uncertain versions, respectively. $tc(S)$ is defined in Eq. (15.1) and $R(S)$ is defined in Eq. (15.16). For $tc(S)$, the optimal values are marked in bold. As mentioned before, MAENS and RTS can both reach the optimal solutions for the static version of the instances. Therefore, the best $tc(S)$ obtained by them were marked in bold for all the instances.

From Table 15.2, it is observed that $R(S)$ is not proportional to $tc(S)$. For MAENS, the solution with the lowest $tc(S)$ and the solution with the lowest $R(S)$ are different on 9 out of the total 23 instances. For RTS, such a phenomenon occurs also on 9 instances. Another interesting observation is that for the lowest $tc(S)$ obtained by MAENS and RTS, although their values are the same and optimal, the corresponding $R(S)$ of the solution can be very different (e.g., in *Dgdb*12, the solutions with lowest $tc(S)$ obtained by the two algorithms have their $R(S)$'s of 642.03 and 603.54, respectively). Based on the above observation, one can conclude that for a static CARP instance, there often exist multiple global optima. However, their robustness in the corresponding dynamic versions may be quite different. When looking at the best-known solutions with respect to $R(S)$, it is seen that for 16 out of the total 23 instances, the best-known solutions have non-optimal $tc(S)$'s. This implies that the global optimum (in terms of robustness) in a dynamic CARP instance may be quite far away from the global optimum in its static counterpart. Comparing with the results obtained by MAENS and RTS, the $R(S)$ values of the best-known solutions are much smaller than the lowest $R(S)$'s obtained by the two algorithms, not to mention the $R(S)$ of the solutions with lowest $tc(S)$. Therefore, we can conclude that when solving *Dgdb* instances by applying algorithms to the *gdb* counterparts, it is difficult to achieve highly robust solutions.

One possible reason that the algorithms for the static CARP cannot perform well when applied to the dynamic CARP may be explained as follows: the algorithms ignore the possibility that the routes may be cut at certain intermediate positions due to the violation of the capacity constraint during the implementation. If the cut position is distant from the depot, the repaired solution will have a much larger total cost. Since the cut position depends on the allocation of the task services, one solution to address this issue is to modify the allocation of the task services so that the possible cut positions are close to the depot.

In summary, the following observations can be drawn:

Table 15.2 The experimental results of MAENS and RTS on the *Dgdb* set. The optimal $tc(S)$'s are marked in bold.

Name	MAENS				RTS				RBK	
	tcBest		RBest		tcBest		RBest			
	$tc(S)$	$R(S)$	$tc(S)$	$R(S)$	$tc(S)$	$R(S)$	$tc(S)$	$R(S)$	$tc(S)$	$R(S)$
1	**316**	380.63	**316**	380.63	**316**	401.95	323	387.32	323	349.49
2	**339**	436.69	345	401.24	**339**	417.96	**339**	417.96	353	383.72
3	**275**	331.19	**275**	331.19	**275**	323.81	**275**	323.81	296	307.00
4	**287**	350.40	**287**	350.40	**287**	345.77	**287**	345.77	287	328.32
5	**377**	492.38	383	472.44	**377**	492.00	**377**	492.00	395	437.79
6	**298**	353.69	**298**	353.69	**298**	369.63	310	367.10	319	342.18
7	**325**	380.06	**325**	380.06	**325**	400.50	**325**	400.50	325	356.09
8	**348**	470.18	354	456.70	**348**	464.33	356	449.56	362	443.87
9	**303**	404.34	309	391.77	**303**	394.75	**303**	394.75	337	385.86
10	**275**	306.72	**275**	306.72	**275**	325.48	**275**	325.48	283	291.61
11	**395**	431.86	**395**	431.86	**395**	442.19	**395**	442.19	409	419.06
12	**458**	642.03	468	595.33	**458**	603.54	468	601.58	474	587.33
13	**536**	598.03	554	593.66	**536**	623.52	552	603.43	544	569.82
14	**100**	118.44	**100**	118.44	**100**	118.47	**100**	118.47	**100**	107.90
15	**58**	60.76	**58**	60.76	**58**	58.91	**58**	58.91	**58**	58.09
16	**127**	146.53	129	145.73	**127**	146.97	**127**	146.97	129	133.43
17	**91**	94.36	**91**	94.36	**91**	96.17	**91**	96.17	**91**	92.32
18	**164**	180.75	**164**	180.75	**164**	181.48	168	179.16	**164**	170.90
19	**55**	67.49	**55**	67.49	**55**	64.24	**55**	64.24	**55**	63.04
20	**121**	139.06	125	135.89	**121**	141.43	122	134.82	123	126.01
21	**156**	171.60	**156**	171.60	**156**	177.74	158	174.79	158	165.41
22	**200**	217.01	**200**	217.01	**200**	218.59	202	217.35	204	210.17
23	**233**	263.60	235	256.52	**233**	260.97	**233**	260.97	235	252.35

- The robustness of solution in the dynamic CARP is conflicting with the absolute performance, and the two measures can hardly reach optimum at the same time;
- Static CARP instances often have multiple global optimal solutions. However, their robustness in the corresponding dynamic version may be quite different;
- Only utilizing the expected information cannot lead to highly robust solutions.
- The algorithms for the static CARP cannot perform well for the dynamic CARP because the possible cut position and the additional cost induced by the cut is not considered. To address this issue, one can estimate the probability of the cut position and adjust the task services so as to reduce the expected additional cost caused by the cut.

15.6 Conclusion

In this chapter, a general dynamic CARP is defined and investigated with evolutionary computation. The dynamic CARP model includes the following four stochastic factors: (1) presence of tasks; (2) demand of tasks; (3) presence of paths between vertices and (4) deadheading costs between vertices. These stochastic factors exist in both the objective and constraints of the problem, and thus influence both the performance and feasibility of solution. Unlike the static CARP the outputs of the dynamic CARP include a solution and a repair operator to modify the solution so that all the constraints can be satisfied during the implementation process. Besides, the objective of the dynamic CARP is to optimize the robustness rather than the absolute quality in a specific environment.

The dynamic CARP has the difficulties caused by the complicatedness of CARP and the dynamic environment. To address the two issues, two competitive approaches for CARP, i.e., RTS and MAENS, are introduced, and a repair operator is designed under the practical considerations.

Then, to investigate the fitness landscape of the dynamic CARP, RTS and MAENS were tested on the *Dgdb* benchmark set, which is extended from the *gdb* static CARP benchmark set. It is found that, although the two algorithms showed excellent performance for static CARP, they were not able to find robust solutions for the dynamic CARP. Therefore, the future work is to design new algorithms that can find more robust solutions by taking advantage of more information. One possible direction is to select the solutions in which the adjacent tasks can be connected by multiple paths with nearly the same lengths to avoid the additional cost induced by the absence of edges.

Although no effective approach has been proposed for the dynamic CARP in this chapter, the formal definition of the problem and the generated benchmark provide a solid foundation of further research work, and the analysis and discussions about the problem characteristics give some guidelines for the algorithm design. The future work includes developing algorithms by taking these analytical results into account.

Acknowledgements. This work was supported by the National Natural Science Foundation of China (NSFC) grant No. 61028009.

References

[1] Agarwal, H.: Reliability based design optimization: Formulations and methodologies. Ph.D. thesis, University of Notre Dame, South Bend, IN, USA (2004)

[2] Agarwal, H., Renaud, J.: Reliability based design optimization using response surfaces in application to multidisciplinary systems. Engineering Optimization 36(3), 291–311 (2004)

[3] Allen, J., Seepersad, C., Choi, H., Mistree, F.: Robust design for multiscale and multidisciplinary applications. Journal of Mechanical Design 128(4), 832–843 (2006)

[4] Amberg, A., Domschke, W., Voß, S.: Multiple center capacitated arc routing problems: A tabu search algorithm using capacitated trees. European Journal of Operational Research 124(2), 360–376 (2000)

[5] Bertsimas, D.: Probabilistic combinatorial optimization problems. Ph.D. thesis, Massachusetts Institute of Technology, Department of Mathematics (1988)

[6] Bertsimas, D.: A vehicle routing problem with stochastic demand. Operations Research, 574–585 (1992)

[7] Birge, J., Louveaux, F.: Introduction to stochastic programming. Springer (1997)

[8] Brandão, J., Eglese, R.: A deterministic tabu search algorithm for the capacitated arc routing problem. Computers and Operations Research 35(4), 1112–1126 (2008)

[9] Branke, J.: Creating robust solutions by means of evolutionary algorithms. In: Eiben, A.E., Bäck, T., Schoenauer, M., Schwefel, H.-P. (eds.) PPSN 1998. LNCS, vol. 1498, pp. 119–128. Springer, Heidelberg (1998)

[10] Branke, J.: Evolutionary optimization in dynamic environments. Kluwer Academic Pub. (2002)

[11] Campbell, J., Langevin, A.: Roadway snow and ice control. In: Arc Routing: Theory, Solutions and Applications, pp. 389–418. Kluwer, Boston (2000)

[12] Chan, K., Skerlos, S., Papalambros, P.: Monotonicity and active set strategies in probabilistic design optimization. Journal of Mechanical Design 128(4), 893–900 (2006)

[13] Christiansen, C., Lysgaard, J., Wøhlk, S.: A Branch-and-Price Algorithm for the Capacitated Arc Routing Problem with Stochastic Demands. Operations Research Letters 37(6), 392–398 (2009)

[14] Chu, F., Labadi, N., Prins, C.: A scatter search for the periodic capacitated arc routing problem. European Journal of Operational Research 169(2), 586–605 (2006)

[15] Das, I.: Robustness optimization for constrained nonlinear programming problems. Engineering Optimization 32(5), 585–618 (2000)

[16] De Rosa, B., Improta, G., Ghiani, G., Musmanno, R.: The arc routing and scheduling problem with transshipment. Transportation Science 36(3), 301–313 (2002)

[17] Dijkstra, E.: A note on two problems in connexion with graphs. Numerische Mathematik 1(1), 269–271 (1959)

[18] Dror, M., Laporte, G., Trudeau, P.: Vehicle routing with stochastic demands: Properties and solution frameworks. Transportation Science 23(3), 166–176 (1989)

[19] Du, X., Wang, Y., Chen, W.: Methods for robust multidisciplinary design. Tech. rep., American Institute of Aeronautics and Astronautics (2000)

[20] El Ghaoui, L., Lebret, H.: Robust solutions to least-squares problems with uncertain data. SIAM Journal on Matrix Analysis and Applications 18(4), 1035–1064 (1997)

[21] Fleury, G., Lacomme, P., Prins, C.: Evolutionary algorithms for stochastic arc routing problems. In: Raidl, G.R., et al. (eds.) EvoWorkshops 2004. LNCS, vol. 3005, pp. 501–512. Springer, Heidelberg (2004)

[22] Gendreau, M., Laporte, G., Séguin, R.: Stochastic vehicle routing. European Journal of Operational Research 88(1), 3–12 (1996)

[23] Ghiani, G., Guerriero, F., Laporte, G., Musmanno, R.: Tabu search heuristics for the arc routing problem with intermediate facilities under capacity and length restrictions. Journal of Mathematical Modelling and Algorithms 3(3), 209–223 (2004)

[24] Ghiani, G., Improta, G., Laporte, G.: The capacitated arc routing problem with intermediate facilities. Networks 37(3), 134–143 (2001)

[25] Golden, B., DeArmon, J., Baker, E.: Computational experiments with algorithms for a class of routing problems. Computer and Operations Research 10(1), 47–59 (1983)

[26] Golden, B., Wong, R.: Capacitated arc routing problems. Networks 11(3), 305–316 (1981)

[27] Gunawan, S., Papalambros, P.: A Bayesian Approach to Reliability-Based Optimization With Incomplete Information. Journal of Mechanical Design 128(4), 909–918 (2006)

[28] Handa, H., Chapman, L., Yao, X.: Robust route optimization for gritting/salting trucks: a CERCIA experience. IEEE Computational Intelligence Magazine 1(1), 6–9 (2006)

[29] Handa, H., Chapman, L., Yao, X.: Robust Salting Route Optimization Using Evolutionary Algorithms. In: Yang, S., Ong, Y.S., Jin, Y. (eds.) Evolutionary Computation in Dynamic and Uncertain Environments. SCI, vol. 51, pp. 497–517. Springer, Heidelberg (2007)

[30] Herrmann, J.: A genetic algorithm for minimax optimization problems. In: Proceedings of the 1999 Congress on Evolutionary Computation, vol. 2, pp. 1099–1103. Citeseer (1999)

[31] Jaillet, P.: Probabilistic traveling salesman problems. Ph.D. thesis, Massachusetts Institute of Technology, Department of Civil Engineering (1985)

[32] Kalsi, M., Hacker, K., Lewis, K.: A comprehensive robust design approach for decision trade-offs in complex systems design. Journal of Mechanical Design 123(1), 1–10 (2001)

[33] Lacomme, P., Prins, C., Ramdane-Cherif, W.: Evolutionary algorithms for periodic arc routing problems. European Journal of Operational Research 165(2), 535–553 (2005)

[34] Laporte, G., Musmanno, R., Vocaturo, F.: An Adaptive Large Neighbourhood Search Heuristic for the Capacitated Arc-Routing Problem with Stochastic Demands. Transportation Science 160(1), 139–153 (2009)

[35] Lewis, A.: Robust regularization. Tech. rep., Simon Fraser University, Vancouver, Canada (2002)

[36] Li, M., Azarm, S., Boyars, A.: A new deterministic approach using sensitivity region measures for multi-objective robust and feasibility robust design optimization. Journal of Mechanical Design 128(4), 874–883 (2006)

[37] McIlhagga, M., Husbands, P., Ives, R.: A comparison of search techniques on a wing-box optimisation problem. In: Ebeling, W., Rechenberg, I., Voigt, H.-M., Schwefel, H.-P. (eds.) PPSN 1996. LNCS, vol. 1141, pp. 614–623. Springer, Heidelberg (1996)

[38] Mei, Y., Tang, K., Yao, X.: A Global Repair Operator for Capacitated Arc Routing Problem. IEEE Transactions on Systems, Man, and Cybernetics, Part B: Cybernetics 39(3), 723–734 (2009)

[39] Mourão, M., Amado, L.: Heuristic method for a mixed capacitated arc routing problem: A refuse collection application. European Journal of Operational Research 160(1), 139–153 (2005)

[40] Papadrakakis, M., Lagaros, N., Plevris, V.: Design optimization of steel structures considering uncertainties. Engineering Structures 27(9), 1408–1418 (2005)

[41] Polacek, M., Doerner, K., Hartl, R., Maniezzo, V.: A variable neighborhood search for the capacitated arc routing problem with intermediate facilities. Journal of Heuristics 14(5), 405–423 (2008)

[42] Tagmouti, M., Gendreau, M., Potvin, J.: Arc routing problems with time-dependent service costs. European Journal of Operational Research 181(1), 30–39 (2007)

[43] Taguchi, G.: Introduction to quality engineering. Asian Productivity Organization, Tokyo (1990)

[44] Tang, K., Mei, Y., Yao, X.: Memetic Algorithm with Extended Neighborhood Search for Capacitated Arc Routing Problems. IEEE Transactions on Evolutionary Computation 13(5), 1151–1166 (2009)

[45] Tillman, F.: The multiple terminal delivery problem with probabilistic demands. Transportation Science 3(3), 192–204 (1969)

[46] Ulusoy, G.: The fleet size and mix problem for capacitated arc routing. European Journal of Operational Research 22(3), 329–337 (1985)

[47] Waters, C.: Vehicle-scheduling problems with uncertainty and omitted customers. The Journal of the Operational Research Society 40(12), 1099–1108 (1989)

[48] Weise, T., Podlich, A., Gorldt, C.: Solving Real-World Vehicle Routing Problems with Evolutionary Algorithms. In: Chiong, R., Dhakal, S. (eds.) Natural Intelligence for Scheduling, Planning and Packing Problems. SCI, vol. 250, pp. 29–53. Springer, Heidelberg (2009)

[49] Weise, T., Podlich, A., Reinhard, K., Gorldt, C., Geihs, K.: Evolutionary Freight Transportation Planning. In: Giacobini, M., et al. (eds.) EvoWorkshops 2009. LNCS, vol. 5484, pp. 768–777. Springer, Heidelberg (2009)

[50] Wiesmann, D., Hammel, U., Back, T.: Robust design of multilayer optical coatings by means of evolutionary algorithms. IEEE Trans. Evol. Comput. 2(4), 162–167 (1998)

Chapter 16
Evolutionary Algorithms for the Multiple Unmanned Aerial Combat Vehicles Anti-ground Attack Problem in Dynamic Environments

Xingguang Peng, Shengxiang Yang, Demin Xu, and Xiaoguang Gao

Abstract. This chapter aims to solve the online path planning (OPP) and dynamic target assignment problems for the multiple unmanned aerial combat vehicles (UCAVs) anti-ground attack task using evolutionary algorithms (EAs). For the OPP problem, a model predictive control framework is adopted to continuously update the environmental information for the planner. A dynamic multi-objective EA with historical Pareto set linkage and prediction is proposed to optimize in the planning horizon. In addition, Bayesian network and fuzzy logic are used to quantify the bias value to each optimization objective so as to intelligently select an executive solution from the Pareto set. For dynamic target assignment, a weapon target assignment model that considers the inner dependence among the targets and the expected damage type is built up. For solving the involved dynamic optimization problems, an environment identification based memory scheme is proposed to enhance the performance of estimation of distribution algorithms. The proposed approaches are validated via simulation with a scenario of suppression of enemy air defense mission.

Xingguang Peng · Demin Xu
School of Marine Engineering, Northwestern Polytechnical University, Xi'an 710072, China
e-mail: pxg0510@gmail.com, xudm@nwpu.edu.cn

Shengxiang Yang
Centre for Computational Intelligence (CCI), School of Computer Science and Informatics, De Montfort University, The Gateway, Leicester LE1 9BH, U.K.
e-mail: syang@dmu.ac.uk

Xiaoguang Gao
School of Electronics and Information, Northwestern Polytechnical University, Xi'an 710129, China
e-mail: xggao@nwpu.edu.cn

S. Yang and X. Yao (Eds.): *Evolutionary Computation for DOPs*, SCI 490, pp. 403–431.
DOI: 10.1007/978-3-642-38416-5_16   © Springer-Verlag Berlin Heidelberg 2013

16.1   Introduction

The application of unmanned aerial combat vehicles (UCAVs) for various military missions has received a growing attention in the last decade. Apart from the obvious advantage of not placing human life at risk, the lack of a human pilot enables significant weight savings and lower costs. UCAVs also provide an opportunity for new operational paradigms. To realize these advantages, UCAVs must have a high level of autonomy and preferably work cooperatively in groups.

In general, there are three layers in the design of autonomy of UCAVs, which are the strategic layer (path planning, task allocation, search patterns, and human mission command, etc), the tactical layer (target observation, path following, communication and cooperation, and human monitor interaction, etc), and the dynamic/control layer (flying control, formation control, and navigation, etc). Since each of the three layers may involve dynamic optimization problems (DOPs), the study on dynamic optimization methods and their application to autonomous control of UCAVs has become a key issue to improve the combat effectiveness.

Evolutionary algorithm (EAs) are now an established research field at the intersection among artificial intelligence, computer science, and operations research. However, most research in EAs has focused on static optimization problems. The main problem of using traditional EAs for DOPs lies in that they eventually converge to an optimum and thereby lose their population diversity which is necessary for efficiently exploring the search space, which consequently deprives them of the ability to adapt to the changes in the environment. To enhance EAs to solve DOPs, over the past two decades, a number of researchers have developed many methods to maintain diversity for traditional EAs to continuously adapt to the changing environment. Most of these methods can be categorized into the following four types of approaches: (1) increasing the diversity after a change, such as the hyper-mutation method [7]; (2) maintaining the diversity throughout the run, such as the random immigrants scheme [10], sharing or crowding mechanisms [5], and the thermodynamical genetic algorithm (GA) [14]; (3) memory-based approaches [1, 19, 26]; and (4) multi-population approaches, such as the self-organizing scouts GA [3], the multi-national GA [23], and the shift balance GA [24]. Comprehensive surveys on EAs applied to dynamic environments can be found in [2, 13, 15, 18, 25].

This chapter focuses on applying EAs to design the autonomy of UCAVs. We will formulate the DOPs involved in anti-ground attack of a fleet of UCAVs and design corresponding EAs to solve them, respectively. In general, the process of anti-ground attack can be summarized as follows. A feet of UCAVs safely fly to the targets via an intelligent tactic flight to coordinately attack single or multiple targets. There are two key DOPs involved in the anti-ground attack. The first one is the intelligent tactic flight, which is an integration of information gathering, situation assessment, and path planning. The leader of the fleet is required to intelligently assess the dangerous level of the hostile environment like a pilot according to the information gathered from onboard sensors or data links. Then, the leader needs to plan an optimal path dynamically for the fleet to minimize the flight time and maximize the safety simultaneously. The second one is target assignment, which is also required to be solved dynamically since some factors of the assignment may be time-variant. For example, the importance values of the targets may be time-variant since the working states of the targets are changeable.

The outline of the rest of this chapter is briefly given as follows. The problems of dynamic online path planning (OPP) and dynamic target assignment are formulated in Sect. 16.2 and Sect. 16.3, respectively. In each section, a problem-solving dynamic EA is designed and the simulation of a suppression of enemy air-defense (SEAD) mission that integrates the above two DOPs is given in Sect. 16.4 to validate the proposed approaches. Finally, conclusions and discussions on the relevant future work are given in Sect. 16.5.

16.2 Intelligent Online Path Planning (OPP)

The intelligent flight is the basic issue for UCAVs to carry out any complicated mission. When the environment is static and known beforehand, the flight path can be well-designed offline [6, 17]. However, when the environment is changeable or there is no exact knowledge about the environment, an UCAV needs to intelligently plan its path online.

In general, path planning involves multiple optimal objectives. For instance, the maximal safety and minimal energy cost are the two common objectives. In some research in the literature, multi-objective optimization problems (MOPs) are transformed to single-objective ones by weighting objectives and summing them up [28]. This requires that the bias to objectives of the optimizer is known beforehand. Unfortunately, sometimes it is difficult to achieve this knowledge. Therefore, from the nature of dynamic MOPs (DMOPs), we optimize all the objectives simultaneously. Moreover, considering the fact that no exact information about the environment is known beforehand and the environment may change during the mission, the objectives involved in path planning are time variant. Accordingly, we need a dynamic multi-objective optimization method to deal with the problem at hand.

In this section, based on the work in [20], we propose a multi-objective EA (MOEA), called a dynamic MOEA with Pareto set linking and prediction and denoted LP-DMOEA, which stores and analyzes the historic information to enhance its performance on DMOPs, to solve the OPP problem. Within LP-DMOEA, the historic Pareto solutions are first linked to construct several time series, and then a prediction method is employed to anticipate the Pareto set of the next problem. Benefiting from such anticipation, the initial population for the new problem can be heuristically generated to accelerate the convergence in the new environment. It is noteworthy that the obtainment of Pareto optimal solutions is not the end of solving an MOP. One solution should be selected from the Pareto set by the decision-maker (DM). As for UCAVs, there are no actual DMs to deal with such decision mission. They should intelligently make such a decision without interacting with human beings. To this end, we employ a Bayesian network (BN) to model the inference process of a pilot when he is assessing the dangerous level of environment. In addition, the fuzzy logic is used to quantify the decision bias reference to inference results.

Fig. 16.1 Illustration of the
online path planning (OPP)
problem

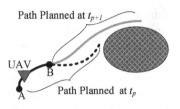

16.2.1 Formulation of the OPP Problem

From the practice point of view, an offline global path planning is likely to be invalid
when the environment is uncertain and time-variant. A UCAV has to independently
plan its online path referring to the local information detected by its onboard sensors. Pongpunwattana has proposed an OPP scheme in the sense of model predictive
control (MPC) in [21]. As shown in Fig. 16.1, suppose the UCAV has planned a path
starting from point A at t_p. Instead of executing the whole planning result, a partial
path (e.g., the path between A and B) will be executed. While flying from A to B,
it is planning a new path that starts from B. When it arrives at B at time t_{p+1}, this
new path will be used. Of course, part of the new path will be executed. Hence,the
OPP can be achieved by iteratively executing the steps above and the environmental
information can be continuously updated to adapt the planner for a changing environment. It is obviously that the path between A and B is the executing horizon and
the paths planned at A or B are the planning horizon. The optimization problem in
each time window can be time-variant. Hence, the OPP problem is indeed a DOP.

In a 2-D case, we formulate the OPP problem as follows:

$$\begin{cases} \min f = \{f_1(\mathbf{u},t), f_2(\mathbf{u},t)\} \\ f_1(\mathbf{u},t) = \prod_{i=1}^{n} p_{kill}\left(\mathbf{x}(t) + \sum_{j=1}^{i} g(u_j)\right) \\ f_2(\mathbf{u},t) = \left\|\mathbf{x}(t) + \sum_{i=1}^{n} g(u_i) - \mathbf{T}\right\|_2 \\ u_i \in \mathbf{u}(i=1,\dots,n) \end{cases} \tag{16.1}$$

where $\mathbf{u}$ is the sequential control input vector of a UCAV from t to $t + \Delta T_s(\Delta T_s = n \times \Delta t)$, $g(u_i)$ denotes the Euclidean deviation of the UCAV caused by the control
input u_i, $\mathbf{x}(t)$ and $\mathbf{T}$ are the vectors of position values of UCAV and destination,
respectively. $p_{kill}(\mathbf{x}(t))$ means the probability of being destroyed. So, the first optimization objective is to minimize the destroy probability when the UCAV flies along
the planning path. The second optimization objective is easy to understand: it aims
at minimizing the distance between the UCAV (at the termination of a path segment) and its goal. For the convenience of simulation, we use a probabilistic threat
exposure map [27] to model the battle field. The probability of becoming disabled
by the i-th threat is characterized by the multi-dimensional Gaussian law.

Fig. 16.2 Follow chart of
the LP-DMOEA

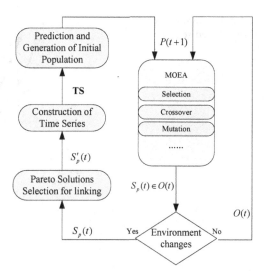

$$p^i_{kill}(t) = \frac{1}{2\pi \sqrt{\det(\mathbf{K}_i)}} \exp\left[-\frac{1}{2}(\mathbf{x}(t) - \boldsymbol{\mu_i})^T \mathbf{K}_i^{-1} (\mathbf{x}(t) - \boldsymbol{\mu_i})\right] \quad (16.2)$$

where $\boldsymbol{\mu_i} = [\mu_{x,i}, \mu_{y,i}]$ and $\mathbf{K}_i = \begin{bmatrix} \sigma^2_{x,i} & 0 \\ 0 & \sigma^2_{y,i} \end{bmatrix}$. Here, $\boldsymbol{\mu_i}$ denotes the position of a threat
and $\mathbf{K}_i$ determines its acting range.

16.2.2   Problem-Solving Approach: LP-DMOEA

In order to deal with the DMOP at hand, we propose the LP-DMOEA. The main
idea of LP-DMOEA lies in heuristically generating the initial population for a new
problem by making prediction from the historical information. Historical Pareto solutions are linked to construct several time series, and then a prediction method is
employed to anticipate the Pareto set of the next problem. At last, the initial population for the new problem can be heuristically generated to accelerate the convergence speed for the new problem.

As shown in Fig. 16.2, suppose the environment changes (a new problem arrives)
at t, the Pareto set $S_p(t)$ in the offspring $O(t)$ is used to generate the next population $P(t+1)$. There are three major steps in LP-DMOEA. Firstly, LP-DMOEA will
select some representative Pareto solutions $S'_p(t)$ from $S_p(t)$ considering both the
computational complexity of prediction and the diversity of Pareto front. Secondly,
$S'_p(t)$ and its historical counterparts are used to construct several time series **TS**. At
last, an anticipating method will be applied to **TS** to predict the location of the new
Pareto set, and the initial population can then be generated base on the prediction
results.

Fig. 16.3 Diagram of the hyper-box based selection (HBS)

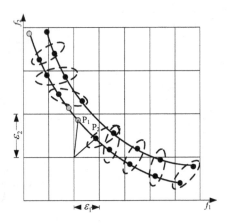

16.2.2.1 Selecting Pareto Solutions for Linking

As for selecting Pareto solutions for linking, there are two major ideas in the literature. The first one accomplishes this by considering the feature of each candidate in the objective space. In [11, 12], the anchor points and closest-to-ideal point of a Pareto front in the objective space are chosen as the key feature points, and the corresponding solutions in the decision space are used to make anticipation. This approach uses two or three key points to characterize the Pareto front. However, this approach may be invalid when the front is concave or very complex. As for the other idea, the candidates are selected directly in the decision space under a special principle. In [29], all the Pareto solutions got before an environmental change are used to anticipate. This approach is more direct because the factors involved in the time series construction are only in the context of the decision space. However, each Pareto solution will be linked to a time series, which may lead to a large number of time series and the computational complexity may be enormous.

In this paper, we follow the second idea. The difference to [29] lies in that we use a hyper-box based selection (HBS) to construct time series from the Pareto sets. As shown in Fig. 16.3, suppose there are two decision values and the range of each value is divided into many sections by a preset parameter ε_1 or ε_2. Thus, the whole decision space is divided into lots of hyper-boxes. Instead of the whole Pareto set, partial solutions will be selected for the time series construction. The HBS allows a hyper-box being occupied by only one Pareto solution in the Pareto set. If there are two or more solutions in a hyper-box, only the one that is the closest to the left-bottom corner of the hyper-box is the winner. For example, P_1 and P_2 are in the same hyper-box, P_2 is the winner and is selected to construct the time series since P_2 is closer to the left-bottom corner of the hyper-box.

16.2.2.2 Construction of Time Series

As for constructing time series, the key issue is how to identify the relationship between the solutions selected by HBS. In this study, we use the minimal distance principle to identify the relationship between two solutions. Each Pareto solution $x_S(i,t)$ in $S'_p(t)$ will be added to the end of a time series and the number of **TS** is equal to the number of $x_S(i,t)$. Suppose $TS(j) \in$ **TS** is a time series constructed previously and x_T is its last element. If the Euclidean distance from $x_S(i,t)$ to x_T is the shortest, then $x_S(i,t)$ should be added to $TS(j)$, i.e.

$$TS(i) = \arg \min_{x_s \in S'_p(t), x_T \in TS(i)} \|x_s - x_T\|_2 \tag{16.3}$$

Considering the limitation of memory and computation resource, we use a preset parameter K to control the maximal order number of a time series. This means there are at most K elements in a time series. If the length of a time series is shorter than K, then $x_S(t)$ will be added to the end of the time series directly. Otherwise, the elements in the time series will follow the first-in-first-out principle.

16.2.2.3 Prediction and Generation of the Initial Population

Many methods can be used to analyze the time series constructed above. In this study, the following simple linear model is adopted:

$$\tilde{x}_{t+1} = x_t + (x_t - x_{t-1}). \tag{16.4}$$

Considering the forecasting error caused by the inaccuracy of the forecasting model and the searching algorithm, the prediction results may not be directly used to initialize the new population. The diversity should be maintain to some degree. Here we maintain diversity in two aspects:

  a) *Only generating part of the initial population based on the prediction results:* A preset parameter α is used to control the rate of the individuals which will be initialized refereing to the prediction and the rest individuals in the initial population will be randomly generated.

  b) *Variation with a noise:* Similar to [29], we bring in a Gaussian noise λ to improve the chance of the initial population to cover the true Pareto set. This noise is added to the predicted result of each decision value. The standard deviation δ of noise is estimated by looking at the changes occurred before:

$$\delta^2 = \frac{1}{4n} \|x_t - x_{t-1}\|_2^2 \tag{16.5}$$

16.2.2.4 Chromosome Representation

The chromosome is the bridge between the optimization problem and the search space. For a path section, it is straightforward to code it as a series of consecutive line segments. However, this coding method is ambiguous for the control system

Fig. 16.4 Diagram of the
chromosome representation

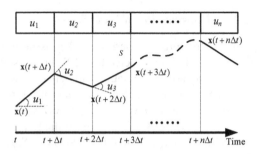

since the control system can not get an explicit input signal from such a chromosome. In this study, we use a series of consecutive yaw angle changing values to code a chromosome. In each time step (i.e., Δt), a UCAV flies at a corresponding yaw angle and the resulting path can be obtained. Suppose the UCAV cruises at a constant velocity and only the 2-D case is considered. As shown in Fig. 16.4, the UCAV makes a change of u_1 in its yaw angle at the moment of t, and a line segment from $\mathbf{x(t)}$ to $\mathbf{x}(t + \Delta t)$ can be geometrically calculated using the kinematics of the UCAV.

16.2.3 Decision-Making on the Selection of Executive Solution

Although a set of Pareto solutions can be dynamically obtained using the LP-DMOEA, the OPP problem is not solved until one feasible solution is selected out for executing. In this section, we focus on how to select a solution referring to the bias of the DM and how to intelligently make such a decision.

16.2.3.1 Methodology to Select Solutions from the Pareto Set

In this work, we use the weighted stress function method (WSFM) proposed in [9] to integrate the DM preference after the search process has been made. For a problem with M optimization objectives, the WSFM converts it to a single-objective optimization problem as follows:

$$\mathbf{x} = \arg \min_{\mathbf{x} \in S} \left(\sum_{1 \leq i < j \leq M} |\gamma_i (f_i(\mathbf{x})) - \gamma_j (f_j(\mathbf{x}))| \right) \quad (16.6)$$

where $\mathbf{x}$ and S denote the decision value vector and the decision space, respectively, $\gamma_i (f_i(\mathbf{x}))$ denotes the "stress" associated to the corresponding objective according to the weight (ω_i ($\sum_i^M w_i = 1$), given by the DM) contributed to each objective, see [9] for more information. Since the equation above considers the maximization problem and each objective should be normalized, we rewrite the original objective function (Eq. (16.1)) as follows:

$$\begin{cases} \min f = \{f_1(\mathbf{u},t), f_2(\mathbf{u},t)\} \\ f_1(\mathbf{u},t) = 1 - \prod_{i=1}^{n} p_{kill}\left(\mathbf{x}(t) + \sum_{j=1}^{i} g(u_j)\right) \\ f_2(\mathbf{u},t) = 1 - \dfrac{\left\|\mathbf{x}(t) + \sum_{i=1}^{n} g(u_i) - \mathbf{T}\right\|_2 - (\|\mathbf{x}(t) - \mathbf{T}\|_2 - nV\Delta T)}{2nV\Delta T} \\ u_i \in \mathbf{u}(i = 1, \ldots, n) \end{cases} \quad (16.7)$$

16.2.3.2   Intelligent Situation Assessment via Bayesian Network (BN)

Now the problem turns to how to set ω_i intelligently. The weights ω_1 and ω_2 reflect the DM's bias to safety (i.e., f_1) and path length (i.e., f_2), respectively. In this study, the BN is employed to assess the dangerous level of the battle filed. The construction process of a BN, including the structure and parameters, is indeed the integration of the DM's knowledge. The resulting BN will make an intelligent inference instead of the DM. In this work, the enemy air defense (i.e., threats for UCAVs) consists of two types of anti-air weapons: the anti-air guns (AAGuns) and the surface-to-air missiles (SAMs). The threat type is written as TT for short. A threat may work in one of the following states: No targets are found and the system is inactive (IA), surveillance radar detects the targets (Surv), targets have been intercepted by radars (Intercept), targets are being traced by radars (Trace) and open fire (Fire). There are five environmental dangerous levels (EDLs): very dangerous (VD), dangerous (D), medium (M), safe (S), and very safe (VS). If the working states of the threats (TS) could be known, the EDL can be easily inferred.

However, the TS is difficult to be known directly and a UCAV has to infer such information referring to the local information collected by its onboard sensors. Suppose there are two major onboard sensors assembled on a UCAV: the missile launching detector (MLD) and the radar warning receiver (RWR). For these sensors, there are two working states: active (A) and inactive (IA). When the RWR works in state A, the UCAV has been detected or traced by enemy radars and the anti-air weapons may be launched in a short future. The matter is worse when the MLD is working in state A which means the UCAV is under attack.

In addition to the sensors' information, the distance between a threat and the UCAV is also a key factor that impacts the EDL. Suppose there are five range (R) scales: R1 (0~1 km), R2 (1~2 km), R3 (2~4 km), R4 (4~6 km), and R5 (larger than 6 km). The BN structure is shown in Fig. 16.5 and the corresponding conditional probability tables (CPTs) are given in Table 16.1 and Table 16.2, respectively.

16.2.3.3   Quantification of Environmental Assessment Results

Since the BN is a qualitative inference tool, we need to quantify the inference results (i.e., the EDLs) to obtain the weight values associated to each optimization objective. In this work, we use a fuzzy logic to set the value of ω_1 and the fuzzy rules are given in Fig. 16.6. The triangular-shaped membership function (as shown in Fig. 16.7) and the center of gravity defuzzification are adopted. After the quantification of ω_1, ω_2 can be easily calculated since $\omega_1 + \omega_2 = 1$.

Fig. 16.5 The BN structure of the environmental assessment model

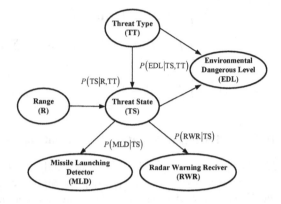

Table 16.1 The conditional probability tables P(RWR | TS) and P(MLD | TS) of the BN of the environmental assessment model

RWR	TS					MLD	TS				
	IA	Surv	Intercept	Trace	Fire		IA	Surv	Intercept	Trace	Fire
A	0	1	1	1	0	A	0	0	0	0	1
IA	1	0	0	0	1	IA	1	1	1	1	0

Table 16.2 The conditional probability tables P(TS | R, TT) and P(EDL | TS, TT) of the BN of the environmental assessment model

TT	TS	R					TT	EDL	TS				
		R1	R2	R3	R4	R5			R1	R2	R3	R4	R5
	IA	0	0	0.05	0.95	1		VD	0	0	0	0.95	1
	Surv	0	0.05	0.9	0.05	0		D	0	0	0.05	0.05	0
AAGun	Intercept	0	0.9	0.05	0	0	AAGun	M	0	0.1	0.9	0	0
	Trace	0.6	0.05	0	0	0		S	0	0.8	0.05	0	0
	Fire	0.4	0	0	0	0		VS	1	0.1	0	0	0
	IA	0	0	0	0.05	0.95		VD	0	0	0	0.1	1
	Surv	0	0	0.05	0.9	0.05		D	0	0	0.2	0.8	0
SAM	Intercept	0	0.05	0.9	0.05	0	SAM	M	0	0.2	0.7	0.1	0
	Trace	0.05	0.9	0.05	0	0		S	0	0.7	0.1	0	0
	Fire	0.95	0.05	0	0	0		VS	1	0.1	0	0	0

Fig. 16.6 Fuzzy rules between EDL and ω_1

> IF EDL is VD, THEN w1 is very high;
> IF EDL is D, THEN w1 is high;
> IF EDL is M, THEN w1 is medium;
> IF EDL is S, THEN w1 is low;
> IF EDL is VD, THEN w1 is very low.

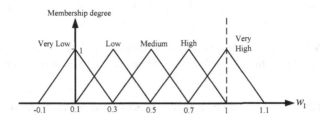

Fig. 16.7 The triangular-shaped membership function. Suppose the value range of ω_1 is [0.1, 1].

16.3   Dynamic Target Assignment

In this section, we consider the following dynamic target assignment problem: a fleet of UCAVs are assigned to attack an anti-air defense site which is composed of several combat units including a command center, surveillance radars, SAMs or AAGuns. These units work cooperatively and their functions are coupled. This means a unit works depending on the states of other units. For example, an AAGun can not work without the target information and commands provided by the radars and the control center, respectively. Similarly, the control center runs depending on the target information from the radars, and the radars should work under the control of the control center. In addition to the inner dependence, the states of the targets are time-variant according to the opposed nature of the battlefield. This may change the importance of the corresponding targets (the targets whose states have changed and the targets that work depending on the state-changed targets). For example, considering the surveillance task and the threat from anti-radiation missiles (a class of missiles which are designed to detect enemy radio emission sources, e.g., radars), networked surveillance radars may keep active (search) in turn. If a radar changes its state from "inactive (radio quiescence)" to "active (search)", the importance of the radar and the SAM launchers or AAGuns that work depending on it will increase. Consequently, the target assignment problem should be achieved dynamically to adapt to the time-variant importance of the targets.

The goal of dynamic target assignment is to assign enemy targets to a group of UCAVs. We achieve this in two steps. Firstly, by dynamically solving the weapon target assignment (WTA) problem, each onboard air-to-ground missile (AGM) of the UCAVs will engage a target. Secondly, the result of the WTA will be transformed to a target list for each UCAV via the weapon-UCAV mapping.

16.3.1   Formulation of the Dynamic WTA Problem

As for an anti-air defense site, the importance values of the combat units are different from each other according to their roles. These units work according to there inner dependence. As for the attacking side, an expected damage should be given by operators as their mission goal. The definitions of the importance vector, inner dependence matrix, and expected damage are given as follows.

Definition 16.1. $A_{N \times N}$ is an inner dependence matrix of N members, where a_{ij} indicates the probability that the j-th member goes fail if the i-th member becomes invalid.

Definition 16.2. W is the vector of weight values of N members, where W_i indicates the importance value of the i-th member and $\sum_{i=1}^{N} W_i = 100$.

Definition 16.3. E is the vector of expected damage types of the targets, where $e_i = \{1, 2, 3\}$ indicates the expected damage type (1-destroyed, 2-disabled, or 3-suppressed) of the i-th target.

Suppose M weapons are assigned to attack N targets. The WTA problem is a kind of combinatorial optimization problem whose objective is to maximize the utility function F_{WTA} of a set of weapon-to-target pairs $\{(k_1, l_1), (k_2, l_2), \dots, (k_i, l_i)\}$, $i = 1, 2, \dots M$, where l_i indicates the target of weapon k_i. The WTA problem can be formulated as follows:

$$
\min F_{WTA} = \sum_{i=1}^{M} \Delta(k_i, l_i)
$$
$$
\forall k_i \in \{1, 2, \dots, M\}, k_i \neq k_j, \textbf{if } i \neq j \qquad (16.8)
$$
$$
\forall l_i \in \{1, 2, \dots, N\}
$$

where $\Delta(k_i, l_i)$ is the utility increment of assigning weapon k_i to target l_i. In the rest of this section, we will show how to calculate $\Delta(k_i, l_i)$.

In general, $\Delta(k_i, l_i)$ is the difference value between gains increment ($\Delta^+(k_i, l_i)$) and cost increment ($\Delta^-(k_i, l_i)$). In other words, $\Delta(k_i, l_i)$ is calculated as follows:

$$
\Delta(k_i, l_i) = \Delta^+(k_i, l_i) - \Delta^-(k_i, l_i) \qquad (16.9)
$$

$\Delta^+(k_i, l_i)$ can be considered as the incremental loss of potential combat capability of the enemy. Similarly, $\Delta^-(k_i, l_i)$ is the incremental loss of potential combat capability of the attacking side. Suppose $i-1$ missiles have engaged their targets, at the $i-$th assignment step, $\Delta^+(k_i, l_i)$ can be calculated as follows:

$$
\Delta^+(k_i, l_i) = F_i^+ - F_{i-1}^+
$$
$$
= \sum_{j=1}^{N} W_{i,j} \left(1 - \prod_{k=1}^{N} (1 - P_i(k) a_{kj}) \right) \qquad (16.10)
$$
$$
- \sum_{j=1}^{N} W_{i-1,j} \left(1 - \prod_{k=1}^{N} (1 - P_{i-1}(k) a_{kj}) \right)
$$

where $W_{i,j}$ and $P_i(k)$ are calculated iteratively as follows. $P_i(k)$ is the invalid probability of the k-th target after the i-th AGM has engaged its target. $Q_i(k) = 1 - P_i(k)$ is the corresponding valid probability of the k-th target.

$$P_i(k) = \begin{cases} P_{i-1}(k), k \neq l_i \\ 1 - \gamma(k_i,k)Q_{i-1}(k), k = l_i \end{cases} \tag{16.11}$$

$$W_{i,j} = W_{i-1,j}\frac{1 - P_i(l_i)a_{l_ij}}{1 - P_{i-1}(l_i)a_{l_ij}} \tag{16.12}$$

where $\gamma(k_i,l_i) = 1 - \xi(k_i,l_i)$ and $\xi(k_i,l_i)$ is the kill probability that AGM k_i damages target l_i to the expected damage type (i.e., destroyed, disabled, or suppressed). $\xi(k_i,l_i)$ can be calculated as follows:

$$\xi(k_i,l_i) = \frac{p_{hit}(k_i,l_i)(1 - p_a)}{OM(k_i,l_i)} \tag{16.13}$$

where $p_{hit}(k_i,l_i)$ is the hit probability of AGM k_i, p_a is the probability that AGM k_i is intercepted by the anti-air defense, and $OM(k_i,l_i)$ denotes the required number of AGMs for destroying target l_i to the expected damage type, i.e., e_{l_i}. Note that those AGMs are of the same kind to AGM k_i.

As for $\Delta^-(k_i,l_i)$, it can be calculated as follows:

$$\Delta^-(k_i,l_i) = F_i^- - F_{i-1}^-$$
$$= \sum_{j=1}^{N} W_{i,j}\left(1 - \frac{1 - (1 - Q_{i-1}(l_i)\lambda_{i-1}(l_i))a_{l_ij}}{1 - \left(1 - \frac{Q_{i-1}(l_i)\lambda_{i-1}(l_i)}{\gamma(k_i,l_i)}\right)a_{l_ij}} \times \prod_{k=1}^{N}\left(1 - \left(1 - \frac{Q_{i-1}(k)\lambda_{i-1}(k)}{\gamma(k_i,l_i)}\right)a_{kj}\right)\right)$$
$$- \sum_{j=1}^{N} W_{i-1,j}\left(1 - \prod_{k=1}^{N}\left(1 - a_{kj}(1 - \lambda_{i-1}(k)Q_{i-1}(k))\right)\right) \tag{16.14}$$

where $\lambda_{i-1}(k) = \prod_{v \in N_T} \gamma(v,k)$ and N_T is the set of the remaining weapons after $i-1$ assignment steps.

According to the above formulation of WTA, the dynamic version of WTA can be easily formulated. In this study, as mentioned above, we consider the case that surveillance radars of the anti-air defense keep active (working in the state of search) in turn to avoid being locked by anti-radiation missiles. This will lead to the change of the importance values of the targets. Therefore, the weight vector W is no longer static but time-variant, we write it as $W(t)$. Accordingly, the dynamic version of Eq. (16.8) can be formulated as follows.

$$\begin{aligned} \min F_{WTA}(t) &= \sum_{i=1}^{M} \Delta(k_i,l_i,W(t)) \\ &\forall k_i \in \{1,2,\ldots,M\}, k_i \neq k_j, \textbf{ if } i \neq j \\ &\forall l_i \in \{1,2,\ldots,N\} \end{aligned} \tag{16.15}$$

16.3.2 Problem-Solving Approach: Memory-Based Estimation of Distribution Algorithm with Environment Identification

In order to solve the dynamic combinational optimization problem which is formulated above, a memory-based estimation of distribution algorithm (EDA) [19] is adopted considering the fact that the changing tendency of the radars' working states is cyclic to some extent.

EDAs are a class of probability model based EAs, where the processes of learning and sampling the probability model replace the genetic operations (e.g., crossover and mutation) in conventional GAs. A probability model indicates the joint probability distribution of high-performance solutions. That is, it characterizes the set of good solutions. If the historic information could be stored as probability models, we would not only save the memory space but also simplify the memory management scheme. Consequently, EDAs are suitable for being extended to be memory-enhanced EAs to solve DOPs.

To this end, an environment identification based memory management scheme (EI-MMS) is applied in this study. Within this scheme, a probability model is regarded as the learning result of the probability distribution of high-performance solutions in an environment. A probability model together with the best individual in the solutions from which the probability model is learnt are stored as a memory element. In order to retrieve the memory elements in EI-MMS, an environment identification method is proposed to select the suitable element according to a special environment. The EI-MMS can be used to extend any binary-encoded static EDA to its dynamic version and we name the corresponding algorithm as EDA with environment identification based memory scheme (EI-MEDA).

16.3.2.1 The EI-MMS

The EI-MMS scheme uses an additional memory and the elements stored are the probability models learnt from the population. Before giving the details, we assume that the environmental changes are detectable. In all the algorithms studied below, the environmental change is detected in each generation by checking whether there is at least one memory element whose evaluation value has changed.

In order to utilize the intervals between every two environmental changes to learn a high-quality probability model, EI-MEDA updates its memory just after the environment changes. As shown in Fig. 16.8, the whole dynamic optimization process is divided into many static optimization processes. In each static process, EDA searches the optimum in the conventional manner. When the e-th environment comes at generation t, EI-MMS manages its memory $\mathbf{M}$ in three major steps. First, it stores the probability model obtained from the generation just before the environmental change, i.e., $PM(t-1)$, into the memory. Then, it finds a memory element $M(k_e)$ ($k_e = 1, 2, \ldots, m$) which best fits the new environment to retrieve using an environment identification method. Finally, the probability model of this memory element, i.e., $mPM(k_e)$, is sampled to generate the first generation of population in the new environment.

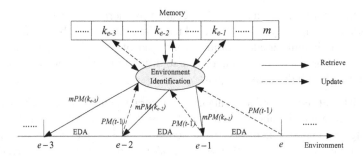

Fig. 16.8 Illustration of the EI-MMS

16.3.2.2   The Environment Identification Method

The environment identification method is very important due to its role of linking between the memory and the dynamic environment. A key aspect of EI-MMS is to find the suitable memory element to retrieve according to the new environment. An intuitive way of achieving this is to consider the average fitness of the solutions sampled from a special element. Considering the computational complexity and the accuracy, we propose a samples averaging plus best individual (SA + BI) method to evaluate the elements in the memory and select the suitable one.

The Samples Averaging (SA) Method. The idea of this method is to evaluate a memory element by averaging the fitness of solutions sampled from it. For each memory element $M(k)$ $(k = 1, 2, \ldots, m)$, N_S solutions are sampled from it and the average fitness of these sampled solutions is calculated in the current environment as the evaluation value of $M(k)$ as follows.

$$f_M(k) = \frac{1}{N_S} \sum_{i=1}^{N_S} f_{ind(i)}^k \tag{16.16}$$

where $f_{ind(i)}^k$ denotes the fitness of the i-th solution sampled from the probability model, i.e. $mPM(k)$, of $M(k)$. This method is the most intuitive way to evaluate a memory element but its computational complexity is high.

The Best Individual (BI) Method. In this method, each memory element consists of two parts: a probability model and the best individual of the population from which the probability model is learnt. Here we denote the memory element by $M(k) = \langle BM(k), mPM(k) \rangle$ $(k = 1, 2, \ldots, m)$, where $BM(k)$ denotes the best individual. The evaluation of the memory element $M(k)$ is defined as follows:

$$f_M(k) = f(BM(k)) \tag{16.17}$$

where $f(BM(k))$ denotes the fitness of $BM(k)$ in the current environment. This method is similar to the method used in [15, 16].

In contrast with the SA method, the accuracy is sacrificed for the sake of the computational complexity. The BI method uses the fitness of the best individual to evaluate the probability model learning from a set of individuals. This may lead to inaccuracy. For example, it is impossible to differentiate two elements when the fitness of their best individuals is equal.

The SA+BI Method. In order to balance the accuracy and the computational complexity, we combine the above two methods, resulting in the SA+BI method. For comparing two memory elements, if the fitness of the best individuals are different, the BI method is applied; otherwise, the SA method is applied to differentiate the memory elements. The pseudo-code of the proposed EI-MEDA with the SA+BI method is shown in Fig. 16.9, where N_{pop} denotes the population size and p_s denotes the truncation selection rate (i.e., for each generation, the $p_s \times N_{pop}$ best samples generated from the current model $PM(t)$ are selected to build up the model $PM(t+1)$ for the next generation).

```
begin
  PM(0,i) := 0.5, mPM(j,i) := 0.5, i ∈ [1,l], j ∈ [1,m];
  Randomly generate BM(j), t := 0, k := 0;
  Randomly initialize first population P₀;
  repeat
    if an environment change is detected then
      ⟨BM(k), mPM(k)⟩ := ⟨B(t−1), PM(t−1)⟩;
      Update the fitness value of each BM(j), j ∈ [1,m];
      Select a proper mPM(k) using the SA+BI method;
      PM(t) := mPM(k);
      Sample PM(t) to generate Pₜ;
    else
      Evaluate Pₜ;
      Select the pₛ × N_pop best individuals from Pₜ
      Learn PM(t);
      Compensate the diversity for PM(t) by the method described in Sect. 16.3.2.3;
      Sample PM(t) to generate the t-th offspring Oₜ;
      Evaluate Oₜ and replace the worst individuals in Pₜ;
      B(t) := BestIndvidual(Pₜ);
      t := t+1;
  until terminated = true;
end;
```

Fig. 16.9 Pseudocode for the proposed EI-MEDA

16.3.2.3   Diversity Loss and Counteracting Methods

As mentioned above, EI-MMS is in fact a diversity maintaining method according to the environmental changes. In addition to diversity maintaining in dynamic landscapes, it is also important to counteract the diversity loss in static EDA which searches the optimum in each environment.

It is well known that the variance of a sample of size N has an expected value of $\sigma^2(1 - 1/N)$ where σ^2 is the variance in the parent distribution. Most EDAs do not compensate for this. When the new probability model is produced, it attempts to model the new population and therefore has a reduced variance. When this is iterated repeatedly, the variance of the sampled population gets smaller and smaller and decays to zero. The probability model evolves to one which can only generate identical configurations. In [22] Shapiro analyzed the dynamics of EDAs in terms of Markov chains and declared that the general EDAs can not satisfy two necessary conditions for being effective search algorithms. Hence, we must counteract the diversity loss to improve the efficiency of an EDA. In this study, we adopt a diversity compensation method that combines the loss correction and boundary correction methods [4], denoted LC+BC in this study.

The Loss Correction (LC) Method. Let l be the length of a chromosome and γ_i $(i = 1, \cdots, l)$ be the probability that the allele of the i-th gene is equal to 1, γ_i is transformed to γ_i' to counteract the diversity loss as follows:

$$\gamma_i' = \begin{cases} \dfrac{1 - \sqrt{1 - 4(1 - \gamma_i)/L_s}}{2}, & \gamma_i \leq \dfrac{1}{2}(1 - \sqrt{1 - L_S}) \\ \dfrac{1 + \sqrt{1 - 4(1 - \gamma_i)/L_s}}{2}, & \gamma_i \geq \dfrac{1}{2}(1 + \sqrt{1 - L_S}) \\ 0.5, & \text{otherwise} \end{cases} \tag{16.18}$$

where $L_S = \dfrac{p_s \times N_{pop} - 1}{p_s \times N_{pop} - p_s}$.

The Boundary Correction (BC) Method. For the BC method, γ_i is transformed to γ_i' to counteract the diversity loss as follows:

$$\gamma_i' = \begin{cases} \beta, & \gamma_i < \beta \\ 1 - \beta, & \gamma_i > 1 - \beta \\ \gamma_i, & \text{otherwise} \end{cases} \tag{16.19}$$

where β is a preset parameter to prevent the distribution from converging to 1 or 0. To guarantee the minimal diversity level, β is set to $1/l$ in this study unless stated otherwise.

The LC+BC Method. For the LC+BC method, LC and BC are applied in turn. In other words, LC is first applied to γ_i, then the resulting γ_i' is taken as the input to the BC method.

16.3.3   Chromosome Representation

A binary-encoded chromosome is used to represent a solution of the WTA problem at a certain moment. As shown in Fig. 16.10, the decimal number of a target is coded into a binary string. The whole chromosome is composed of the target numbers of all weapons, encoded in binary strings.

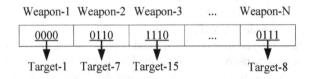

Fig. 16.10 Chromosome representation of the WTA problem

16.3.4   Weapon-UCAV Mapping

One can observe that the attacking sequence to the targets has been considered to some extent in the procedure of WTA. The high-value targets whose working states can affect the validity of other targets will be attacked in priority and will be allocated more AGMs. However, the importance of the independent targets (whose working states can not affect other targets) can not be reflected by the sequence in which they are attacked. In order to plan a sequential target list for each UCAV, we firstly refine the WTA result according to the following additional principles.Then, the refined WTA result is mapped to the target lists for the UCAVs according to the owner-member relationship between the AGMs and the UCAVs.

1. If there is no dependence between two targets, the one with a higher weight will be attacked in priority.
2. If there is no dependence between two targets and their weight values are the same, the one that is closer to the attacker will be attacked in priority.

16.4   Simulation Results and Analysis

16.4.1   Simulation Scenario

In this section, we will validate and test the proposed approaches via the simulation with a SEAD mission scenario. Four UCVAs (red side) are assigned to attack the anti-air defense equipments of the blue side. The brief procedure can be described as following: The UCAVs maneuver according to the intelligent tactic flight so as to dynamically avoid the threats in partly known or fully unknown battlefield. Meanwhile, the targets will be dynamically engaged by the UCAVs according to the changing working states of the targets. This procedure keeps on going until the fire condition is satisfied and then the AGMs are released.

Table 16.3 Weapon configuration of the UCAVs

	UCAV-1	UCAV-2	UCAV-3	UCAV-4
Weapon station-1	No.1 (AGM-A)	No.2 (AGM-B)	No.3 (AGM-A)	No.4 (AGM-B)
Weapon station-2	No.5 (AGM-A)	No.6 (AGM-B)	No.7 (AGM-A)	No.8 (AGM-B)
Weapon station-3	No.9 (AGM-A)	No.10 (AGM-B)	No.11 (AGM-A)	No.12 (AGM-B)
Weapon station-4	No.13 (AGM-A)	No.14 (AGM-B)	No.15 (AGM-A)	No.16 (AGM-B)

16.4.1.1 Red Side

Among the four UCAVs, UCAV-1 is the leader who will be in charge of the intelligent OPP problem. The other three wing UCAVs will maneuver to follow their leader in a desired formation[1]. The initial position of UCVA-1 is (0,10) km and the initial yaw angle is zero (in east-north-height coordination). The UCAVs are equipped electronic warning equipments to report the event that being traced by enemy radars or being locked by enemy anti-air missiles. Each UCAV carries four AGMs of the same class. There are two classes of AGMs: AGM-A and AGM-B. The hit probability (p_{hit}) of AGM-A is 0.85 and that of AGM-B is 0.7, the maximal attack range of the two classes of AMGs is 10 km. The weapon configuration of the UCAVs is shown in Table 16.3.

16.4.1.2 Blue Side

The air-defense of the blue side is composed of sixteen combat units: one command center, three surveillance radars, and twelve AAGuns. Detailed parameters (positions, weight values, and expected damage types) of the blue units are shown in Table 16.4. In order to reduce the risk of being locked by anti-radiation missiles, the surveillance radars will keep active (searching) for two seconds in turn. The weight value of an active surveillance radar is sixteen and any AAGun that works depending on this active radar will have a weight value of five. The weight value of an inactive radar and the related AAGuns are nine and two, respectively.

The inner dependence matrix is shown in Table 16.5. Table 16.6 shows the required number of AGM-A and AGM-B for destroying a certain target to the expected damage type. The interception probability p_a could be calculated by Eq. (16.20) according to the quantity of AGMs (i.e., M).

---

[1] Formation control is a well-studied issue in multiagent cooperative control and is not in the scope of this study. We assume the formation can be kept well in this simulation and the relative distance and angle are 100 m and 60° respectively

Table 16.4 Parameters of the air-defense units

No.	Role	Position (km)	Weight value[a]	Expected damage type
1	Command center	(21.23, -10.67)	30	$e_1 = 1$ (destroyed)
2	Surveillance radar	(19.50, -9.67)	16	$e_2 = 1$ (destroyed)
3	AAGun	(21.23, -8.67)	5	$e_3 = 2$ (disabled)
4	AAGun	(22.96, -9.67)	5	$e_4 = 2$ (disabled)
5	AAGun	(20.50, -9.67)	5	$e_5 = 2$ (disabled)
6	AAGun	(18.50, -9.67)	5	$e_6 = 2$ (disabled)
7	Surveillance radar	(19.50, -8.67)	9	$e_7 = 1$ (destroyed)
8	AAGun	(19.50, -10.67)	2	$e_8 = 2$ (disabled)
9	AAGun	(22.23, -8.67)	2	$e_9 = 2$ (disabled)
10	AAGun	(20.23, -8.67)	2	$e_{10} = 2$ (disabled)
11	AAGun	(21.23, -7.67)	2	$e_{11} = 2$ (disabled)
12	Surveillance radar	(21.23, -9.67)	9	$e_{12} = 1$ (destroyed)
13	AAGun	(23.96, -9.67)	2	$e_{13} = 2$ (disabled)
14	AAGun	(21.96, -9.67)	2	$e_{14} = 2$ (disabled)
15	AAGun	(22.96, -8.67)	2	$e_{15} = 2$ (disabled)
16	AAGun	(22.96, -10.67)	2	$e_{16} = 2$ (disabled)

[a] Initial weight values that will change according to the time-variant working states of the targets.

Table 16.5 The inner dependence matrix

	1	2	3	4	5	6	7	8	9	10-	11	12	13	14	15	16
1	1.0	0.7	0.8	0.8	0.8	0.8	0.7	0.8	0.8	0.8	0.8	0.7	0.8	0.8	0.8	0.8
2	0.33	1	0.9	0.9	0.9	0.9	0	0.1	0.1	0.1	0.1	0	0.1	0.1	0.1	0.1
3	0	0	1	0	0	0	0	0	0	0	0	0	0	0	0	0
4	0	0	0	1	0	0	0	0	0	0	0	0	0	0	0	0
5	0	0	0	0	1	0	0	0	0	0	0	0	0	0	0	0
6	0	0	0	0	0	1	0	0	0	0	0	0	0	0	0	0
7	0.33	0	0.1	0.1	0.1	0.1	1	0.9	0.9	0.9	0.9	0	0.1	0.1	0.1	0.1
8	0	0	0	0	0	0	0	1	0	0	0	0	0	0	0	0
9	0	0	0	0	0	0	0	0	1	0	0	0	0	0	0	0
10	0	0	0	0	0	0	0	0	0	1	0	0	0	0	0	0
11	0	0	0	0	0	0	0	0	0	0	1	0	0	0	0	0
12	0.33	0	0.1	0.1	0.1	0	0.1	0.1	0.1	0.1	1	0.9	0.9	0.9	0.9	
13	0	0	0	0	0	0	0	0	0	0	0	0	1	0	0	0
14	0	0	0	0	0	0	0	0	0	0	0	0	0	1	0	0
15	0	0	0	0	0	0	0	0	0	0	0	0	0	0	1	0
16	0	0	0	0	0	0	0	0	0	0	0	0	0	0	0	1

Table 16.6 The required number of AGM-A and AGM-B for destroying a certain target to the expected damage type

	Destroyed		Disabled		Suppressed	
	AGM-A	AGM-B	AGM-A	AGM-B	AGM-A	AGM-B
Command center	2.5	2	1.8	1.5	1.1	1.0
Surveillance radar	2.0	1.8	1.3	1.2	1.0	1.0
AAGun	2.0	1.8	1.3	1.2	1.0	1.0

Table 16.7 Parameters of the threats

	$(\mu_x, \mu_y)/km$	(σ_x, σ_y)	Type
Threat-1	(5,5)	(1.4,1)	SAM
Threat-2	(5,-5)	(2,2)	SAM
Threat-3	(15,-11)	(1.4,1.4)	AAGun
Threat-4	(13,2)	(1.0,1.0)	SAM

$$
p_a = \begin{cases}
0.16M/5 \ , \ M \leq 5 \\
0.077M/10 \ , \ M \leq 10 \\
0.05M/20 \ , \ M \leq 20 \\
0.04M/30 \ , \ M \leq 30 \\
0.03M/40 \ , \ M \leq 40 \\
0.02 \ , \ M > 40
\end{cases}
\tag{16.20}
$$

In addition to the targeted anti-air defense site, there are another four independent threats in the battlefield. The parameters of these threats are given in Table 16.7.

16.4.2 Results and Analysis on the Intelligent OPP Problem

In the following experiments, we chose NSGA2 [8] as the basic MOEA. We apply LP-DMOEA and random restart methods to NSGA2 and the resulting dynamic MOEAs are written as LP-DNSGA2 and R-DNSGA2, respectively. For comparison study, the OPP algorithms that adopt LP-DNSGA2 and R-DNSGA2 are denoted as OPP-A and OPP-B, respectively.

1) Parameters involved in NSGA2: the population size was set to 100, the length of a chromosome was set to 20, the probabilities of simulated binary crossover (SBX) and polynomial mutation (PM) were set to 0.9 and 1/20, respectively, and the special parameters of SBX and PM are 10 and 20, respectively.

2) Parameters involved in LP-DNSGA2: The rate of the heuristically generated individuals was set to 50% (i.e. $\alpha = 0.5$), the maximal order number of a time series was set to 5 (i.e. $K = 5$), both ε_1 and ε_2 were set to 0.1.

3) Parameters involved in OPP: The time step Δt was set to 1s, executing horizon was set to one time step (i.e., 1s) and the number of time steps (planning horizon) of a sequential control input was set to 20 (i.e., the chromosome length $n = 20$).

16.4.2.1   Validation of LP-DMOEA Based OPP Algorithm

In order to test the validity of the LP-DMOEA based OPP algorithm (i.e., OPP-A), we consider a moving-threat case. In this case, two enemy moving threats patrol in the mission field. A UCAV should dynamically plan its flying path to avoid being detected. Fig. 16.11 shows the simulation snapshots. It can be seen that the UCAV can successfully keep its stealth. In other words, in the 215 s simulation, the UCAV successfully keeps its survival probability equal to 1.0.

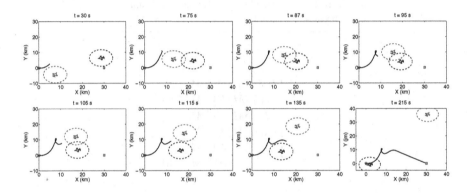

Fig. 16.11 A set of simulation snapshots of online path planning for a UCAV. There are two moving threats (e.g. enemy aircrafts) with $\sigma_x = \sigma_y = 2$ flying at a constant speed of 150 m/s. The speed of the UCAV is 200 m/s, $\omega_1 = 0.7$, and $\omega_2 = 0.3$.

16.4.2.2   Comparison of Two OPP Algorithms

We compare OPP-A with OPP-B to show the advantage of the proposed dynamic MOEA over the random restart method. Besides, we would like to test the validity of WSFM. Therefore, in the following experiments, the intelligent decision-making method will not be used. Two OPP algorithms will be tested in unknown environment (no information about the four threats are known in advance) with fixed bias weight values ($\omega_1 = 0.7$ and $\omega_2 = 0.3$, or $\omega_1 = 0.3$ and $\omega_2 = 0.7$).

As shown in Fig. 16.12, the paths obtained by OPP-A (solid line) are more reasonable and smoother than that of OPP-B (dotted line). This can be explained by Fig. 16.13 where the curves of the yaw angle are compared. It can be seen that the

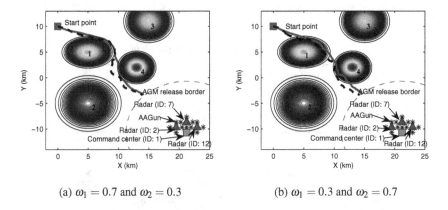

(a) $\omega_1 = 0.7$ and $\omega_2 = 0.3$ (b) $\omega_1 = 0.3$ and $\omega_2 = 0.7$

Fig. 16.12 Planned pathes of OPP-A (solid line) and OPP-B (dotted line)

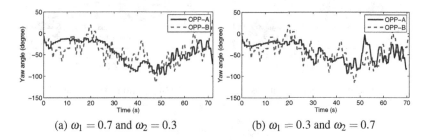

(a) $\omega_1 = 0.7$ and $\omega_2 = 0.3$ (b) $\omega_1 = 0.3$ and $\omega_2 = 0.7$

Fig. 16.13 The curves of yaw angle of OPP-A and OPP-B

yaw angle curve calculated by OPP-B fluctuates more violently, which is harmful for the control mechanism of an actual UCAV.

Besides, the results of the survival probability and flight time are given in Table 16.8. It is obvious that OPP-A outperforms OPP-B again. This is because LP-DNSGA2 employed by OPP-A can effectively improve the dynamic optimization performance in contrast to R-DNSGA2. In addition, the paths considering the DM bias of $\omega_1 = 0.7$ and $\omega_2 = 0.3$ are more likely to keep away from the threats in contrast to the bias of $\omega_1 = 0.3$ and $\omega_2 = 0.7$. This reveals that the WSFM can effectively integrate the DM bias into the automatic planner.

16.4.2.3 Validation of Intelligent Behavior against A Pop-up Threat

In the following experiments, we will show the validity of the intelligent OPP which is the combination of OPP-A and the intelligent environment assessment. Here, we assume that Threat-4 does not appear or work until the UCAV flies for 50 seconds. In such case, the UCAV should assess the environment and react to the pop-up threat intelligently.

Table 16.8 Result values of the objective functions, i.e., the survival probability and flight time of the leader UCAV

	Survival probability	Flight time (s)
OPP-A, $w_1 = 0.7, w_2 = 0.3$	0.987	72
OPP-B, $w_1 = 0.7, w_2 = 0.3$	0.979	77
OPP-A, $w_1 = 0.3, w_2 = 0.7$	0.821	69
OPP-B, $w_1 = 0.3, w_2 = 0.7$	0.808	72

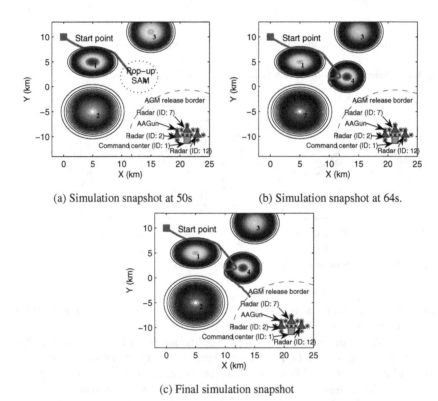

(a) Simulation snapshot at 50s   (b) Simulation snapshot at 64s.

(c) Final simulation snapshot

Fig. 16.14 Path planned by intelligent OPP algorithm proposed in this paper

As shown in Fig. 16.14, in the first 50 seconds, the UCAV has successfully evaded the threats and would straightly fly to-ward to its goal if the hostile environment would not change. Then, a pop-up threat (Thread-4) suddenly appears at 50s. Fortunately, as seen from the dotted path, the UCAV can react to this change by flying away from the pop-up threat as quickly as possible. At this moment, the intelligent environmental assessment works effectively to increase the probability of $P(EDL = VD)$ accordingly and the weight value (ω_1) associated to the safety is set to a higher one. As seen from Fig. 16.15(a), the probability $P(EDL = VD)$ raises

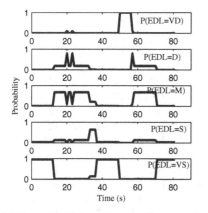
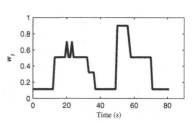

(a) The probabilities of each EDL versus time   (b) The value of ω_1 versus time

Fig. 16.15 Results of environment assessment

at 50s and goes down again (at about 58s) when the UCAV has flied away from the pop-up threat. Accordingly, as shown in Fig. 16.15(b), the value of ω_1 is set to 0.9 from 50s to 58s and goes down when the environment seems safe again.

16.4.3   Results and Analysis on the Dynamic WTA Problem

We apply EI-MEDA to univariate marginal distribution algorithm (UMDA) [16], named EI-MUMDA in this set of experiments. The corresponding parameters are given as follows. The population size was set to 200, the memory size was 10, and the truncation selection rate of the UMDA was set to 0.5.

The results of the last thirty seconds on the dynamic WTA problem are shown in Fig. 16.16 and the last three times of UCAV-to-target assignment are given in Table 16.9. It can be seen that the command center is the most important priority target to be attacked and is allocated more attacking resources comparing with the radars and AAGuns. After the survival probability of the command center has gone down to a certain level, its importance value will significantly decrease. This will make the WTA process adapt its priority target to the active anti-air units (i.e., the active surveillance radar and its relative AAGuns). Among the combat units of an anti-air site, the AAGuns work according to the target information from the surveillance radar. Thus, the surveillance radar is the priority target in an active anti-air site. It is noteworthy that the priority of an active radar is only reflected in the quantity of the allocated AGMs. The attacking sequence to the surveillance radars is not considered since there is no dependence among them. In order to assign each UCAV a sequential target list, the principles in Sect. 16.3.4 are used to refine the WTA results. As can be seen from Table 16.9, each UCAV has its reasonable target list and the coordinated target assignment is achieved.

$t = 40$ s: $1 - 1 - 1 - 7 - 12 - 2 - 1 - 12 - 2 - 7 - 1 - 2 - 12 - 12 - 12$
$t = 42$ s: $1 - 1 - 1 - 7 - 1 - 1 - 2 - 1 - 2 - 2 - 2 - 7 - 7 - 2 - 12 - 12$
$t = 44$ s: $1 - 1 - 1 - 2 - 1 - 1 - 12 - 1 - 7 - 7 - 12 - 12 - 2 - 7 - 2 - 9$
$t = 46$ s: $7 - 1 - 1 - 7 - 12 - 2 - 1 - 12 - 2 - 7 - 1 - 1 - 2 - 12 - 12 - 12$
$t = 48$ s: $1 - 1 - 1 - 7 - 1 - 1 - 2 - 1 - 2 - 2 - 2 - 7 - 7 - 2 - 12 - 12$
$t = 50$ s: $1 - 1 - 1 - 2 - 1 - 1 - 12 - 1 - 7 - 7 - 12 - 12 - 2 - 7 - 2 - 9$
$t = 52$ s: $1 - 1 - 1 - 7 - 12 - 2 - 2 - 12 - 2 - 7 - 1 - 1 - 12 - 12 - 12 - 12$
$t = 54$ s: $1 - 1 - 1 - 7 - 1 - 1 - 2 - 1 - 2 - 2 - 2 - 7 - 7 - 2 - 12 - 12$
$t = 56$ s: $1 - 1 - 1 - 1 - 1 - 1 - 12 - 2 - 7 - 7 - 12 - 12 - 2 - 7 - 2 - 10$
$t = 58$ s: $1 - 1 - 1 - 7 - 12 - 2 - 1 - 12 - 2 - 7 - 1 - 7 - 2 - 12 - 12 - 12$
$t = 60$ s: $1 - 1 - 1 - 7 - 1 - 1 - 2 - 1 - 2 - 2 - 2 - 7 - 7 - 2 - 12 - 12$
$t = 62$ s: $1 - 1 - 1 - 2 - 1 - 1 - 12 - 1 - 7 - 7 - 12 - 12 - 2 - 7 - 2 - 9$
$t = 64$ s: $1 - 1 - 1 - 7 - 12 - 2 - 7 - 12 - 2 - 7 - 1 - 1 - 2 - 12 - 12 - 12$
$t = 66$ s: $1 - 1 - 1 - 7 - 1 - 1 - 2 - 1 - 2 - 2 - 2 - 7 - 7 - 2 - 12 - 12$
$t = 68$ s: $1 - 1 - 1 - 2 - 1 - 1 - 12 - 1 - 7 - 7 - 12 - 12 - 2 - 7 - 2 - 10$
$t = 70$ s: $1 - 1 - 1 - 7 - 12 - 2 - 7 - 12 - 2 - 7 - 1 - 1 - 2 - 12 - 12 - 12$
$t = 72$ s: $1 - 1 - 1 - 7 - 1 - 1 - 2 - 1 - 2 - 2 - 2 - 7 - 7 - 2 - 12 - 12$
$t = 74$ s: $1 - 1 - 1 - 2 - 1 - 1 - 12 - 1 - 7 - 7 - 12 - 12 - 2 - 7 - 2 - 10$
$t = 76$ s: $1 - 1 - 1 - 7 - 12 - 1 - 1 - 12 - 2 - 7 - 2 - 1 - 2 - 12 - 12 - 12$
$t = 78$ s: $1 - 1 - 1 - 7 - 1 - 1 - 2 - 1 - 2 - 2 - 2 - 7 - 7 - 2 - 12 - 12$
$t = 80$ s: $1 - 1 - 1 - 2 - 1 - 1 - 12 - 1 - 7 - 7 - 12 - 12 - 2 - 7 - 2 - 9$

Fig. 16.16 The results of dynamic WTA in the last 30 seconds

Table 16.9 The last three times of UCAV-to-target assignment

UCAV number	Target list at $t = 76s$	Target list at $t = 78s$	Target list at $t = 80s$
UCAV-1	$1 - 12 - 2 - 2$	$1 - 1 - 2 - 7$	$1 - 1 - 7 - 2$
UCAV-2	$1 - 1 - 12 - 7$	$1 - 1 - 2 - 2$	$1 - 1 - 7 - 7$
UCAV-3	$1 - 1 - 12 - 2$	$1 - 2 - 2 - 12$	$1 - 2 - 12 - 12$
UCAV-4	$12 - 12 - 1 - 7$	$7 - 1 - 7 - 12$	$2 - 1 - 12 - 9$

For comparison, the random immigrant method [10] is applied to the UMDA, denoted RUMDA, to solve the dynamic WTA problem. Fig. 16.17 shows the comparison of the fitness of the best individuals of EI-MUMDA and RUMDA. One can see that the proposed EI-MMS can significantly improve the performance of UMDA and thereby the target assignment result.

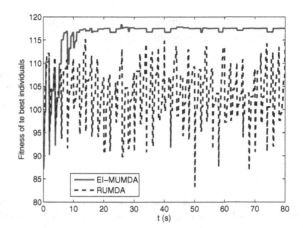

Fig. 16.17 Fitness of the best individuals of EI-MUMDA and RUMDA. The immigrant rate for RUMDA is 0.1.

16.5   Conclusions and Future Work

Usually, the mission area of UCAVs is unknown and might change arbitrarily. Therefore, the involved optimization problems should be solved dynamically. The major contribution of this chapter is to solve DOPs in the UCAVs anti-ground attack with EAs. Firstly, we have formulated the MPC-like OPP as a dynamic multi-objective real-valued optimization problem and a dynamic MOEA with the idea of Pareto set linkage and prediction (LP-DMOEA) has been proposed to solve this problem. This dynamic MOEA was then used to optimize the planning horizon of the MPC. The intelligent behavior is achieved by adopting the BN and fuzzy logic to model and quantify the environmental assessment which is accomplished by a pilot in a manned aircraft. Secondly, the dynamic WTA problem has been formulated with respect to the inner dependence among the targets and the expected damage type. Then, a memory-based EDA with environment identification (EI-MEDA) has been proposed to solve the dynamic combinational optimization problem. At last, the proposed approaches have been tested and validated through a computational simulation with a SEAD scenario.

The experimental results show that the LP-DMOEA works more effectively for the OPP in contrast to the restart method due to the positive impact on heuristically initializing the population. The intelligent methods for selecting executive solutions can automatically assess the changing environment and adapt the path planner. The formulation of WTA is practical to model a group of anti-air defense units and the expected damage intention of the attacking side. The EI-MEDA can effectively solve the dynamic WTA problem and significantly outperforms the algorithm with random immigrants.

In the future work, we plan to extend our intelligent OPP algorithm to the 3-D case and pay more attention to the kinetic and dynamic models of a real UCAV. As for the dynamic WTA problem, we hope to extend our research to the case where targets are detected one by one and assigned dynamically.

Acknowledgements. This work was supported by the National Nature Science Foundation of China under Grant 61105068 and the NPU Foundation for Fundamental Research under Grant JCY20130110, and partially by the State Key Laboratory of Synthetical Automation for Process Industries, Northeastern University, China.

References

[1] Branke, J.: Memory enhanced evolutionary algorithms for changing optimization problems. In: Proc. 1999 IEEE Congr. on Evol. Comput., vol. 3, pp. 1875–1882 (1999)

[2] Branke, J.: Evolutionary optimization in dynamic environments. Kluwer Academic Pub. (2002)

[3] Branke, J., Kaußler, T., Schmidt, C., Schmeck, H.: A multi-population approach to dynamic optimization problems. In: Proc. 4th Int. Conf. Adaptive Comput. Des. Manuf., pp. 299–308 (2000)

[4] Branke, J., Lode, C., Shapiro, J.: Addressing sampling errors and diversity loss in umda. In: Proc. 9th Annual Conf. Genetic and Evol. Comput., pp. 508–515 (2007)

[5] Cedeno, W., Vemuri, V.: On the use of niching for dynamic landscapes. In: Proc. 1997 IEEE Congr. on Evol. Comput., pp. 361–366 (1997)

[6] Chandler, P.R., Rasmussen, S., Pachter, M.: Uav cooperative path planning. In: AIAA Guidence, Navigation, and Control Conference and Exhibit (2000)

[7] Cobb, H.: An investigation into the use of hypermutation as an adaptive operator in genetic algorithms having continuous, time-dependent nonstationary environments. Tech. Rep. AIC-90-001, Naval Research Lab, Washington, D.C. (1990)

[8] Deb, K.: A fast and elitist multiobjective genetic algorithm: NSGA-II. IEEE Trans. on Evol. Comput. 6(2), 182–197 (2002)

[9] Ferreira, J.C., Fonseca, C.M., Gaspar-Cunha, A.: Methodology to select solutions from the pareto-optimal set: A comparative study. In: Proc. 9th Annual Conf. Genetic and Evol. Comput., pp. 789–796 (2007)

[10] Grefenstette, J.: Genetic algorithms for changing environments. In: Proc. 2nd Int. Conf. Parallel Problem Solving from Nature, pp. 137–144 (1992)

[11] Hatzakis, I., Wallace, D.: Dynamic multi-objective optimization with evolutionary algorithms: A forward-looking approach. In: Proc. 8th Annual Conf. Genetic and Evol. Comput., pp. 1201–1208 (2006)

[12] Hatzakis, I., Wallace, D.: Topology of anticipatory populations for evolutionary dynamic multi-objective optimization. In: 11th AIAA/ISSMO Multidisciplinary Analysis and Optimization Conf. (2006)

[13] Jin, Y., Branke, J.: Evolutionary optimization in uncertain environments-a survey. IEEE Trans. on Evol. Comput. 9(3), 303–317 (2005)

[14] Mori, N., Kita, H., Nishikawa, Y.: Adaption to a changing environment by means of the thermodynamical genetic algorithm. In: Ebeling, W., Rechenberg, I., Voigt, H.-M., Schwefel, H.-P. (eds.) PPSN 1996. LNCS, vol. 1141, pp. 513–522. Springer, Heidelberg (1996)

[15] Morrison, R.: Designing evolutionary algorithms for dynamic environments (2004)

[16] Mühlenbein, H., Paass, G.: From recombination of genes to the estimation of distributions i. Binary parameters. In: Ebeling, W., Rechenberg, I., Voigt, H.-M., Schwefel, H.-P. (eds.) PPSN 1996. LNCS, vol. 1141, pp. 178–187. Springer, Heidelberg (1996)

[17] Nikolos, I.K., Valavanis, K.P., Tsourveloudis, N.C., Kostaras, A.N.: Evolutionary algorithm based offline/online path planner for uav navigation. IEEE Trans. on Syst., Man, and Cybern.–Part B: Cybern. 33(6), 898–912 (2003)

[18] Nguyen, T.T., Yang, S., Branke, J.: Evolutionary dynamic optimization: A survey of the state of the art. Swarm and Evol. Comput. 6, 1–24 (2012)

[19] Peng, X., Gao, X., Yang, S.: Environment identification based memory scheme for estimation of distribution algorithms in dynamic environments. Soft Computing 15(2), 311–326 (2011)

[20] Peng, X., Xu, D., Yan, W.: Intelligent flight for uav via integration of dynamic moea, bayesian network and fuzzy logic. In: Proc. 50th IEEE Conf. on Decision and Control and European Control Conf., pp. 3870–3875 (2011)

[21] Pongpunwattana, A., Rysdyk, R.: Evolution-based dynamic path planning for autonomous vehicles. In: Chahl, J.S., Jain, L.C., Mizutani, A., Sato-Ilic, M. (eds.) Innovations in Intelligent Machines - 1. SCI, vol. 70, pp. 113–145. Springer, Heidelberg (2007)

[22] Shapiro, J.: Drift and scaling in estimation of distribution algorithms. Evol. Comput. 13(1), 99–123 (2005)

[23] Ursem, R.: Multinational gas: Multimodal optimization techniques in dynamic environments. In: Proc. 2000 Genetic and Evol. Comput. Conf., pp. 19–26 (2000)

[24] Wineberg, M.O.F.: Enhancing the ga's ability to cope with dynamic environments. In: Proc. 2000 Genetic and Evol. Comput. Conf., pp. 3–10 (2000)

[25] Yang, S., Jiang, Y., Nguyen, T.T.: Metaheuristics for dynamic combinatorial optimization problems. IMA J. of Management Mathematics (2012), doi:10.1093/imaman/DPS021

[26] Yang, S., Yao, X.: Population-based incremental learning with associative memory for dynamic environments. IEEE Trans. on Evol. Comput. 12(5), 542–561 (2008)

[27] Zengin, U., Dogan, A.: Dynamic target pursuit by uavs in probabilistic threat exposure map. In: AIAA 3rd "Unmanned Unlimited" Technical Conference, Workshop and Exhibit, Chicago, Illinois (2004)

[28] Zheng, C., Li, L., Xu, F., Sun, F., Ding, M.: Evolutionary route planner for unmanned air vehicles. IEEE Trans. on Robotics 21(4), 609–620 (2005)

[29] Zhou, A., Jin, Y., Zhang, Q., Sendhoff, B., Tsang, E.: Prediction-based population reinitialization for evolutionary dynamic multi-objective optimization. In: Obayashi, S., Deb, K., Poloni, C., Hiroyasu, T., Murata, T. (eds.) EMO 2007. LNCS, vol. 4403, pp. 832–846. Springer, Heidelberg (2007)

Chapter 17
Advanced Planning in Vertically Integrated Wine Supply Chains

Maksud Ibrahimov, Arvind Mohais, Maris Ozols, Sven Schellenberg,
and Zbigniew Michalewicz

Abstract. This chapter gives detailed insights into a project for transitioning a wine manufacturing company from a mostly spreadsheet driven business with isolated silo-operated planning units into one that makes use of integrated and optimised decision making by use of modern heuristics. We present a piece of the puzzle – the modelling of business entities and their silo operations and optimizations, and pave the path for a further holistic integration to obtain company-wide globally optimised decisions. We argue that the use of "Computational Intelligence" methods is essential to cater for dynamic, time-variant and non-linear constraints and solve today's real-world problems exemplified by the given wine supply chain.

17.1 Introduction

In managing supply chain thousands of individual decisions have to be made at regular intervals. These decisions are of different scope and significance

Maksud Ibrahimov
School of Computer Science, University of Adelaide, South Australia 5005, Australia
e-mail: maksud.ibrahimov@adelaide.edu.au

Arvind Mohais · Maris Ozols
SolveIT Software, Pty Ltd., 99 Frome Street, Adelaide, SA 5000 Australia
e-mail: {am,mo}@solveitsoftware.com

Sven Schellenberg
SolveIT Software, Pty Ltd., Level 2, 198 Harbour Esplanade, Docklands, VIC 3008, Australia. School of Computer Science, RMIT University, Building 14, Level 8 City campus, 414-418 Swanston Street, Melbourne, Vic 3000 Australia
e-mail: sven.schellenberg@rmit.edu.au

Zbigniew Michalewicz
School of Computer Science, University of Adelaide, South Australia 5005, Australia, Institute of Computer Science, Polish Academy of Sciences, ul. Ordona 21, 01-237 Warsaw, Poland, Polish-Japanese Institute of Information Technology, ul. Koszykowa 86, 02-008 Warsaw, Poland
e-mail: zbigniew.michalewicz@adelaide.edu.au

S. Yang and X. Yao (Eds.): *Evolutionary Computation for DOPs*, SCI 490, pp. 433–463.
DOI: 10.1007/978-3-642-38416-5_17   © Springer-Verlag Berlin Heidelberg 2013

(e.g., operational, tactical, strategic); there are also countless trade-offs between various decisions. Further, supply chain networks are extremely complex and they are set in constrained and dynamic environment, with many (usually conflicting) objectives. Any successful implementation of Advance Planning System (APS) for such complex network should be capable of addressing several key issues; these include dealing with huge search spaces, many objectives, variable constraints, and variability and uncertainty.

Due to the high level of complexity, it becomes virtually impossible for deterministic systems or human domain experts to find an optimal solution – not to mention that the term 'optimal solution' loses its meaning in multi-objective environment, as often we can talk only about trade-offs between different solutions. Moreover, the manual iteration and adjustment of scenarios (what-if scenarios and trade-off analysis), which is needed for strategic planning, becomes an expensive, if not unaffordable, exercise. Many texts on Advance Planning and Supply Chain Management (SCM) (e.g., [29]) describe several commercial software applications (e.g., AspenTech – aspenONE, i2 Technologies – i2 Six.Two, Oracle – JDEdwards EnterpriseOne Supply Chain Planning, SAP – SCM, and many others), which emerged mainly in 1990s. However, it seems that the areas of supply chain management in general, and advanced planning in particular, are ready for a new genre of applications which are based on Computational Intelligence methods. Many supply chain related projects run at large corporations worldwide failed miserably (projects that span a few years and cost many millions). In [29], the authors wrote:

> "In recent years since the peak of the e-business hype Supply Chain Management and especially Advanced Planning Systems were viewed more and more critically by industry firms, as many SCM projects failed or did not realize the promised business value."

The authors also identified three main reasons for such failures:

- the perception that the more you spend on IT (e.g., APS) the more value you will get from it,
- an inadequate alignment of the SCM concept with the supply chain strategy, and
- the organizational and managerial culture of industry firms.

While it is difficult to argue with the above points, it seems that the forth (and unlisted) reason is the most important: maturity of technology. Small improvements and upgrades of systems created in 1990s do not suffice any longer for solving companies' problems of 21st Century. A new approach is necessary which would combine seamlessly the forecasting, simulation, and optimization components in a new architecture. Further, many existing applications are not flexible enough in the sense that they cannot cope with any exceptions, i.e., it is very difficult, if not impossible, to include some problem-specific features – and most businesses have some unique features which need to be included in the underlying model, and are not adequately captured by off-the-shelf standard applications. Thus the results are often not realistic and the team of operators return to their spreadsheets and whiteboards rather than to rely on unrealistic recommendations of the software.

Several studies have investigated optimization techniques for various supply chain components, including the job-shop scheduling problem [8, 11, 32], planning and cutting problems [20, 21], routing problems [31], allocation and distribution problems [10, 37], to name a few. Many algorithms were developed to solve various supply chain components in a single silo environment [22, 33, 36]. However, optimization of each silo without considering relationships with other silos usually does not lead to a globally optimal solution of the whole supply chain. Because of that, large businesses started to become more interested in optimization of their whole system rather than the optimization of single components of the system. This approach is commonly referred in the literature as coordinated supply chain management. Davis, one of the experts in supply chain optimization, described this situation as follows [12]:

> "...companies should be able to use solutions to optimise across their sourcing and procurement, production and distribution processes all at the same time. Instead, supply chain solutions tend to break those functions out into separate modules, each of which runs separately from the others. The answer you get when you first do production and then do distribution or any of these functions independently is not as good as the answer you get when you do them together."

And Ackoff [1] correctly identified this issue over 30 years ago:

> "Problems require holistic treatment. They cannot be treated effectively by decomposing them analytically into separate problems to which optimal solutions are sought."

In this chapter a wine supply chain of five component is described. The current state of implementation models these components as individual silos tackling some of the dynamic challenges and business objectives. We describe one part of the journey towards a fully integrated company-wide decision support system.

The rest of the chapter is organized as follows. The next section provides a review of relevant literature of supply chain management and Computational Intelligence methods. Section 17.3 discusses briefly different components of the wine supply chain. Three of those components are described in more detail in the following sections respectively. Section 17.4 explains vintage intake planning, the process from harvesting grapes to their processing in a winery. Section 17.5 presents tank farm processing. Section 17.6 discusses the bottling wine problem. The last section concludes the chapter.

17.2 Literature Review

In this section, a brief literature review on coordinated supply chain management is provided, together with some discussion on time-varying constraints and the importance of Computational Intelligence methods.

17.2.1 Supply Chain Management

Supply chain management is a field that combines management of various business functions with a broad range of theoretical domains such as systems theory, inventory control, optimization, logistics, mathematical and computer modelling and other. The Global Supply Chain Forum defines Supply Chain Management as the integration of key business processes across the supply chain for the purpose of creating value for customers and stakeholders as mentioned in [17]. The term supply chain management was first mentioned in 1982 by [24], however, the concept of supply chain management was born long before, in the early 20th century, with the creation of the assembly line. In 1960s appearance of IBM computers helped raising interest to modern supply chain management.

In recent years, there has been an increased interest in solving supply chain management problems using Computational Intelligence methods. Naso *et al.* [23] looked at the problem of coordination of just-in-time production and transportation in a network of partially independent facilities of ready-mixed concrete. They optimized the network of independent and distributed production centres serving a number of customers distributed across a certain geographical area. This problem, which has high complexity and strict time delivery constraints, was approached with a meta-heuristic based on a hybrid genetic algorithm with combined constructive heuristics. Altiparmak *et al.* [3] proposed algorithms using mixed-integer, non-linear programming model for multi-objective optimization of a supply chain network based on the real world problem of a company that produces plastic products. They compared three approaches to find the set of Pareto-optimal solutions and discuss the pros and cons of each of them. Zhou *et al.* [36] presented a novel genetic algorithm to solve bi-criteria, multiple warehouse allocation problem. The proposed method finds the Pareto-front of wide range of non-dominated solutions without the arbitrary determination of weighting coefficients. In [33], an evolutionary algorithm was developed for dealing with the coordination of supply chain flows between different members of the chain. The authors of that paper recognized the importance of an overarching algorithm that optimises the whole system. Their work looked at the flow of materials in terms of supply and demand, but did not consider the intricacies of production within each business silo. Lee and Choi [18] applied genetic algorithms to solve a single machine scheduling problem with distinct dates and attempt to minimize all penalties. This method produces near optimal solutions, which they proved by comparison with an exact algorithm. Lee *et al.* [19] addressed the problem of inventory management of a refinery that imports several types of crude oil and proposes a mixed-integer linear programming model. Martin *et al.* [22] created a linear programming model to optimise flat glass production.

Large businesses typically split their business into operational components such as purchasing, production, distribution, etc. In the past, organisations have concentrated their research and efforts on these single operational components of the supply chain. As optimization of each individual silo in isolation may not lead to the global optimum, large businesses started to become more interested in optimization of their whole system rather than optimization of single components of the system. This approach is commonly referred in the literature as coordinated supply chain management. Although the field of supply chain management emerged not a long time ago, the idea of coordinated planning was already proposed in 1960 by Clark and Scarf [9]. They investigated multi-echelon inventory problem based on sequential and tree-like models – and they approached the problem using recursive decomposition method. Vidal and Goetschalckx [34] provides a review of strategic production-distribution systems with global integration. They stressed that very little research exists that addresses optimization of the whole supply chain rather than its single components. They mainly concentrated on mixed integer programming models. Aikens [2] presented a problem of optimal location of warehouses starting with simple uncapacitated single-commodity case with zero echelons and ending with more complex capacitated multicommodity and multi-echelon facility location models. Review by [30] defines three categories of operational coordination: Buyer-Vendor coordination, Production-Distribution coordination, and Inventory-Distribution coordination. The model described later in this chapter belongs to the second category. Within the production-distribution coordination category, several researchers approached it with dynamic programming heuristics, mixed integer programming, and Markov chains. Authors underline the complexity of problems of this category and discuss strategic planning supply chain models, the majority of the which are mixed integer programming based models.

In one of the first papers on production-distribution coordination [13] the author considered several models with interacting silos and stochastic demands and discusses an analytical approach to find optimum inventory levels. Production orders are processed in weekly batches. Pyke and Cohen in [27] considers a single-product, three silo supply chain that include a factory, finished goods stockpile and a distribution centre. The whole model is demand driven and based on the Markov chain. The finished goods stockpile in this model is used more like a buffer between production and distribution silos. Authors present near-optimal algorithms to determine batch size, normal reorder point, and expedite reorder point. The companion paper [28] upgrades their system to support multi-product situations. Pankaj and Marshall in [7] developed a two component demand driven model where products are first produced at the production plant and then distributed by the fleet of trucks to a number of retail outlets (also known as vehicle routing problem). Then, the authors compared two approaches, in the first one they solve the production and vehicle routing problems separately, and then together combined under a single optimization algorithm. The production planning problem is solved optimally and the vehicle routing problem is solved using heuristics. Their results show an advantage of using the second approach. However, the authors used relatively small datasets for their problem. Multi-echelon inventory optimization systems have been studied by various researchers. Caggiano *et al.* [5] used a greedy algorithm for his multi-item, multi-echelon service parts distribution system. Caglar *et al.* [6] solved his multi-item, two-echelon problem with Lagrangian decomposition. Wong *et al.* [35] approached the same problem with a greedy algorithm.

17.2.2   Time-Varying Constraints

The dynamic nature of real-world optimization problems is well known, but a closer examination reveals a few different aspects of the problem that can be described as *dynamic*. In this section, we introduce a classification of dynamic optimization problems into three categories:

1. Time-varying objective functions.
2. Time-varying input variables.
3. Time-varying constraints.

There is a large body of research literature that addresses dynamic property of such optimization problems, and we will undertake a brief review of work in each of the areas. It will be noticed that while there is an abundance of work on problems fitting into the categories of time-varying objective functions and time-varying input variables, there are relatively few published reports dealing with dynamic optimization problems of the 'third kind', that is one that deal with time-varying constraints.

These are problems in which the constraints of the environment within which a solution must be found change from day to day. These varying constraints add an additional level of complexity to the problem because a good approach must be able to operate equally well regardless of how many of these constraints are in place, and in what particular combination they occur. An example of a varying constraint is the requirement that a new production schedule be created that meshes seamlessly with the existing schedule once a number of days of the current production schedule have been fixed. In general, we do not know what the current schedule might be at any given moment in time, yet we must create an algorithm that produces a new solution that matches up with the existing one, and still produces an optimal result for the long term.

Good example of the types of problems that can be categorized as time-varying constraints can be found in [15], wherein the authors described circumstances that require the generation of regular new production schedules due to uncertainties (both expected and unexpected) in the production environment. They touch on typical examples such as machine breakdowns, increased order priority, rush orders arrival and cancellations. All of these issues are also considered in this article, from the perspective of an integrated evolutionary algorithms-based software solution. There are many other example of works recognizing this type of problem, for example [14, 16, 25], considering other Computational Intelligence methods.

Based on our experience in solving real-world optimization problems for commercial organisations, we have found that the type of problems commonly experienced in industry are those that belong to the second and third categories. Interestingly we have found that the vast majority of published research in evolutionary algorithms addresses the first category and to a lesser extent the second category. However, dynamic optimization of the 'third kind', i.e., where the problem involves time-varying constraints, although well-recognised in other domains, has been the subject of relatively few investigations in the literature. This observation is especially true when we extend our search to fully-fledged application of a

Computational Intelligence method to a dynamic real-world problem. In the same time, Computational Intelligence methods are very well suited for powering software applications for addressing these operational issues; we discuss them briefly in the following subsection.

17.2.3   Computational Intelligence

Computational Intelligence is considered an alternative to classical artificial intelligence and it relies on heuristic algorithms (such as in fuzzy systems, neural networks and evolutionary computation). In addition, Computational Intelligence also embraces techniques such as swarm intelligence, fractals and chaos theory, artificial immune systems, and others. Computational Intelligence techniques often combine elements of learning, adaptation, evolution and fuzzy logic to create programs that are, in some sense, intelligent.

An interesting question, which is being raised from time to time, asks for guidance on the types of problems for which Computational Intelligence methods are more appropriate than, say, standard Operation Research methods. From our perspective, the best answer to this question is given in a single phrase: complexity. Let us explain. Real-world problems are usually difficult to solve for several reasons, and include the following:

- The number of possible solutions is so large as to forbid an exhaustive search for the best answer.
- The evaluation function that describes the quality of any proposed solution is noisy or varies with time, thereby requiring not just a single solution but an entire series of solutions.
- The possible solutions are so heavily constrained that constructing even one feasible answer is difficult, let alone searching for an optimum solution.

Naturally, this list could be extended to include many other possible obstacles. For example, we could include noise associated with our observations and measurements, uncertainly about given information, and the difficulties posed by problems that have multiple and possibly conflicting objectives (which may require a set of solutions rather than a single solution). All these reasons are just various aspects of the complexity of the problem.

Note that every time we solve a problem we must realise that we are in reality only finding the solution to a model of the problem. All models are a simplification of the real world - otherwise they would be as complex and unwieldy as the natural setting itself. Thus, the process of problem solving consists of two separate general steps: (i) creating a model of the problem, and (ii) using that model to generate a solution:

$$Problem \rightarrow Model \rightarrow Solution.$$

Note that the "solution" is only a solution in terms of the model. If our model has a high degree of fidelity, we can have more confidence that our solution will be meaningful. In contrast, if the model has too many unfulfilled assumptions and rough approximations, the solution may be meaningless, or worse. So, in solving real-world problems there are at least two ways to proceed:

1. we can try to simplify the model so that traditional methods might return better answers, or
2. we can keep the model with all its complexities, and use non-traditional approaches, to find a near-optimum solution.

In other words, the more complex the problem (e.g., size of the search space, evaluation function, noise, constraints), the more appropriate it is to use a non-traditional method, e.g., Computational Intelligence method. Anyway, it is difficult to obtain a precise solution to a problem because we either have to approximate a model or approximate the solution. And a large volume of experimental evidence shows that this latter approach can often be used to practical advantage: many Computational Intelligence methods have already been incorporated into software applications that handle levels of supply chain complexity that is unapproachable by traditional methods.

17.3 Wine Supply Chain

Wine supply chain is a complex multi-component system with various kinds of inter- and intra- company dependencies. A typical supply chain may consist of components such as maturity models, vintage intake planning, crushing, tank farm, bottling, supply of dry goods (for example bottles, corks, bottle labels, etc), various distribution components (such as distribution depots and hubs), storage components (warehouses), external sourcing of raw materials and other. In [29], the authors defined that

> "...supply chain consists of two or more legally separated organizations, being linked by material, information and financial flows."

In order to reduce the complexity of this enormous supply chain, a company may concentrate on the subset of the supply chain that belongs to this company. This type of management control where several consecutive stages of supply chain operated by a single company called *vertical integration* and a supply chain with this style of management is called *vertically integrated supply chain*. In Fig. 17.1 dashed boxes represent external components to the company, for example dry goods suppliers, distribution companies, storages, etc. The components represent the vertically integrated wine supply chain, all components of which belong to the given company. Material flow goes from left to right on this figure, whereas information flows from right to left. The vertically integrated supply chain on this figure is represented by five components: Maturity Models, Vintage Intake Planning, Crushing, Tank Farm and Bottling.

From the broader perspective, these entities may not be legally separated, especially in vertically integrated supply chains, as they are within jurisdiction of a single wine company. In more generic sense a supply chain consists of generally separate

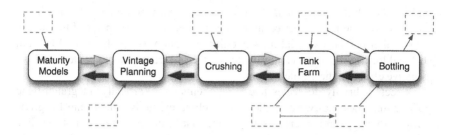

Fig. 17.1 Wine Supply Chain. Blue arrows on this diagram represent a material flow, black arrows represent information flow. The dashed boxes illustrate external components of the company (e.g., suppliers or data feeds). A screen shot of the applications can be found in subsequent sections describing the application components.

entities (within one structure) involved in production and distribution of a certain commodity with possibly different key performance indicators (KPIs), being linked by material, information and financial flows. This vertically integrated system can have two types of KPIs: *global*, which reflect a measure of the supply chain as a whole and *local*, which shows the performance of individual component.

Concentrating on the vertically integrated supply chain which is a core element in the whole system for the company opens many benefits from the financial and business points of view. A software application has been developed for each of the components. All these software applications were based on the paradigms of Computational Intelligence. When deployed together and later on integrated, these applications have the potential to optimise all planning and scheduling activities across a winery's entire supply chain and can help wineries to optimise various business processes, predict and deal with unexpected events, and address key operational issues, such as:

- Creating optimal harvest schedules that can accommodate last-minute changes.
- Maximising the utilisation of crushers, pressers, and fermenters.
- Optimising tank farm transfers and activities.
- Dealing with sudden changes (e.g., delayed transport, demand spikes, equipment failure, extreme weather conditions, etc.)
- Improving resource allocation under risk and uncertainty.
- Minimising transportation and other logistics costs.
- Improving the accuracy of demand forecasts.

Having a system that controls the whole supply chain becomes even more essential because of the dynamic nature of the market and constantly coming and changing customer orders with many, often conflicting, objectives. Apart from the finding the globally optimal solution of how to run the supply chain in a relatively short period of time, this vertically integrated system has additional benefits such as the *ability to quickly react on dynamic changes to the schedule* which gives a company a highly competitive advantage, *a quick identification of bottlenecks and problems* which significantly decreases the chance of having a human error. One of the most significant benefits is the *global what-if scenario analysis* which is the ability to quickly analyse the impact of certain changes in one or several parts of the supply chain on other parts is essential for schedulers and analysts. Looking at KPIs (both global and local) schedulers can evaluate whether a certain change made a positive or negative effect on the whole system.

This section briefly describes five components of a vertically integrated wine supply chain. These software applications include predictive modelling for grape maturity, vintage planning, crush scheduling, tank farm optimization, bottling-line sequencing, and demand forecasting. In the later sections, due to the space limitations, only three of these components are described in details.

This represents the current state of the project. Later on, all systems are coordinated by a *global module* which facilitates the cooperation between the components for the whole vertically integrated supply chain.

17.3.1 Maturity Models

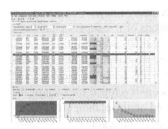

Vintage intake plans are heavily dependent on the prediction of expected grape maturity dates. It is possible to export the prediction dates from some external system that functions as a black box providing only one date when it believes the harvesting should occur. However, limited visibility into the prediction process often prompts requests to revisit the prediction functionality of this process. So the maturity models deploy a new prediction module that provides improved prediction dates and visibility of the prediction-calculation provided.

Grape maturity can be defined as the physiological age of the berry on the vine. It is important to define the optimal grape maturity for wine production and to develop clear chemical or biochemical traits that can be used to define the peak of ripeness. The definition of optimal maturity will vary depending upon the style of wine being made; the working definition of quality; varietal; rootstock; site; interaction of varietal, rootstock and site; seasonal specific factors; viticultural practices; and downstream processing events and goals. If a clear descriptive analysis of the quality target exists, then the time of harvest can be optimised to meet those goals. Several grape and cluster characteristics have been used to assess ripeness (e.g., sugar, pH, acidity, berry metabolites, berry proteins, taste). There are, of course, other non-compositional factors that influence the decision to harvest, including labour availability; seasonal changes such as rainfall; heat waves; tank space limitations; and other factors beyond the winemaker's control.

A "black box" prediction approach provides no audit capability for the user making it difficult to detect and promptly address issues related to accuracy of prediction. These factors can easily cause errors in forecasting maturity dates to go unnoticed and unrectified for prolonged periods of time. Decisions on when to book certain grapes for harvesting and crushing are relying heavily on the experience of the personnel involved in the process, and may result in a non-optimal allocation of harvesting and crushing resources. Each of these situations could result in higher costs for harvesting, transportation, and crushing, and reduction in grape quality.

17.3.2   Vintage Intake Planning

Vintage intake planning manages the grape supply intake from the "vineyard to the weighbridge". The functionality of this module supports the creation and maintenance of vintage intake plans that satisfy capacity constraints and facilitate the harvesting of grapes during periods of time when the quality is the highest. This stage is described in detail in Section 17.4.

17.3.3   Crushing

Crushers are used to process wine grapes and are often connected to different types of pressing machines, The optimal processing capacity of the crushing machines is about 40-45 tons per hour. However, but if necessary, it may be increased to 60-80 tons per hour. The most important limiting factor is the capacity of the pressing machines and fermentation containers. The processing capacity for the pressing machines ranges from 4 to 12 tonnes per hour depending on the type of grapes. It is important to generate optimal schedules for all crushers over some time horizon. However, the generated weekly schedule may incur frequent changes due to contractual influences, weather conditions, seasonal influences, and daily production variances. When the changes occur, The Crush Scheduler re-optimises and generates alternative schedules to fill available capacity. Also, a variety of constraints are present in this part of the wine supply chain, including:

- constraints in time (e.g., not processing fruit during the hottest part of the day).
- constraints in the throughput of presses.
- constraints in the throughput of fermentation vessels.
- the throughput of trucks via the crusher to be a continuous flow.
- scheduled repairs and maintenance of equipment.
- scheduled changeover and clean up (white to red, or lower grade grape-to-higher grade grape).
- special demand-fulfilling variety shortages to address meeting capacity needs.

17.3.4 Tank Farm

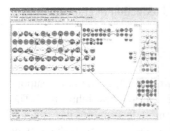

Wineries make daily decisions on the processing and storage of wines and juices in their tank farms, with major wineries having several hundred tanks that have differing capacities and attributes. These tanks may be insulated or refrigerated, for instance, and could have an agitator. Some of these tanks might be used to store juice after the grapes are crushed, fermented, and pressed, while others might be primarily designated for blending. Different types of juices may also require specific tank attributes for processing, such as refrigeration jackets or agitators. This stage is described in details in Section 17.5.

17.3.5 Bottling

The primary task of the Bottling-Line Scheduler is to generate optimal production schedules for the wineries' bottling operations. The software uses advanced optimization techniques for generating optimal production schedules. Opportunities for optimization include manipulating the sequencing order, selecting which bottling lines to use, consolidating similar orders within the planning horizon, and suggesting changes to the requested dates that improve the overall schedule. Some of the key objectives are to maximise export and domestic service levels (i.e., DIFOT), maximising production efficiency, and minimising cost. This stage is described in details in section 17.6.

17.4 Vintage Intake Planning

This section describes Vintage Intake Planning component of the wine supply chain in details.

17.4.1 Description of the Problem

The aim of the vintage intake is to harvest grape blocks from vineyards, carry the grapes into wineries where they will be further processed. Vintage intake planning in the scope of this chapter describes the planning of all necessary actions of bringing wine grapes from vineyards into wineries and the immediate subsequent grape processing steps within the winery. The vintage intake planning finishes once the juice's fermentation is over and the wine will be refined in the tank farm (17.5).

Parties that are involved in the intake process are harvesting and carrier companies and the winery personnel.

While wineries and the winery equipment such as crushers, presses, centrifuges and fermenters are owned by the wine producing company, the harvesting and transportation of grapes into the winery is carried out by contractors. Those contractors usually provide their services to many wine producers. Due to the nature of the ripening process of grapes and the inherent variations caused by different grape varieties, locations of the blocks and constantly changing weather conditions during the maturity process, contractors such as harvester or truck providers' availability fluctuates significantly and is often unknown until very close to the actual date their services become necessary.

The uncertainty around logistics resources and the inability to control them by the vintage planners demand a constant update of the plans. The further uncertainty imposed by sudden unpredicted weather changes which impacts on the grape maturity and thus the preferred wine intake day limits the planning horizon to only one or two weeks. These short-term plans are created in the week before the planning week commences, but do not get locked in until two days prior. At this point in time, the logistics manager for a particular region finished the process of requesting the required capacities to harvest and carry the grapes, and the service providers responded with their offer of available resources that meet or undercut the necessary requirements. This process will continue until all resource requirements are met or if none of the available providers within the region has any capacities left. The resources will be booked and locked in to the plan. After locking in the available resources, only unexpected events such as technical problems or human error can compromise the plan.

Harvesters and Trucks are not the only resources to take into consideration when drafting a vintage intake plan. The intake does not finish at the arrival at the winery, but after the wine grapes have been processed. At the winery the processing of the grapes begins:

- Grapes get weighted on weighbridges.
- Depending on which process for the harvested grapes to follow, the grapes will be crushed (usually, only red grapes get crushed).
- Some crushed grapes may require to stay together with the stalks and skins for a few days.
- Grapes get pressed (usually whites).
- The resulting press cut juices gets centrifuged.
- Finally, the juice is fermented in fermenters for about 1-2 weeks.

These processing steps are by no means complete, but they represent steps that pose major constraints on the grape intake. They represent the constraints human planners would take into account when assessing the winery intake capacity for drafting an intake plan.

17.4.2 Constraints

The logistic constraints during peak season of the vintage are probably the most pressing capacity constraints a vintage intake planner faces. Not only do they take into account preferred harvesters of growers, but also do they compete with other wine producing companies in allocating and booking carrier capacities.

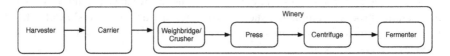

Fig. 17.2 Grape flow at intake. Note that not all grape processing steps within a winery are mandatory.

The first resource at the winery that poses a capacity bottleneck is the weighbridge as in most cases it can only be used sequentially. In some instances, it may be connected with the crushers in which case grapes will be loaded into special bins which incorporate a weighing machine. The filled bin will be subsequently tipped into the crushers which grind the grapes. Prior to the crushing, the winemaker undertakes a final assessment of the quality of the grapes and classifies the grapes into their final usage. This final usage determines the processes the grapes go through the manufacturing network all the way to the final bottling of the product.

Depending on the quality of the grapes and the intended final product they contribute to, the grapes will be routed through different processing stages, the *process*.

Presses pose the next potential bottleneck in the grape processing pipeline. Presses are also used to extract juice from grapes, but they deploy a much finer extraction technique than crushers. Through presses, the winemaker is able to extract different quality grades from the grapes, the so-called *press cuts* or *pressings*. They are constrained by the capacity of grapes they can be filled with (typically around 40 Tonnes) and the actual time it takes to carry out the pressing (approximately six hours, depending on the load and type of press).

Further clarification prior to fermentation is carried out in the centrifuges. Unlike presses, centrifuges are able to process a constant stream of juice. Depending on the type of centrifuge and the state of the infeed, centrifuges can reach a rate of $60,000L$ of juice per hour [4].

The last step of the intake is the fermentation of juice. Fermenters can be filled over a period of days. It takes approximately 10-14 days until the fermentation is finished and the fermentation vessel is available for new juice. In some instances, however, the fermentation tank is used for storage also so that it will not return to the pool of empty tanks. This may be the case when capacity shortages in the tank farm occur which may have causes the planner cannot anticipate.

The above mentioned winery equipment has a physical processing capacity limit. Those limits vary only rarely so that planning of their usage and anticipated completion time can be easily determined. Issues arise if winery equipment is not functional

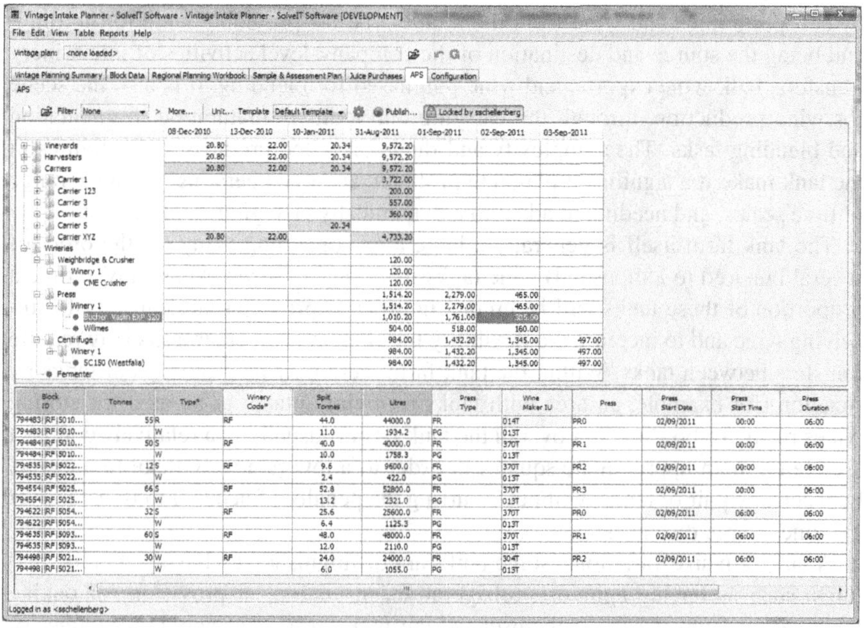

Fig. 17.3 Screenshot of the vintange intake planning system

(unplanned outages) or personnel that operates the equipment is not available. In these cases subsequent assignments become infeasible as they rely on the operation of the now unavailable equipment. The next weekly plan has to take into account and adopt to the change in operational resources.

Figure 17.3 shows the screenshot of the Vintage Intake Planning system illustrating the flow of material and the conversion from grapes to juice to wine by various processing steps. The figures in the table represent the quantity of grapes being harvesters, carried and crushed. From the press stage on downwards, the values are shown in litres. Note that constraint violations are highlighted in the upper overview table (red background cell colour). A selection of the violating cell, will bring up the below detailed table with all necessary information on the root cause of the violation.

17.5 Tank Farm

This section describes Tank Farm component of the wine supply chain in details.

17.5.1 Description of the Problem

The tank farm of a large winery is a major and integral part of the successful and efficient wine manufacturing enterprise. It interfaces to all the other major processes, by accepting juice and wine from the intake process; providing wine for bottling; and being the source and destination of the company-level activities of interwinery transfers, bulk wine exports, and wine purchased for blending. It is also the scene for wine production, through the carrying out additive, maturation, fermentation and blending tasks. The complexity and range of varied products and activities in the tank make it a significant planning problem, and one operating at a wide range of time scales, and needing to adapt to a dynamically changing environment.

The tank farm itself is generally a large area containing many (in the order of several hundred to a thousand) wine tanks in a range of sizes. At any given time a proportion of these tanks will be empty, in order to provide spare capacity for receiving wine and to increase the flexibility for blending and other tasks may require transfers between tanks. Within the tank farm there may be groupings of tanks by function (for example, an area with tanks receiving intake; or an area for storing wines ready for bottling) but overall the tank farm is usually in a relatively compact area (e.g., in an approximate square) in order to allow relatively flexible transfers between any pair of tanks, without incurring the penalty of needing to transfer over extremely large distances.

Apart form size variations, some tanks are distinguished by having specific features, such as having agitators, refrigeration, insulation, or provision for oaking wine. Apart from the tanks themselves, the tank farm may include a number of additional elements. These include various types of equipment (such as filters, centrefuges or heat exchangers); some of this equipment is portable, but some are at fixed locations within tank farm. In addition, the tank farm may have a a fixed piping and manifold system for transferring wine between tanks, supplemented with portable ground pipes for the same purpose, providing more flexibility, but with higher labour and setup costs.

17.5.1.1 Tank Selection

Overall there are two main optimization tasks for organising activities within the tank farm, and these can be simply described as relating to *which tank(s)* should be chosen, and *when* a specific operation on the tanks (such as a transfer) should occur. The first of these activities can be divided between the two quite distinct tasks of *Source Tank Selection* and *Destination Tank Selection*.

Obviously the primary constraint in source tank selection concerns the actual content of a tank from which we plan to draw. If we need to bottle of order for 5,000 litres of 2009 Shiraz, then we must be drawing from source tanks with the correct variety and vintage year. This constraint is common to any system of warehousing or stocking of multiple types of product. What is specific to wine industry, and the tank farm situation in particular, is the need to do this in a way that avoids oxidation of wine through contact with the air. If wine is left in a partially filled tank, then, over time, the wine will degrade and significantly lose value. It is such a serious problem to be avoided in the wine industry, that it has specific terms: *ullage* is the proportion of air within a wine tank, and *being on ullage* refers to a tank and its content having non-zero ullage.

For destination tank selection, we are frequently dealing with empty tanks. An important rule for optimising the global behaviour is to ensure that specialist tanks are not selected if not needed - for example, fermentation tanks should not be selected as a destination for a generic transfer (particularly during the vintage season) unless other tanks are not available.

Other rules, that relate equally to source and destination tank selection, revolve around reducing the costs associated with the execution of the transfers. A transfer over a longer distance is more expensive, due to increased setup times (for example to lay out a ground pipe) and the additional wastage and water use that is required for cleaning out the longer pipe. In a similar way, having many transfers to or from many smaller tanks will be more expensive than a single transfer involving a single pair of tanks.

17.5.1.2   Labour and Resource Scheduling

Once the operations of the tank farm, and the selection of tank. have been decided, there remains the scheduling issue of when the operation will occur, by whom, and with what specific equipment resources.

Each operation can be divided into a number of subtasks, such as setup, startup, execution, and finish-up. For each such task there are fixed amounts of time, labour, and potentially equipment, that are required. With the execution task, such as for a transfer between tanks, there may also be a variable cost in time, dependent on the number of litres being transferred.

For routine planning, a standard schedule of times that various shifts of tank farm workers is defined, and the various types and numbers of equipment can be acquired. But in the dynamic real world, this is not sufficient: workers can call in sick, and pieces of equipment can become unavailable for maintenance or repair. Such variations need to be handled by a real-world system.

17.5.2   Functionality

The Tank Farm Optimiser is designed to deal with all the tank farm issues of tank selection, and resource scheduling, across the full range of scenarios, including receiving vintage intake, wine manufacturing operations, servicing bottling requests, as well as blend planning, and management of bulk wine transfers (interwinery, export, and purchasing.) We describe some major features, with particular emphasis on optimization and dealing with dynamic changes.

The Tank Farm Optimiser is arranged via a series of tabs that allow access the main planning activities (intake, operation, bottling, bulk wine movement and blending) as well more generic informational screens (tank farm maps and schedule, capacity plan, the vintage plan, and the long term production plan). This arrangement allows customising the interface for specific user roles: for example, an intake planner may just have the vintage plan, intake, and tank farm map/schedule tabs.

The Tank Farm map and table give two distinct visual representations of the contents and events of the tanks in the tank farm (see Fig. 17.4 and 17.5).

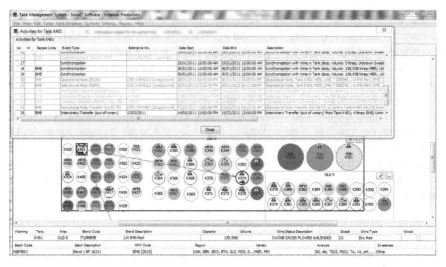

Fig. 17.4 Tank farm map view screenshot. The circles represent physical layout and contents (by colour). An Activities popup shows the transactions on a single tank.

In Fig. 17.4 the map view gives a graphical layout of the tanks on a specific date. The sizes of the circles tanks reflect the total capacity of the tanks; the colours denote the type of content (this is customisable, but might represent the grape variety, such as Shiraz or Chardonnay); and the proportions that are coloured indicate how full the tanks are: tanks that all full or all white (empty) would represent a situation where no tank is left on ullage. Arrows between tank circles represent scheduled transfers, whilst warning markers indicate the presence of issues at some point in the tank's timeline (such as possible constraint violations).

Simply clicking on a tank circle reveals other basic facts about the contents of the tank, displayed in the lower part of Fig. 17.4, including the precise varieties, vintages years, exact volume, status of the wine (e.g., ready for bottling) and a succinct representation of analysis results (such as current measurements of the wine sweetness and alcohol levels). By opening up any tank, an Activities screen for the tank is displayed (visible in the upper part of the figure) which displays the sequence of all transactions that relate to that tank in the model, such as transfers of fluids in or out, operations (such as additives), and synchronisations with updated content information.

The Table view (see Fig. 17.5) is similar to the Map view, but the tanks are listed in a tablular form, rather than arranged by geographical location. Each row presents data for a single tank, and the various textual and numerical columns (visible on the left of the figure) give various attributes of the tank and its content for a specific date (these columns are particularly useful for sorting and filtering for tanks having specific properties.) On the right of Fig. 17.5 we see the timeline of the content of the tank represented, providing a graphical view of the dynamic changes of the tank, both historical and future planned. The colour of each bar represents the type

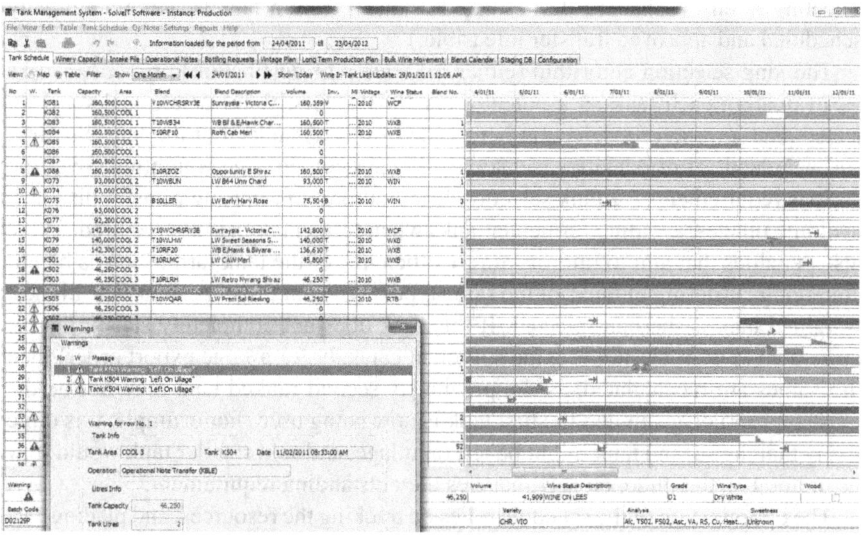

Fig. 17.5 Tank farm table view Screenshot. Timelines for the tanks are visible on the right, coloured by the wine variety.

of content (using the same colour mapping as used in the Map view colouring of the circles) and the height of the bar represent the proportion that the tank is filled. Transfers are represented as icons on the bars. Scrolling the right side of this table allows the timeframe of these timelines to be altered. A wide range of filters can be applied to both views, and the selection context and filtering is maintained as the user switches between Map and Table views.

In the lower left of Fig. 17.5 a warning icon on a row has been opened to a warning details panel, giving information about contraints that have been violated on the tank on that row. These constraints may be volume violations (e.g., attempting to transfer more from a tank than it contains at the time that transfer is scheduled) or contents violations (e.g., mixing content of one tank into another tank, which would break configured rules of what varieties may be mixed).

The underlying model of tank transfer events is an effective way of dealing with the dynamic temporal aspects. For example, the tank model may show that a 20,000 litre tank T100 has 10,000 litres of 2009 Shiraz, and suppose this is precisely what is required for us to service a bottling request for the next two days. A manual operator may be tempted to use this for the bottling; but instead, the optimiser might suggest to only take 4,000l from T100, whilst taking the remaining 6,000l from a small 6,000 litre tank, full of 2009 Shiraz. It seems like a poor solution, since T100 now continues to be left on ullage, with 4,000 litres. The explanation is that there is already a booking in the system that will take 6,000 from T100 in several days time. The optimiser is looking ahead at existing scheduled transfer bookings, and selecting to ensure that we do not take too much from a tank (leaving an already scheduled and approved transfer infeasible.)

The tank selection algorithm (either for source or destination) primarily works by evaluating each tank as a potential candidate. Firstly, infeasible tanks are rejected, and a reason for rejection is recorded for such tanks. For the feasible tank, and weighted set of penalties is calculated and recorded. The weighting cover the various requirements for tank selection listed before: minimising transfer distance, avoiding ullages, avoiding fermenter tanks when fermentation is not required. There also penalties that aim to improve future efficiency of the tank farm, such as by encouraging clustering of tanks of the same type of wine in the same area. The highest weighted tank is the one selected. If this tank does not completely satisfy the requirement (e.g., it did not provide sufficient capacity for a large export order) then the values are recalculated, as the previously second-ranked tank may not be the most appropriate given that the first tank is now being used. For example, it is quite likely that the second tank might be left on ullage, and so a smaller tank would now be preferable, that more closely matches the outstanding requirement.

The optimization of the schedule relies on tracking the resources, and placing earliest due tasks earlier than later tasks. However there can occur task order inversion, on the basis of availability of specific resources. Furthermore the algorithm assesses the costs of varied resources, aiming to use lower cost resources in preferences.

17.5.3   Results

Both the tank selection and labour and resource optimization features of the Tank Farm Optimiser system provide effective and timely solutions, and adapt well the real world, dynamical changes. An important aspect of the utility of the system is how it interacts with the workflows of the users and with their individual planning.

17.5.3.1   Tank Selection

As new orders and requests enter the system, via a batch loading system, they are automatically optimised together providing a level of integration towards a more global optimum. This provides new tank selection allocations, without disrupting existing allocations. This is important to help provide stability of the tank farm for the users, and to preserve the additional planning information already added for existing allocations. When requests are revised for requests, particularly for volume, this is highlighted to users visually, but does not result in an automated re-optimization. This is because the users may still wish to use existing tank allocations, and will merely add or remove tanks from the existing assignment.

A particularly important aspect of the use of the tank selection, has been the presentation of the penalties and constraints involved with tank selection, when demanded by the user. Users have needed to be able to refer to such decision making data, in order to gain trust in the choices made by the Tank Farm Optimiser. We believe that this may partly due to the traditional based aspect of the wine making process, where individual winemakers have been used to considerable personal control and responsibility with all aspects of the process, and hence a relative unwillingness to give up aspects of this, particularly if the optimization appears as a black box. Another importance in presenting the optimization reasoning, is so that users can deal better with the effects of dynamic changes. For example, a user who is allocated a tank which appears "inferior" to that allocated to someone who books later than them, can sense an aspect of unfairness - and perhaps suboptimality - in the system. However, once the optimization records are presented, they can see that at the time of their booking's optimization, their selection was indeed optimal.

Since the incremental approach of allocating tank is not guaranteed to be globally optimal, users also make use of other features that give extra flexibility. These include reoptimising selections that occurred previously, and manually stipulating tank selection, by-passing the optimiser. One of the noted results of manual bypassing of the optimiser is that the users' deliberate choices can work to undermine the global goals of the optimization program - such as reducing the clustering of tanks for the same or similar wines.

17.5.3.2 Tank Farm Resources Scheduling

The labour and resource scheduling is carried out automatically, at the same time as the tank selection, when tank operation records are loaded into the system.

Interestingly, although the Tank Farm Optimiser does not provide a rationale for its resource scheduling, this does not seem to be a major concern to users. This may be because they have less inherent interest in at which particular hour, and by which cellar hand, the work will be carried out; or it may be because in a visual presentation of the schedule, the interlinked and cross-dependent complexity of the problem is more evident, leading to less obvious substitute solutions.

As a result, the changes that are made to the schedule appear more in the nature of refinements and improvements, constructively using the existing schedule as a basis. Examples of such changes arise from the users noting that the setup time for a following task may be reduced, or even eliminated, since it may be able to use the equipment setup up the preceding task - and thus reflect a global optimization that is not yet accessible to the Tank Farm Optimiser.

17.6 Bottling

This section describes the Bottling component of the wine supply chain in detail. It is based on a real-world wine bottling system for mass-production environment. The problem and its various challenges are discussed, particularly concerning the issue of time-varying constraints. Methods to overcome these difficulties are introduced. This system has been deployed into fully live production environment and is in daily use.

Before getting to the point where wine is bottled into a finished product, the liquid would have gone through a series of fermentation and other processing steps. We will assume that we are at the point where the liquid is in a finished, consumable state, and is residing in a bulk storage tank, which ranges in capacity between several tens of thousands to one million litres. This bulk liquid remains in storage awaiting the bottling process wherein it is pumped into a bottling factory and put into consumer size bottles, with a typical volume of less than one litre.

A bottling factory houses several bottling *lines*. These are machines that are connected to intermediate feeding tanks that contain finished wine, and are used to transfer the wine into bottles. The bottling lines also take care of related tasks such as capping the bottle with a screw cap or a cork, applying a label to the bottle, and packaging the bottles into cartons. Each bottling line is capable of bottling a subset of the types of finished wine products manufactured by the wine company (e.g., some lines are only capable of producing only red wines).

These two elements, the bulk wine liquid, and the bottling lines, constitute the basic working elements of the wine bottling problem. The bottling process is illustrated in Fig. 17.6. Which bulk wines are put into which bottles, and when that is done, are determined by the demand. The wine company receives orders for particular finished goods from their clients and it is those orders that must be carefully considered in order to determine the best way of running the bottling plant. In an ideal situation, customers place their orders with sufficient lead time to ensure timely bottling of their goods.

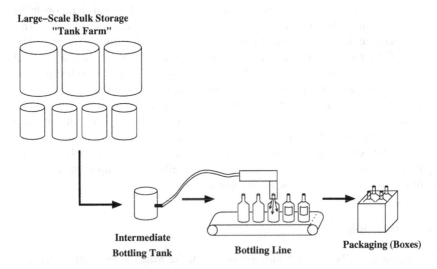

Fig. 17.6 The wine bottling process

The problem is to determine a sequence of orders to be carried out on each bottling line, hence classifying it as a scheduling problem, such that optimal use is make of the company's resources, from the point of view of making maximum profit, and also maximising customer satisfaction. Hence a good schedule will minimise production costs, and at the same time ensure that orders are produced in full, sufficiently before their due dates. This is the fundamental part of the wine bottling problem, that is deciding how to schedule production so as to make the best use of limited resources.

Before getting into the business intricacies that give rise to the time-varying constraints, we will first consider some more basic issues that affect the scheduling problem:

- *Due dates*: When a customer places an order, the sales department will assign it a due date that is acceptable to the customer, and which should also be realistic taking into account the size of the order and available resources at the bottling plant.
- *Bulk wine availability*: Some orders may need to be inevitably delayed due to the fact that the bulk wine needed to fill the bottles may not yet be ready.
- *Dry Materials*: In addition to the liquid wine, there are a few dry goods that are required to produce a bottle of wine.
- *Job run lengths*: It is inefficient to have machines frequently changing from one type of bottling job to another because this incurs set-up and take-down time and reduces the overall utilization of the machine. Hence, the scheduling algorithm must attempt to group similar orders for sequential execution so this type of inefficiency is avoided.
- *Wine changeovers*: Wines are categorized broadly in terms of their color, there is red, white, and rose (pink). When a bottling line finishes working with one type of wine and switches over to another type, this is referred to as a *wine changeover*. Bottling line must be cleaned during changeovers, to varying degrees, depending on the nature of the change.
- *Other changeovers*: Although wine changeovers incur the most time, there are a variety of other changeovers that can happen, even within a run of the same color wine.
- *Bottling line availability*: Some industries use machines that are kept in operation continuously. This is sometimes the case in large-scale wine companies, but for only limited periods of time.
- *Routings*: Each product can be bottled on a number of different lines.

Our work on the wine bottling problem resulted in a full-featured piece of software integrated around a core evolutionary algorithm that deals with all of the above listed issues, and creates feasible, optimal schedules for satisfying customer orders for wine. The application was launched for a major global wine company, and it experiences heavy daily usage at international sites. It is a cornerstone of their scheduling department, and we think it is a good example of an integrated evolutionary algorithm serving robustly in prime time.

17.6.1 Time-Varying Challenges in Wine Bottling

In this subsection, we will look at some business requirements that led to time-varying constraints that had to be addressed in the software. Here we will only consider the issues, and their actual solution, including algorithmic details, will be covered in a later section.

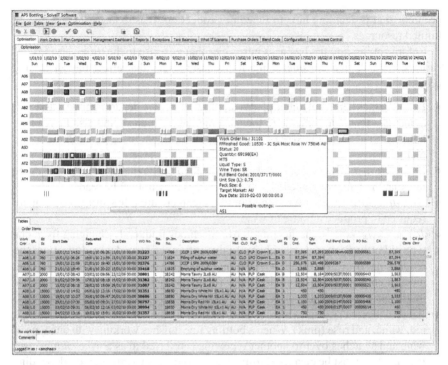

Fig. 17.7 Visualisation of a schedule on multiple production lines in the software application

- *Manual assignments*: There are various scenarios in which the human scheduler would need the ability to override the schedule. For instance, it might be that a very important customer makes a late, but urgent request for a large quantity of wine. Even if this causes severe disruptions to the smooth running of the bottling plant, this type of request if usually accommodated due to the high value placed on some customers, and the importance maintaining a good business relationship with them. The software application that was developed had to be flexible enough to allow its built-in algorithm to work in interactive way that it could actively seek out an optimal solution that satisfies the constraints described before, but at the same time allowed inefficient manual overrides dictated by a human operator to co-exist with the otherwise optimal solution.
- *Machine breakdowns*: From time to time, a bottling line will break down and become unavailable for use. It may be that a solution found by the optimiser previously would have been planned around that machine being available during a period of time that has now become unavailable due to the breakdown. The software must be flexible enough to repair the previous solution to take into account the breakdown.
- *Freeze periods*: Once the schedule is accepted and saved to an internal database, subsequent daily use of the optimiser to schedule newly arriving customer orders must be done in a controller manner so that there a buffer period at the

beginning of the schedule that remains the same as it was yesterday. This unchanging portion of the schedule is referred to as a *freeze period*.

- *Modified orders*: Once a schedule has been created and saved, there might be a situation in which the next time the software package is opened, it realizes that there was a modification made to one of the scheduled orders in the database. This might be for example a change in quantity. More, or less bottles of wine may be required, and by adjusting the scheduled order, the start and end times of all subsequent orders on the same bottling line become affected.
- *Poor/Excellent Job Execution*: Due to a number of factors, the efficiency of a bottling machine may be better or worse on any given day, and the factory manager would expect the scheduling optimiser to take this into account when creating a new schedule, or when adjusting an existing one.

17.6.2   Objective

The problem is to determine the best sequence of orders to be scheduled on the bottling lines in such a way that minimum number of constraints mentioned in the previous section are violated. Each constraint has a penalty coefficient associated with if violated. So the total penalty of the solution is

$$TotalPenalty = \sum_i k_i P_i$$

where k_i is the coefficient for penalty P_i.

17.6.3   The Algorithm

In this subsection, we will look at the structural, algorithmic and programmatic details required to solve the problem using an evolutionary algorithm.

17.6.3.1   Representation

For the wine bottling scheduling problem, the core of the problem was conceptualized as having a number of orders that must be placed on a fixed number of bottling machines in an efficient sequence. Hence the natural representation to use is one of a mapping of lists of orders to machines. This can be visualized as in Fig. 17.8. The representation illustrated in this diagram is quite similar to the final schedule presented visually in Fig. 17.7 above.

The representation shown in Fig. 17.8 is stored programmatically as a map of machines to variable-length lists of orders. The list on any given machine is sorted chronologically in terms of which orders will be carried out first. The individual is constructed in such a way that the assignment of orders to machines is always valid, in other words the assigned orders always respects product routings.

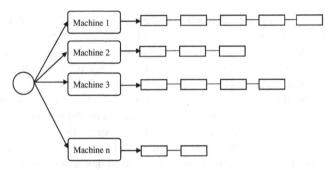

Fig. 17.8 Scheduling individual representation

17.6.3.2 Decoding

In order to manage the decoding process, we employed a concept we referred to as *time blocks*. Each such block keeps track of a start time, an end time and an activity that is performed during that time, allow for the possibility that nothing is done actually done, in which case the time block is referred to as an *available* time block. The link to the activity performed during a time block may point to an external data structure that contains any information on any level of detail required to accurately model a scenario.

Decoding begins with a series of multiply-linked-lists of time block nodes associated with each machine. Each machine has a list of available time blocks, and occupied time blocks. At the outset, before anything is placed on a machine, it would only contain a list of available time blocks, each representing a chunk of time during which the machine is available for use.

Decoding proceeds by going through all machine-job pairs found in the individual representation and proceeding to the corresponding machine, finding an available time block node, and marking it as occupied for the corresponding job. Some jobs may not be able to fit in the first available time block, and may need to be split into multiple parts. How this is done, indeed, if it is permissible at all depends on the policies of the business for which the scheduling application is being created. In the case of the wine bottling application that we are considering, orders were split across adjacent available time blocks.

17.6.3.3 Operators

A number of operators were used to manipulate the representation given above. To avoid the problem of having to perform extensive repairs based on invalid representation states, crossover-type operators were avoided. The operators listed below, which are typical examples from the set used, may all be considered as mutation operators.

- *Routing Mutation*: This operator modifies the machine that was selected to execute a job. An alternative is randomly chosen from the set of possible options.

- *Load Balancing Mutation*: There is a variation to the routing mutation operator which, instead of merely randomly choosing an alternative machine for an order, could choose from a subset of machines that are under-loaded. Such an approach would help with load-balancing of the machines.
- *Grouping*: This operator groups orders based on some common characteristic, such as wine colour, bottle type, or destination export country. There are several variants of the grouping operators. Some operate quite randomly, looking for a group of orders based on some characteristic and then looking left or right for a similar group and then merging the two. Other examples are given below.
- *Recursive Grouping*: A more directed variation on grouping is recursive grouping. This operator seeks out an existing group of jobs based on wine colour, then within that group perform random grouping based on some other characteristic, such as bottle size. This process may then be repeated inside one of the subgroups.
- *Outward Grouping*: Outward grouping is a term we use to describe the process of identifying groups in a list of jobs based on a primary characteristic, then randomly selecting one of them and from that location looking left and right for another group with a common secondary characteristic and finally bringing together the two.
- *Order Prioritization*: Orders that may be showing up as being produced late after decoding an individual are stochastically prioritized by moving them left in the decoding queue. Again, this type of operator could be made more intelligent than random by moving groups of jobs along with the one that is identified as being late. That way the job gets prioritized, but at the same time disruption to grouping is minimized. This operator works on one of the most important objectives of the solution, to maximize DIFOT.

17.6.3.4   Addressing the Dynamic Issues

We will now look at how we dealt with the time-varying issues that were identified in section 17.6.1. As will be observed, the problems were addressed by a combination of modifications made to the initial time block node linked-lists, to the decoding process, by altering the input variables to the optimiser, and by introducing a step between the optimiser and the human user call *solution re-alignment*.

- *Addressing Manual Overrides*: This problem was solved by applying a constraint to the decoding process, and indirectly affecting the fitness function. The software application allows the user to select a particular order, and specify which machine it should be done on, as well as the date and time of assignment. This becomes a timetable constraint for the genotype decoder. When a candidate individual is being decoded, the initial state of the machine time block usages includes the manually assigned orders as part of the list of occupied time blocks.

   With this approach, it is primarily the fitness function that would guide the search to a relatively good solution built around the manual assignment. However it is also important to de-emphasize some of the more structured and aggressive operators such as recursive grouping and outward grouping, which were designed with a clean slate in mind. They would still contribute toward the search for a good solution in parts of the time period that do not contain manual assignments, but the main evolutionary algorithm loop should keep track of the performance of these operators and adjust their probability of application accordingly.

- *Addressing Freeze Periods*: The application allows the user to select a freeze date and time on each machine used in the bottling plant. During an optimization run, all assignments in the existing schedule that are before the freeze period cutoff are internally marked as manual assignments, and therefore behave in exactly the same way as described above for user-defined manual assignments.
- *Addressing Machine Breakdowns, Modified Orders, and Poor/Excellent Job Execution*: These three problems were solved using a similar approach. Take Modified Orders for example. When the application re-loads and realizes that an existing order has been modified, for example its quantity has been increased or decreased, then the necessary action to be taken is at the level of modifying the existing solution, prior to it being fed back into the next run of the evolutionary algorithm. This was accomplished using a process call *solution re-alignment*.

As we already mentioned earlier in paper, the system has been deployed into live production environment at a few locations and is in daily use.

17.7 Conclusion

The previous sections described five applications modelling functional units of a vertically integrated wine supply chain. These five applications integrate operations and decision support that was previously carried out in isolation and often without visibility of peer's decisions. We consider this as first step in the roadmap towards an integrated company-wide decision support system that leverages the ability to 'see' across fragmented business units to find a more globally optimal solution. The benefits can already be materialised through the integration of sub processes into each of the five modules, but their full potential will be realised by an even higher degree of integration by providing an environment in which these five applications can cooperate. The benefits are:

- Better use of available production capacity.
- reduced risk of late deliveries due to production capacity issues, supply issues, or scheduling errors, and better visibility of potential risk.
- Schedulers can quantify the relative merits of different schedules, make informed decisions as to which scheduled to choose.
- Higher confidence in production schedules may allow running at lower inventory levels.
- Long-term planning from sales forecasts (e.g., assist with production capacity planning and production smoothing, supply planning for long lead-time items, and inventory planning, what-if scenarios for strategic and operational planning, testing the impact of changes on business rules, infrastructure investment, over-time or extra shifts).

- Reporting on production (e.g., identification of capacity problems, identification of production or supply bottlenecks, high-level overview as well as information on specific orders).
- Reduction in overall transfers, leading to less water consumed in the wine production process, a smaller carbon footprint, less spillage, less plant maintenance, and increased safety.
- Reduction in labour requirements through labour balancing, a reduction in labour-intensive operations, and a reduction in overall transfers.
- Reduction in "free working space" on the tank farm, leading to increased tank utilisation, capacity, and throughput.
- Process improvement in the area of work order handling, by reducing paper handling and data duplication.
- Provision of centralised applications that are maintained and supported.
- Provision of a scalable platform for future extensions.
- Straightforward integration with other applications for prediction and optimization.
- Provision of integrated views (carrier, winery, etc.)
- Provision of integrated inputs (e.g., for Grower Liaison Officers and Logistics Coordinators).
- Provision of optimised capacity planning (i.e., automated "smoothing").
- Automatic generation of robust, optimised production schedules that maximise service levels and utilisation, while minimising cost.
- Faster feedback to production planners, management, sales, and other interested parties for placing orders or requesting changes.

There are also a number of flow-on benefits, e.g., planners require less time to produce bottling plans, less chance of human error, identification of potential data problems, ability to handle dynamic changes to the schedule, whilst minimising the impact on existing Work Orders near their production date.

To achieve such an ambitious goal, a *global module* would need to interact between the sub modules in a cooperative manner, such as in [26]. Another possibility is to employ an agent-based system with each of the sub modules acting as competing or cooperating units negotiating for resources and capacities. The options for further integration are manifold and we will explore them in a later publication.

One statement, however, from the beginning of this chapter can already be confirmed: To realise all these benefits, Computational Intelligence method must play the central role in the development of the software. It would be very difficult, for example, to build a linear model of such whole supply chain, representing all objectives, constraints, and dependencies – further, most standard software packages also failed in this complex environment over the last 15 years. As we have already indicated in the Introduction, a new genre of applications (based on Computational Intelligence methods) is necessary to address complex issues of advanced planning in the supply chain management – and this section illustrated this point.

Acknowledgements. This work was partially funded by the ARC Discovery Grant DP0985723 and by grants N 516 384734 and N N519 578038 from the Polish Ministry of Science and Higher Education (MNiSW).

References

[1] Ackoff, R.L.: The future of operational research is past. J. Oper. Res. Soc. 30, 93–104 (1979)

[2] Aikens, C.H.: Facility location models for distribution planning. Europ. J. Oper. Res. 22(3), 263–279 (1985)

[3] Altiparmak, F., Gen, M., Lin, L., Paksoy, T.: A genetic algorithm approach for multi-objective optimization of supply chain networks. Comput. Ind. Eng. 51(1), 196–215 (2006)

[4] Boulton, R.B., Singleton, V.L., Bisson, L.F., Kunkee, R.E.: Principles and Practices of Winemaking. Springer (1998)

[5] Caggiano, K.E., Jackson, P.L., Muckstadt, J.A., Rappold, J.A.: Optimizing Service Parts Inventory in a Multiechelon, Multi-Item Supply Chain with Time-Based Customer Service-Level Agreements. Oper. Res. 55(2), 303–318 (2007)

[6] Caglar, D., Li, C.L., Simchi-Levi, D.: Two-echelon spare parts inventory system subject to a service constraint. IIE Transactions 36(7), 655–666 (2004)

[7] Chandra, P., Fisher, M.L.: Coordination of production and distribution planning. Europ. J. Oper. Res. 72(3), 503–517 (1994)

[8] Cheng, R., Gen, M., Tsujimura, Y.: A tutorial survey of job-shop scheduling problems using genetic algorithms—i: representation. Comput. Ind. Eng. 30(4), 983–997 (1996)

[9] Clark, A.J., Scarf, H.: Optimal policies for a multi-echelon inventory problem. Manage. Sci. 50(12 suppl.), 1782–1790 (2004)

[10] Coit, D.W., Smith, A.E.: Solving the redundancy allocation problem using a combined neural network/genetic algorithm approach. Comput. Oper. Res. 23(6), 515–526 (1996)

[11] Davis, L.: Job shop scheduling with genetic algorithms. In: Proc. 1st Int. Conf. Genetic Algorithms, pp. 136–140 (1985)

[12] Davis, L.: Embracing complexity. Toward a 21st century supply chain solution (2008), Web-resource http://sdcexec.com/online/printer.jsp?id=9012

[13] Hanssmann, F.: Optimal inventory location and control in production and distribution networks. Oper. Res. 7(4), 483–498 (1959)

[14] Holthaus, O.: Scheduling in job shops with machine breakdowns: an experimental study. Comput. Ind. Eng. 36(1), 137–162 (1999)

[15] Jain, A.K., Elmaraghy, H.A.: Production scheduling/rescheduling in flexible manufacturing. Int. J. Prod. Res. 35(1), 281–309 (1997)

[16] Kutanoglu, E., Sabuncuoglu, I.: Routing-based reactive scheduling policies for machine failures in dynamic job shops. Int. J. Prod. Res. 39(14), 3141–3158 (2001)

[17] Lambert, D.M.: Supply chain management: Implementation issues and research opportunities. Int. J. of Logistics Management 9, 1–20 (1998)

[18] Lee, C.Y., Choi, J.Y.: A genetic algorithm for job sequencing problems with distinct due dates and general early-tardy penalty weights. Comput. Oper. Res. 22(8), 857–869 (1995)

[19] Lee, H., Pinto, J.M., Grossmann, I.E., Park, S.: Mixed-integer linear programming model for refinery short-term scheduling of crude oil unloading with inventory management. Ind. Eng. Chem. Res. 35(5), 1630–1641 (1996)

[20] Levine, J., Ducatelle, F.: Ant colony optimization and local search for bin packing and cutting stock problems. J. Oper. Res. Soc. 55(7), 705–716 (2004)

[21] Liang, K.H., Yao, X., Newton, C., Hoffman, D.: A new evolutionary approach to cutting stock problems with and without contiguity. Comput. Oper. Res. 29(12), 1641–1659 (2002)

[22] Martin, C.H., Dent, D.C., Eckhart, J.C.: Integrated production, distribution, and inventory planning at libbey-owens-ford. Interfaces 23(3), 68–78 (1993)

[23] Naso, D., Surico, M., Turchiano, B., Kaymak, U.: Genetic algorithms for supply-chain scheduling: A case study in the distribution of ready-mixed concrete. Europ. J. Oper. Res. 177(3), 2069–2099 (2007)

[24] Oliver, R.K., Webber, M.D.: Supply-chain management: Logistics catches up with strategy. In: Logistics. Chapman and Hall (1982) (Reprint from Outlook)

[25] Petrovic, D., Alejandra, D.: A fuzzy logic based production scheduling/rescheduling in the presence of uncertain disruptions. Fuzzy Sets and Systems 157(16), 2273–2285 (2006)

[26] Potter, M.: The design and analysis of a computational model of cooperative coevolution. Ph.D. Thesis, George Mason University (1997)

[27] Pyke, D.F., Cohen, M.A.: Performance characteristics of stochastic integrated production-distribution systems. Europ. J. Oper. Res. 68(1), 23–48 (1993)

[28] Pyke, D.F., Cohen, M.A.: Multiproduct integrated production–distribution systems. Europ. J. Oper. Res. 74(1), 18–49 (1994)

[29] Stadtler, H., Kilger, C.: Supply Chain Management and Advanced Planning. Springer (2008)

[30] Thomas, D.J., Griffin, P.M.: Coordinated supply chain management. Europ. J. Oper. Res. 94(1), 1–15 (1996)

[31] Toth, P., Vigo, D.: The Vehicle routing problem. Society for Industrial and Applied Mathematics (2001)

[32] Van Laarhoven, P.J.M.: Job shop scheduling by simulated annealing. Oper. Res. 40, 113 (1992)

[33] Vergara, F.E., Khouja, M., Michalewicz, Z.: An evolutionary algorithm for optimizing material flow in supply chains. Comput. Ind. Eng. 43(3), 407–421 (2002)

[34] Vidal, C.J., Goetschalckx, M.: Strategic production-distribution models: A critical review with emphasis on global supply chain models. Europ. J. Oper. Res. 98(1), 1–18 (1997)

[35] Wong, H., Kranenburg, B., van Houtum, G., Cattrysse, D.: Efficient heuristics for two-echelon spare parts inventory systems with an aggregate mean waiting time constraint per local warehouse. OR Spectrum 29(4), 699–722 (2007)

[36] Zhou, G., Min, H., Gen, M.: A genetic algorithm approach to the bi-criteria allocation of customers to warehouses. Int. J. Prod. Econ. 86(1), 35–45 (2003)

[37] Zielinski, K., Weitkemper, P., Laur, R., Kammeyer, K.D.: Parameter study for differential evolution using a power allocation problem including interference cancellation. In: Proc. 2006 IEEE Congr. Evol. Comput., pp. 1857–1864 (2006)

Author Index

Subject Index

Printed in the United States
By Bookmasters